MARGO
MULTIPROXY APPROACH FOR THE RECONSTRUCTION OF THE GLACIAL OCEAN SURFACE

The cover image shows sea ice in the Beaufort Sea, Arctic Ocean. Photograph taken onboard CCGS Sir Wilfried Laurier in July 2002. Image courtesy of Dr Kate Darling.

MARGO
MULTIPROXY APPROACH FOR THE RECONSTRUCTION OF THE GLACIAL OCEAN SURFACE

Edited by

M. KUCERA

Institute for Geosciences, Eberhard-Karls University of Tübingen
D-72076 Tübingen, Germany

R. SCHNEIDER

Institute for Geosciences, University of Kiel
D-24098 Kiel, Germany

M. WEINELT

Institute for Geosciences, University of Kiel
D-24098 Kiel, Germany

ELSEVIER

Amsterdam - Boston - Heidelberg - London - New York - Oxford - Paris
San Diego - San Francisco - Singapore - Sydney – Tokyo

ELSEVIER B.V.
Radarweg 29
P.O. Box 211, 1000 AE Amsterdam
The Netherlands

ELSEVIER Inc.
525 B Street, Suite 1900
San Diego, CA 92101-4495
USA

ELSEVIER Ltd
The Boulevard, Langford Lane
Kidlington, Oxford OX5 1GB
UK

ELSEVIER Ltd
84 Theobalds Road
London WC1X 8RR
UK

First edition 2005

Reprinted from Quaternary Science Reviews, Volume 24/7-9

Printed and bound by CPI Group (UK) Ltd, Croydon, CR0 4YY

Transferred to Digital Print 2012

ISBN-13: 9780080447025
ISBN-10: 0080447023

♾ The paper used in this publication meets the requirements of ANSI/NISO Z39.48-1992 (Permanence of Paper). Printed in Great Britain.

Contents

Quaternary Science Reviews 24 (2005) 813–819

Multiproxy approach for the reconstruction of the glacial ocean surface (MARGO)

Michal Kucera[a],*, Antoni Rosell-Melé[b], Ralph Schneider[c],
Claire Waelbroeck[d], Mara Weinelt[e]

[a]Department of Geology, Royal Holloway University of London, Egham TW20 0EX, UK
[b]ICREA and Institute of Environmental Science and Technology, Universitat Autònoma de Barcelona, 08193 Bellaterra, Catalonia, Spain
[c]Département de Géologie et Océanographie, UMR5805-EPOC, CNRS/Université de Bordeaux1, 33405 Talence Cedex, France
[d]Laboratoire des Sciences du Climat et de l' Environnement, Laboratoire mixte CNRS-CEA, F-91198 Gif sur Yvette, France
[e]Institute for Geosciences, University of Kiel, Olshausen Strasse 40, D-24098 Kiel, Germany

Received 19 July 2004; accepted 23 July 2004

The oceans and the atmosphere are the two main reservoirs of greenhouse gases and latent heat on Earth. These reservoirs interact through the ocean surface, and the dynamics of this interaction is a major determinant of global climate. Accurate reconstructions of the physical state of the global ocean are therefore critical to the understanding of past climate changes. This is in turn required to assess the significance of instrumentally observed climate variability, and for the forcing and validation of global circulation models, which are used to predict future climate change.

Systematic instrumental measurements of sea surface properties exist for only a few decades, with the longest regional records rarely extending beyond the 19th century. Yet, it is only with the aid of climate records spanning thousands of years and encompassing dramatically different climatic states of the planet that one can truly understand the dynamics of the ocean–atmosphere interface and perform meaningful and useful tests of global climate models. Information on the state of the planet in the past, and the amplitude, frequency and mechanisms of its changes is of paramount importance to our society, as it is used to inform and guide long-term environmental policies and planning and to predict impact of climate change on land, our habitat.

Any effort to provide past climate records of sufficient extent and time range will have to resort to the use of indirect information: proxies based on biological, chemical and physical signals preserved in ancient geological materials. In open-ocean settings, the organic

*Corresponding author. Tel.: +44-1784-443586; fax: +44-1784-471780.

E-mail address: m.kucera@gl.rhul.ac.uk (M. Kucera).

doi:10.1016/j.quascirev.2004.07.017

and mineral remains of marine microplankton in deep-sea sediments provide the most comprehensive source of such proxies (Mix et al., 2001; Henderson, 2002). Unfortunately, it is scientifically and technically not possible at present to aim to reconstruct a continuous record in space and time of past climate variation. Instead, a discrete and distinct time interval is needed to focus research efforts of the scientific community. To be useful for validation of climate models, this time interval ought to represent a period of climate markedly different from that of today, yet not too distant from the present so that the basic assumptions and parameters of the climate models need not be modified.

With respect to past ocean surface conditions, particularly since the pioneering effort of the CLIMAP group (climate long-range investigation, mapping, and prediction; CLIMAP, 1976, 1981) which started 30 years ago, the time of the last glacial maximum (LGM) has served as common target for climate modelling experiments and palaeoenvironmental proxy reconstructions. The LGM interval, around 21,000 years ago, represents the nearest of a series of past climatic extremes characterising the waxing and waning of Quaternary ice ages and as such serves as an excellent testing ground for assessment of sensitivity of the Earth's climatic system. Since the CLIMAP project conclusions were published, a large amount of new sediment material has been recovered, its age determined using ever-improving dating techniques, and its palaeoclimatic significance assessed with an ever-expanding battery of proxies. Yet, a global synthesis of this material is still lacking. In 1999, an international initiative was launched by the scientific community, with the aim to facilitate a new synthesis of the last ice age Earth's surface. The environmental processes of the ice age: land, ocean, glaciers (EPILOG) initiative commenced by the IMAGES-PAGES program (the international marine past global changes study-past global changes; core project of the international geosphere-biosphere programme) provided an updated review of the progress in palaeoclimatic reconstructions since CLIMAP. It summarised the salient points and obstacles in the way of a new synthesis, and set a series of benchmarks to allow a precise definition of the LGM chronozone (Mix et al., 2001).

Following EPILOG, and with the above advancements in mind, the MARGO working group was launched in September 2002, when 33 scientists from 13 countries met at the Hanse Institute for Advanced Studies in Delmenhorst, Germany, to initiate the "multiproxy approach for the reconstruction of the glacial ocean surface" (MARGO). MARGO acts as an open international project involving data gathering, sharing and harmonisation, with the aim of producing a new synthesis of sea-surface temperature (SST) and sea-ice extent of the glacial ocean. The overall MARGO objective is to collate and harmonise all the available proxy data and transfer function techniques, and place them into a common framework for a multi-proxy global glacial ocean reconstruction. However, prior to this global synthesis, huge efforts have been put by the MARGO working group members into the assembly of new regional or proxy-specific SST reconstructions which are reported in this volume or in recently published studies, like the GLAMAP reconstruction (Sarnthein et al., 2003; and references therein) and the TEMPUS compilation (Rosell-Melé et al., 1998, 2004). The contributions presented in this issue form the first phase of MARGO: (i) compilation of quality-assessed and harmonised proxy-specific calibration datasets and LGM reconstructions, (ii) documentation of individual proxies and techniques and (iii) an outline of possible methods of the presentation of the final synthesis.

A selection of 8 papers out of 13 herein provides a series of compilations for different SST proxies (Fig. 1), including Mg/Ca ratios of planktonic foraminifera (Barker et al., 2005; Meland et al., 2005) and various transfer function approaches based on census counts of assemblages of planktonic foraminifera (Barrows and Juggins, 2005; Chen et al., 2005; Hayes et al., 2005; Kucera et al., 2005), diatoms and radiolaria (Gersonde et al., 2005) as well as dinoflagellate cysts (de Vernal et al. 2005). Moreover, a new Holocene oxygen isotope data synthesis based on planktonic foraminifera is presented (Waelbroeck et al., 2005) and a stimulating contribution by Morey et al. (2005) discusses the distribution of planktonic foraminifer assemblages in surface sediments as a function of multiple environmental variables rather than exclusively related to SST. The series of papers addressing glacial surface ocean conditions is followed by a study describing the termination of the last glacial period in the Pacific (Kiefer and Kienast, 2005) and this MARGO special issue ends with two contributions dealing with the issues of mapping techniques for sparsely and non-homogeneously distributed proxy data (Schäfer-Neth et al., 2005) as well as their comparison with results from climate model experiments (Paul and Schäfer-Neth, 2005).

This preface briefly summarises the recommendations and common standards agreed at two MARGO workshops which form the innovative guidelines for the new compilations presented herein, and which will serve as the internationally agreed base for the overall multi-proxy synthesis of the last glacial SST reconstruction. For further information on the MARGO guidelines, their application to the new compilations and the individual data sets the reader is referred to the MARGO website (http://www.pangaea.de/projects/MARGO) and to the individual articles in this volume. These guidelines evolved from the major aims of MARGO, as formulated at the first meeting in 2002.

Distribution of MARGO Last Glacial Maximum SST proxy records

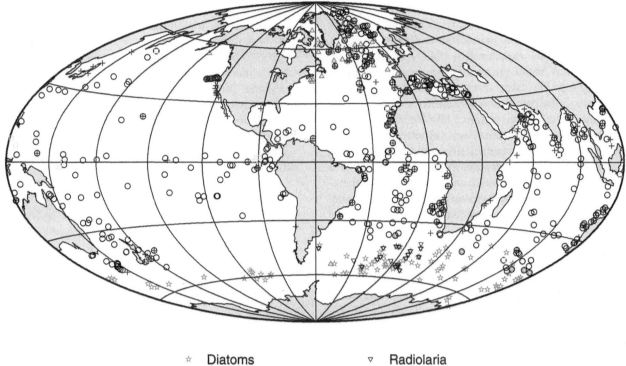

☆ Diatoms	▽ Radiolaria
△ Dinoflagellates	☐ Mg/Ca
○ Foraminifera	+ U_{37}^k

Fig. 1. The location of proxy records of LGM SST included in the MARGO reconstructions. The alkenone data are from Rosell-Melé et al. (2004). All other proxy data are presented in this volume.

- Compilation of new calibration datasets using consistent criteria for sample quality, and assignment and definition of modern SST values to each calibration sample.
- Compilation of new LGM SST reconstructions for each proxy, based on the new calibrations, providing age quality assessment for each sample using harmonised criteria.
- Assessment of the feasibility of a single, multi-proxy, LGM SST reconstruction.

In parallel, the oxygen isotopic composition of planktonic foraminifera has been included among the MARGO target proxies of glacial surface ocean conditions. Although one cannot use oxygen isotope data alone to deduce palaeotemperatures, compiling a new dataset for this proxy has numerous advantages

- Combining planktonic oxygen isotope values with independent SST estimates makes it possible to derive the isotopic composition of surface water, which is related to surface salinity and thus yields information on the hydrological cycle (e.g., Duplessy et al., 1991; Lea et al., 2000);

- Oxygen isotopes give a first approximation of the surface density of the ocean as the oxygen isotopic composition of planktonic foraminifera is a function of salinity and temperature;
- An increasing number of ocean circulation numerical models compute the isotopic composition of calcite explicitly (e.g., Paul et al., 1999), and these models can be directly validated by comparison of their output with foraminifer isotopic composition;
- Finally, a large amount of oxygen isotope data is currently available on planktonic foraminifera from recent and glacial sediments (e.g., Duplessy et al., 1981, 1991; Billups and Schrag, 2000; Schmidt and Mulitza, 2002), which provides a unique opportunity to test the consistency of the regional SST estimates based on different methods and proxies.

The MARGO oxygen isotope dataset now consists of over 2100 measurements from recent sediments (Wael-broeck et al., 2005) and 410 data points from LGM sediments with thorough age control that have been checked for internal consistency. The LGM dataset will be analysed in conjunction with the envisaged revised global LGM SST compilation.

According to the MARGO recommendations, seasonal SST reconstructions as well as annual temperatures are provided wherever possible. This is the case with all transfer function techniques using planktonic foraminifera, dinoflagellate cysts, radiolarian and diatom assemblages, while the geochemical SST reconstructions provide annual mean estimates (alkenones) or summer SST (Mg/Ca). To cater for a proper comparison of single proxy reconstructions we agreed on common use of World Ocean Atlas version 2 (WOA, 1998; 1° grid version) as modern reference and calibration data. The WOA98 dataset has benefited from error reduction as compared to previous versions while no significant new data were included. Seasonal and/or annual temperatures for the sample sites were extracted for 10 m water depth using a common data extraction tool (http://www.palmod.uni-bremen.de/~csn/woasample.html). For seasonal temperatures three-month averages of January–March (northern winter) and July–September (northern summer), and 12-months average for annual SST are used. Temperature at sample sites is computed as the area-weighted average of the four WOA temperature points surrounding the sample location; WOA data points marked as land were omitted from the averaging.

Sea-ice extent is one of the most elusive properties of the ocean in terms of the prospect of its accurate reconstruction from geological records, yet it is a crucial parameter of climate models and its knowledge is essential for assessment of the different oceanographic mechanisms that could be at play in a given region. The MARGO group has, therefore, recommended this variable should be reconstructed as far as is possible and encourage further research in this field. In this issue, Gersonde et al. (2005) present an updated reconstruction of patterns of summer and winter sea-ice extent in the Southern Ocean reconstructed from the distribution patterns of diatom sea-ice indicator species. In the Northern Hemisphere, deVernal et al. (2005) and Kucera et al. (2005) produced updated estimates of glacial sea-ice extent in the Nordic Seas based on dinocyst and planktonic foraminifer assemblages; both indicated ice-free summers in the Norwegian Sea.

MARGO adopted the same definition of the LGM interval or chronozone as in Mix et al. (2001): 19–23 cal kyr BP (ka). The definitions of the levels of certainty are identical as well:

- *LGM Chronozone Level 1*: Chronologic control based either on annually counted layers extending through the LGM chronozone, or two radiometric dates within the interval, such as U/Th dates or reservoir-corrected 14 C-yr dates adjusted to the calendar scale using the CALIB software (Stuiver et al., 1998).
- *LGM Chronozone Level 2*: Chronologic control based on two bracketing radiometric dates of any kind

within the interval 12–30 ka (i.e., within marine oxygen-isotope stage 2), or by correlation of non-radiometric data to similar regional records that have been dated to match the level 1 protocol (for example, $\delta^{18}O$ stratigraphy).
- *LGM Chronozone Level 3*: Chronologic control based on other stratigraphic constraints (for example, a regional lithologic index such as %CaCO$_3$) that are correlated to similar records dated elsewhere to match the level 2 protocol.

In addition, it was recommended to label samples with no stratigraphic control as level 4.

For the purpose of improved calibration of transfer functions and calibration equations used in MARGO with respect to WOA98, the MARGO working group also attempted to better constrain the quality of "modern" or "core-top" samples. Therefore, the Late Holocene chronozone was defined in a similar way: 0–4 cal kyr BP, with the following levels of certainty:

- *LH Chronozone Level 1, 0–2 ka*: Chronologic control based either on annually counted layers covering the last 2 ky, or one radiometric date (such as U/Th dates or reservoir-corrected 14 C-yr dates converted into calendar age) within the interval 0–2 ka.
- *LH Chronozone Level 2, 0–4 ka*: Chronologic control based on one radiometric date of any kind within the interval 0–4 ka, or stained benthic foraminifera with sedimentation rate higher than 5 cm/ky.
- *LH Chronozone Level 3, 0–4 ka*: Chronologic control based on one radiometric date of any kind within the interval 4–8 ka or specific stratigraphic control (for e.g., % *G. hirsuta* left coiling) indicating that the sample belongs to the interval 0–4 ka.
- *LH Chronozone Level 4, 0–4 ka*: Chronologic control based on other stratigraphic constraints (for e.g., $\delta^{18}O$ stratigraphy, or a regional lithologic index such as %CaCO$_3$) indicating that the sample belongs to the interval 0–4 ka.

In addition, it was recommended to label samples with no stratigraphic control as Level 5, and to report the 0–4 ka and 0–2 ka intervals separately when Level 1 is achieved.

Table 1
Conversion of MARGO/EPILOG Holocene and LGM Chronozone boundary ages using the INTCAL98 calibration curve from Stuiver et al. (1998)

Calendar age	Reservoir age–corrected ^{14}C age
2 ka $\pm$ 0.1 kyr	2.05 ^{14}C kyr BP $\pm$ 0.1 kyr
4 ka $\pm$ 0.1 kyr	3.6 ^{14}C kyr BP $\pm$ 0.1 kyr
19 ka $\pm$ 0.1 kyr	15.9 ^{14}C kyr BP $\pm$ 0.35 kyr
23 ka $\pm$ 0.1 kyr	19.4 ^{14}C kyr BP $\pm$ 0.4 kyr

The definition of the Holocene and LGM Chronozone intervals in terms of reservoir age-corrected ^{14}C ages is shown in Table 1. In high latitudes (beyond 40°N and 40°S), reservoir ages significantly increase during the LGM with respect to modern values (Sikes et al., 2000; Waelbroeck et al., 2001). As the LGM reservoir age is generally not well defined, this induces very large uncertainties on calendar dates derived from radiocarbon dates. The reliability of the radiocarbon dates from LGM high-latitude samples is thus lower than that for dates derived from low- or mid-latitudes samples.

Given the perspective for a single multi-proxy reconstruction for the glacial surface ocean conditions which should be the final product of MARGO and considering the individual proxy compilations presented in this volume, we would like to end this preface with some general thoughts addressing the challenge that the compilation of a unique multi-proxy SST/SSS/sea-ice data set or map presents, and on the possible strategies to do so. Given the current array of proxies available to reconstruct SST it might appear that the reconstruction of these parameters could be achieved with more certainty than ever before. This would have been the case if coinciding SST estimates were obtained by all proxies, but unfortunately this is not what always happens. There are only a few comprehensive studies on the comparison of different proxy SST estimates. One of these is the study carried out by Bard (2001). One of its chief conclusions is that overall SST proxies agree on the amplitude of changes at low and mid-latitudes. However, a level of disagreement between proxies must be expected because each approach reflects different past environmental conditions. The estimates depend on the ecology and biology of each source organism as well as the statistical approaches used to calibrate the proxy. The uncertainties are intrinsic to each approach, as each calibration is empirically derived, based on data sets of different size and with contrasting spatial coverage. The sedimentary data is usually calibrated against "modern conditions" but there is an incomplete knowledge of the ecology and biology of the source organisms and incomplete information on oceanographic conditions to derive "modern SST" as registered in the sediments. A proxy measured in a sediment sample also represents an integrated signal over time and space of the sedimentation of the chemical or microfossil parameters on which the approach is based. This will also be different for each proxy given that the remains of each source organism will sediment differently as a function of density and size of particles, among other factors.

In addition, the environmental information inferred from each approach may relate to more than one climatic parameter. This may be a general property of the proxy, as in the case of $\delta^{18}O$ in calcareous tests of foraminifera, or occurring just in specific circumstances, which means that certain approaches are geographically constrained. For instance, $U^K_{37'}$ is not reliable in low-salinity environments, foraminiferal transfer functions are questioned in upwelling regions, diatom and radiolarian-based approaches are vulnerable in areas where sediments are undersaturated in silica, dinoflagellate cysts are not found at present in many open ocean environments where they were common in the last glacial.

Each approach also has key uncertainties that must be resolved to clarify the meaning of the temperature estimates inferred in each case. In the case of the alkenones this possibly relates to the depth and time of production of the signal, and the role of sedimentary processes in laterally mixing the alkenone inputs. In the case of the use of microfossils abundances the Achilles heel is in the understanding of the ecology of each fossil group, its precise relationship to the desired environmental parameter and the validity of this relationship in space and through time. For the Mg/Ca measurement in planktonic foraminifera the key pending issue is probably on the development of worldwide valid calibration, the role of vital effects in the calibration, and the imprint of calcite dissolution in the deep ocean. Finally, for $\delta^{18}O$ in calcareous tests the challenge is to establish in each studied region the relative importance of the environmental factors that influence the isotopic signature. In conclusion, all proxy approaches provide slightly or substantially different SST estimations, and the meaning of what is "sea surface" and "temperature", and for which season in each case, is still a matter of debate.

The compilation of global SST reconstruction maps which summarise the information from all approaches available may thus appear too much of a daunting challenge at present. It is a key issue to decide if any single approach should be given more credibility than others. But given that all proxies are fraught with uncertainties taking this a priori assumption on a general basis seems unjustified. A multi-proxy compilation of SST for the same depth and season is also impossible to obtain at present. Maps like those produced by CLIMAP are only feasible on a single-proxy basis. Once accepted that no single proxy is "right" and "right everywhere", the joint interpretation of multiple maps to infer SST during the last glacial is not trivial. We cannot provide the solution here, but in our effort to advance in updating CLIMAP reconstructions, it may be useful to think about the purpose of generating the compilations and who will be the end-users of such products. Given the current shortage of resources it may be more efficient to focus efforts on some strategies rather than others to achieve short-term progress whilst maintaining long-term goals in sight. Rather than waiting for years to provide a consensus multi-proxy map of SST for the LGM, some intermediate products could be assembled that still represent a genuine advance in our understanding of the LGM climate, and be of use to the community. For instance, why is it necessary, in contrast to being desirable, to provide a single multi-proxy SST map? If multi-proxy maps of SST cannot be derived, what could still be useful to advance our understanding of climate change? If multiple maps are provided for different approaches, how may they be interpreted by different users?

The second challenge in such an interdisciplinary research field is the flow of information between different communities, such as data producers and users like climate modellers. It is our perception that issues that may be taken for granted among some groups may be ignored by others even if the general field of work is the same, in our case climate research. Few individuals work in the transition between disciplines and the flow and processing of information between research areas is not always smooth and unbiased by each individual's perception. We decided to find out as objectively as possible, as far as we were able to, the perception and opinion on some of the issues outlined above by constructing a questionnaire directed to climate modellers. Its aim was to dispel some doubts on how the palaeo-maps were perceived, their possible uses, and obtain suggestions on how to provide useful mapping outputs that would represent an improvement on the state of the art and be valuable to other users than those in the MARGO project. Of course, we all knew, or thought we knew, the answers to some issues, but the answers were not always the same and were based on our restricted circle of contacts. The questionnaire was publicised among the palaeoclimate modelling community and was returned by North American, European and Japanese modellers from 11 groups, which represent more than half of the participants in the Palaeoclimate Modelling Intercomparison Project (http://www-lsce.cea.fr/pmip2/). These groups are engaged in almost all cases in 3D general circulation modelling, and although those that provided a reply identified themselves, it was our commitment to maintain the anonymity of the participants. A summary and discussion of the results of the exercise are shown on the MARGO web site (http://www.pangaea.de/projects/MARGO). Perhaps, the key (and not unexpected) message that one can draw from the questionnaire replies is that palaeomaps will be more useful to modellers, the simpler they are in their graphical representations and the better explained the proxy approaches are. The former means that fancy coloured maps are good for wall displays and as teaching aids, but that raw and gridded data are more useful for science. The latter means that the better described the constraints in the interpretation of the proxies, and the advantages and shortcomings of each approach, are the better use will be made of the data by those that need them.

Acknowledgements

First of all we would like to express our most sincere thanks to the contributors to this volume for their tremendous efforts in assembling their databases into the MARGO archive, for their contributions to the discussions held at the two MARGO workshops, and finally for their willingness to reassess all their data sets and reshape their compilations according to the MARGO guidelines which they agreed on beforehand. R.R.S greatly acknowledges the financial support by the IMAGES-PAGES program and the HANSE Wissenschaftskolleg (HWK) at Delmenhorst to hold the first MARGO workshop at the HWK and also to the DEKLIM program of the BMBF Germany and the European 5th framework projects (CESOP, EVR1-2001-40018, and MOTIF, EESD-ESD-3 (JO 2000/C 324/09)) for data compilation into the PANGAEA archive. A.R.-M. is grateful to the Fundacio Abertis and the IMAGES-PAGES program for providing technical and financial resources for the second MARGO workshop at the castle de Castellet de Foix. Finally, we would like to thank all reviewers who helped to profoundly improve the articles presented in this volume. Without their willingness to provide timely reviews this volume would not exist.

References

Bard, E., 2001. Comparison of alkenone estimates with other paleotemperature proxies. Geochemistry Geophysics Geosystems 2, art. no. 2000GC000050.

Barker, S., Cacho, I., Benway, H.M., Tachikawa, K., 2005. Planktonic foraminiferal Mg/Ca as a proxy for past oceanic temperatures: a methodological overview and data compilation for the last glacial maximum. Quaternary Science Reviews, this issue, doi:10.1016/j.quascirev.2004.07.016.

Barrows, T.T., Juggins, S., 2005. Sea-surface temperatures around the Australian margin and Indian Ocean during the last glacial maximum. Quaternary Science Reviews, this issue, doi:10.1016/j.quascirev.2004.07.020.

Billups, K., Schrag, D.P., 2000. Surface ocean density gradients during the last glacial maximum. Paleoceanography 15, 110–123.

Chen, M.-T., Huang, C.-C., Pflaumann, U., Waelbroeck, C., Kucera, M., 2005. Estimating glacial western Pacific sea-surface temperature: methodological overview and data compilation of surface sediment planktonic foraminifer faunas. Quaternary Science Reviews, this issue, doi:10.1016/j.quascirev.2004.07.013.

CLIMAP Project Members, CLIMAP, 1976. The surface of the ice-age Earth. Science 191, 1131–1137.

CLIMAP Project Members, CLIMAP, 1981. Seasonal reconstructions of the Earth's surface at the last glacial maximum. Geological Society of American Map and Chart Series, MC-36, Geological Society of America, Boulder, CO.

de Vernal, A., Eynaud, F., Henry, M., Hillaire-Marcel, C., Londeix, L., Mangin, S., Matthiessen, J., Marret, F., Radi, T., Rochon, A., Solignac, S., Turon, J.-L., 2005. Reconstruction of sea-surface conditions at middle to high latitudes of the Northern Hemisphere during the last glacial maximum (LGM) based on dinoflagellate cyst assemblages. Quaternary Science Reviews, this issue, doi:10.1016/j.quascirev.2004.06.04.

Duplessy, J.C., Bé, A.W.H., Blanc, P.L., 1981. Oxygen and carbon isotopic composition and biogeographic distribution of planktonic foraminifera in the Indian Ocean. Palaeogeography Palaeoclimatology Palaeoecology 33, 9–46.

Duplessy, J.-C., Labeyrie, L., Juillet-Leclerc, A., Maitre, F., Duprat, J., Sarnthein, M., 1991. Surface salinity reconstruction of the North Atlantic Ocean during the last glacial maximum. Oceanologica Acta 14, 311–324.

Gersonde, R., Crosta, X., Abelmann, A., Armand, L., 2005. Sea surface temperature and sea ice distribution of the last glacial Southern Ocean—A circum-Antarctic view based on siliceous microfossil records. Quaternary Science Reviews, this issue, doi:10.1016/j.quascirev.2004.07.015.

Hayes, A., Kucera, M., Kallel, N., Sbaffi, L., Rohling, E.J., 2005. Glacial Mediterranean sea surface temperatures based on planktonic foraminiferal assemblages. Quaternary Science Reviews, this issue, doi:10.1016/j.quascirev.2004.02.018.

Henderson, G.M., 2002. New oceanic proxies for paleoclimate. Earth and Planetary Science Letters 203, 1–13.

Kiefer, T., Kienast, M., 2005. Patterns of deglacial warming in the Pacific Ocean: a review with emphasis on the time interval of Heinrich event 1. Quaternary Science Reviews, this issue, doi:10.1016/j.quascirev.2004.02.021.

Kucera, M., Weinelt, Mara, Kiefer, T., Pflaumann, U., Hayes, A., Weinelt, Martin, Chen, M.-T., Mix, A.C., Barrows, T.T., Cortijo, E., Duprat, J., Juggins, S., Waelbroeck, C., 2005. Reconstruction of the glacial Atlantic and Pacific sea-surface temperatures from assemblages of planktonic foraminifera: multi-technique approach based on geographically constrained calibration datasets. Quaternary Science Reviews, this issue, doi:10.1016/j.quascirev.2004.07.017.

Lea, D.W., Pak, D.K., Spero, H.J., 2000. Climate impact of late Quaternary equatorial Pacific sea surface temperature variations. Science 289, 1719–1724.

Meland, M.Y., Jansen, E., Elderfield, H., 2005. Constraints on SST estimates for the northern North Atlantic/Nordic Seas during the LGM. Quaternary Science Reviews, this issue, doi:10.1016/j.quascirev.2004.05.011.

Mix, A.C., Bard, E., Schneider, R., 2001. Environmental processes of the ice age: land, oceans, glaciers (EPILOG). Quaternary Science Reviews 20, 627–657.

Morey, A.E., Mix, A.C., Pisias, N.G., 2005. Planktonic foraminiferal assemblages preserved in surface sediments correspond to multiple environmental variables. Quaternary Science Reviews, this issue, doi:10.1016/j.quascirev.2004.09.011.

Paul, A., Schäfer-Neth, C., 2005. How to combine sparse proxy data and coupled climate models. Quaternary Science Reviews, this issue, doi:10.1016/j.quascirev.2004.05.010.

Paul, A., Mulitza, S., Pätzold, J., Wolff, T., 1999. Simulation of oxygen isotopes in a global ocean model. In: Fischer, G., Wefer, G. (Eds.), Use of Proxies in Paleoceanography. Springer, Berlin, pp. 655–686.

Rosell-Melé, A., Bard, E., Emeis, K.C., Farrimond, P., Grimalt, J., Müller, P., Schneider, R.R., 1998. Project takes a new look at past sea surface temperatures. EOS, transactions, American Geophysical Union 79, 393–394.

Rosell-Melé, A., Bard, E., Emeis, K.-C., Grieger, B., Hewitt, C., Müller, P.J., Schneider, R.R., 2004. Sea surface temperature anomalies in the oceans at the LGM estimated from the alkenone-$U_{37'}^K$ index: comparison with GCMs. Geophysical Research Letters 31, L03208.

Sarnthein, M., Gersonde, R., Niebler, S., Pflaumann, U., Spielhagen, R., Thiede, J., Wefer, G., Weinelt, M., 2003. Overview of glacial Atlantic ocean mapping (GLAMAP 2000). Paleoceanography 18 (2), 1030 doi:10.1029/2002PA000769.

Schäfer-Neth, C., Paul, A., Mulitza, S., 2005. Perspectives on mapping the MARGO reconstructions by variogram analysis/kriging and objective analysis. Quaternary Science Reviews, this issue, doi:10.1016/j.quascirev.2004.06.017.

Schmidt, G.A., Mulitza, S., 2002. Global calibration of ecological models for planktonic foraminifera from coretop carbonate oxygen-18. Marine Micropalaeontology 44, 125–140.

Sikes, E.L., Samson, C.R., Guilderson, T.P., Howard, W.R., 2000. Old radiocarbon ages in the southwest Pacific Ocean during the last glacial period and deglaciation. Nature 405, 555–559.

Stuiver, M., Reimer, P.J., Bard, E., Beck, J.W., Burr, G.S., Hughen, K.A., Kromer, B., McCormac, F.G., van der Plicht, J., Spurk, M., 1998. INTCAL98 radiocarbon age calibration, 24,000–0 cal BP. Radiocarbon 40, 1041–1083.

Waelbroeck, C., Duplessy, J.C., Michel, E., Labeyrie, L., Paillard, D., Duprat, J., 2001. The timing of the last deglaciation in North Atlantic climate records. Nature 412, 724–727.

Waelbroeck, C., Mulitza, S., Spero, H., Dokken, T., Kiefer, T., 2005. A global compilation of Late Holocene planktonic foraminiferal $\delta^{18}O$: relationship between surface water temperature and $\delta^{18}O$. Quaternary Science Reviews, this issue, doi:10.1016/j.quascirev.2003.10.014.

WOA, 1998. World ocean atlas 1998, version 2, http://www.nodc.noaa.gov/oc5/woa98.html. Technical Report, National Oceanographic Data Center, Silver Spring, Maryland.

Quaternary Science Reviews 24 (2005) 821–834

Planktonic foraminiferal Mg/Ca as a proxy for past oceanic temperatures: a methodological overview and data compilation for the Last Glacial Maximum

Stephen Barker[a],*, Isabel Cacho[b], Heather Benway[c], Kazuyo Tachikawa[d]

[a]*Department of Earth Sciences, University of Cambridge, Downing Street, Cambridge CB2 3EQ, UK*
[b]*CRG Marine Geosciences, Department of Stratigraphy, Paleontology and Marine Geosciences, University of Barcelona,*
C/Martí i Franqués, s/n, E-08028 Barcelona, Spain
[c]*College of Oceanic and Atmospheric Sciences, 104 COAS, Administration Bldg., Oregon State University, Corvallis, OR 97331, USA*
[d]*CEREGE, Europole de l'Arbois BP 80, 13545 Aix en Provence, France*

Abstract

As part of the Multi-proxy Approach for the Reconstruction of the Glacial Ocean (MARGO) incentive, published and unpublished temperature reconstructions for the Last Glacial Maximum (LGM) based on planktonic foraminiferal Mg/Ca ratios have been synthesised and made available in an online database. Development and applications of Mg/Ca thermometry are described in order to illustrate the current state of the method. Various attempts to calibrate foraminiferal Mg/Ca ratios with temperature, including culture, trap and core-top approaches have given very consistent results although differences in methodological techniques can produce offsets between laboratories which need to be assessed and accounted for where possible. Dissolution of foraminiferal calcite at the sea-floor generally causes a lowering of Mg/Ca ratios. This effect requires further study in order to account and potentially correct for it if dissolution has occurred. Mg/Ca thermometry has advantages over other paleotemperature proxies including its use to investigate changes in the oxygen isotopic composition of seawater and the ability to reconstruct changes in the thermal structure of the water column by use of multiple species from different depth and or seasonal habitats. Presently available data are somewhat limited to low latitudes where they give fairly consistent values for the temperature difference between Late Holocene and the LGM (2–3.5 °C). Data from higher latitudes are more sparse, and suggest there may be complicating factors when comparing between multi-proxy reconstructions.
© 2004 Elsevier Ltd. All rights reserved.

1. Introduction

Planktonic foraminiferal Mg/Ca thermometry is a relatively recent addition to the expanding set of proxies used for reconstructing past changes in sea surface temperature (SST). Mg^{2+} is one of several divalent cations which may substitute for Ca during the formation of biogenic calcium carbonate. Its incorporation into foraminiferal calcite is influenced by the temperature of the surrounding seawater during growth such that foraminiferal Mg/Ca ratios increase with increasing temperature. The temperature sensitivity of foraminiferal Mg/Ca ratios was first reported by Chave (1954) and Blackmon and Todd (1959) using X-ray diffraction. Later studies by Kilbourne and Sen Gupta (1973) using atomic absorption analysis and by Duckworth (1977) using electron microprobe analysis reinforced this finding. Cronblad and Malmgren (1981) produced down-core records of foraminiferal Mg and Sr content variability and suggested the potential paleoclimatic significance of minor element composition

*Corresponding author. Lamont Doherty Earth Observatory, 202 New Core Laboratory, P.O. Box 1000, 61 Route 9W Palisades, New York 10964 USA Columbia University, USA. Tel.: 1-845-365-8866; fax: 1-845-365-8154.

E-mail addresses: sbarker@ldeo.columbia.edu (S. Barker), icacho@ub.edu (I. Cacho), hbenway@coas.oregonstate.edu (H. Benway), kazuyo@cerege.fr (K. Tachikawa).

0277-3791/$ - see front matter © 2004 Elsevier Ltd. All rights reserved.
doi:10.1016/j.quascirev.2004.07.016

variation. More recently Mg/Ca thermometry has been greatly refined and applied almost routinely to paleoceanographic questions concerning temperature variability through time (e.g. Nürnberg et al., 1996; Hastings et al., 1998; Lea et al., 1999, 2000, 2002; Mashiotta et al., 1999; Elderfield and Ganssen, 2000; Rosenthal et al., 2000; Koutavas et al., 2002; Stott et al., 2002; Pahnke et al., 2003; Rosenthal et al., 2003; Visser et al., 2003). Mg/Ca ratios of other calcifying organisms have also been shown to be sensitive to changes in temperature. These include benthic foraminifera (e.g. Izuka, 1988; Rathburn and De Deckker, 1997; Rosenthal et al., 1997; Lear et al., 2000; Toyofuku et al., 2000; Martin et al., 2002; Billups and Schrag, 2003), ostracods (e.g. Chivas et al., 1986; Dwyer et al., 1995; Wansard, 1996; De Deckker et al., 1999), coccoliths (Stoll et al., 2001) and corals (e.g. Mitsuguchi et al., 1996; Wei et al., 1999; Watanabe et al., 2001).

Mg/Ca thermometry has distinct advantages over other temperature proxies. The oceanic residence times for Ca and Mg are relatively long (10^6 and 10^7 years, respectively) therefore the Mg/Ca ratio of seawater may be considered to be constant over glacial/interglacial timescales. This assertion removes considerable uncertainty when reconstructing paleotemperatures using foraminiferal Mg/Ca ratios. In contrast, foraminiferal $\delta^{18}O$, which is also sensitive to changes in temperature, is strongly influenced by changes in the isotopic composition of seawater (δw). Since δw varies both as a function of global ice volume and local salinity differences, the direct interpretation of foraminiferal $\delta^{18}O$ is not straightforward. However, because these two proxies are attained from a single medium (i.e. foraminiferal calcite) they can be combined in order to reconstruct variations in δw over time (e.g. Mashiotta et al., 1999; Elderfield and Ganssen, 2000; Lea et al., 2002). A similar approach has been to combine foraminiferal $\delta^{18}O$ with other SST proxies such as alkenones and faunal counting (e.g. Rostek et al., 1993; Cayre et al., 1999; Arz et al., 2003). However, only Mg/Ca ratios coupled with $\delta^{18}O$ can guarantee a common source of the signal, averaging the same environmental conditions (season and spatial habitat). This is one of the greatest contributions of Mg/Ca paleothermometry since errors and uncertainties for δw reconstructions can be substantially minimized. Further, since temperature estimates based on Mg/Ca ratios are specific to the species employed they may be used to reconstruct temperatures from different depths in the water column depending on the species' habitat preferences. Measurement of foraminiferal Mg/Ca ratios is quite straightforward with modern techniques of elemental analysis and as a result, high resolution records may be attained in a relatively short time.

Foraminiferal Mg/Ca thermometry is still a relatively new proxy with respect to some of the more established methods of paleotemperature estimation. The aim of this report is to bring together existing published and previously unpublished Mg/Ca measurements and temperature estimates from several species of planktonic foraminifera of both Late Holocene and Last Glacial Maximum (LGM) age. This synthesis of available Mg/Ca data contributes to the aims of the Multi-proxy Approach for the Reconstruction of the Glacial Ocean (MARGO) working group. In order to make a direct comparison of Mg/Ca data and inferred temperatures from different studies it is first necessary to review some of the various technical aspects of Mg/Ca thermometry. These include laboratory procedures, temperature calibrations, use of multiple species, dissolution effects, discussion of current data and comparison with other temperature proxies. This work does not aim to reinterpret or evaluate previous work by other authors. It is most importantly a synthesis of published data and summary of considerations applicable to Mg/Ca thermometry.

2. Calibration of planktonic foraminiferal Mg/Ca versus temperature

Studies on inorganic precipitation of calcite in seawater predict a positive temperature control on Mg incorporation (e.g. Kinsman and Holland, 1969; Katz, 1973; Mucci, 1987; Oomori et al., 1987). As discussed by Rosenthal et al. (1997) and Lea et al. (1999) thermodynamic considerations predict that the temperature dependence of Mg uptake into calcite should be exponential, with a sensitivity of around 3% per °C increase in temperature. This prediction is bourn out by the work of Oomori et al. (1987) on inorganic calcite precipitates (Fig. 1a). The temperature dependence of Mg uptake into foraminiferal calcite is also generally accepted to be exponential, with a sensitivity of about 10% per °C (Fig. 1b). Measured Mg/Ca ratios of natural foraminiferal calcite are generally 1–2 orders of magnitude lower than those predicted for inorganic precipitates (Fig. 1). This suggests that biological influences play an important role in biogenic calcification. The offset in Mg/Ca between inorganic precipitates and foraminiferal calcite, together with differences between individual species of foraminifera (e.g. Lea et al., 1999) (Fig. 1c), stresses the importance for empirical determination of foraminiferal Mg/Ca thermometry calibrations.

Various approaches have been made to calibrate Mg/Ca versus temperature for several species of planktonic foraminifera. These include core-top, culture and sediment trap studies, each of which may have its advantages and disadvantages although as it turns out, very good agreement has been achieved between all three approaches (Table 1). Culture-based calibrations

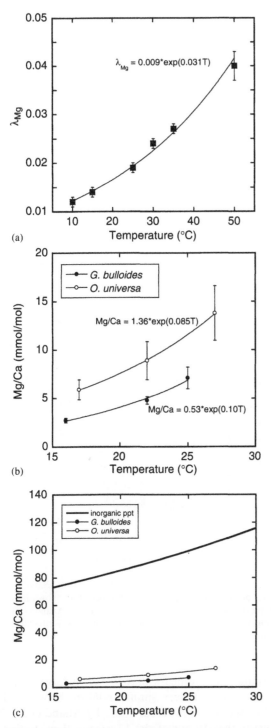

Fig. 1. (a) Results of an inorganic precipitation experiment by Oomori et al. (1987) showing an exponential relation between temperature and the distribution coefficient (λ_{Mg}) of Mg^{2+} ions between calcite and solution ($\lambda_{Mg} \approx (Mg/Ca)_{calcite}/(Mg/Ca)_{solution}$). The observed sensitivity is about 3% per °C. (b) Culturing experiments also suggest an exponential relation between foraminiferal Mg/Ca and calcification temperature with a sensitivity of about 9–10% per °C (modified after Lea et al., 1999). Offsets between species highlight the benefit of species specific calibrations. The importance of empirical calibrations is also highlighted by the difference between measured foraminiferal Mg/Ca ratios (Lea et al., 1999) and those predicted by inorganic precipitation experiments (Oomori et al., 1987) (c).

such as by Nürnberg et al. (1996) and Lea et al. (1999) have the distinct advantage that temperature (T) is constrained during the experiment i.e. it is an independent variable. This is not the case for core-top or plankton tow or trap studies where T must be estimated using either a climatological atlas or some derivation of the calcification temperature attained from foraminiferal $\delta^{18}O$ and δw (e.g. Elderfield and Ganssen, 2000; Anand et al., 2003). In these cases, T itself becomes a dependent variable and may introduce greater error into the calibration than the measurement of Mg/Ca (Anand et al., 2003). A potential disadvantage of culture calibrations is that laboratory conditions may not realistically reproduce the natural environment sufficiently to ensure natural chamber growth. Further, since foraminiferal reproduction cannot be achieved in the laboratory, juvenile specimens must be collected from their natural habitat before culturing. Controlled conditions and subsequent chemical analysis can therefore only be applied to later stages of test formation i.e. those chambers formed during culture. Routine determination of foraminiferal Mg/Ca ratios for paleotemperature reconstruction is performed on whole tests representing the entire period of test growth. Studies on foraminiferal test chemistry from various size fractions of foraminifera have shown that Mg/Ca ratios as well as other properties (e.g. Sr/Ca, $\delta^{13}C$ and $\delta^{18}O$) are controlled by test size (Oppo and Fairbanks, 1989; Ravelo and Fairbanks, 1995; Spero and Lea, 1996; Elderfield et al., 2002). Uncertainty in culture calibrations may therefore arise if differences in Mg/Ca during growth are caused by changing biological controls rather than solely by changes in temperature.

Calibrations determined using core-top material such as those by Elderfield and Ganssen (2000) are valuable since they are based on material that will eventually form the sedimentary record, having gone through a complete life cycle including gametogenesis and any secondary calcite formation. However, complications in calibrating the results may arise if the samples used have undergone post-depositional alteration. As discussed later, partial dissolution of foraminiferal calcite tends to cause a decrease in Mg/Ca. Since the solubility of carbonate tends to increase with depth at any location, core-top material from greater depths will be prone to alteration of Mg/Ca. Dekens et al. (2002) produced a core-top calibration using samples from the tropical Pacific and Atlantic. Their calibration included a correction term for dissolution effects on Mg/Ca in foraminifera which we discuss in Section 5.

Calibrations based on water column samples have the great advantage that the season of growth is known and therefore better constraints can be made on the specific temperatures used in calibrations. Trap material also has the advantage that it most closely represents the material entering the sedimentary record without

Table 1
Summary of published temperature calibrations for single and multiple species of planktonic foraminiferal Mg/Ca

Species	Source	Temperature	B	A	Reference
N. pachyderma	Coretops	0–200 m water depth	0.46	0.088	Nürnberg et al., 1995
G. sacculifer	Coretops	Annual SST	0.37	0.09	Dekens et al., 2002
G. sacculifer	Culture	Fixed	0.39 (±0.06)	0.089 (±0.008)	Nürnberg et al., 1996
G. sacculifer	Sediment trap	δ^{18}O derived	0.35 (±0.01)	0.090	Anand et al., 2003
G. bulloides	Culture	Fixed	0.53 (±0.17)	0.102 (±0.008)	Lea et al., 1999
G. bulloides	Coretop/culture	Annual SST/fixed	0.47 (±0.03)	0.107 (±0.003)	Mashiotta et al., 1999
G. ruber (w)	Coretop	Annual SST	0.30 (±0.06)	0.089 (±0.007)	Lea et al., 2000
G. ruber (w)	Coretop	Annual SST	0.38	0.09	Dekens et al., 2002
G. ruber (w)	Sediment trap	δ^{18}O derived	0.34 (±0.08)	0.102 (±0.01)	Anand et al., 2003
Mixed	Coretop	δ^{18}O derived	0.52	0.100	Elderfield and Ganssen, 2000
Mixed	Sediment trap	δ^{18}O derived	0.38 (±0.02)	0.090 (±0.003)	Anand et al., 2003
Cibicidoides ssp	Coretop	BWT	0.87 (±0.05)	0.109 (±0.007)	Lear et al., 2002

The data of Nürnberg et al. (1995, 1996) have been recalibrated using the general form Mg/Ca = B exp (A × T). G. ruber (w) is the white variety picked from the 250–350 μm size fraction. The benthic calibration of Lear et al. (2002) is given for comparison. SST = sea surface temperature, BWT = bottom water temperature.

actually reaching the sediment surface. If a reliable temperature estimate can be made for water column samples, for example by δ^{18}O derived temperature calculations (Anand et al., 2003), this may provide one of the most robust approaches to Mg/Ca thermometry calibration.

The points outlined above (specifically, problems in determining a reliable calcification temperature and partial dissolution) could potentially lead to differences between various calibrations. Further uncertainties may be caused by differences in analytical procedures and or sample preparation between different laboratories (see later). Nevertheless, as mentioned previously, results attained using very different approaches are remarkably consistent. Modern calibrations may be expressed by an exponential relation of the form:

$$Mg/Ca = B \exp(A \times T), \tag{1}$$

where T is the calcification temperature in °C. A and B are constants and dependent on species. The value of A is consistently found to be around 0.09–0.1 which reflects a temperature sensitivity in Mg/Ca of about 10% per °C increase in T. A summary of published calibrations for single and multiple species is given in Table 1. Calibrations for Mg/Ca in benthic foraminifera also suggest a similar temperature sensitivity to planktonic species (e.g. Rosenthal et al., 1997; Lear et al., 2002, Table 1).

The point has been made that empirically based calibrations are important since biological differences between species may cause significant offsets in Mg/Ca for a given temperature. However, for species where a narrow temperature range prohibits statistically meaningful calibration (see Anand et al., 2003) or for species now extinct, this requirement presents obvious difficulties. Fortunately, thanks to the growing number of calibration studies on multiple species of planktonic

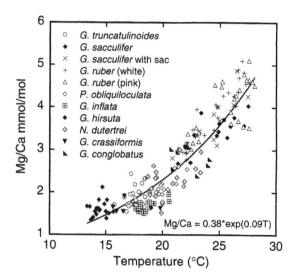

Fig. 2. Mg/Ca calibration results of Anand et al. (2003) for several species of planktonic foraminifera. Temperatures shown here are the isotopically derived calcification temperatures of Anand et al. (2003). A single temperature equation may be used to describe all data ($r = 0.93$). Modified after Anand et al. (2003).

foraminifera it appears that in fact a large number of species display a very similar relation between Mg/Ca and temperature. A recent study by Anand et al. (2003) suggests that 10 out of the 12 species they studied may be described by a single relation (Table 1, Fig. 2). The uncertainty associated with temperatures estimated using a generic equation will naturally be greater than by use of a species specific relationship for a given species. The deviation between the two estimates will depend on the species employed. For example, based on data from Anand et al. (2003), the average deviation of calculated temperature from isotopically derived "calcification" temperature using the species specific calibration for *Globigerinoides ruber* (w) (Anand et al., 2003) is

0.6 °C. This compares with 1.0 °C when using the generic calibration. The combined average deviation between calculated and isotopically derived temperatures for all species used in the generic calibration of Anand et al. (2003) increases from 0.9 to 1.5 °C for specific and generic calibrations, respectively. On the other hand, for *G. aequilateralis*, the average deviation increases from 1.0 to 3.8 °C. The increased uncertainty in this case is because this species has particularly high Mg/Ca values for a given temperature (Anand et al., 2003). Therefore although species specific calibrations are statistically preferable, where this is not possible, and with certain exceptions, some confidence may be ascribed to the use of a generic temperature calibration.

3. Intra- and inter-specific variability in Mg/Ca ratios

Planktonic foraminifera commonly migrate vertically throughout their life cycle, often forming calcite at deeper depths as they mature. This depth/temperature migration results in heterogeneity of Mg/Ca ratios within the tests of individual foraminifera (Duckworth, 1977; Nürnberg, 1995; Elderfield and Ganssen, 2000; Jha and Elderfield, 2000; Benway et al., 2003). Further intra-test variability occurs in many species due to the secretion of a secondary crust of calcite at the time of gametogenesis. Laboratory culture experiments have provided a unique source of information in this direction (e.g. Bé, 1980; Hemleben et al., 1985). For example, it has been observed that many planktonic species add a calcite crust (or "gametogenic crust") in deeper, colder water immediately prior to reproduction. In the case of *G. sacculifer* gametogenic calcite can represent up to 30% of the test weight (Bé, 1980). This crust may have a Mg/Ca ratio distinct from other regions of the test, even when formed under similar temperature conditions, as a result of differing bio-physiological controls on Mg uptake (e.g. Nürnberg et al., 1996). Thus individual fossil tests may comprise a range of compositions that reflect the changing conditions during an individual's lifetime. Recent developments in Laser Ablation, LA–ICP–MS techniques provide a unique opportunity to accurately monitor intra-test variability in trace element concentrations and understand better the distinct signature of foraminiferal migration in calcite secretion (Eggins et al., 2003; Hathorne et al., 2003). Early results suggest a migration signal in the assimilation of Mg/Ca by individual chambers of *Neogloboquadrina dutertrei* while other species like *G. ruber* (w) appear to give rather constant Mg/Ca values (Eggins et al., 2003).

It is clear that measured bulk test Mg/Ca ratios should be considered as an integration of the temperatures experienced during calcification. The Mg/Ca ratio of 'whole-test' samples therefore represents the integra-tion of various depth habitats and (potentially) biological controls. As a result, the temperature derived from foraminiferal Mg/Ca ratios is the temperature of calcification at whatever depth range and season that particular species grew. This definition is distinct from the strict notion of sea surface temperature (SST) in that many species of planktonic foraminifera live at depths greater than 50 m or so (e.g. Erez and Honjo, 1981; Deuser and Ross, 1989; Anand et al., 2003). This issue does not seem to be critical for surface and mixed layer species such as *G. sacculifer* and *G. ruber* (w) as demonstrated by the close agreement between the calibrations for *G. ruber* (w) reported by Lea et al. (2000) using SST and Anand et al. (2003) using isotopically derived calcification temperatures (Table 1). However, the distinction becomes more important when dealing with deeper dwelling species.

Inter-species differences in Mg/Ca, due to differences in seasonal and or depth habitats, could be viewed as a limitation of the Mg/Ca method; for instance it is clear that the temperature derived from a thermocline dwelling species will not reflect sea surface temperature. However, application of Mg/Ca thermometry to multiple species of planktonic foraminifera within a single core provides a potentially powerful tool for reconstructing changes in water column structure. For instance, *G. ruber* (w) may be used to reconstruct surface and mixed layer temperatures while *N. dutertrei* generally represents thermocline conditions (Spero et al., 2003). By measuring Mg/Ca ratios in both species from a single core it is possible to reconstruct temporal changes in upper water column thermal gradient. A multi-species approach has been applied in several studies based on the stable isotope compositions of planktonic foraminifera ($\delta^{18}O$ and $\delta^{13}C$) (e.g. Mulitza et al., 1997; Spero et al., 2003; Simstich et al., 2003). These studies demonstrate that with a suitable selection of species, it is possible to reconstruct past oceanographic conditions for both the thermocline and mixed layer. But such an approach has not yet been broadly used for Mg/Ca studies. Elderfield and Ganssen (2000) analysed several different planktonic foraminifers in a site from the tropical Atlantic and observed that Mg/Ca temperature estimates provide consistently warmer or colder temperatures in accordance with the depth-habitat preferences of each of the species considered. They also showed that glacial–interglacial thermal variability evolved differently between the different habitats i.e. the upper water column thermal structure changed between glacial and interglacial times. Stott et al. (2002) studied two surface-dwelling species that they interpret to reflect different seasonal conditions for the region studied (western tropical Pacific). These early results open up new and promising perspectives for paleoceanographic reconstructions. The ability to reconstruct vertical and possibly seasonal temperature gradients is an advantage over other geochemical proxies such as alkenone based SST reconstructions which strictly monitor the conditions of the euphotic layer in which the particular phytoplankton grow.

4. Methodology

Methods of preparing and analysing foraminiferal samples for Mg/Ca measurement vary between laboratories; there is no *standardised* method per se. Differences in methodology and analysis could lead to discrepancies between laboratories and as such it is important to know what differences exist and how these might affect the determination of Mg/Ca ratios. These questions have recently been addressed by an inter-laboratory comparison study by Rosenthal et al. (2004) in which several foraminiferal samples and standard solutions were run by a number of laboratories throughout Europe and the USA in order to assess the relative accuracy, precision and reproducibility of their methodologies. Results from this work suggest that analytical techniques used by a selection of laboratories produced generally good precision but poor inter-lab consistency. This was probably due to inaccuracies introduced during preparation of calibration standard solutions within particular laboratories and may be addressed in the future by use of a universal accuracy standard. The findings also point towards more significant disparities caused by differences in sample preparation (see below).

Modern analytical techniques for analysing bulk foraminiferal Mg/Ca ratios include Inductively Coupled Plasma Mass Spectrometry (ICP-MS) (e.g. Rosenthal et al., 1999) and ICP-AES (Atomic Emission Spectrometry) (e.g. de Villiers et al., 2002). Both of these techniques provide high precision measurements of Mg/Ca ratios of the order ±0.5% (Rosenthal et al., 2004). An alternative approach to Mg/Ca analysis is by use of electron microprobe (e.g. Nürnberg, 1995) or Laser Ablation, LA–ICP–MS techniques (e.g. Eggins et al., 2003; Reichart et al., 2003). These sorts of technique allow investigation into heterogeneity in Mg/Ca throughout a single test which is of value to studies concerning the systematics of Mg incorporation during test formation or removal during dissolution. Another approach to investigating Mg/Ca heterogeneity is that of Haley and Klinkhammer (2002) using a sequential leaching technique (this method is discussed in more detail later).

Most modern techniques of sample preparation for measurement of foraminiferal Mg/Ca ratios follow the general procedures developed by Boyle (1981), Boyle and Keigwin (1985) and Lea and Boyle (1991) for the determination of foraminiferal Cd/Ca and Ba/Ca ratios. The general requirements for cleaning foraminiferal

calcite prior to Mg/Ca analysis are the removal of silicate phases (predominantly clays from within the test), organic matter and potentially a Mn-oxide coating. There has been discussion over the absolute necessity of the last of these steps (e.g. Brown and Elderfield, 1996; Haley and Klinkhammer, 2002; Martin and Lea, 2002; Barker et al., 2003). This has resulted in the evolution of essentially two variants of the Mg-cleaning protocol; one with a reductive step for removal of an oxide coating, the other without (the "Cd" and "Mg" versions as described by Rosenthal et al., 2004). The effect of reductive cleaning is to lower Mg/Ca. This is highlighted by a systematic difference (of about 8%) between laboratories using the "Cd" and "Mg" cleaning protocols (Rosenthal et al., 2004) (Fig. 3). Reductive cleaning involves bathing the sample in a hot buffered solution of hydrazine (Boyle and Keigwin, 1985) and causes some dissolution of foraminiferal carbonate. Since partial dissolution is known to cause a lowering of Mg/Ca ratios (see later), the reduction in Mg/Ca during reductive cleaning may be due either to removal of a contaminant phase or to the partial dissolution that occurs as a side-effect of reductive cleaning. Barker et al. (2003) point out that core-top samples having initially very low Mn/Ca ratios (presumably with only a very small amount of oxide coating) still show a significant (~10%) decrease in foraminiferal Mg/Ca after reductive treatment and suggest that this probably reflects loss through partial dissolution. It is also possible that the lowering of Mg/Ca during reductive cleaning may be due to the removal of some contaminating phase which is not necessarily the Mn-oxide coating but may itself be trapped by such a coating and only released by its

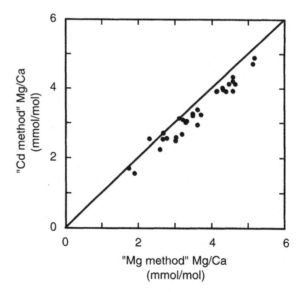

Fig. 3. Comparison of two cleaning procedures ("Cd" versus "Mg") used on splits of the same foraminiferal samples. Results using the "Cd method" are almost consistently lower than by use of the "Mg method"—see text for details. Data are from Rosenthal et al. (2004).

removal. This issue is yet to be resolved but it is clear that if contaminant Mg is present, it should be removed.

Intra-lab reproducibility of measured foraminiferal Mg/Ca ratios using a consistent cleaning protocol can be better than ±2% (or about 0.2 °C) (Barker et al., 2003). Use of a consistent method of sample preparation between laboratories would presumably lead to greater inter-lab reproducibility and may therefore be desirable. If reductive cleaning is deemed unnecessary for Mg/Ca work, it seems logical to omit this step. But as pointed out by Rosenthal et al. (2004) those laboratories wishing to simultaneously measure other trace metal ratios such as Cd/Ca will presumably continue with the "Cd" method. Although the difference in measured Mg/Ca ratios attained by use of the two versions is systematic, it is also pointed out by Barker et al. (2003) that the offset (equivalent to perhaps 1 °C) is approximately equal to the uncertainty quoted for Mg/Ca temperature calibrations of 1.3 °C (Lea et al., 1999) and 1.2 °C (Anand et al., 2003). It may also be suggested that samples should be cleaned by the same method as that employed for constructing the temperature calibration chosen for their interpretation. Once again, since calibrations made using the "Cd" and "Mg" methods are almost identical within error, it appears that differences may not be too problematic. Some further work is required to establish firmly whether differences produced by the two cleaning protocols result in different interpretations of temperature changes through time. It should also be mentioned that samples from certain locations and or time intervals may demand a more or less rigorous cleaning procedure than others. For instance, sediments from strongly reducing environments may be more prone to contamination from Mg associated with ferromanganese phases and therefore reductive cleaning may be necessary. Similarly, sediments containing significant amounts of coarse-grained terrigenous material will require extra care to remove such material from foraminiferal samples.

5. Effects of partial dissolution

Several studies have shown that foraminiferal Mg/Ca ratios systematically decrease through post-depositional dissolution under the influence of undersaturated bottom-waters or pore-waters (Fig. 4) (Lorens et al., 1977; Rosenthal and Boyle, 1993; Russell et al., 1994; Brown and Elderfield, 1996; Rosenthal et al., 2000). The cause of this reduction is thought to be the preferential dissolution of high Mg/Ca regions of the test i.e. those formed in warmer waters (Brown and Elderfield, 1996; Rosenthal et al., 2000). The effect of partial dissolution is therefore a biasing of mean test Mg/Ca values towards colder temperatures.

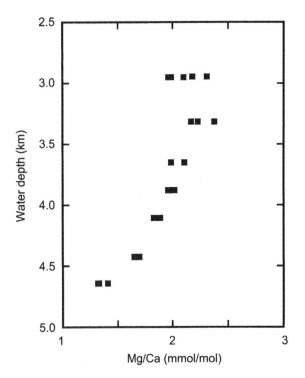

Fig. 4. Coretop Mg/Ca data for *G. tumida* from the Atlantic Ocean show a strong relationship with water depth, demonstrating the effect of partial dissolution on Mg/Ca (after Russell et al., 1994).

If the interpretation of foraminiferal Mg/Ca ratios is influenced by post-depositional alteration, it is necessary to estimate the uncertainty due to such alteration and if possible, correct for it. There have been various methods suggested for accounting for Mg/Ca variations due to dissolution. Lea et al. (2000) show that Mg/Ca ratios in *G. ruber* (w) from core-top samples across the Ontong Java Plateau (OJP) decrease by about 12% per kilometre increase in water depth. They use this finding to suggest a potential uncertainty in their glacial paleo-temperature reconstructions of about +0.5 °C taking into account published records of carbonate preservation from that region (Farrell and Prell, 1989; Le and Shackleton, 1992). This uncertainty may be compared with their calculated glacial-interglacial temperature changes of up to 5 °C over the last 450 ka. Dekens et al. (2002) approach the problem by incorporating a dissolution term (in this case water depth or bottom-water carbonate ion concentration, $[CO_3^=]_{bw}$, into the calibration of Mg/Ca with temperature. Using $[CO_3^=]_{bw}$ rather than water depth allows use of a single equation for both the Atlantic and Pacific Ocean basins since the saturation state of bottom waters at a particular depth differs markedly between these ocean basins. An interesting observation made by Dekens et al. (2002) is that different species show varying responses to dissolution with respect to their Mg/Ca ratios. These responses are not necessarily related to the overall susceptibility of a particular species to dissolution. For example, on the OJP N. dutertrei showed a decrease in Mg/Ca of 20% per km increase in water depth (Dekens et al., 2002). This may be compared to a response in G. sacculifer of 5% per km. Berger (1970) ranked these two species 16 and 5, respectively, out of 22 in his dissolution index (1 being the most susceptible to dissolution) i.e. Mg/Ca in N. dutertrei appears to be more sensitive to dissolution than in G. sacculifer. Observations of this sort may become important when comparing multi-species records from sites where temporal changes in preservation are suspected.

One of the greatest obstacles in correcting for dissolution effects on foraminiferal Mg/Ca ratios is quantifying the extent of dissolution undergone by a particular sample. For core-top samples, correction may be made since bottom water conditions are known but for reliable correction of paleo-Mg/Ca ratios it would be desirable to be able to quantify dissolution by some independent dissolution indicator, ideally one coupled with the foraminiferal sample in question. One such approach is by the use of foraminiferal test weights. Based on the work of Lohmann (1995) Broecker and Clark (2001a) demonstrated that weight loss in the tests of planktonic foraminifera, caused by partial dissolution, in core-top samples could be calibrated with $[CO_3^=]_{bw}$. They subsequently used these calibrations to reconstruct paleo-$[CO_3^-]_{bw}$ in the glacial oceans (e.g. Broecker and Clark, 2001b; Broecker et al., 2001). Rosenthal and Lohmann (2002) demonstrated that the pre-exponential constant, B, in the Mg/Ca temperature calibration equation (Eq. (1)) varied as a function of test weight during dissolution for the tropical species G. ruber (w) and G. sacculifer. Using this relationship Mg/Ca data may be corrected for dissolution by incorporating test weight variability into temperature reconstructions. Tachikawa et al. (2003) applied this approach to a core from the Arabian Sea, which is characterised by highly variable productivity. After applying a dissolution correction based on foraminiferal test weights, Mg/Ca-based temperatures were in good agreement with alkenone-based temperatures. This result suggests that the correction of Rosenthal and Lohmann (2002) is valuable for dissolution caused by metabolic respiration within sediments. A possible limitation of this method is the observation that initial test weights of several species of planktonic foraminifera vary both regionally and temporally (Barker and Elderfield, 2002). Shell weights are thought to be sensitive to changes in the $[CO_3^-]$ of seawater in which they grow and hence may vary in parallel with changes in atmospheric CO_2 through time (Spero et al., 1997; Bijma et al., 1999; Barker and Elderfield, 2002). Obviously the initial test weight must be known in order to calculate weight loss due to dissolution.

Another potential method for quantifying foraminiferal dissolution is by the measurement of so-called 'crystallinity' (Barthelemy-Bonneau, 1978; Bonneau et al., 1980). Crystallinity is quantitatively the peak width (at half maximum height) of the 104 calcite diffraction peak measured by X-ray diffraction (XRD). It is essentially a measure of how 'perfect' the calcite lattice is; a perfect calcite crystal would comprise a single crystallised domain of effectively infinite dimension and would have a very narrow 104 diffraction peak. As a foraminiferal test dissolves at the sea-floor, "poorly crystallized" regions of calcite within the test tend to dissolve preferentially, thereby causing the measured 104 calcite peak of the sample to narrow. Hence foraminiferal crystallinity is sensitive to increasing water depth (Barthelemy-Bonneau, 1978; Bassinot et al., 2004) and as such may potentially be used to correct for dissolution in foraminiferal Mg/Ca ratios if initial environmental influences have negligible effects (Bassinot et al., 2001).

During dissolution, decreasing Mg/Ca is thought to represent the preferential removal of calcite formed at warmer temperatures (Brown and Elderfield, 1996; Rosenthal et al., 2000). The region of calcite within a foraminiferal test with the highest Mg/Ca ratio probably reflects the warmest temperature experienced by that individual and therefore may give the best impression of near surface temperatures. Benway et al. (2003) describe a method addressing the issues of dissolution and heterogeneity using the flow-through leaching technique described by Haley and Klinkhammer (2002) which allows continuous measurement of Mg/Ca during dissolution. The results show how Mg/Ca ratios of sequential leaches decrease during dissolution, presumably because higher Mg/Ca calcite is dissolved first. The high Mg/Ca ratios of the earliest stages of dissolution are interpreted to reflect calcification temperatures of near surface waters. Providing some of the high Mg/Ca calcite remains, this method may provide the opportunity to assess calcification temperatures even after partial dissolution at the seafloor.

6. Mg/Ca temperature reconstructions for the LGM

Mg/Ca paleothermometry is still a relatively new proxy and as such there are a limited number of published records relative to other, more established paleotemperature proxies. Within the framework of the MARGO project, we have developed a compilation of Mg/Ca data representing surface ocean temperature conditions during the LGM (implementing the EPI-LOG-defined LGM chronozone: 19–23 kyr BP Mix et al., 2001) and Late Holocene (MARGO-defined LH chronozone: 0–4 cal. kyr BP) periods. We have included Mg/Ca data from the mixed layer-dwelling foraminiferal

Table 2
Summary of current Mg/Ca reconstructions covering the LGM listed according to the annotation of Fig. 5. (http://www.pangaea.de/Projects/MARGO/)

Map #	Source reference	Core	Calibr. used	Species	LGM Mg/Ca temp	LH-LGM Mg/Ca temp dif.
1	Banakar (unpublished data)	SK129/CR2	6	G. sacculifer	27.6	0.8
2	Barker and Elderfield, 2002	NEAP 8K	8	G. bulloides	9.7	1.2
3	Barker (this manuscript)	16867-2	2	G. ruber	17.7	(*)
		13289-2	2	G. ruber	16.9	1.6
		13289-2	2	G. bulloides	14.6	2.9
		15637-1	2	G. ruber	14.6	4.3
		15637-1	2	G. bulloides	13.9	(*)
		KF09	2	G. ruber	15.4	(*)
		KF13	2	G. ruber	15.0	(*)
		15612-2	2	G. ruber	13.3	2.9
4	Cacho et al. (in prep.)	MD95-2043	2	G. bulloides	14.3	3.7
5	Elderfield and Gassen, 2000	BOFS 31K	2	G. ruber	15.4	0.7
		BOFS 31K	2	G. bulloides	17.1	1.9
6	Hastings et al., 1998	EN066-17GGC	6	G. sacculifer	22.7	1.4
		TT9108-1GC	6	G. sacculifer	23.2	1.8
		CP6001-4PC	6	G. sacculifer	23.1	2
7	Kiefer et al., 2003	ODP883	2	N. pachyderma (s)	5.6	−0.6
8	Koutavas et al., 2002	V21-30	6	G. sacculifer	21.3	1.0
9	Lea et al., 2000	ODP806B	4	G. ruber	26.2	2.4
		TR163-22	4	G. ruber	22.3	3.1
		TR163-20B	4	G. ruber	21.1	2.4
		TR163-19	4	G. ruber	23.3	2.9
		TR163-18	4	G. ruber	23.3	2.4
10	Lea et al., 2003	PL07-39PC	1	G. ruber	23.9	2.9
11	Mashiotta et al., 1999	RC11-120	5	G. bulloides	7.9	3.8
		E11-2	5	N. pachyderma (s)	2.8	(*)
12	Nürnberg et al., 2000	GeoB 1112	6	G. sacculifer	24.1	1.7
		GeoB 1105	6	G. sacculifer	22.5	2.1
13	Pahnke et al., 2003	MD97-2120	5	G. bulloides	8.3	4.3
14	Palmer and Pearson, 2003	ERDC-92	1	G. sacculifer	26.7	1.7
15	Rosenthal et al., 2003	MD97-2141	7	G. ruber	26.7	2.3
16	Skinner and Elderfield (subm.)	MD99-2334K	2	G. bulloides	15.7	1.0
17	Stott et al., 2002	MD98-2181	6	G. ruber	27.0	2.5
18	Visser et al., 2003	MD982162	3	G. ruber	25.7	3.4

Calib: (1) Dekens et al., 2002. (2) Elderfield and Ganssen, 2000. (3) Hastings et al., 2001. (4) Lea et al., 2000. (5) Mashiotta et al., 1999. (6) Nürnberg et al., 1996. (7) Rosenthal and Lohman, 2002. (8) Barker and Elderfield, 2002.
(*) No data available for the Late Holocene Chronozone.

species G. ruber (w), G. sacculifer, G. bulloides, and N. pachyderma (s). The data summarize results from 33 sites (Table 2), and are available on the MARGO website (http://www.pangaea.de/Projects/MARGO/). Most of these sites are concentrated in the equatorial-tropical band (Fig. 5), but more studies are now focussing on higher latitudes which, until very recently, were a barren territory for Mg/Ca studies. Further LGM Mg/Ca reconstructions are presented in Meland et al. (2004) for the North Atlantic and Nordic Sea, also included in the MARGO website.

SST reconstructions in the tropics have been a controversial issue due to the ambiguity of reconstructions based on different proxies from both terrestrial and marine sources (Broecker and Denton, 1989). The CLIMAP reconstruction (CLIMAP, 1976) suggested little change in tropical SSTs between LGM and modern. This finding is in conflict with terrestrial and

coral derived SST reconstructions which suggest low latitude LGM temperatures were up to 5–6 °C colder than the present day (e.g. Clapperton, 1993; Guilderson et al., 1994; Stute et al., 1995). Mg/Ca paleothermometry has provided insight to this paleoclimate problem; several Mg/Ca-based temperature reconstructions have yielded tropical LGM temperatures that are 2.0–3.5 °C colder than modern. This range is more in line with modelling results which suggest an LGM cooling of about 2.5 °C in equatorial regions (Crowley, 2000).

Mg/Ca records from the Western Pacific warm pool suggest a LH-LGM difference of 2.3 °C and 2.5 °C for the Sulu Sea and the nearby Mindanao Sea respectively, based on G. ruber (w) and G. sacculifer (Stott et al., 2002; Rosenthal et al., 2003). These values are comparable to the 2.4 °C LH-LGM temperature change estimate for the Ontong Java Plateau (Lea et al., 2000), and the 1.7 °C LH-LGM temperature difference

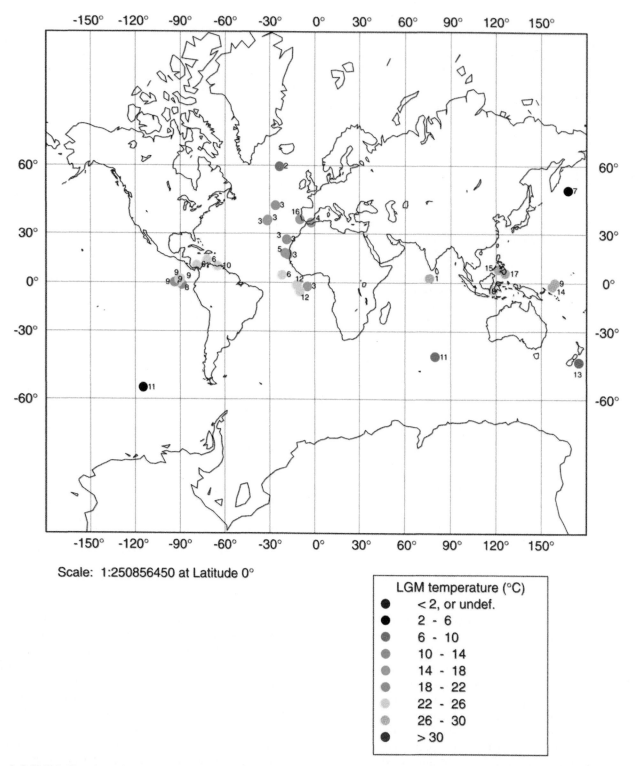

Scale: 1:250856450 at Latitude 0°

Fig. 5. LGM Mg/Ca estimated temperatures from planktonic foraminifers, data available in the MARGO data set (http://www.pangaea.de/Projects/ MARGO/). (1) Banakar (unpublished data); (2) Barker and Elderfield, 2002; (3) Barker (this manuscript); (4) Cacho et al. (in prep.); (5) Elderfield and Ganssen, 2000; (6) Hastings et al., 1998; (7) Kiefer et al., in press; (8) Koutavas et al., 2002; (9) Lea et al., 2000; (10) Lea et al., 2003; (11) Mashiotta et al., 1999; (12) Nürnberg et al., 2000; (13) Pahnke et al., 2003; (14) Palmer and Pearson, 2003; (15) Rosenthal et al., 2003; (16) Skinner (unpublished data); (17) Stott et al., 2002; (18) Visser et al., 2003.

obtained further south, also in the western tropical Pacific (Palmer and Pearson, 2003). A recent Mg/Ca temperature reconstruction from the Indo-Pacific warm pool shows a slightly larger LH-LGM temperature change of 3.4 °C (Visser et al., 2003). These results agree with alkenone-based SST estimates from the same region (Pelejero et al., 1999; Kienast et al., 2001). In the eastern tropical Pacific just north of the equator, Lea et al. (2000) estimated a LH-LGM temperature change of ~2.8 °C based on four Mg/Ca records from *G. ruber* (w). However, a Mg/Ca reconstruction based on *G. sacculifer* from the equatorial cold tongue region yielded a LH-LGM temperature change of only ~1 °C (Koutavas et al., 2002). These results may collectively suggest an intensified LGM cross-equatorial gradient (Koutavas and Lynch-Stieglitz, 2003).

In the tropical Atlantic Ocean results are less numerous but they are consistent with those from the Pacific Ocean. Results from *G. sacculifer* indicate that the LH-LGM SST difference was about 2 °C in the Caribbean Sea and in the eastern equatorial Atlantic Ocean (Hastings et al., 1998; Nürnberg et al., 2000) and nearly 3 °C for the Cariaco basin based on *G. ruber* (w) (Lea et al., 2003). *G. ruber* (w) results from the eastern tropical Atlantic also show a LH-LGM difference of 2 °C (Elderfield and Ganssen, 2000; this manuscript) but the amplitude is larger (4 °C) further north in the eastern tropical Atlantic (this manuscript).

When we move to higher latitudes *G. ruber* (w) and *G. sacculifer* become less dominant within the foraminiferal assemblage, particularly for glacial times. Therefore, the species chosen for mid- to high-latitude studies is commonly *G. bulloides*. *N. pachyderma* (s) has also been used in those studies from colder regions. High latitude Mg/Ca reconstructions are too sparse to provide conclusive results regarding LGM-LH temperature changes. However, a synthesis of the available data suggests that the situation may be rather complex, particularly for those sites from the Northern Hemisphere. Results for *G. bulloides* in the North Atlantic show little to no temperature difference for LH-LGM at 60 °N (Barker and Elderfield, 2002). At this latitude, results from different SST proxies are conflicting. Alkenone-based SSTs suggest that the LGM was warmer than the LH, whereas foraminiferal assemblages suggest that the LGM was ~5 °C cooler than the LH (see the MARGO database). Nordic sea Mg/Ca reconstructions (Meland et al., this issue) also show conflicting results, particularly for the central part of the Nordic Sea where LGM temperature reconstructions are unrealistically warm. But Mg/Ca reconstructions for the eastern and southern part of the Nordic Sea show consistent results with those estimated from $\delta^{18}O$ reconstructions, suggesting 0.5 °C colder temperatures for the LGM. Moving southward, the LGM-LH contrast is ~4 °C at 45 °N (this manuscript), 1 °C at 39 °N near the western Iberian margin (Skinner and Elderfield, submitted), and 4 °C at 36 °N in the western Mediterranean Sea (Cacho et al., in prep). Alkenone-based SST reconstructions from comparable locations suggest larger LH-LGM temperature differences of 6, 4, and 7.5 °C, respectively (Cacho et al., 2001; Calvo et al.,

2001; Pailler and Bard, 2002). The discrepancies between these two SST proxies in the Meditteranean may reflect changes in the seasonal production of *G. bulloides*. Identification of seasonal variability in foraminiferal species abundances is important since it may affect both Mg/Ca and stable isotope reconstructions.

In the Southern Hemisphere a Mg/Ca record for *G. bulloides* indicates a LH-LGM cooling of 3.8 °C for the sub-Antarctic Indian Ocean (Mashiotta et al., 1999). This is in general agreement with warm season temperature estimates based on radiolarian analyses. A recent study in the Southwest Pacific, also based on *G. bulloides*, suggests a LH-LGM SST temperature difference of 4.3 °C (Pahnke et al., 2003) which is comparable to alkenone results (3.5 °C) from a near by core (Sikes et al., 2002).

7. Summary and conclusions

We have reviewed the present status of planktonic foraminiferal Mg/Ca thermometry and presented a compilation of published and unpublished Mg/Ca temperatures for the LGM. We have highlighted the potential advantages of this method over other paleo-temperature proxies and outlined some issues regarding analytical methods, and dissolution. These issues should be addressed before more overreaching interpretation of the available data may be made. For example, results attained from deeper core sites may need to be corrected for dissolution once this has been quantified. Results originating from different laboratories should be compared with caution over the possible effects of differing techniques. Nevertheless, Mg/Ca thermometry has distinct strengths as a paleotemperature proxy; firstly Mg/Ca measurements may be combined with stable isotopes in order to reconstruct variations in $\delta^{18}O_{seawater}$. Secondly the use of multiple species, representing environmental conditions at different depths and/or seasons can provide more hydrographic information than is possible from SST reconstructions alone.

The Mg/Ca data currently available for the LGM are still very sparse. However, this proxy has already provided a significant contribution to the question of tropical cooling during the LGM. Most Mg/Ca reconstructions are concentrated in tropical regions and results consistently suggest LGM temperatures 2.0–3.5 °C cooler than the Late Holocene for these regions. Results from northern latitudes are very scarce and some of them show inconsistencies with other proxies, particularly in those areas in which the temperature change associated with the last deglaciation is believed to be particularly large (>4 °C). Further data are required before any firm conclusions can be made regarding the origin of these disparities.

Acknowledgements

S.B. would like to thank Franck Bassinot for describing the theory behind foraminiferal crystallinity. We also thank the MARGO organising committee. This work was funded in part by the Comer Fellowship for the study of abrupt climate change to S.B. and I.C.

References

Anand, P., Elderfield, H., Conte, M.H., 2003. Calibration of Mg/Ca thermometry in planktonic foraminifera from a sediment trap time series. Paleoceanography 18.

Arz, H.W., Pätzold, J., Müller, P.J., Moammar, M.O., 2003. Influence of Northern Hemisphere climate and global sea level rise on the restricted Red Sea marine environment during termination I. Paleoceanography 18.

Barker, S., Elderfield, H., 2002. Foraminiferal calcification response to glacial-interglacial changes in atmospheric CO_2. Science 297, 833–836.

Barker, S., Greaves, M., Elderfield, H., 2003. A study of cleaning procedures used for foraminiferal Mg/Ca paleothermometry. Geochemistry Geophysics Geosystems 4.

Barthelemy-Bonneau, M.-C., 1978. Dissolution expérimentale et naturelle de foraminifères planctoniques—Approches morphologique, isotopique et cristallographique. Ph.D. Thesis, Université Pierre et Marie Curie, Paris.

Bassinot, F., Melieres, F., Levi, C., Ghelen, M., Labeyrie, L., Elderfield, H., 2001. Improving Mg/Ca paleothermometer: correction of dissolution effects on Mg/Ca ratio of foraminifera shells. Paper presented at 7th International Conference on Paleoceanography, Sapporo, Japan.

Bassinot, F., Melieres, F., Gehlen, M., Levi, C., Labeyrie, L., 2004. Crystallinity of foraminifer shells: a proxy to reconstruct past bottom water CO_3^- changes? Geochemistry Geophysics Geosystems 5, doi:10.1029/2003GC000668.

Bé, A.W.H., 1980. Gametogenic calcification in a spinose planktonic foraminifer, *Globigerinoides sacculifer* (Brady). Marine Micropaleontology 5, 283–310.

Benway, H.M., Haley, B.A., Klinkhammer, G., Mix, A., 2003. Adaptation of a flow-through leaching procedure for Mg/Ca paleothermometry. Geochemistry Geophysics Geosystems 4.

Berger, W.H., 1970. Planktonic foraminifera: selective solution and the lysocline. Marine Geology 8, 111–138.

Bijma, J., Spero, H., Lea, D.W., 1999. Reassessing foraminiferal stable isotope geochemistry: impact of the oceanic carbonate system (Experimental Results). In: Fischer, G., Wefer, G. (Eds.), Uses of Proxies in Paleoceanography: Examples from the South Atlantic. Springer, Berlin, Heidelberg, pp. 489–512.

Billups, K., Schrag, D.P., 2003. Application of benthic foraminiferal Mg/Ca ratios to questions of Cenozoic climate change. Earth and Planetary Science Letters 209, 181–195.

Blackmon, P.D., Todd, R., 1959. Mineralogy of some foraminifera as related to their classification and ecology. Journal of Paleontology 33, 1–15.

Bonneau, M.C., Melieres, F., Vergnaud-Grazzini, C., 1980. Variations isotopiques (oxygène et carbone) et cristallographiques chez des espèces actuelles de foraminifères planctoniques en fonction de la profondeur de dépôt. Bulletin de la Societe Geologique de France 22, 791–793.

Boyle, E.A., 1981. Cadmium, zinc, copper, and barium in foraminifera tests. Earth and Planetary Science Letters 53, 11–35.

Boyle, E.A., Keigwin, L.D., 1985. Comparison of Atlantic and Pacific paleochemical records for the last 215,000 years: Changes in deep ocean circulation and chemical inventories. Earth and Planetary Science Letters 76, 135–150.

Broecker, W., Clark, E., 2001a. An evaluation of Lohmann's foraminifera weight dissolution index. Paleoceanography 16, 531–534.

Broecker, W.S., Clark, E., 2001b. Glacial-to-Holocene redistribution of carbonate ion in the deep sea. Science 294, 2152–2155.

Broecker, W.S., Denton, G.H., 1989. The role of ocean-atmosphere reorganizations in glacial cycles. Geochimica et Cosmochimica Acta 53, 2465–2501.

Broecker, W.S., Anderson, R., Clark, E., Fleisher, M., 2001. Record of seafloor $CaCO_3$ dissolution in the central equatorial Pacific. Geochemistry Geophysics Geosystems 2, U1–U7.

Brown, S.J., Elderfield, H., 1996. Variations in Mg/Ca and Sr/Ca ratios of planktonic foraminifera caused by postdepositional dissolution: evidence of shallow Mg-dependent dissolution. Paleoceanography 11, 543–551.

Cacho, I., Grimalt, J.O., Canals, M., Sbaffi, L., Shackleton, N.J., Schönfeld, J., Zahn, R., 2001. Variability of the western Mediterranean Sea surface temperatures during the last 25,000 years and its connection with the northern hemisphere climatic changes. Paleoceanography 16, 40–52.

Cacho, I., Shackleton, N.J., Elderfield, H., Grimalt, J.O., in prep. Geochemical approach on multi-foraminiferal species to reconstruct past rapid variability in the hydrography of the Alboran Sea (W. Mediterranean Sea).

Calvo, E., Villanueva, J., Grimalt, J.O., Boelaert, A., Labeyrie, L.D., 2001. New insights into the glacial latitudinal temperature gradients in the North Atlantic. Results from $U_{37}^{K'}$-sea surface temperatures and terrigenous inputs. Earth and Planetary Science Letters 188, 509–519.

Cayre, O., Lancelot, Y., Vicent, E., Hall, M.A., 1999. Paleoceanographic reconstructions from planktonic foraminifera off the Iberian Margin: temperature, salinity, and Heinrich events. Paleoceanography 14 (3), 384–396.

Chave, K.E., 1954. Aspects of the biogeochemistry of magnesium 1. Calcareous marine organisms. Journal of Geology 62, 266–283.

Chivas, A.R., De Deckker, P., Shelley, J.M.G., 1986. Magnesium content of non-marine ostracod shells: a new palaeosalinometer and palaeothermometer. Paleogeography Paleoclimatology Paleoecology 54, 43–61.

Clapperton, C.M., 1993. Nature of Environmental-Changes in South-America at the last glacial maximum. Paleogeography Paleoclimatology Paleoecology 101, 189–208.

CLIMAP Project Members, 1976. The Surface of the Ice-Age Earth. Science 191, 1131–1137.

Cronblad, H.G., Malmgren, B.A., 1981. Climatically controlled variation of Sr and Mg in Quaternary planktonic foraminifera. Nature 291, 61–64.

Crowley, T.J., 2000. CLIMAP SSTs re-revisited. Climate Dynamics 16, 0241–0255.

De Deckker, P., Chivas, A.R., Shelley, J.M.G., 1999. Uptake of Mg and Sr in the euryhaline ostracod Cyprideis determined from in vitro experiments. Paleogeography Paleoclimatology Paleoecology 148, 105–116.

de Villiers, S., Greaves, M., Elderfield, H., 2002. An intensity ratio calibration method for the accurate determination of Mg/Ca and Sr/Ca of marine carbonates by ICP-AES. Geochemistry Geophysics Geosystems 3, doi:10.1029/2001GC000169.

Dekens, P.S., Lea, D.W., Pak, D.K., Spero, H.J., 2002. Core top calibration of Mg/Ca in tropical foraminifera: refining paleotemperature estimation. Geochemistry Geophysics Geosystems 3, U1–U29.

Deuser, W.G., Ross, E.H., 1989. Seasonally abundant planktonic foraminifera of the Sargasso Sea: succession, deep-water fluxes, isotopic compositions, and paleoceanographic implications. Journal of Foraminiferal Research 19, 268–293.

Duckworth, D.L., 1977. Magnesium concentration in the tests of the planktonic foraminifer *Globorotalia truncatulinoides*. Journal of Foraminiferal Research 7, 304–312.

Dwyer, G.S., Cronin, T.M.P., Baker, A., Raymo, M.E., Buzas, J.S., Correge, T., 1995. North-Atlantic deep-water temperature-change during late pliocene and late quaternary climatic cycles. Science 270, 1347–1351.

Eggins, S., DeDeckker, P., Marshall, A.T., 2003. Mg/Ca variation in planktonic foraminfera tests: implications for reconstructing palaeo-seawater temperature and habitat migration. Earth and Planetary Science Letters 212, 291–306.

Elderfield, H., Ganssen, G., 2000. Past temperature and delta-^{18}O of surface ocean waters inferred from foraminiferal Mg/Ca ratios. Nature 405, 442–445.

Elderfield, H., Vautravers, M., Cooper, M., 2002. The relationship between shell size and Mg/Ca, Sr/Ca, δO18, and δC13 of species of planktonic foraminifera. Geochemistry Geophysics Geosystems 3.

Erez, J., Honjo, S., 1981. Comparison of isotopic composition of planktonic foraminifera in plankton tows, sediment traps and sediments. Paleogeography Paleoclimatology Paleoecology 33, 129–156.

Farrell, B.F., Prell, W., 1989. Climatic change and CaCO$_3$ preservation; an 800,000 year bathymetric reconstruction from the Central Equatorial Pacific Ocean. Paleoceanography 4, 447–466.

Guilderson, T.P., Fairbanks, R.G., Rubenstone, J.L., 1994. Tropical temperature variations since 20,000 years ago: modulating interhemispheric climate change. Science 263, 663–665.

Haley, B.A., Klinkhammer, G.P., 2002. Development of a flow-through system for cleaning and dissolving foraminiferal tests. Chemical Geology 185, 51–69.

Hastings, D.W., Russell, A.D., Emerson, S.R., 1998. Foraminiferal magnesium in *Globigerinoides sacculifer* as a paleotemperature proxy. Paleoceanography 13, 161–169.

Hathorne, E.C., Alard, O., James, R.H., Rogers, N.W., 2003. Determination of intratest variability of trace elements in foraminifera by laser ablation inductively coupled plasma-mass spectrometry. Geochemistry Geophysics Geosystems 4.

Hemleben, C., Spindler, M., Breitinger, I., Deuser, W.G., 1985. Field and laboratory studies on the ontogeny and ecology of some *Globorotaliid* species from the Sargasso Sea off Bermuda. Journal of Foraminiferal Research 15, 254–272.

Izuka, S.K., 1988. Relationship of magnesium and other minor elements in tests of Cassidulina subglobosa and Cassidulina oriangulata to physical oceanic properties. Journal of Foraminiferal Research 18, 151–157.

Jha, P., Elderfield, H., 2000. Variation of Mg/Ca and Sr/Ca in planktonic and benthic foraminifera from single test chemistry. Eos Trans. AGU 81 (48) Fall Meet. Suppl., Abstract OS11C-13.

Katz, A., 1973. The interaction of magnesium with calcite during crystal growth at 25–90degC and one atmosphere. Geochimica et Cosmochimica Acta 37, 1563–1586.

Kiefer, T., Sarnthein, M., Elderfield, H., Erlenkeuser, H., Grootes, P., in press. Warmings in the far northwestern Pacific support pre-Clovis immigration to America during Heinrich event 1.

Kienast, M., Steinke, S., Stattegger, K., Calvert, S.E., 2001. Synchronous tropical South China Sea SST change and Greenland warming during deglaciation. Science 291, 2132–2134.

Kilbourne, R.T., Sen Gupta, B.K., 1973. Elemental composition of planktonic foraminiferal tests in relation to temperature-depth habitats and selective solution. Geological Society of America Abstracts with Programs 5, 408–409.

Kinsman, D.J.J., Holland, H.D., 1969. The co-precipitation of cations with CaCO$_3$-IV. The co-precipitation of Sr^{2+} with aragonite between 16° and 96 °C. Geochimica et Cosmochimica Acta 33, 1–17.

Koutavas, A., Lynch-Stieglitz, J., 2003. Glacial-interglacial dynamics of the eastern equatorial Pacific cold tongue-Intertropical Convergence Zone system reconstructed from oxygen isotope records. Paleoceanography 18, 1089.

Koutavas, A., Lynch-Stieglitz Jr, J.T.M.M., Sachs, J.P., 2002. El Niño-Like Pattern in Ice Age Tropical Pacific Sea Surface Temperature. Science 297, 226–230.

Le, J., Shackleton, N.J., 1992. Carbonate dissolution fluctuations in the western Equatorial Pacific during the late Quaternary. Paleoceanography 7, 21–42.

Lea, D.W., Boyle, E.A., 1991. Barium in planktonic foraminifera. Geochimica et Cosmochimica Acta 55, 3321–3331.

Lea, D.W., Mashiotta, T.A., Spero, H.J., 1999. Controls on magnesium and strontium uptake in planktonic foraminifera determined by live culturing. Geochimica et Cosmochimica Acta 63, 2369–2379.

Lea, D.W., Pak, D.K., Spero, H.J., 2000. Climate impact of late quaternary equatorial Pacific sea surface temperature variations. Science 289, 1719–1724.

Lea, D.W., Martin, P.A., Pak, D.K., Spero, H.J., 2002. Reconstructing a 350 ky history of sea level using planktonic Mg/Ca and oxygen isotope records from a Cocos Ridge core. Quaternary Science Review 21, 283–293.

Lea, D.W., Pak, D.K., Peterson, L.C., Hughen, K.A., 2003. Synchroneity of Tropical and High-Latitude Atlantic Temperatures over the Last Glacial Termination. Science 301, 1361–1364.

Lear, C.H., Elderfield, H., Wilson, P.A., 2000. Cenozoic deep-sea temperatures and global ice volumes from Mg/Ca in benthic foraminiferal calcite. Science 287, 269–272.

Lear, C.H., Rosenthal, Y., Slowey, N., 2002. Benthic foraminiferal Mg/Ca-paleothermometry: a revised core-top calibration. Geochimica et Cosmochimica Acta 66, 3375–3387.

Lohmann, G.P., 1995. A model for variation in the chemistry of planktonic foraminifera due to secondary calcification and selective dissolution. Paleoceanography 10, 445–457.

Lorens, R.B., Williams, D.F., Bender, M.L., 1977. The early non-structural chemical diagenesis of foraminiferal calcite. Journal of Sedimentary Petrology 47, 1602–1609.

Martin, P.A., Lea, D.W., 2002. A simple evaluation of cleaning procedures on fossil benthic foraminiferal Mg/Ca. Geochemistry Geophysics Geosystems 3 art. no.-8401.

Martin, P.A., Lea, D.W., Rosenthal, Y., Shackleton, N.J., Sarnthein, M., Papenfuss, T., 2002. Quaternary deep sea temperature histories derived from benthic foraminiferal Mg/Ca. Earth and Planetary Science Letters 198, 193–209.

Mashiotta, T.A., Lea, D.W., Spero, H.J., 1999. Glacial-interglacial changes in Subantarctic sea surface temperature and d^{18}O-water using foraminiferal Mg. Earth and Planetary Science Letters 170, 417–432.

Meland, M., Jansen, E., Elderfield, H., 2004. Constraints on SST estimates for the northern North Atlantic/Nordic Seas during the LGM. Quaternary Science Review, this issue (doi:10.1016/j.quascirev.2004.05.011).

Mitsuguchi, T., Matsumoto, E., Abe, O., Uchida, T., Isdale, P.J., 1996. Mg/Ca thermometry in coral-skeletons. Science 274, 961–963.

Mix, A.C., Bard, E., Schneider, R., 2001. Environmental processes of the ice age: land, oceans, glaciers (EPILOG). Quaternary Science Review 20, 627–657.

Mucci, A., 1987. Influence of temperature on the composition of magnesium calcite overgrowths precipitated from seawater. Geochimica et Cosmochimica Acta 51, 1977–1984.

Mulitza, S., Durkoop, A., Hale, W., Wefer, G., Niebler, H.S., 1997. Planktonic foraminifera as recorders of past surface-water stratification. Geology 25, 335–338.

Nürnberg, D., 1995. Magnesium in tests of *Neogloboquadrina pachyderma* (sinistral) from high northern and southern latitudes. Journal of Foraminiferal Research 25, 350–368.

Nürnberg, D., Bijma, J., Hemleben, C., 1996. Assessing the reliability of magnesium in foraminiferal calcite as a proxy for water mass temperatures. Geochimica et Cosmochimica Acta 60, 803–814.

Nürnberg, D., Müller, A., Schneider, R.R., 2000. Paleo-sea surface temperature calculations in the equatorial east Atlantic from Mg/Ca ratios in planktic foraminifera: a comparison to sea surface temperature estimates from $U^{K'}_{37}$, oxygen isotopes, and foraminiferal transfer function. Paleoceanography 15, 124–134.

Oomori, T., Kaneshima, H., Maezato, Y., 1987. Distribution coefficient of Mg^{2+} ions between calcite and solution at 10–50 °C. Marine Chemistry 20, 327–336.

Oppo, D.W., Fairbanks, R.G., 1989. Carbon isotope composition of tropical surface waters during the past 22,000 years. Paleoceanography 4, 333–351.

Pahnke, K., Zahn, R., Elderfield, H., Schulz, M., 2003. 340,000-Year Centennial-Scale Marine Record of Southern Hemisphere Climatic Oscillation. Science 301, 948–952.

Pailler, D., Bard, E., 2002. High frequency palaeoceanographic changes during the past 140000 yr recorded by the organic matter in sediments of the Iberian Margin. Paleogeography Paleoclimatology Paleoecology 181, 431–452.

Palmer, M.R., Pearson, P.N., 2003. A 23,000-Year Record of Surface Water pH and PCO_2 in the Western Equatorial Pacific Ocean. Science 300, 480–482.

Pelejero, C., Grimalt, J.O., Heilig, S., Kienast, M., Wang, L., 1999. High resolution U^{K}_{37} temperature reconstructions in the South China Sea over the last 220 kyrs. Paleoceanography 14, 224–231.

Rathburn, A.E., DeDeckker, P., 1997. Magnesium and strontium compositions of recent benthic foraminifera from the Coral Sea, Australia and Prydz Bay, Antarctica. Marine Micropaleontology 32, 231–248.

Ravelo, A.C., Fairbanks, R.G., 1995. Carbon isotopic fractionation in multiple species of planktonic foraminifera from core-tops in the Tropical Atlantic. Journal of Foraminiferal Research 25, 53–74.

Reichart, G.J., Jorissen, F., Anschutz, P., Mason, P.R.D., 2003. Single foraminiferal test chemistry records the marine environment. Geology 31, 355–358.

Rosenthal, Y., Boyle, E.A., 1993. Factors controlling the fluoride content of planktonic foraminifera: an evaluation of its paleoceanographic applicability. Geochimica et Cosmochimica Acta 57, 335–346.

Rosenthal, Y., Lohmann, G.P., 2002. Accurate estimation of sea surface temperatures using dissolution-corrected calibrations for Mg/Ca paleothermometry. Paleoceanography 17.

Rosenthal, Y., Boyle, E.A., Slowey, N., 1997. Temperature control on the incorporation of magnesium, strontium, fluorine, and cadmium into benthic foraminiferal shells from Little Bahama Bank: prospects for thermocline paleoceanography. Geochimica et Cosmochimica Acta 61, 3633–3643.

Rosenthal, Y., Field, M.P., Sherrell, R.M., 1999. Precise determination of element/calcium ratios in calcareous samples using sector field inductively coupled plasma mass spectrometry. Analytical Chemistry 71, 3248–3253.

Rosenthal, Y., Lohmann, G.P., Lohmann, K.C., Sherrell, R.M., 2000. Incorporation and preservation of Mg in *Globigerinoides sacculifer*: implications for reconstructing the temperature and O-18/O-16 of seawater. Paleoceanography 15, 135–145.

Rosenthal, Y., Oppo, D.W., Linsley, B.K., 2003. The amplitude and phasing of climate change during the last deglaciation in the Sulu Sea, western equatorial Pacific. Geophysical Research Letters 30.

Rosenthal, Y., Perron-Cashman, S., Lear, C.H., Bard, E., Barker, S., Billups, K., Bryan, M., Delaney, M.L., Demenocal, P., Dwyer, G.S., Elderfield, H., German, C.R., Greaves, M., Lea, D.,

Marchitto, T., Pak, D., Ravelo, A.C., Paradis, G.L., Russell, A.D., Schneider, R.R., Scheindrich, K., 2004. Laboratory inter-comparison study of Mg/Ca and Sr/Ca measurements in planktonic foraminifera for paleoceanographic research. Geochemistry Geophysics Geosystems, doi:10.1029/2003GC000650.

Rostek, F., Ruhland, G., Bassinot, F.C., Muller, P.J., Labeyrie, L.D., Lancelot, Y., Bard, E., 1993. Reconstructing sea-surface temperature and salinity using $\delta^{18}O$ and alkenone records. Nature 364, 319–321.

Russell, A.D., Emerson, S., Nelson, B.K., Erez, J., Lea, D.W., 1994. Uranium in foraminiferal calcite as a recorder of seawater uranium concentrations. Geochimica et Cosmochimica Acta 58, 671–681.

Sikes, E.L., Howard, W.R., Neil, H.L., Volkman, J.K., 2002. Glacial-interglacial sea surface temperature changes across the subtropical front east of New Zealand based on alkenone unsaturation ratios and foraminiferal assemblages. Paleoceanography 17.

Simstich, J., Sarnthein, M., Erlenkeuser, H., 2003. Paired delta O-18 signals of *Neogloboquadrina pachyderma* (s) and *Turborotalita quinqueloba* show thermal stratification structure in Nordic Seas. Marine Micropaleontology 48, 107–125.

Skinner, L., Elderfield, H., submitted. Constraining ecological and biological bias in planktonic foraminiferal Mg/Ca and $\delta^{18}O_{cc}$: an "ex post" approach to proxy calibration testing.

Spero, H.J., Lea, D.W., 1996. Experimental determination of stable isotope variability in *Globigerina bulloides*: implications for paleoceanographic reconstructions. Marine Micropaleontology 28, 231–246.

Spero, H.J., Bijma, J., Lea, D.W., Bemis, B.E., 1997. Effect of seawater carbonate concentration on foraminiferal carbon and oxygen isotopes. Nature 390, 497–500.

Spero, H.J., Mielke, K.M., Kalve, E.M., Lea, D.W., Pak, D.K., 2003. Multispecies approach to reconstructing eastern equatorial Pacific thermocline hydrography during the past 360 kyr. Paleoceanography 18 art. no.-1022.

Stoll, H.M., Encinar, J.R., Alonso, J.I.G., Rosenthal, Y., Probert, I., Klaas, C., 2001. A first look at paleotemperature prospects from Mg in coccolith carbonate: cleaning techniques and culture measurements. Geochemistry Geophysics Geosystems 2 art. no. 2000GC000144.

Stott, L., Poulsen, C., Lund, S., Thunell, R., 2002. Super ENSO and Global Climate Oscillations at Millennial Time Scales. Science 297, 222–226.

Stute, M., Forster, M., Frischkorn, H., Serejo, A., Clark, J.F., Schlosser, P., Broecker, W.S., Bonani, G., 1995. Cooling of Tropical Brazil (5 °C) during the Last Glacial Maximum. Science 269, 379–383.

Tachikawa, K., Vidal, L., Sepulcre, S., Bard, E., 2003. Size-normalised test weights, Mg/Ca and Sr/Ca of planktonic foraminifera from the Arabian Sea. Goldschmidt 2003, A464.

Toyofuku, T., Kitazato, H., Kawahata, H., Tsuchiya, M., Nohàra, M., 2000. Evaluation of Mg/Ca thermometry in foraminifera: comparison of experimental results and measurements in nature. Paleoceanography 15, 456–464.

Visser, K., Thunell, R.C., Stott, L.D., 2003. Magnitude and timing of temperature change in the Indo-Pacific warm pool during deglaciation. Nature 421, 152–155.

Wansard, G., 1996. Quantification of paleotemperature changes during isotopic stage 2 in the La Draga continental sequence (NE Spain) based on the Mg/Ca ratio of freshwater ostracods. Quatenary Science Review 15, 237–245.

Watanabe, T., Winter, A., Oba, T., 2001. Seasonal changes in sea surface temperature and salinity during the Little Ice Age in the Caribbean Sea deduced from Mg/Ca and $^{18}O/^{16}O$ ratios in corals. Marine Geology 173, 21–35.

Wei, G.J., Li, X.H., Nie, B.F., Sun, M., Liu, H.C., 1999. High resolution Porites Mg/Ca thermometer for the north of the South China Sea. China Science Bulletin 44, 273–276.

Quaternary Science Reviews 24 (2005) 835–852

Constraints on SST estimates for the northern North Atlantic/Nordic Seas during the LGM

Marius Y. Meland[a,b,*], Eystein Jansen[a,b], Henry Elderfield[c]

[a]Bjerknes Centre for Climate Research, University of Bergen, Allégaten 55, 5007 Bergen, Norway
[b]Department of Earth Sciences, University of Bergen, Allégaten 41, 5007 Bergen, Norway
[c]Department of Earth Sciences, University of Cambridge, Downing Street, Cambridge CB2 3EQ, UK

Received 24 October 2003; accepted 7 May 2004

Abstract

A map of estimated calcification temperatures of the planktic foraminifer *Neogloboquadrina pachyderma* sinistral (T_{Nps}) for the Nordic Seas and the northern North Atlantic for the Last Glacial Maximum was produced from oxygen isotopes with support of Mg/Ca ratios. To arrive at the reconstruction, several constraints concerning the plausible salinity and δ^{18}O-fields were employed. The reconstruction indicates inflow of temperate waters in a wedge along the eastern border of the Nordic Seas and at least seasonally ice-free waters. The reconstruction from oxygen isotopes shows similarities with Mg/Ca based paleotemperatures in the southern and southeastern sector, while unrealistically high Mg/Ca values in the central Nordic Seas prevent the application of the method in this area. The oxygen isotope based reconstruction shows some agreement with temperature reconstructions based on the modern analogue technique, but with somewhat lower temperatures and a stronger internal gradient inside the Nordic Seas. All told, our results suggest a much more ice-free and dynamic high latitude ocean than the CLIMAP reconstruction.
© 2004 Elsevier Ltd. All rights reserved.

1. Introduction

The state of the sea surface in the North Atlantic and Nordic Seas provide important constraints on the climate in Europe and Eurasia, and strongly influence the strength of the meridional overturning circulation. The reliability of reconstructions for past climate states are crucial for developing a physical understanding of past climate changes, their dynamics, magnitude and underlying driving mechanisms. Reliable sea surface reconstructions are also critical as boundary conditions for atmospheric General Circulation Model (GCM) experiments of the Last Glacial Maximum (LGM) and as data for validation of experiments with fully coupled GCMs.

The problem of acquiring reliable SST reconstructions in the low temperature end is a long-standing issue. Due to the problem of low planktic foraminifer diversity in Arctic and Polar water masses, SST estimates based on planktic foraminiferal transfer functions are unreliable when summer temperatures are below 4–5 °C (e.g. Pflaumann et al., 1996, 2003). Other approaches also have their inherent problems: diatoms are often absent in LGM samples, the alkenone and dinocyst assemblage methods have difficulties in the low temperature end (Rosell-Melé and Comes, 1999; de Vernal et al., 2000), and the sensitivity of Mg/Ca ratios to temperature change is less at low temperatures than at higher (e.g. Elderfield and Ganssen, 2000; and see below).

Despite its wide usage, it has been clear for some time that the reconstruction of CLIMAP (1981) (Fig. 1a) based on the transfer function method of Imbrie and Kipp (1971) may be unreliable in the high latitude North Atlantic/Nordic Seas. In the CLIMAP reconstruction there is a permanent sea-ice cover over most of the area,

*Corresponding author. Bjerknes Centre for Climate Research, University of Bergen, Allégaten 55, 5007 Bergen, Norway. Tel.: +47-55-58-98-06; fax: +47-55-58-43-30.

E-mail address: marius.meland@bjerknes.uib.no (M.Y. Meland).

0277-3791/$ - see front matter © 2004 Elsevier Ltd. All rights reserved.
doi:10.1016/j.quascirev.2004.05.011

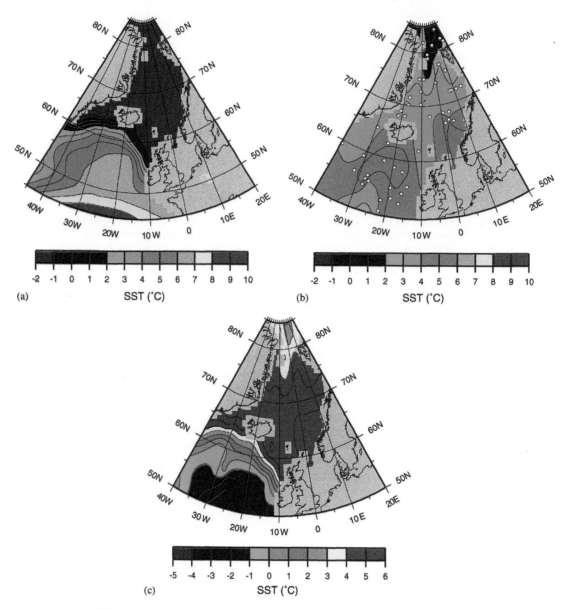

Fig. 1. (a) Summer (August) SST reconstruction for the LGM of CLIMAP (1981). CLIMAP placed the summer sea ice margin along the 0 °C isoline, with a perennial sea-ice cover inside this isoline. (b) Summer SST reconstruction for the LGM, based on gridding and contouring of the published SST data set of GLAMAP (Pflaumann et al., 2003). The SSTs were calculated from foraminiferal transfer functions using the SIMMAX-28 modern analogue technique. The white dots show the data locations. (c) Difference in SST between the GLAMAP and CLIMAP reconstructions, calculated as: $\Delta SST = SST_{GLAMAP}-SST_{CLIMAP}$. Grey coloured areas indicate land areas during the LGM (Peltier, 1994).

although newer evidence clearly point at least to seasonally open waters (e.g. Figs. 1b, 1c; Veum et al., 1992; Hebbeln et al., 1994; Wagner and Henrich, 1994; Sarnthein et al., 1995; Weinelt et al., 1996). The diversity problem of the transfer function approach, i.e. that there is only one dominant species, *Neogloboquadrina pachyderma* (sinistral coiled), in the Polar water, make these alternate reconstructions at least partially unreliable in the sense that they provide a more qualitative reconstruction rather than an accurate or realistic SST field.

The difference between a perennially frozen ocean and an open or seasonally open ocean has, however, wide climatic implications, for example in the possibility for the ocean to steer storm tracks and provide moisture supply to the high latitude ice sheets, to interact with marine based ice sheets and not least to constrain the location of deep water formation and the possible strength and northward extent of the meridional overturning circulation. Renewed attempts to better constrain the LGM state of the Nordic Seas and the northern North Atlantic should therefore be pursued.

Here we develop constraints on the plausible range of SST reconstructions for the LGM, primarily based on planktic foraminiferal oxygen isotopes. Although the oxygen isotopic composition of foraminiferal calcite depends on the oxygen isotopic composition of the ambient water (i.e. salinity), besides being a function of temperature, it is possible to evaluate the plausible salinity ranges constraining the data, and thus obtain estimates of SST from the oxygen isotopic composition by employing these constraints. We then compare these results with results using the Mg/Ca method and recent (post-CLIMAP) transfer function SST-estimates to identify the degree to which these different approaches provide a coherent picture.

2. Chronology

GLAMAP 2000 used two LGM time slices, comprising the intervals of 18.0–21.5 cal. ka (15–18 ^{14}C ka) and 19.0–22.0 cal. ka (16–19 ^{14}C ka) (Vogelsang et al., 2000). The first of these time slices is used in this work to temporally constrain the LGM, which in oxygen isotope records from the Nordic Seas is known to be a period of stability and minimum meltwater influx (Sarnthein et al., 1995). Since in some southern Atlantic records deglacial warming is already recorded in the younger part of the GLAMAP LGM time slice, the EPILOG working group (Mix et al., 2001) suggested a slightly

different definition of the LGM: 19.0–23.0 cal. ka (16.0–19.5 ^{14}C ka). In northern Atlantic records, the early warming is not apparent in influencing oxygen isotope records (Sarnthein et al., 1995; Vogelsang et al., 2000; Pflaumann et al., 2003), which means that choice of time slice has limited importance, as long as the most important aspect is to cover a time slice of relatively stable climatic conditions. Since previous work on the LGM in the North Atlantic area mainly used the definition of 18.0–21.5 ka (e.g. Sarnthein et al., 1995, 2000, 2003; Weinelt et al., 1996; Nørgaard-Pedersen et al., 2003; Pflaumann et al., 2003), from which our data partly is drawn from, we find it most practical to use this definition of the LGM here.

The Mg/Ca data are from the time slice 19.0–21.5 ka, a period covered of both the GLAMAP and EPILOG time scales, but this slight difference has no influence on the results.

3. LGM data sets

3.1. Oxygen isotopes of N. pachyderma (sin.) ($\delta^{18}O_{Nps}$)

The LGM time slice is based on a total of 158 cores, covering the northern North Atlantic and the Nordic Seas (Fig. 2a). Table 1 shows the data sources. Fifty-three of these cores are dated by Accelerator Mass Spectrometry (AMS) ^{14}C within the range of the LGM

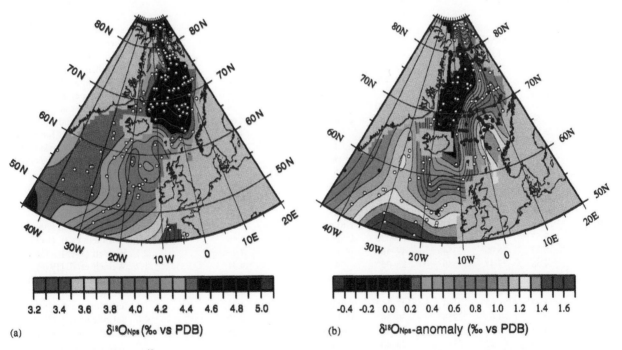

Fig. 2. (a) Oxygen isotope values ($\delta^{18}O_{Nps}$) for the LGM, based on the planktonic foraminifer *N. pachyderma* (sin.). The gridding, contouring and colouring are based on 158 cores, marked as white circles (Location in Table 1) (b) $\delta^{18}O_{Nps}$-anomaly between the LGM and modern values, calculated from the gridded plots as: plot($\delta^{18}O_{Nps(anomaly)}$) = plot($\delta^{18}O_{Nps(LGM)}$) − plot($\delta^{18}O_{Nps(modern)}$) − 1.1, where 1.1 is the ice volume effect. Black dots mark cores which contain $\delta^{18}O_{Nps(modern)}$ values, while white dots mark cores which contain $\delta^{18}O_{Nps(LGM)}$ values. The $\delta^{18}O_{Nps(modern)}$ values are from Simstich et al. (2003). Grey coloured areas indicate land areas during the LGM (Peltier, 1994).

Table 1
Cores used for studies of oxygen isotope ratios in the tests of *N. pachyderma* (sinistral). The isotopic temperatures (T_{Nps}) are determined by use of Eqs. (1)–(3) in text

Core no.	Longitude	Latitude	Water depth (m)	LGM level 18–21.5 ka depth (cm)	Number of averaged $\delta^{18}O_{Nps}$ measurements	Averaged $\delta^{18}O_{Nps}$	T_{Nps}	Data source for $\delta^{18}O_{Nps}$
BOFS 5 K	−21.87	50.68	3547	77–94[a]	9	4.17	5.1	Maslin (1992)
BOFS 8 K	−22.04	52.50	4045	78–110	13	4.21	4.3	Maslin (1992)
BOFS 14 K	−19.44	58.62	1756	34–44	5	3.94	5.3	Maslin (1992)
BOFS 16 K	−23.14	59.28	2502	42–48	2	4.26	3.7	Vogelsang (1990)
BOFS 17 K	−16.50	58.00	1150	68–95	13	4.17	4.9	Maslin (1992)
CH 67-19	−3.95	45.75	1982	380	1	3.57	7.4	Labeyrie and Duplessy (1985)
CH 69-12	−4.69	46.02	3642	Unknown	Unknown	3.65	6.9	Duplessy et al. (1991)
CH 69-32	−5.18	45.40	4777	Unknown	Unknown	3.56	7.5	Duplessy et al. (1991)
CH 72-101	−8.56	47.47	2428	280	1	3.51	7.7	Labeyrie and Duplessy (1985)
CH 72-104	−8.08	46.90	4590	380	1	3.25	8.9	Labeyrie and Duplessy (1985)
CH 73-108	−10.73	58.08	2032	Unknown	Unknown	4.18	4.5	Duplessy et al. (1991)
CH 73-110	−8.93	59.50	1365	102–110	Unknown	4.00	5.1	Weinelt (1993)
CH 73-136	−14.47	55.57	2201	120	1	4.18	4.7	Keigwin and Boyle (1989)
CH 73-139	−16.35	54.63	2209	160–190[a]	4	3.99	5.6	Bard et al. (1987)
CH 73-141	−16.52	52.86	3489	Unknown	Unknown	4.00	5.7	Duplessy et al. (1991)
CH 77-07	−10.52	66.60	1487	145–200	Unknown	4.62	1.0	Ruddiman and McIntyre (1981)
DSDP 609	−24.00	50.00	3884	88–101[a]	>5	4.25	4.6	Bond et al. (1993)
ENAM 93-21	−4.00	62.74	1020	220–250[a]	7	4.48	2.9	Rasmussen et al. (1996)
FRAM 1/4	−8.95	84.50	3820	15–23[a]	6	4.41	−1.9	Zahn et al. (1985)
FRAM 1/7	−6.96	83.88	2990	26–40	4	4.61	−1.8	Zahn et al. (1985)
HM 25-09	4.79	63.05	600	254–264	2	4.49	2.4	Jansen and Erlenkeuser (1985)
HM 31-33	4.78	63.63	1580	112	1	4.44	3.0	Jansen and Erlenkeuser (1985)
HM 31-36	0.53	64.25	2620	29–34	2	4.75	1.9	Jansen and Erlenkeuser (1985)
HM 52-43	0.73	64.25	2781	83–90[a]	3	4.51	2.7	Veum et al. (1992)
HM 57-07	−13.53	68.25	1668	34–42	Unknown	4.48	1.6	Sarnthein et al. (1995)
HM 71-12	−13.87	68.43	1547	44–48	3	4.69	1.6	Sarnthein et al. (1995)
HM 71-14	−18.08	69.98	1624	64	1	4.65	0.4	Sarnthein et al. (1995)
HM 71-19	−9.51	69.48	2210	42	1	4.81	0.9	Vogelsang (1990)
HM 80-30	1.60	71.78	2821	36–44	5	4.57	2.2	Sarnthein et al. (1995)
HM 80-42	−9.23	72.25	2416	54–64	6	4.44	0.3	Sarnthein et al. (1995)
HM 80-60	−11.86	68.90	1869	68–76	5	4.69	1.4	Sarnthein et al. (1995)
HM 94-13	−1.62	71.63	1946	44–52	5	4.66	1.6	Sarnthein et al. (1995)
HM 94-18	5.70	74.50	2469	25–35	2	4.58	1.9	Sarnthein et al. (1995)
HM 94-25	1.32	75.60	2469	38–47[a]	4	4.68	1.6	Sarnthein et al. (1995)
HM 94-34	−2.54	73.77	3004	50–53[a]	2	4.71	1.2	Sarnthein et al. (1995)
HM 100-7	−4.72	61.67	1125	63–100	38	4.39	3.4	Sarnthein et al. (1995)
K 11	1.60	71.78	2900	30–56	Unknown	4.66	1.9	Ruddiman and McIntyre (1981)
KN 708-1	−23.75	50.00	4053	100	1	4.11	5.2	Ruddiman and McIntyre (1981)
KN 708-6	−29.57	51.57	2469	60	1	4.46	3.0	Keigwin and Boyle (1989)
KN 714-15	−25.78	58.77	2598	415–419	1	4.23	3.7	Keigwin and Boyle (1989)
M 17045	−16.65	52.43	3663	80–110[a]	15	4.01	5.8	Winn et al. (1991)
M 17048	−18.16	54.30	1859	71–96	Unknown	4.02	5.3	Sarnthein et al. (1995)
M 17049	−26.73	55.28	3331	150–163[a]	14	4.29	2.9	Jung (1996)
M 17051	−31.98	56.17	2300	200–209[a]	3	4.34	2.7	Jung (1996)
M 17701	11.68	68.53	1421	92–141	Unknown	4.34	2.2	Sarnthein et al. (1995)
M 17719	12.57	72.15	1823	260–320	Unknown	4.46	2.5	Sarnthein et al. (1995)
M 17724	8.33	76.00	2354	47–58[a]	6	4.63	2.1	Weinelt (1993)
M 17725	4.58	77.47	2580	25–35[a]	4	4.41	1.7	Weinelt et al. (1996)
M 17728	3.95	76.52	2485	12–17	Unknown	4.69	1.0	Sarnthein et al. (1995)
M 17730	7.31	72.05	2769	132–147[a]	5	4.60	2.2	Weinelt (1993)
M 17732	4.23	71.62	3103	137–142	Unknown	4.75	1.7	Sarnthein et al. (1995)
M 23041	0.22	68.68	2258	30–36	Unknown	4.70	2.0	Sarnthein et al. (1995)
M 23043	−3.35	70.27	2133	30–38	Unknown	4.56	2.0	Sarnthein et al. (1995)
M 23055	4.10	68.42	2311	35–40	5	4.76	1.9	Vogelsang (1990)
M 23056	3.83	68.50	2665	27–35[a]	2	4.68	2.0	Weinelt et al. (1996)
M 23057	3.31	68.40	3157	24–29	Unknown	4.70	2.1	Sarnthein et al. (1995)
M 23059	−3.12	70.30	2283	29–32	4	4.72	1.4	Vogelsang (1990)
M 23062	0.16	68.73	2244	30–35	4	4.73	1.9	Vogelsang (1990)
M 23063	0.00	68.75	2299	31	1	4.76	1.8	Vogelsang (1990)

Table 1 (continued)

Core no.	Longitude	Latitude	Water depth (m)	LGM level 18–21.5 ka depth (cm)	Number of averaged $\delta^{18}O_{Nps}$ measurements	Averaged $\delta^{18}O_{Nps}$	T_{Nps}	Data source for $\delta^{18}O_{Nps}$
M 23064	0.33	68.67	2571	32–35	Unknown	4.66	2.2	Sarnthein et al. (1995)
M 23065	0.81	68.50	2804	29–34[a]	6	4.72	2.0	Vogelsang (1990)
M 23068	1.50	67.83	2230	61–75	2	4.74	2.0	Vogelsang (1990)
M 23071	2.93	67.08	1308	98–147[a]	5	4.73	2.7	Vogelsang (1990)
M 23074	4.92	66.67	1157	125–350[a]	19	4.62	2.3	Vogelsang (1990)
M 23254	9.63	73.12	2273	80–95	Unknown	4.70	1.7	Sarnthein et al. (1995)
M 23256	10.95	73.18	2061	70–90	Unknown	4.73	1.5	Sarnthein et al. (1995)
M 23258	13.98	75.00	1768	710–840	Unknown	4.52	2.1	Sarnthein et al. (1995)
M 23259	9.25	72.03	2518	140–170	7	4.68	1.8	Weinelt (1993)
M 23260	11.46	72.13	2089	110–140	5	4.71	1.6	Weinelt (1993)
M 23261	13.11	72.17	1628	192–260	10	4.60	1.9	Weinelt (1993)
M 23262	14.43	72.23	1130	200–320[a]	10	4.33	2.9	Weinelt (1993)
M 23269	0.68	71.45	2872	43–49	2	4.83	1.2	Weinelt (1993)
M 23294	−10.59	72.37	2224	110–123[a]	3	4.71	0.1	Weinelt (1993)
M 23323	5.93	67.77	1286	Unknown	Unknown	4.42	3.0	Sarnthein et al. (1995)
M 23351	−18.21	70.36	1672	55–78	6	4.40	−0.4	Voelker (1999)
M 23354	−10.63	70.33	1747	70–80[a]	2	4.50	1.2	Voelker (1999)
M 23415	−19.20	53.17	2472	135–164[a]	12	4.13	5.2	Jung (1996)
M 23419	−19.74	54.97	1491	40–42	1	3.96	5.5	Jung (1996)
M 23519	−29.60	64.80	1893	58–84	3	4.53	1.6	Hohnemann (1996)
MD 2010	4.56	66.68	1226	217–373[a]	47	4.61	2.3	Dokken and Jansen (1999)
MD 2011	7.64	66.97	1048	972–1361[a]	195	4.42	2.9	Dreger (1999)
MD 2012	11.43	72.15	2094	262–468[a]	37	4.58	2.0	Dreger (1999)
MD 2284	−0.98	62.37	1500	800–1050[a]	25	4.34	3.5	Jansen and Meland (2001)
MG 123	0.81	79.27	3050	58–72	2	4.65	0.9	Morris (1988)
NA 87-22	−14.57	55.50	2161	370–419[a]	20	4.06	4.8	Duplessy et al. (1992)
NO 77-14	−20.42	62.45	1531	Unknown	Unknown	4.65	2.5	Duplessy et al. (1991)
NO 79-06	−36.89	54.52	2734	290	1	4.45	1.8	Labeyrie and Duplessy (1985)
NO 79-25	−27.28	46.98	2826	Unknown	Unknown	4.10	6.3	Duplessy et al. (1991)
NP 90-12	9.42	78.41	628	220–275	7	4.65	1.1	Dokken (1995)
NP 90-36	9.94	77.62	1360	330–360	5	4.60	1.5	Dokken (1995)
NP 90-39	9.90	77.26	2119	127–155	8	4.48	2.0	Dokken (1995)
OD 41:4:1	11.24	84.03	3344	13–17[a]	3	4.76	−1.4	Nørgaard-Pedersen et al. (2003)
PS 1171	−18.07	68.20	935	22–50[a]	12	4.49	1.5	Lackschewitz et al. (1994)
PS 1230	−4.78	78.86	1235	20–28[a]	9	4.28	0.1	Nørgaard-Pedersen et al. (2003)
PS 1294	5.37	78.00	2668	35–55	5	4.75	0.3	Hebbeln and Wefer (1997)
PS 1295	2.43	78.00	3112	36–43	4	4.63	0.6	Jones and Keigwin (1989)
PS 1308	−4.83	80.02	1444	20–32	Unknown	3.95	−0.5	Nørgaard-Pedersen et al. (2003)
PS 1314	4.50	80.00	1382	21–28	Unknown	4.22	1.1	Nørgaard-Pedersen et al. (2003)
PS 1533	15.18	82.03	2030	60–85[a]	7	4.60	0.1	Köhler (1992)
PS 1535	1.85	78.75	2557	32–47[a]	8	4.60	1.5	Köhler (1992)
PS 1730	−17.7	70.12	1617	103–110[a]	1	4.29	−0.3	Stein et al. (1996)
PS 1894	−8.30	75.81	1975	36–125[a]	43	4.39	−0.4	Nørgaard-Pedersen et al. (2003)
PS 1919	−11.90	75.00	1876	13–33[a]	2	4.39	−0.7	Stein et al. (1996)
PS 1922	−8.77	75.00	3350	117–149[a]	2	4.40	−0.4	Stein et al. (1996)
PS 1927	−17.12	71.50	1734	65–95[a]	4	4.35	−0.6	Stein et al. (1996)
PS 1951	−20.82	68.84	1481	70–102[a]	4	4.48	1.4	Stein et al. (1996)
PS 2122	7.55	80.39	705	159–200	Unknown	4.20	1.4	Knies (1994)
PS 2123	9.86	80.17	571	187–228	Unknown	4.43	1.6	Knies (1994)
PS 2129	17.47	81.37	861	20–40[a]	Unknown	4.68	0.8	Knies (1999)
PS 2206	−2.51	84.28	2993	10–13	1	4.60	−1.8	Stein et al. (1994)
PS 2208	4.60	83.64	3681	15–25	Unknown	4.60	−1.1	Stein et al. (1994)
PS 2210	10.70	83.04	3702	12–14	1	4.65	−0.9	Stein et al. (1994)
PS 2212	15.85	82.07	2550	80–100	Unknown	4.26	0.0	Vogt (1997)
PS 2423	−5.45	80.04	829	20–100	Unknown	3.80	−0.6	Notholt (1998)
PS 2424	−5.74	80.04	445	390–433	Unknown	4.20	−0.6	Notholt (1998)
PS 2613	−0.48	74.18	3259	30–40[a]	3	4.71	1.2	Voelker (1999)
PS 2644	−21.77	67.87	778	111–129[a]	19	4.50	0.5	Voelker (1999)
PS 2837	2.38	81.23	1023	389–397[a]	9	4.72	0.0	Nørgaard-Pedersen et al. (2003)
PS 2876	−9.43	81.91	1976	9–62[a]	>20	4.55	−1.5	Nørgaard-Pedersen et al. (2003)

Table 1 (*continued*)

Core no.	Longitude	Latitude	Water depth (m)	LGM level 18–21.5 ka depth (cm)	Number of averaged $\delta^{18}O_{Nps}$ measurements	Averaged $\delta^{18}O_{Nps}$	T_{Nps}	Data source for $\delta^{18}O_{Nps}$
PS 2887	−4.61	79.60	1411	35–52[a]	18	3.33	−0.1	Nørgaard-Pedersen et al. (2003)
PS 16396	−11.25	61.87	1145	198–540[a]	38	4.08	4.5	Sarnthein et al. (1995)
PS 16397	−11.18	61.87	1145	4–98[a]	13	4.00	4.8	Sarnthein et al. (1995)
PS 21291	8.70	78.00	2400	85–105	Unknown	4.54	1.6	Weinelt (1993)
PS 21736	−5.17	74.33	3460	28	1	4.65	0.1	Jünger (1993)
PS 21842	−16.52	69.45	982	27–39[a]	2	4.44	0.6	Sarnthein et al. (1995)
PS 21900	−2.32	74.53	3538	25–33	2	4.45	0.7	Jünger (1993)
PS 21906	−2.15	76.93	2990	21–30[a]	9	4.25	0.6	Nørgaard-Pedersen et al. (2003)
PS 21910	1.32	75.62	2454	35–40	Unknown	4.50	1.6	Weinelt (1993)
PS 23199	5.24	68.38	1968	74–103	4	4.76	1.9	Vogelsang (1990)
PS 23205	5.76	67.62	1411	50–90	5	4.62	2.2	Vogelsang (1990)
PS 23243	−6.54	69.38	2715	55	1	4.71	1.2	Vogelsang (1990)
PS 23246	−12.86	69.40	1858	45	1	4.58	1.2	Vogelsang (1990)
RC 9-225	−15.40	54.89	2334	135	1	3.99	5.6	Keigwin and Boyle (1989)
SO 82-5	−30.9	59.18	1416	95–120[a]	6	4.43	1.9	van Kreveld et al. (2000)
SU 90-32	−22.42	61.78	2200	140–170[a]	Unknown	3.98	4.8	Sarnthein et al. (1995)
SU 90-33	−22.08	60.57	2370	Unknown	Unknown	4.22	3.9	Cortijo et al. (1997)
SU 90-39	−21.93	52.57	2900	81–90	Unknown	4.25	4.5	Cortijo (1995)
SU 90-106	−39.45	59.98	1615	19–23	3	4.40	2.2	Weinelt et al. (1996)
SU 90-107	−28.08	63.08	1625	17–19	Unknown	4.17	3.5	Sarnthein et al. (1995)
V 23-23	−44.55	56.08	3292	115	1	4.34	2.0	Mix and Fairbanks (1985)
V 23-42	−27.92	62.18	1514	70	1	4.52	2.5	Keigwin and Boyle (1989)
V 23-81	−16.14	54.03	2393	236–304[a]	4	3.82	5.4	Jansen and Veum (1990)
V 23-82	−21.93	52.59	3974	110	1	4.34	4.2	Keigwin and Boyle (1989)
V 23-83	−24.26	49.87	3971	100	1	4.31	4.5	Keigwin and Boyle (1989)
V 27-17	−37.31	50.08	4054	35	1	4.42	3.1	Keigwin and Boyle (1989)
V 27-19	−38.79	52.10	3466	35	1	4.46	1.9	Keigwin and Boyle (1989)
V 27-60	8.58	72.17	2525	188–203	2	4.72	1.7	Labeyrie and Duplessy (1985)
V 27-86	1.12	66.60	2900	50–67	2	4.72	2.1	Labeyrie and Duplessy (1985)
V 27-114	−33.07	55.05	2532	291	1	4.42	2.2	Keigwin and Boyle (1989)
V 27-116	−30.33	52.83	3202	60	1	4.52	2.3	Keigwin and Boyle (1989)
V 28-14	−29.58	64.78	1855	149–170	3	4.60	1.2	Shackleton (1974)
V 28-38	−4.40	69.38	3411	184–192	Unknown	4.82	1.0	Keigwin and Boyle (1989)
V 28-56	−6.12	68.03	2941	60–80	4	4.67	1.5	Kellogg et al. (1978)
V 29-180	−23.87	45.30	3049	57	1	3.80	7.5	Keigwin and Boyle (1989)
V 29-183	−25.50	49.14	3629	38	1	4.10	5.4	Keigwin and Boyle (1989)
V 29-206	−29.28	64.90	1624	170	1	4.37	2.2	Keigwin and Boyle (1989)
V 30-108	−32.50	56.10	3171	85	1	4.52	2.0	Keigwin and Boyle (1989)
V 30-164	8.97	69.83	2901	Unknown	Unknown	4.83	1.3	Duplessy et al. (1991)

[a]The depths of the LGM levels are determined with use of AMS ^{14}C. Where no asterisk is shown, the LGM levels are determined by use of isotope stratigraphy.

level. In the other cores the LGM level is found based on the planktic oxygen isotope curves. Single $\delta^{18}O_{Nps}$-values from different low-resolution cores (see Table 1) are also included in the data set. In the publications from which the data was gathered, the authors simply defined the LGM as the event with the youngest distinct $\delta^{18}O_{Nps}$-maximum prior to Termination 1. We assume here that these values are inside the GLAMAP definition of the LGM, but admit that a higher robustness, resolution and more AMS ^{14}C dates are preferable. When integrating these data into the data set, we observe, however, that the spatial pattern is not changed, indicating that these data points are coherent with neighbouring data points that are better constrained by actual dates.

3.2. Mg/Ca ratios of N. pachyderma (sin.)

Foraminiferal Mg/Ca ratios are increasingly used as indicators of past ocean temperatures (e.g. Lea et al., 2000, 2002; Elderfield and Ganssen, 2000; Rosenthal et al., 2000). To add this information to the temperature information from the oxygen isotopes we applied Mg/Ca paleothermometry to a subset of samples from the LGM time slice and to a set of modern (core top) sediment samples.

The LGM part of the work is based on a total of 31 cores that provide a broad coverage of the North Atlantic north of 50°N, and the Nordic Seas. Table 2 shows the data. We measured the Mg/Ca ratios of the polar species *N. pachyderma* (sin.) in 1–6 samples from each core covering the LGM time slice, and note that some of the cores are not AMS ^{14}C dated (see Table 2). In these cases the LGM time slice definition was based on the planktic oxygen isotope records as described above. When choosing samples for Mg/Ca analyses, samples with planktic oxygen isotope values close to the average for the whole of the LGM level were given the highest priority. This was done to avoid potential internal fluctuations and obtain a best possible fit to oxygen isotope data.

Sediment surface samples from the Nordic Seas were also analysed for Mg/Ca content, with the purpose to investigate the correlation between the core top Mg/Ca ratios of *N. pachyderma* (sin.) and modern sea surface temperatures (SST). The data are shown in Table 3, and all these samples have Fe/Ca ratios below 0.1 mmol/mol, indicating that contamination should not influence the Mg/Ca ratios significantly (Barker et al., 2003). All samples are from box cores, covering the upper 0.5 cm of the sediment (Johannessen, 1992).

The samples were cleaned following the cleaning procedure of Barker et al. (2003), and were measured using the inductively coupled plasma-atomic emission spectroscopy (ICP-AES) method, with a relative precision of <0.3% (de Villiers et al., 2002). All cleaning and measurements were performed at the Department of Earth Sciences, University of Cambridge, UK.

4. Constraints on the temperatures derived from $\delta^{18}O$

The oxygen isotopic composition of foraminiferal calcite depends on both the temperature and the oxygen isotopic composition of the ambient seawater, $\delta^{18}O_w$. Since we here are primarily interested in the temperature, we need to address the factors that influence the $\delta^{18}O_w$, which are: the ice volume effect which results from the storage of water depleted in ^{18}O in ice sheets,

Table 2
Core data used for studies of Mg/Ca ratios in the tests of *N. pachyderma* (sinistral). The Mg/Ca ratios are analysed in this study

Core no.	Longitude	Latitude	Water depth (m)	LGM level 19–21.5 ka depth (cm)	Number of averaged trace element measurements	Average Mg/Ca (mmol/mol)	Reference LGM-level
BOFS 5 K	−21.87	50.68	3547	78–94[a]	4	0.87	Maslin (1992)
BOFS 7 K	−22.54	51.75	2429	38–40	2	0.91	Manighetti et al. (1995)
BOFS 8 K	−22.04	52.50	4045	85–110	2	0.81	Maslin (1992)
BOFS 10 K	−20.65	54.67	2761	108–114	4	0.93	Manighetti et al. (1995)
BOFS 14 K	−19.44	58.62	1756	38–44	3	0.92	Maslin (1992)
BOFS 17 K	−16.50	58.00	1150	74–95	4	1.16	Maslin (1992)
ENAM 94-09	−9.43	60.34	1286	543–583	3	0.89	Lassen et al. (2002)
HM 100-7	−4.72	61.67	1125	71–100	4	0.83	Sarnthein et al. (1995)
HM 52-43	0.73	64.25	2781	85–90[a]	2	0.81	Veum et al. (1992)
HM 71-12	−13.87	68.43	1547	44–48	3	0.96	Sarnthein et al. (1995)
HM 71-15	−17.43	70.00	1547	38–41[a]	4	0.71	Roe (1998)
HM 71-19	−9.51	69.48	2210	42	1	1.22	Vogelsang (1990)
HM 79-26	−5.93	66.90	3261	60	1	1.05	Sarnthein et al. (1995)
HM 80-30	1.60	71.78	2821	38–44	4	0.88	Sarnthein et al. (1995)
HM 80-42	−9.23	72.25	2416	57–64	4	0.90	Sarnthein et al. (1995)
HM 80-60	−11.86	68.90	1869	70–76	3	1.11	Sarnthein et al. (1995)
HM 94-13	−1.62	71.63	1946	46–52	4	0.98	Sarnthein et al. (1995)
HM 94-18	5.70	74.50	2469	28–35	2	0.96	Sarnthein et al. (1995)
HM 94-25	1.32	75.60	2469	38–47[a]	4	0.95	Sarnthein et al. (1995)
HM 94-34	−2.54	73.77	3004	50	1	0.79	Sarnthein et al. (1995)
M 23357	−5.53	70.96	1969	57	1	1.29	Goldschmidt (1994)
MD 2010	4.56	66.68	1226	261–373[a]	6	0.73	Dokken and Jansen (1999)
MD 2011	7.64	66.97	1048	1066–1361[a]	5	0.80	Dreger (1999)
MD 2284	−0.98	62.37	1500	870–1050	5	0.74	Jansen and Meland (2001)
MD 2289	4.21	64.66	1395	610–690	4	0.72	Berstad (pers. com.)
MD 2304	9.95	77.62	853	420–512[a]	5	0.74	Fevang (2001)
NEAP 8 K	−23.54	59.47	2419	179–192	4	0.94	Vogelsang et al. (2000)
PS 21842	−16.52	69.45	982	31–39[a]	5	0.85	Sarnthein et al. (1995)
SO 82-4	−30.48	59.10	1503	181–188[a]	1	0.89	Moros et al. (1997)
SU 90-32	−22.42	61.78	2200	140–170	1	0.80	Sarnthein et al. (1995)
V 23-81	−16.14	54.03	2393	260–304[a]	3	0.95	Jansen and Veum (1990)

[a]The depths of these LGM levels are determined with use of AMS ^{14}C.

Table 3
Core top data used for studies of Mg/Ca ratios of *N. pachyderma* (sin.). The Mg/Ca ratios are analysed in this study. All core tops are from box cores, upper 0.5 cm of the sediment (Johannessen, 1992)

Core no.	Longitude	Latitude	Water depth (m)	Observed Mg/Ca (mmol/mol)	$\delta^{18}O_{Nps}$ ratios	T JAS 50 m depth[a]	Calcification temperature[b]	"Calculated" Mg/Ca (mmol/mol)[c]	Mg/Ca deviation (mmol/mol)[d]
HM 16132	−0.72	64.57	2798	1.00	2.11	8.2	8.5	1.27	−0.28
HM 49-14	−3.23	65.42	2863	1.06	2.46	6.6	6.8	1.08	−0.02
HM 49-15	−0.36	66.34	3260	1.16	2.41	7.4	7.2	1.12	0.04
HM 52-34	−8.28	65.35	1101	1.14	3.00	5.9	5.1	0.91	0.23
HM 57-14	−6.21	67.00	3005	1.21	2.68	4.3	5.1	0.91	0.30
HM 57-20	1.67	62.65	750	1.13	2.40[e]	9.2	7.3	1.13	0.00
HM 94-12	−3.92	71.53	1816	0.99	3.25[e]	3.8	3.6	0.78	0.20
HM 94-16	5.61	73.38	2356	0.89	3.13[e]	2.8	4.3	0.84	0.05
HM 94-25	1.32	75.60	2469	1.12	3.68[e]	1.6	1.7	0.65	0.47
HM 94-42	−22.5	68.75	1339	0.98	3.10[e]	1.2	2.0	0.67	0.31

[a] Calculated from Levitus and Boyer (1994).

[b] Calcification temperatures are calculated from the oxygen isotope ratios given in table, and Eqs. (1) and (3) in text.

[c] "Calculated" Mg/Ca ratios are the ratios which should be expected from the calcification temperatures, using the equation of Nürnberg (1995) (Eq. (4) in text).

[d] Calculated in this manner: Mg/Ca deviation = observed Mg/Ca − "calculated" Mg/Ca.

[e] $\delta^{18}O_{Nps}$ measurements are not performed for these core tops. These values are interpolated $\delta^{18}O_{Nps}$ data calculated from nearby core tops with $\delta^{18}O_{Nps}$ measurements from Simstich et al. (2003).

and the salinity and the slope of the $\delta^{18}O_w$–salinity relationship.

4.1. Constraints on the ice volume effect

Before temperature calculations can be made, the effect that storage of isotopically depleted water in continental ice sheets has on $\delta^{18}O_w$ between the LGM and today must be assessed. Fairbanks (1989) proposed that ice volume indicated a change in $\delta^{18}O_w$ of 1.2–1.3‰ between LGM and today. Schrag et al. (1996), on the other hand, suggested an amplitude of 1.0‰, based on pore water measurements in deep sea sediments. For this work we have chosen an intermediate value of 1.1‰ for the ice volume effect at the LGM. The range of the different estimates of the ice volume effect indicated above, would impose an uncertainty on the temperature reconstructions of about ±0.5–1 °C on the absolute values, but do not influence the spatial patterns we reconstructed.

4.2. Oxygen isotope distribution of N. pachyderma (sin.) during the LGM

During the LGM, the northern North Atlantic and the Nordic Seas were characterized by a marked contrast in the foraminiferal isotopic composition between the areas south of the Iceland–Scotland Ridge and the Nordic Seas (Fig. 2a). In the eastern North Atlantic the $\delta^{18}O_{Nps}$-values were 3.8–4.4‰, while $\delta^{18}O_{Nps}$-values in the central Nordic Seas were quite homogenous with values about 4.6–4.8‰. In the western and eastern flanks of the Nordic Seas the $\delta^{18}O_{Nps}$-values were lower, with values averaging about 4.2–4.5‰.

By subtracting an ice volume effect of 1.1‰ from the raw data, the pattern in Fig. 2b is derived, showing $\delta^{18}O_{Nps}$-anomalies in the Nordic Seas. These anomalies suggest that the ice volume corrected $\delta^{18}O_{Nps}$-values in the Greenland and Iceland Seas for the LGM were not very different from today, indicating that similar surface environments as today prevailed in the LGM in the present cold water part of the Nordic Seas. The largest changes are evident south of the Iceland–Scotland Ridge and in the eastern Norwegian Sea, indicating that the meridional heat transport was much weaker during the LGM compared with today.

4.3. Constraints on the relationship between salinity (S) and $\delta^{18}O_w$ for the LGM

Constraining this relationship (or mixing line) for the Nordic Seas and the northern North Atlantic is not a straightforward work, since the water masses have different characteristics. $\delta^{18}O_w$ and salinity data from the database of Schmidt et al. (1999), show that the oxygen isotope values of the fresh water end member ($\delta^{18}O_0$) vary from −16‰ south of the Greenland–Scotland Ridge, −23‰ in the Norwegian Sea, −33‰ in Greenland–Iceland Seas and −35‰ in the Arctic Ocean. Simstich et al. (2003) have published relationships with $\delta^{18}O_0$ of −12‰ in the eastern and central Nordic Seas, and $\delta^{18}O_0$ of −17‰ in the East Greenland Current (EGC).

For the LGM the salinity/$\delta^{18}O$ relationship is not known. The Greenland deep ice cores show glacial oxygen isotope ratios down to less than −40‰ (Dansgaard and Oeschger, 1989), which are substantially below the modern values around −30‰, suggesting a steeper mixing line than today. On the other hand, more intensive sea-ice formation during the LGM may produce sea-ice with an isotopic signature of, e.g. typically $-2‰$, the same as in the surrounding sea water, thus producing a fresh water source with isotopically high values that effectively lowers the salinity/$\delta^{18}O$ slope. It is therefore not obvious that the LGM slope was steeper than the modern. Based on an Earth system model of intermediate complexity, CLIMBER-2, Roche et al. (2004) found that no drastic changes occurred in the $\delta^{18}O_w$:S relationship. In the absence of conclusive evidence for a shift in this relationship, we suggest that the modern mixing line for both the North Atlantic and the Nordic Seas seems most appropriate. We therefore use the equation of GEOSECS (1987) obtained from $\delta^{18}O_w$ and salinity for water samples collected in the upper 250 m of the Atlantic Ocean during the GEOSECS expedition:

$$\delta^{18}O_w = -19.264 + 0.558 \times \text{salinity}. \tag{1}$$

4.4. Constraints on the LGM salinity

Since we have no direct method of assessing the salinity, we need to constrain the possible range of salinities from other criteria. In this case we need to consider:

- The homogenous $\delta^{18}O_{Nps}$-values in the Nordic Seas (Fig. 2a), which indicate that the salinity (and temperature) range within the Nordic Seas was quite low.
- Oxygen isotope change between modern data and the LGM (Fig. 2b), which can be used to assess the possible range of salinity changes in certain areas (see below).
- The general high latitude cooling of the LGM implies that inflow of saline waters from the North Atlantic into the Nordic Seas was reduced compared with today, due to the fact that warmer waters in general are more saline.
- Possible existence of deep convection, which requires salinity above certain thresholds.
- Higher glacial salinities in the Fram Strait, due to less melting of sea ice, reduced river runoff, and no inflow of low saline water through Bering Strait due to the low sea level.

From plotting the $\delta^{18}O_{Nps}$-anomalies between the recent data and the LGM (Fig. 2b), we find that the smallest anomaly between LGM and modern values exist in an elongated area in the Iceland Sea and the Greenland Sea, where open-ocean convection in Arctic water masses occurs today. This similarity indicates similar oceanographic conditions in this area during the

LGM, with SSSs of 34.4–34.9‰ (the global salinity rise of $+1‰$ due to the ice volume effect is not added here and in the following discussion) and summer-SSTs of 0–6 °C, which is typical for Arctic water masses (Swift and Aagard, 1981). Considering that most of the Nordic Seas had similar isotope values, this is an indication for presence of Arctic water masses, with SSS- and SST-values as mentioned above.

In the northernmost cores in our compilation, in the Fram Strait region, the planktic isotope values indicate higher sea surface salinity during the LGM than today, since applying modern salinities for these locations to calculate SSTs from the oxygen isotope paleotemperature equation give LGM SST-values lower than -1.8 °C, which is below the freezing point of sea water. To correct this improbable result, the salinities of the LGM must be set to higher values than now in this area.

In the central and eastern parts of the North Atlantic we believe that salinities must have been lower than today as a result of the reduced northward oceanic heat flux and reduced advection of saline water from the tropics to higher latitudes.

Based on high epibenthic carbon isotope values in the Nordic Seas (Veum et al., 1992) and benthic oxygen isotope values (Dokken and Jansen, 1999), it appears likely that deep-sea ventilation and some element of open ocean convection took place in the Nordic seas in the LGM. Dokken and Jansen (1999) proposed that deep-water formation shifted between open ocean convection and brine-release due to sea-ice formation, and that shifts in the relative importance of the two deep-water formation types accompanied the millennial scale climate shifts of the last glacial. The brine mechanism can operate with lower preformed salinities than is required for open ocean convection. The benthic $\delta^{18}O$-change between LGM and Holocene in the Nordic Seas is 0.3‰ lower than the ice volume effect of 1.1‰ (Fairbanks, 1989; Schrag et al., 1996). This implies that the isotopic composition of the deep water to some extent was influenced by the brine mechanism. Yet the high epibenthic carbon isotope values (Veum et al., 1992), and the generally higher oxygen isotope values compared to many of the stadials of the last glacial would imply that at least some convection in open ocean areas occurred. Open-ocean convection will bring isotopically heavy water to the deep sea, while overturning by the brine-mechanism brings a light isotopic signal originating from densification of fresher waters by sea-ice formation to the deep sea. A mix of these two modes of deep-water formation may have produced an average LGM-Holocene change in benthic oxygen isotopes, which is 0.3‰ less than what would be expected from ice volume alone. Hence the average salinity of surface waters was probably 34.5–34.9‰, i.e. somewhat lower than today, but not by a large extent, as this may have hindered overturning

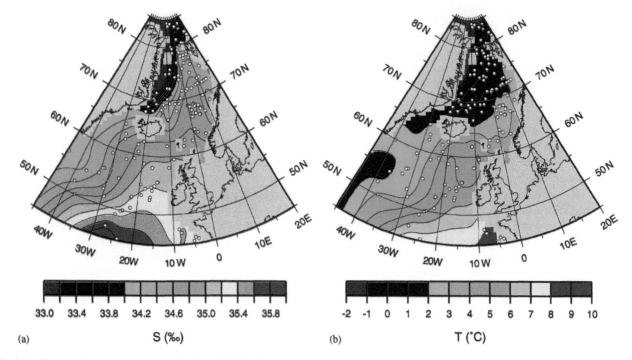

Fig. 3. (a) Suggested sea surface salinities for the LGM, based on the approach described in text (Sections 4.4.1 and 4.4.2). The higher global salinity (+1‰) during the LGM caused by the ice volume effect is not included. (b) Summer SST reconstruction for the LGM computed from the approach described in text (Section 4.5). The gridding, contouring and colouring were based on the estimated SST values (Table 1) calculated from Eq. (3) in text. Grey coloured areas indicate land areas during the LGM (Peltier, 1994).

and ventilation through some amount of open ocean convection.

Based on the discussion above, we have constrained the likely LGM salinity field follows (Fig. 3a):

- Lower salinities in the central and eastern Nordic Seas than today. They were, however, high enough to allow for some convection.
- The most likely open-ocean convection sites are where the highest $\delta^{18}O_{Nps}$-values are found ($>4.7‰$), assuming that convection occurred in the areas with highest $\delta^{18}O_{Nps}$-values, as it does today.
- Higher salinities in the Fram Strait and the border to the Arctic Ocean than today.
- Somewhat lower salinity in the North Atlantic relative to today, based on a lower northward advection of warm waters.
- Generally, lower salinity gradients between different areas compared with today, based on the homogenous LGM $\delta^{18}O_{Nps}$-field.

These constraints were used to calculate a possible salinity field in two steps:

4.4.1. 1st step: Salinities in the coldest water, where $T_{modern} \leqslant 2\,°C$

We have here selected data from the core locations where the summer temperature (at 50 m depth, JAS) today is below 2 °C. It is highly unlikely that the temperatures during the LGM were significantly higher

than today. There was probably partly ice-free water as far north as the Fram Strait (Hebbeln et al., 1994), thus the summer SST were above $-1.8\,°C$. Therefore, we postulate that the SSTs have been between -1.8 and $+2\,°C$ for these cores and that as a first approximation the temperatures at their respective locations were similar to modern day temperatures.

We then used the modern day temperatures, the modern mixing line between salinity and $\delta^{18}O_w$ (Eq. (1)), and the paleotemperature equation of Shackleton (1974) to calculate salinities for these core locations. This results in a correlation of LGM and modern salinities of 0.62 in the northwestern area denoted in Fig. 3a. The best fit between modern and LGM salinities is expressed by the following equation:

$$S_{LGM} = 0.3192^*(S_{modern})^2 - 20.81^*S_{modern} + 371.8),$$
$$R^2 = 0.62. \qquad (2)$$

4.4.2. 2nd step: Salinities for the area where $T_{modern} > 2\,°C$

As a means of assessing the plausible range of salinities, we use Eq. (2) to calculate LGM salinities outside of the coldest end member area. The calculated salinities give similar salinity pattern as today, but with lower gradients between high and low salinity areas, in particular a reduced northward extent of saline surface waters compared with the modern values (Fig. 3a). We stress that this is not based on a strictly objective criterion, yet the pattern agrees with the pattern that we expect to see considering the constraints discussed in the text above.

In favour of this attempt we can argue that the salinities in the coldest region cannot be much different, due to the fact that temperatures cannot be lower than freezing. There must also be a gradient between the low latitudes and the high latitude region as today, but with a smaller influx of saline waters in the Nordic seas areas due to the prevailing colder temperatures in the mid latitude region and the eastern boundary of the area which is consistent in all reconstructions of LGM temperatures. The calculated salinities give a similar salinity pattern as today, but with lower salinity gradients in the study area, as expected (Fig. 3a).

With the applied mixing line for oxygen isotopes vs. salinity an 0.48‰ error in salinity will result in a 1 °C change in the temperature estimate. With a steeper mixing line, e.g. closer to unity, an 0.25‰ salinity error will result in a 1 °C reconstructed temperature change. With the constraint that LGM salinities must be the same or less than modern (ice volume effect not included), and the observation that there must have been some deep convection, we contend that salinities cannot have been more than 0.3–0.5‰ lower than modern over the central and eastern parts of the Nordic Seas. Thus the salinity assumption used for our SST reconstruction produce an error of not more than 1 °C.

4.5. Temperature estimates

In the area where $T_{modern} \leqslant 2.0\,°C$, we have defined that $T_{LGM} = T_{modern}$, based on arguments described in Section 4.4.1. Where $T_{modern} > 2.0\,°C$, the salinities are defined as in Eq. (2) based on the arguments described in Section 4.4.2. $\delta^{18}O_w$ is then calculated using Eq. (1). The calcification temperatures of *N. pachyderma* (sin.) are then calculated using the paleotemperature equation of Shackleton (1974):

$$T = 16.9 - 4.38*(\delta^{18}O_{Nps} - \delta^{18}O_w + 0.27)$$
$$+ 0.10*(\delta^{18}O_{Nps} - \delta^{18}O_w + 0.27)^2. \qquad (3)$$

The resulting temperatures are shown in Fig. 3b.

4.6. Discussion of oxygen isotope results

Assuming that the calcification of the foraminifers is a summer season phenomenon, in particular during the LGM, and that it takes place in the upper 100 m of the water column, at least for Arctic water masses (Johannessen, 1992; Simstich et al., 2003), the results in Fig. 3b can be viewed as an estimate of summer temperature of the surface mixed layer, possibly the lower part of it. Thus the peak summer temperatures of the surface may

have been slightly higher than these estimates. The corridor of somewhat higher temperatures entering the Nordic seas in the SW in Fig. 3b is consistent with other evidence, such as the drift routes of icebergs inferred from sediment patterns of ice rafted constituents with known provenance in the North Sea (Hancock, 1984), and occurrence of significant amounts of subpolar planktic foraminifers in a band on the eastern side of the Nordic Seas as north as Fram Strait (Hebbeln et al., 1994). It is also consistent with the North Atlantic pattern in Fig. 3b, documenting that despite a more pronounced zonal temperature field than today, and a much more reduced northward heat flux in the ocean, there is an element of a meridional transport that is maintained. This meridionality is in conflict with the CLIMAP reconstruction, which had a perennial ice cover over the Nordic Seas at the LGM. The idea of some convection is also consistent with inferences from deep-water benthic foraminifer $\delta^{18}O$ of LGM conditions favourable for open ocean convection in the Nordic Seas (Dokken and Jansen, 1999). The low gradient towards the west is also consistent with the abundance patterns of biogenic carbonate (Hebbeln et al., 1998). The absolute temperatures south of the Greenland–Scotland Ridge are consistent with foraminiferal transfer function estimates (Pflaumann et al., 2003), while the temperatures within the Nordic Seas are slightly lower than those inferred from foraminiferal transfer functions, in particular in the western part of the ocean. This is most likely due to the lack of sensitivity of the transfer functions in cold waters.

The constraints on this reconstruction in terms of the possible range of salinity impose an error range of about $\pm 1\,°C$ on the reconstruction. We conclude that this reconstruction, despite its shortcomings is a more likely LGM reconstruction than the CLIMAP, and probably also the GLAMAP reconstructions.

5. Constraints from Mg/Ca ratios

To further test the oxygen isotope based reconstruction, we also employed Mg/Ca paleothermometry on a subset of samples from the LGM as well as tests on the core top distribution of the Mg/Ca ratio. Applying the method in this type of environment is challenging, both due to the reduced slope of the Mg/Ca relationship vs. temperature in cold waters, thereby reducing its sensitivity and imposing larger error bars, but also due to the enhanced possibility of contamination from detrital minerals in the glacial marine sediments.

5.1. Contamination of clay and organic material

Even if the cleaning procedure of Barker et al. (2003) is used, there may still be significant amounts of clay

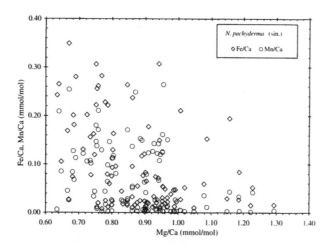

Fig. 4. Scatter of Fe/Ca and Mn/Ca plotted against Mg/Ca for all the LGM data, showing no covariance between these metals.

and/or organic material (hereafter called contaminant) in the analysed carbonate samples, which may influence the measured Mg/Ca values. Contamination will nearly always give higher Mg/Ca values, since the amount of Mg in silicate in nearly all cases is higher than the Mg content in carbonate. Indicators for contamination may be the content of Fe and Mn, documented in the Fe/Ca and Mn/Ca ratios of samples. These secondary metal ratios exhibit no positive correlation with Mg/Ca measured in N. pachyderma (sin.) for the LGM samples, and therefore suggest that contamination is not a significant control on Mg/Ca variability (Fig. 4). If it is assumed that the measured "contaminant" metals are representative of a silicate phase, then for a typical silicate Mg/Fe ratio of ∼1 mol/mol, −30% of the measured Mg/Ca in N. pachyderma (sin.) could be attributed to contamination. However, it is probable that the Fe and Mn contaminant metals are in fact associated with a ferromanganese carbonate overgrowth (Boyle, 1983), particularly given the elevated Fe/Ca and Mn/Ca ratios observed.

5.2. Mg/Ca ratios in the core tops

While foraminiferal Mg/Ca ratios have shown great potential as indicators of past ocean temperatures in tropical and subtropical areas of the ocean (e.g. Lea et al., 2000, 2002; Elderfield and Ganssen, 2000; Rosenthal et al., 2000), the Mg/Ca ratios shown in Fig. 5a indicate a more complicated pattern. There seems to be very little temperature dependency across the basin, contrary to what would be expected. This lack of a temperature trend is due to the anomalously high values in the areas north to northeast of Iceland where the Mg/Ca ratios are much higher than expected if existing Mg/Ca based temperature equations are used (e.g. Elderfield and Ganssen, 2000; Nürnberg, 1995). The Mg/Ca ratios are

not clearly correlated with calcification temperatures. We tested how well the observed core top Mg/Ca ratios fit with the paleotemperature equation of Nürnberg (1995) for the North Atlantic and the Nordic Seas:

$$Mg/Ca = 0.549 * e^{(0.099 \times T)}. \qquad (4)$$

We calculated from this equation the Mg/Ca ratios, which would correspond to the calcification temperatures deduced from core top oxygen isotopes (see Table 3), and found that 4 samples (marked with green circles in Fig. 5a) deviated with less than 0.06 mmol/mol from the expected Mg/Ca values. All these cores are located in Atlantic water masses. 1 sample (marked with a yellow circle in Fig. 5a) shows 0.28 mmol/mol lower than expected measured Mg/Ca ratios, while the samples to the west inside or close to Arctic water masses all have too high Mg/Ca values. The reason for these deviations from expected Mg/Ca values cannot be that the Nürnberg (1995) equation (Eq. (4)) does not work with the Arctic Water, since in this case we would expect that the deviations from sample to sample would be consistent in the whole region. Therefore other factors influence the Mg composition. At the moment, we suggest that Eq. (4) can be used for calculation of Mg/Ca based paleotemperatures in the eastern sector of the Nordic Seas, where 4 of the core tops deviate not more than 0.06 mmol/mol from the expected values based on the calibration curve of Nürnberg (1995). More data and analyses of various factors, such as salinity, dissolution, secondary encrustation, need to be explored.

We do not have core top measurements south of the Greenland–Scotland Ridge, but previous work suggest that the temperature dependency to Mg/Ca works better here (Nürnberg, 1995; Elderfield and Ganssen, 2000).

5.3. LGM results

The Mg/Ca ratios document a clear meridional trend in the eastern sector of the Norwegian Sea for the LGM, with somewhat lower ratios in the eastern Norwegian Sea (0.7–0.8 mmol/mol) than further south, in the North Atlantic west of Ireland (0.8–0.9 mmol/mol) (Fig. 5b). These Mg/Ca ratios were transferred to temperatures using Eq. (4) in the area south of the Greenland–Scotland Ridge and the Norwegian Sea.

5.3.1. Temperature estimates

The temperatures were calculated using Eq. (4), and are shown in Fig. 5c. Note that the Mg/Ca based temperatures are only calculated in the eastern Nordic Seas and south of the Greenland–Scotland Ridge, based on arguments mentioned in Section 5.2.

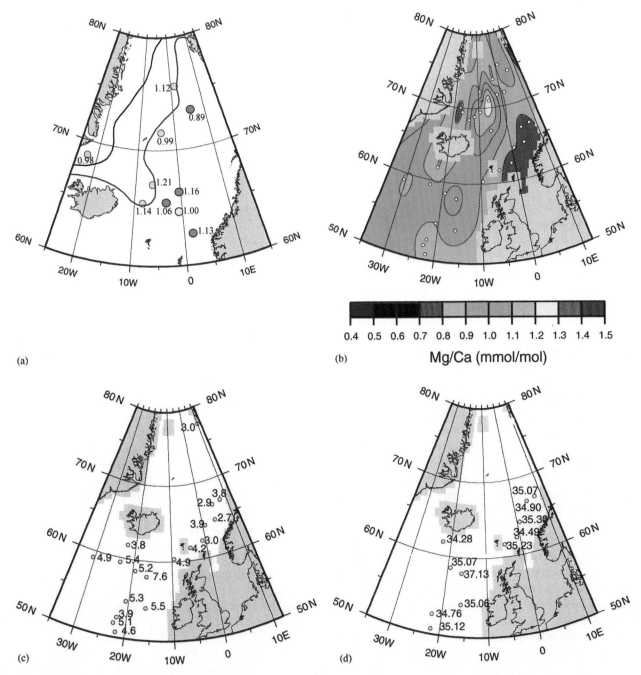

Fig. 5. (a) Mg/Ca ratios (mmol/mol) of core tops, measured on *N. pachyderma* (sin.). The green coloured circles are samples with measured Mg/Ca ratios deviating less than 0.06 mmol/mol from "expected" Mg/Ca ratios, calculated from Eq. (4) in text. The yellow coloured sample has a measured Mg/Ca ratio of 0.28 mmol/mol below the "expected" Mg/Ca ratios. The grey coloured samples show "too high" Mg/Ca ratios. See Table 3 for details. The red line marks the Arctic Front, separating Atlantic and Arctic water masses. The blue line marks the Polar Front, separating Arctic and Polar water masses. (b) Mg/Ca distribution for the LGM, based on averaged Mg/Ca ratios in Table 2 of *N. pachyderma* (sin.). The white dots show the locations of the cores. (c) Mg/Ca calcification temperatures of *N. pachyderma* (sin.), calculated using the Mg/Ca temperature equation of Eq. (4) in text. (d) Calcification salinities of *N. pachyderma* (sin.), calculated using the Mg/Ca temperatures in Fig. 5c and oxygen isotope ratios from Table 1. Not all points from Fig. 5c are shown on the map, since oxygen isotope data are not available for the LGM in some of the cores. The higher global salinity (+1‰) during the LGM is not included. See the text for details on salinity calculation. Grey coloured areas indicate land areas during the LGM (Peltier, 1994).

5.3.2. $\delta^{18}O_w$ and salinity estimates

Values of $\delta^{18}O_w$ and salinities are calculated for the cores in Fig. 5c where we have suggested that Eq. (4) can be used for calculation of Mg/Ca based temperatures. The $\delta^{18}O_w$ can easily be computed by a rearrangement of Eq. (3):

$$\delta^{18}O_w = \delta^{18}O_{Nps} + 0.27 - 21.9 + \sqrt{(310.61 + 10T_{Nps})}.$$

$$(5)$$

By assuming that the mixing line from GEOSECS (1987) (Eq. (1)) can be used, as argued above, we calculate salinities by rearrangement of Eq. (1):

$$S = [(\delta^{18}O_w + 34.52) \times 1.792] - 1. \qquad (6)$$

Note that the global salinity rise of $+1\permil$ due to the ice volume effect is subtracted on the end of this equation for comparison to the modern salinities. The resulting salinities are shown in Fig. 5d.

5.4. Discussion of Mg/Ca results

South of the Greenland–Scotland Ridge and in the eastern parts of the Nordic Seas the temperatures and salinities deviate somewhat from the estimates obtained from oxygen isotopes above, i.e. on average higher temperature and salinity. Yet, the meridional trend in the Eastern sector of the study area is similar to the oxygen isotope temperature trend (Figs. 3b, 5b–d). The Mg/Ca results for the LGM are consistent with a warmer and sea-ice free sector in the Eastern Nordic Seas, as compared with the CLIMAP reconstruction. The main problem with the Mg/Ca results is the higher Mg/Ca ratios to the west (not shown in Fig. 5c). Existing temperature equations (e.g. Eq. (4)) give temperatures about 5–10 °C, which are much higher than the summer SSTs today in this area (0–5 °C). It is clear that there must be an artefact on the ratios in the areas north of Iceland. We note that the same deviation from a temperature relationship is found in this area in the core top data (Fig. 5a). This may point to a specific problem with the method in this sub-area. Since all the Fe/Ca ratios of these samples are less than 0.1 mmol/mol, we find it difficult to ascribe the high Mg/Ca ratios to contamination. A possibility is that the high Mg/Ca ratios in the Greenland and Iceland Seas could be due to high salinity, because of salt rejection via sea-ice formation and brine. This process requires, however, intensive sea-ice formation to increase the salinity in significant amounts, and it is questionable whether there was enough sea-ice formation in the central ocean, since sea-ice formation is largely a shelf process. Another possibility is that sea-ice cover, of reasons not explored, may give high Mg/Ca ratios both for the core tops and the LGM. The area where the green and yellow core tops are marked (Fig. 5a), covers Atlantic water masses, while the area with grey core tops covers an area inside or close to Arctic water masses. The positions where the grey core tops are marked, may in several winters be sea-ice covered. Nürnberg (1995) found that core tops in areas partly covered by sea-ice showed anomalously high Mg/Ca ratios. The same may be true for the LGM time slice. The anomalously high Mg/Ca ratios for the LGM (Fig. 5b) may also reflect water masses, at least partly, covered by sea-ice. It may also be that physiological processes, food supply or gametogenic

calcite play a more important role than temperature in this cold area, where the exponential character of the Mg/Ca temperature curves make the curves flatter in most published temperature equations. Further investigations in this region need to be performed to identify the possible causes for the anomalously high Mg/Ca values. We just note that the values northeast of Iceland are unreasonable in terms of a temperature reconstruction, i.e. both the modern and LGM data from the sub-area are too high.

6. Discussion

We have developed constraints on the range of possible salinity and $\delta^{18}O_w$ fields in the Nordic Seas and the North Atlantic. Given these constraints we obtain an oxygen isotope based reconstruction of summer temperatures of the upper 100 m of the water column that gives a reasonable fit to other observations of the paleoceanography of the region during the LGM (Fig. 3b). It indicates a reduced, but significant meridional advection in the east, i.e. similar to the modern pattern, but with much lower temperatures, and a low gradient between this area and the western area, which was colder. Oxygen isotope derived SSTs in the North Atlantic are consistent with newer transfer function estimates for this region (Pflaumann et al., 2003), while the oxygen isotope based reconstruction for the Nordic Seas is different from the transfer function estimates. The newer transfer function based reconstructions have no discernable east–west gradients, something which appears unlikely based on physical reasoning, since the pattern of inflowing waters in the east, which is seen both as a northward trending tongue of warmer temperatures south of the Iceland–Scotland Ridge and as a wedge of somewhat smaller reconstructed temperatures in the eastern Nordic Seas, would imply that there was a compensating southward flow of colder, Polar waters off east Greenland. Thus the reconstruction of Fig. 3b appears more likely. The mean temperature in the Nordic Seas is also lower in the oxygen isotope based reconstruction than in the transfer function based estimates. We ascribe both these differences to the lack of resolution for transfer functions at the cold end member, and contend that the oxygen isotope based reconstruction is more realistic. The CLIMAP reconstruction is in conflict with evidence from various sources that there must have been seasonally open waters in the Nordic Seas during the LGM (e.g. Fig. 1b; Veum et al., 1992; Hebbeln et al., 1994; Wagner and Henrich, 1994; Sarnthein et al., 1995; Weinelt et al., 1996), thus the oxygen isotope based reconstruction is much more in line with this evidence than the CLIMAP reconstruction, which had the whole Nordic Seas covered by a perennial sea-ice cover.

In agreement with the oxygen isotope based temperatures, the Mg/Ca based temperatures also indicate meridional advection in the east (Fig. 5c). Even if we have observed that the core tops in the eastern Nordic Seas deviate not much from the temperature calibration line of Nürnberg (1995), we suggest that more core top and downcore studies are required to develop a more consistent Mg/Ca temperature equation for the studied area, before the method can be used with sufficient reliability both in the eastern Nordic Seas and the northern North Atlantic. For the central and western Nordic Seas it is obvious that other factors than temperature influence the Mg/Ca ratios. It should also be mentioned that other methods have problems in this partly sea-ice covered area, like the alkenone method (Rosell-Melé and Comes, 1999), the dinocyst assemblage method (de Vernal et al., 2000) and partly the planktic foraminifer assemblage method (Pflaumann et al., 1996, 2003, Fig. 1b). Common for these methods and the Mg/Ca method is that they all suggest unreasonably high temperatures. Also factors like salinity, pH, physiological processes and gametogenic calcite may be important by influencing the Mg/Ca ratios, not only for the western and central Nordic Seas, but also for the eastern Nordic Seas and for some areas laying south of the Greenland–Scotland Ridge. Further studies are needed to elucidate the causes of these high Mg/Ca ratios.

We expect that replacing the commonly used CLIMAP data set for this region with the oxygen isotope based reconstruction (Fig. 3b) as ocean boundary conditions for climate models will change a number of aspects of the modelled dynamics of the LGM climate, such as storm tracks, degree of meridionality/zonality and existence of the well known modes of modern climate variability, e.g. the NAO/AO. The seasonally ice-free waters apparent in our reconstruction may act as a moisture source, consistent with the current understanding of the rapid growth of the Fennoscandian and Barents Ice Sheets, during the LGM (Boulton, 1979; Elverhøi et al., 1995; Mangerud et al., 2002). The growth clearly requires a strong source of winter precipitation, and a degree of meridional circulation. A zonal circulation, which is the result when using the CLIMAP reconstruction as boundary conditions, would probably provide too little winter precipitation to feed rapid growth of ice sheets in the northern portion of the Nordic Seas.

7. Conclusions

(1) The LGM temperatures, based on oxygen isotopes in the planktic foraminifer species *N. pachyderma* (sin.), imply that central and eastern parts of the Nordic Seas were ice-free, at least during the summer.

(2) The LGM temperatures, based on Mg/Ca ratios in *N. pachyderma* (sin.), show some meridionality south of the Greenland–Scotland Ridge and in the eastern parts of the Nordic Seas. More core top and downcore studies are needed before a more consistent Mg/Ca temperature equation for the studied area can be used with sufficient reliability. We thus suggest that the oxygen isotope based reconstruction of Fig. 3b is a better method to constrain LGM temperatures here.

(3) The temperature reconstruction based on oxygen isotopes is probably the most realistic approach for qualitative reconstructions of sea surface conditions in the Nordic Seas during the LGM. The results in this reconstruction appear consistent with dynamics required for the rapid growth of the Fennoscandian and Barents Ice Sheets during the LGM.

Acknowledgements

We thank Mervyn Greaves for patient training on the Mg/Ca cleaning method and for help with the Mg/Ca analyses, at the University of Cambridge. Stephen Barker is thanked for helpful advice on cleaning methods. Rune Søraas and Odd Hansen are thanked for keeping the stable isotope mass spectrometer in good shape. This manuscript was greatly improved by formal reviews from two anonymous reviewers, and by comments and suggestions from Carin Andersson Dahl, Ulysses Ninnemann and Trond Dokken. This project is financed by The Norwegian Academy of Science and Statoil under the VISTA program, the CESOP-project funded by the European Commission (EVR1-2001-40018), and the Bjerknes Centre.

References

Bard, E., Arnold, M., Maurice, P., Duprat, J., Moyes, J., Duplessy, J.-C., 1987. Retreat velocity of the North Atlantic polar front during the last deglaciation determined by ^{14}C accelerator mass spectrometry. Nature 328, 791–794.

Barker, S., Greaves, M., Elderfield, H., 2003. A study of cleaning procedures used for foraminiferal Mg/Ca paleothermometry. Geochemistry, Geophysics, Geosystems 4 (9), 8407.

Bond, G., Broecker, W.S., Johnsen, S., McManus, J., Labeyrie, L., Jouzel, J., Bonani, G., 1993. Correlations between climate records from North Atlantic sediments and Greenland ice. Nature 365, 143–147.

Boulton, G.S., 1979. Glacial history of the Spitsbergen archipelago and the problem of a Barents Shelf ice sheet. Boreas 8, 31–57.

Boyle, E.A., 1983. Manganese carbonate overgrowths on foraminifera tests. Geochimica et Cosmochimica Acta 47, 1815–1819.

CLIMAP, 1981. Seasonal reconstruction of the Earth's surface at the Last Glacial Maximum. Map Chart Series MC-36.

Cortijo, E., 1995. La variabilité climatique rapide dans l'Atlantique Nord depuis 128 000 ans: relations entre les calottes de glace et l'ocean de surface. Ph.D. Thesis, University of Paris, France, 235pp.

Cortijo, E., Labeyrie, L., Vidal, L., Vautravers, M., Chapman, M.R., Duplessy, J.-C., Elliot, M., Arnold, M., Turon, J.-L., Auffret, G., 1997. Changes in sea surface hydrology associated with Heinrich event 4 in the North Atlantic Ocean between 40° and 60°N. Earth and Planetary Science Letters 146, 29–45.

Dansgaard, W., Oeschger, H., 1989. Past environmental long-term records from the Arctic. In: Oeschger, H., Langway, Jr., C.C. (Eds.), The Environmental Record in Glaciers and Ice Sheets. Wiley, New York, pp. 287–318.

de Vernal, A., Hillaire-Marcel, C., Turon, J.-L., Matthiessen, J., 2000. Reconstruction of sea-surface temperature, salinity, and sea-ice cover in the northern North Atlantic during the last glacial maximum based on dinocyst assemblages. Canadian Journal of Earth Sciences 37, 725–750.

de Villiers, S., Greaves, M., Elderfield, H., 2002. An intensity ratio calibration method for the accurate determination of Mg/Ca and Sr/Ca of marine carbonates by ICP-AES. Geochemistry, Geophysics, Geosystems 3 2001GC000169.

Dokken, T., 1995. Last interglacial/glacial cycle on the Svalbard/Barents Sea margin. Ph.D.Thesis, University of Tromsø, Norway, 175pp.

Dokken, T., Jansen, E., 1999. Rapid changes in the mechanism of ocean convection during the last glacial period. Nature 401, 458–461.

Dreger, D., 1999. Decadal-to-centennial-scale sediment records of ice advance on the Barents shelf and meltwater discharge into the north-eastern Norwegian Sea over the last 40 kyr. Ph.D. Thesis, University of Kiel, Germany, 79pp.

Duplessy, J.-C., Labeyrie, L., Juillet-Leclerc, A., Maitre, F., Duprat, J., Sarnthein, M., 1991. Surface salinity reconstruction of the North Atlantic Ocean during the Last Glacial Maximum. Oceanologica Acta 14, 311–324.

Duplessy, J.-C., Labeyrie, L., Arnold, M., Paterne, M., Duprat, J., van Weering, T.C.E., 1992. Changes in surface salinity of the North Atlantic Ocean during the last deglaciation. Nature 358, 485–488.

Elderfield, H., Ganssen, G.M., 2000. Past temperature and $\delta^{18}O$ of surface ocean waters inferred from foraminiferal Mg/Ca ratios. Nature 405, 442–445.

Elverhøi, A., Andersen, E.S., Dokken, T., Hebbeln, D., Spielhagen, R.F., Svendsen, J.I., Sørflaten, M., Rørnes, A., Hald, M., Forsberg, C.F., 1995. The growth and decay of Late Weichselian ice sheet in Western Svalbard and adjacent areas based on provenance studies of marine sediments. Quaternary Research 44, 303–316.

Fairbanks, R.G., 1989. A 17,000-year glacio-eustatic sea level record: influence of glacial melting rates on the Younger Dryas event and deep-ocean circulation. Nature 342, 637–642.

Fevang, A., 2001. Klimaendringer i de nordiske hav i siste glasiale maksimum og Terminasjon 1. Unpublished Cand. Scient. Thesis in Marine Geology, University of Bergen, Norway, 106pp.

GEOSECS Atlantic, Pacific and Indian Ocean Expeditions, 1987. Shorebased data and graphics. In: GEOSECS Executive Committee, Ostlund, H.G., Craig, H., Broecker, S.W.S., Spencer, D. (Eds.), National Science Foundation 7.

Goldschmidt, P.M., 1994. The ice-rafting history in the Norwegian-Greenland Sea for the last two glacial/interglacial cycles. Berichte—Reports, 313, University of Kiel, Germany, 103pp.

Hancock, J.M., 1984. Cretaceous. In: Glennie, K.W. (Ed.), Introduction to the Petroleum Geology of the North Sea. Blackwell, Cambridge, pp. 133–150.

Hebbeln, D., Wefer, G., 1997. Late Quaternary Paleoceanography in the Fram Strait. Paleoceanography 12, 65–78.

Hebbeln, D., Dokken, T., Andersen, E.S., Hald, M., Elverhøi, A., 1994. Moisture supply for northern ice-sheet growth during the Last Glacial Maximum. Nature 370, 357–360.

Hebbeln, D., Henrich, R., Baumann, K.-H., 1998. Paleoceanography of the last glacial/interglacial cycle in the polar North Atlantic. Quaternary Science Reviews 17, 125–153.

Hohnemann, C., 1996. Zur Paläoceanographie der südlichen Dänemarkstraße. Unpublished Dipl. Arb. Thesis, University of Kiel, Germany.

Imbrie, J., Kipp, N.G., 1971. A new micropaleontological method for quantitative micropaleontology: application to a late Pleistocene Caribbean core. In: Turekian, K. (Ed.), Late Cenozoic Glacial Ages. Yale University Press, New Haven, pp. 71–81.

Jansen, E., Erlenkeuser, H., 1985. Ocean circulation in the Norwegian Sea during the last deglaciation: isotopic evidence. Palaeogeography, Palaeoclimatology, Palaeoecology 49, 189–206.

Jansen, E., Meland, M., 2001. Stratigraphy of IMAGES cores in the Faeroese-Shetland area. Technical Report 01-01, Bjerknes Centre for Climate Research, Bergen, Norway.

Jansen, E., Veum, T., 1990. Evidence for two-step deglaciation and its impact on North Atlantic deep-water circulation. Nature 343, 612–616.

Johannessen, T., 1992. Stable isotopes as climatic indicators in ocean and lake sediments. Dr. Scient Thesis, University of Bergen, Norway.

Jones, G.A., Keigwin, L., 1989. Evidence from Fram Strait (78°N) for early deglaciation. Nature 336, 56–59.

Jung, S.J.A., 1996. Wassermassenaustausch zwischen NE-Atlantik und Nordmeer während der letzten 300 000/80 000 Jahre im Abbild stabiler O- und C-isotope. Berichte aus dem Sonderforschungsbereich, 313, University of Kiel, 61, 104pp.

Jünger, B., 1993. Tiefenwassererneuerung in der Grønlandsee während der letzten 340 000 Jahre. Ph.D. Thesis, University of Kiel, Germany.

Keigwin, L., Boyle, E.A., 1989. Late Quaternary paleochemistry of high-latitude surface waters. Palaeogeography, Palaeoclimatology, Palaeoecology 73, 85–106.

Kellogg, T.B., Duplessy, J.-C., Shackleton, N.J., 1978. Planktonic foraminiferal and oxygen isotope stratigraphy and paleoclimatology of Norwegian deep-sea cores. Boreas 7, 61–73.

Knies, J., 1994. Spätquartäre Sedimentation am Kontinentalhang nordwestlich Spitzbergens, Der letzte Glazial/Interglazial-Zyklus. Thesis, Justus-Liebig-University, Giessen, Germany, 95pp.

Knies, J., 1999. Spätquartäre Paläoumweltbedingungen am nördlichen Kontinentalrand der Barents- und Kara-See: Eine Multi-Parameter-Analyse. Berichte Polarforschung, 304. Alfred Wegener Institute, Bremerhaven, Germany, 159pp.

Köhler, S.E.I., 1992. Spätquartäre paläo-ozeanographische Entwicklung des Nordpolarmeeres und Europäischen Nordmeeres anhand von Sauerstoff- und Kohlenstoffisotopenverhältnissen der planktischen Foraminifere *Neogloboquadrina pachyderma* (sin.). Geomar Research Centre for Marine Geosciences, Kiel, Germany, 104pp.

Labeyrie, L., Duplessy, J.-C., 1985. Changes in the oceanic $^{13}C/^{12}C$ ratio during the last 140,000 years: high-latitude surface water records. Palaeogeography, Palaeoclimatology, Palaeoecology 50, 217–240.

Lackschewitz, K.L., Dehn, J., Wallrabe-Adams, H.J., 1994. Volcaniclastic sediments from mid-oceanic Kolbinsey Ridge, north of Iceland: Evidence for submarine volcanic fragmentation processes. Geology 22, 975–978.

Lassen, S., Kuijpers, A., Kunzendorf, H., Lindgren, H., Heinemeier, J., Jansen, E., Knudsen, K.L., 2002. Intermediate water signal leads surface water response during Northeast Atlantic deglaciation. Global and Planetary Change 32, 111–125.

Lea, D.W., Pak, D.K., Spero, H.J., 2000. Climate impact of late quaternary equatorial Pacific sea surface temperature variations. Science 289, 1719–1724.

Lea, D.W., Martin, P.A., Pak, D.K., Spero, H.J., 2002. Reconstructing a 350 ky history of sea level using planktonic Mg/Ca and oxygen isotope records from a Cocos Ridge core. Quaternary Science Reviews 21, 283–293.

Levitus, S., Boyer, T.P., 1994. World Ocean Atlas 1994, vol. 4, Temperature, NOAA Atlas NESDIS 4. US Department of Commerce, Washington, DC, 117pp.

Mangerud, J., Astakhov, V., Svendsen, J.I., 2002. The extent of the Barents-Kara ice sheet during the Last Glacial Maximum. Quaternary Science Reviews 21, 111–119.

Manighetti, B., McCave, I.N., Maslin, M., Shackleton, N.J., 1995. Chronology for climate change: Developing age models for the Biogeochemical Ocean Flux Study cores. Paleoceanography 10, 513–525.

Maslin, M., 1992. A study of the paleoceanography of the N.E. Atlantic in the late Pleistocene. Ph.D. Thesis, University of Cambridge, England.

Mix, A.C., Fairbanks, R.G., 1985. North Atlantic surface-ocean control of Pleistocene deepocean circulation. Earth and Planetary Science Letters 73, 231–243.

Mix, A.C., Bard, E., Schneider, R., 2001. Environmental processes of the ice age: land, oceans, glaciers (EPILOG). Quaternary Science Reviews 20, 627–657.

Moros, M., Endler, R., Lackschewitz, K.L., Wallrabe-Adams, H.J., Mienert, J., Lemke, W., 1997. Physical properties of Reykjanes Ridge sediments and their linkage to high-resolution Greenland Ice Sheet Project 2 ice core data. Paleoceanography 12, 687–695.

Morris, T.H., 1988. Stable isotope stratigraphy of the Arctic Ocean: Fram Strait to Central Arctic. Palaeogeography, Palaeoclimatology, Palaeoecology 64, 201–219.

Notholt, H., 1998. The implication of the "North East Water"—Polynya on the sedimentation by NE-Greenland and late Quaternary paleo-oceanic investigations. Alfred Wegener Institute for Polar and Marine Research, Bremerhaven, Germany, 182pp.

Nürnberg, D., 1995. Magnesium in tests of *Neogloboquadrina pachyderma* sinistral from high northern and southern latitudes. Journal of Foraminiferal Research 25, 350–368.

Nørgaard-Pedersen, N., Spielhagen, R.F., Erlenkeuser, H., Grootes, P.M., Heinemeier, J., Knies, J., 2003. Arctic Ocean during the Last Glacial Maximum: Atlantic and polar domains of surface water mass distribution and ice cover. Paleoceanography 18, 1063.

Peltier, W.R., 1994. Ice Age paleotopography. Science 265, 195–201.

Pflaumann, U., Duprat, J., Pujol, C., Labeyrie, L., 1996. SIMMAX, a Transfer technique to deduce Atlantic Sea Surface Temperatures from planktonic foraminifera—the EPOCH approach. Paleoceanography 11, 15–35.

Pflaumann, U., Sarnthein, M., Chapman, M.R., d'Abreu, L., Funnell, B., Huels, M., Kiefer, T., Maslin, M., Schulz, H., Swallow, J., van Kreveld, S., Vautravers, M., Vogelsang, E., Weinelt, M., 2003. Glacial North Atlantic: sea-surface conditions reconstructed by GLAMAP 2000. Paleoceanography 18, 1065.

Rasmussen, T.L., Thomsen, E., van Weering, T.C.E., Labeyrie, L., 1996. Rapid changes in surface and deep water conditions at the Faeroe Margin during the last 58,000 years. Paleoceanography 11 (6), 757–771.

Roche, D., Paillard, D., Ganopolski, A., Hoffmann, G., 2004. Oceanic oxygen-18 at the present day and LGM: equilibrium simulations with a coupled climate model of intermediate complexity. Earth and Planetary Science Letters 218, 317–330.

Roe, A.B., 1998. Hurtige klimavariasjoner i havområdet øst for Scoresbysund, Grønland—isotoptrinn 3 og 2. Unpublished Cand. Scient. Thesis, University of Bergen, Norway, 93pp.

Rosell-Melé, A., Comes, P., 1999. Evidence for a warm Last Glacial Maximum in the Nordic seas or an example of shortcomings in U^K_{37} and U^K_{37} to estimate low sea surface temperature. Paleoceanography 14, 770–776.

Rosenthal, Y., Lohmann, G.P., Lohmann, K.C., Sherrell, R.M., 2000. Incorporation and preservation of Mg in *Globigerinoides sacculifer*: Implications for reconstructing the temperature and the $^{18}O/^{16}O$ of seawater. Paleoceanography 15, 135–145.

Ruddiman, W.F., McIntyre, A., 1981. The North Atlantic during the last deglaciation. Palaeogeography, Palaeoclimatology, Palaeoecology 35, 145–214.

Sarnthein, M., Jansen, E., Weinelt, M., Arnold, M., Duplessy, J.-C., Erlenkeuser, H., Flatøy, A., Johannessen, G., Johannessen, T., Jung, S.J.A., Koc, N., Labeyrie, L., Maslin, M., Pflaumann, U., Schulz, H., 1995. Variations in Atlantic surface ocean paleoceanography, 50°–80°N: A time-slice record of the last 30,000 years. Paleoceanography 10, 1063–1094.

Sarnthein, M., Stattegger, K., Dreger, D., Erlenkeuser, H., Grootes, P.M., Haupt, B.J., Jung, S.J.A., Kiefer, T., Kuhnt, W., Pflaumann, U., Schäfer-Neth, C., Schulz, H., Schulz, M., Seidov, D., Simstich, J., van Kreveld, S., Vogelsang, E., Völker, A., Weinelt, M., 2000. Fundamental modes and abrupt changes in North Atlantic circulation and climate over the last 60 ky—concepts, reconstruction and numerical modelling. In: Schäfer, P., Ritzrau, W., Schlüter, M., Thiede, J. (Eds.), The Northern North Atlantic: A Changing Environment. Springer, Berlin, pp. 365–410.

Sarnthein, M., Pflaumann, U., Weinelt, M., 2003. Past extent of sea ice in the northern North Atlantic inferred from foraminiferal paleotemperature estimates. Paleoceanography 18, 1047.

Schmidt, G.A., Bigg, G.R., Rohling, E.J., 1999. Global Seawater Oxygen-18 Database. Web page: www.giss.nasa.gov/data/o18data.

Schrag, D.P., Hampt, G., Murray, D.W., 1996. Pore fluid constraints on the temperature and oxygen isotopic composition of the glacial ocean. Science 272, 637–642.

Shackleton, N.J., 1974. Attainment of isotopic equilibrium between ocean water and the benthonic foraminifera genus *Uvigerina*: isotopic changes in the ocean during the Last Glacial. Colloques Internationaux du CNRS 219, 203–209.

Simstich, J., Sarnthein, M., Erlenkeuser, H., 2003. Paired $\delta^{18}O$ signals of *Neogloboquadrina pachyderma* (s) and *Turborotalita quinqueloba* show thermal stratification structure in Nordic Seas. Marine Micropaleontology 48, 107–125.

Stein, R., Schubert, C., Vogt, C., Fütterer, D., 1994. Stable isotope stratigraphy, sedimentation rates and paleosalinity in the latest Pleistocene to Holocene Central Arctic Ocean. Marine Geology 119, 333–355.

Stein, R., Nam, S., Grobe, H., Hubberten, H., 1996. Late Quaternary glacial history and short-term ice rafted debris fluctuations along the East Greenland continental margin. In: Andrews, J.T., Austin, W.E.N., Bergsten, H., Jennings, A.E. (Eds.), Late Quaternary Paleoceanography of the North Atlantic Margins. Geological Society Special Publication, The Geological Society, Oxford, UK, pp. 135–152.

Swift, J.H., Aagard, K., 1981. Seasonal transitions and water mass formation in the Iceland and Greenland Seas. Deep-Sea Research I 28, 1107–1129.

van Kreveld, S., Sarnthein, M., Erlenkeuser, H., Grootes, P.M., Jung, S.J.A., Nadeau, M.J., Pflaumann, U., Voelker, A.H., 2000. Potential links between surging ice sheets, circulation changes, and the Dansgaard–Oeschger cycles in the Irminger Sea. Paleoceanography 15, 425–442.

Veum, T., Arnold, M., Beyer, I., Duplessy, J.-C., 1992. Water mass exchange between the North Atlantic and the Norwegian Sea during the last 28,000 years. Nature 356, 783–785.

Voelker, A.H., 1999. Zur Deutung der Dansgaard–Oeschger Ereignisse in ultrahochauflösenden Sedimentprofilen aus dem Europäischen Nordmeer. Ph.D. Thesis, University of Kiel, Germany, 180pp.

Vogelsang, E., 1990. Paläo-Ozeanographie des Europäischen Nordmeeres an Hand stabiler Kohlenstoff- und Sauerstoffisotope. Ph.D. Thesis, University of Kiel, Germany.

Vogelsang, E., Sarnthein, M., Pflaumann, U., 2000. δ^{18}O Stratigraphy, Chronology, and Sea Surface Temperatures of Atlantic Sediment Records (GLAMAP-2000 Kiel). Berichte—Reports, 13, University of Kiel, Germany.

Vogt, C., 1997. Zur Paläozeanographie und Paläoklima im spätquartären Arktischen Ozean: Zusammensetzung und Flux terrigener und biogener Sedimentkomponenten auf dem Morris Jesup Rise und dem Yermak Plateau. Berichte Polarforschung, 251. Alfred Wegener Institute, Bremerhaven, Germany, 309pp.

Wagner, T., Henrich, R., 1994. Organo- and lithofacies of TOC-lean glacial/interglacial deposits in the Norwegian-Greenland Sea: sedimentary and diagenetic responses to paleoceanographic and paleoclimatic changes. Marine Geology 120, 335–364.

Weinelt, M., 1993. Veränderungen der Oberflächenzirkulation im Europäischen Nordmeer während der letzten 60,000 Jahre – Hinweise aus stabilen Isotopen. Ph.D. Thesis, University of Kiel, Germany, 106pp.

Weinelt, M., Sarnthein, M., Pflaumann, U., Schulz, H., Jung, S.J.A., Erlenkeuser, H., 1996. Ice-free Nordic Seas during the Last Glacial Maximum? Potential sites of deepwater formation. Paleoclimates 1, 283–309.

Winn, K., Sarnthein, M., Erlenkeuser, H., 1991. δ^{18}O stratigraphy and chronology of Kiel sediment cores from the East Atlantic. Berichte—Reports 45, Department of Geology and Paleontology, University of Kiel, Germany, 99p.

Zahn, R., Markussen, B., Thiede, J., 1985. Stable isotope data and depositional environments in the late Quaternary. Nature 314, 433–435.

Quaternary Science Reviews 24 (2005) 853–868

A global compilation of late Holocene planktonic foraminiferal δ^{18}O: relationship between surface water temperature and δ^{18}O

C. Waelbroeck[a,*], S. Mulitza[b], H. Spero[c], T. Dokken[d], T. Kiefer[e], E. Cortijo[a]

[a]*Laboratoire des Sciences du Climat et de l' Environnement, Domaine du CNRS, 91198 Gif-sur-Yvette, France*
[b]*Fachbereich Geowissenschaften, Universität Bremen, D-28359 Bremen, Germany*
[c]*Department of Geology, University of California, Davis, California, USA*
[d]*Bjerknes Centre for Climate Research, University of Bergen, Allegaten 55, Bergen N-5007, Norway*
[e]*Department of Earth Sciences, University of Cambridge, Downing Street, Cambridge CB3 2EQ, UK*

Received 19 July 2003; accepted 2 October 2003

Abstract

We review the different sources of uncertainty affecting the oxygen isotopic composition of planktonic foraminifera and present a global planktonic foraminifera oxygen isotope data set that has been assembled within the MARGO project for the Late Holocene time slice. The data set consists of over 2100 data from recent sediment with thorough age control, that have been checked for internal consistency. We further examine how the oxygen isotopic composition of fossil foraminifera is related to hydrological conditions, based on published results on living foraminifera from plankton tows and cultures. Oxygen isotopic values (δ^{18}O) of MARGO recent fossil foraminifera are 0.2–0.8‰ higher than those of living foraminifera. Our results show that this discrepancy is related to the stratification of the upper water mass and generally increases at low latitudes. Therefore, as stratification of surface waters and seasonality depends on climatic conditions, the relationship between temperature and δ^{18}O established on fossil foraminifera from recent sediment must be used with caution in paleoceanographic studies. Before models predicting seasonal flux, abundance and δ^{18}O composition of a foraminiferal population in the sediment are available, we recommend studying relative changes in isotopic composition of fossil planktonic foraminifera. These changes primarily record variations in temperature and oxygen isotopic composition of sea water, although part of the changes might reflect modifications of planktonic foraminifera seasonality or depth habitat.
© 2004 Elsevier Ltd. All rights reserved.

1. Introduction

Oxygen stable isotopic composition (^{18}O/^{16}O) is one of the most important tools in paleoclimatology. The oxygen isotopic composition of planktonic foraminifera has played a capital role in establishing a high resolution stratigraphy of deep sea sediment (e.g., Shackleton and Opdyke, 1973; Imbrie et al., 1984). However, because the oxygen isotopic composition of foraminifera calcite is a function of ambient temperature and the isotopic composition of the water in which calcification takes place, it also has the potential to yield important information about past environments. For instance:

(1) Combining planktonic oxygen isotope values with independent sea-surface temperature (SST) estimates allows one to derive the isotopic composition of surface water, which is related to surface salinity and thus yields information on the hydrological cycle (e.g., Duplessy et al., 1991; Mashiotta et al., 1999; Lea et al., 2000).
(2) Planktonic oxygen isotopes give a first approximation of the surface density of the ocean as the oxygen isotope composition is a function of salinity and temperature.

*Corresponding author. Tel.: +33-1-69-82-43-27; fax: +33-1-69-82-35-68.
E-mail address: claire.waelbroeck@lsce.cnrs-gif.fr (C. Waelbroeck).

0277-3791/$ - see front matter © 2004 Elsevier Ltd. All rights reserved.
doi:10.1016/j.quascirev.2003.10.014

(3) Since planktonic foraminifera live depth-stratified, the oxygen isotope difference between certain species can be used as a proxy for stratification (e.g., Williams and Healy Williams, 1980; Mulitza et al., 1997).

(4) An increasing number of ocean circulation numerical models compute the isotopic composition of calcite explicitly (e.g., Paul et al., 1999), so that these models can be directly validated by comparison of their output with foraminifera isotopic composition.

(5) Finally, a large amount of oxygen isotope data is currently available on planktonic foraminifera from recent and glacial sediments (e.g., Duplessy et al., 1981a, 1991; Billups and Schrag, 2000; Schmidt and Mulitza, 2002), which provides a unique opportunity to test the consistency of the regional SST estimates based on different methods and proxies.

The use of this large paleoceanographic archive is limited, however, by a number of factors that are critical to the application of oxygen isotopes as a paleoceanographic tool. In the following sections, we first review the various sources of uncertainties in fossil foraminifera oxygen isotope measurements, including those involved in the comparison of data of different origins. We then discuss how oxygen isotopic composition of fossil foraminifera is related to the hydrological conditions that prevailed during the foraminifera life. We summarize the current state of knowledge in this matter, based on published results on living foraminifera from plankton tows and cultures. This allows us to highlight a number of unresolved questions concerning the processes affecting the planktonic isotopic values from recent sediments.

The MARGO data set includes over 2100 planktonic foraminifera oxygen isotopic data from more than 15 laboratories. The main contributing laboratories are University of Bergen (Bergen, Norway), University of Kiel (Kiel, Germany), University of Bremen (Bremen, Germany), LSCE (Gif-sur-Yvette, France), Alfred–Wegener-Institute (Bremerhaven, Germany) and the Free University of Amsterdam (Amsterdam, The Netherlands). The database is available at www.pangaea.de/projects/MARGO.

2. Oxygen isotopes

Oxygen has three stable isotopes: ^{16}O, ^{17}O and ^{18}O. ^{16}O is the most abundant (99.76%) while ^{17}O and ^{18}O only comprise 0.04% and 0.2% of the total oxygen. The $^{18}O/^{16}O$ ratio of a sample is expressed as $\delta^{18}O$:

$$\delta^{18}O(‰) = \left[\frac{(^{18}O/^{16}O)_{sample} - (^{18}O/^{16}O)_{standard}}{(^{18}O/^{16}O)_{standard}} \right] \times 10^3.$$

Carbonate samples are measured relative to the PDB-standard (Cretaceous belemnite from the Pee Dee River formation in North Carolina, USA) and water samples refer to Standard Mean Ocean Water (SMOW). This nomenclature has recently been changed to V-PDB and V-SMOW (Coplen, 1996). To convert from the V-SMOW scale to the V-PDB scale, a correction of 0.27‰ is necessary (Hut, 1987; see Bemis et al. (1998) for additional discussion of this correction for paleoceanographic applications). The basic principles of mass spectrometry and the preparation of carbonate and water samples first established by McCrea (1950) and Epstein and Mayeda (1953), are fully described by Hoefs (1987) and Rohling and Cooke (1999).

A combination of temperature-dependent equilibrium and kinetic isotope fractionations occurs when water undergoes a phase transition. Water evaporating from the sea surface is depleted in heavy isotopes with respect to sea water, while rain precipitating from the atmosphere is enriched in the heavy isotopes with respect to the cloud water vapor. The major source of water vapor is the surface ocean. Poleward transport of this water vapor over land masses results in gradual rainout and a depletion of the remaining moisture in ^{18}O. Hence, the isotopic composition of precipitation shows systematic variations with latitude, altitude and distance from vapor source.

Compared to the atmosphere, the variation in the oxygen isotopic composition of seawater (δ_w) is relatively small. The distribution of salinity and oxygen isotopes in the open ocean is primarily controlled by precipitation and evaporation. For this reason, salinity and δ_w of surface waters are linearly related, with high δ_w and salinity values in regions of high atmospheric pressure such as the central subtropical gyres, and low δ_w and salinity values in regions of low atmospheric pressure such as the western Pacific warm pool, along the equator, and below the westerlies. However, the slope of the δ_w—salinity relationship ranges between ~0.1 for tropical surface waters and ~1 for high-latitude surface waters (e.g., Charles and Fairbanks, 1992; Paul et al., 1999). Variations in slope are controlled by the source of the freshwater end-member. In low latitudes, this end-member is precipitation, whereas in high latitudes such as the North Atlantic, very low ^{18}O Greenland ice sheet meltwater can dominate the freshwater end-member. Isotopic fractionation taking place during the formation of sea ice is so small that the freezing process leads to an increase in salinity, with essentially no observable influence on δ_w (Craig and Gordon, 1965). Finally, the isotopic composition of sea water is also affected by advective mixing of different water masses. In summary, in the modern ocean, δ_w allows us to distinguish between different freshwater sources. In this way, reconstructing variations in δ_w can yield information on changes in the hydrological cycle in the past.

$\delta^{18}O$ of foraminifera calcite depends on both the temperature and the isotopic composition of the water in which the shell grew. The temperature dependence of the equilibrium fractionation of inorganic calcite precipitation has been determined experimentally (e.g., O'Neil et al., 1969; Kim and O'Neil, 1997). The quadratic approximation of the O'Neil et al. (1969) relationship around 16.9 °C is given in Shackleton (1974) as:

$$T = 16.9 - 4.38 \times (\delta_c - \delta_w) + 0.1 \times (\delta_c - \delta_w)^2, \qquad (1)$$

where T is water temperature in °C; δ_c and δ_w, are the $\delta^{18}O$ of the calcite and ambient water, respectively. For paleoceanographic applications, the $\delta^{18}O$ vs. temperature slope varies from 0.21‰/°C at 30 °C to 0.29‰/°C at 0 °C. Urey (1947) first proposed that the oxygen isotopic composition of carbonate fossils could be used to reconstruct temperatures. During the past 20 years, many empirically-derived paleotemperature relationships have been published, which produce results that can differ by several degrees, when applied to the same data set (see Bemis et al., 1998 for review). In the present study, we use Eq. (1) as the reference equilibrium equation for inorganic calcite and discuss how fossil foraminifera $\delta^{18}O$ relates to inorganic calcite $\delta^{18}O$ on the one hand, and to living foraminifera $\delta^{18}O$, on the other hand.

3. Inter-comparison of fossil foraminifera $\delta^{18}O$ measurements

In this section we review the primary sources of uncertainty on fossil foraminifera $\delta^{18}O$ in order to evaluate errors when comparing data from various origins.

3.1. Intra-specific variability

Interpretation of geochemical data from deep sea cores must take into account aspects of the life cycle and habitat preferences of each foraminifera species. Because the life cycle of most planktonic foraminifera is 2–4 weeks (Bijma et al., 1990), the geochemistry of each shell records environmental conditions during a finite period of the year. When the seasonal distribution of a species is considered together with its habitat depth range (Fairbanks and Wiebe, 1980; Fairbanks et al., 1982; Sautter and Thunell, 1989; Sautter and Thunell, 1991), and the influence of bioturbation (Boyle, 1984), it is not surprising that each interval in a core displays a large amount of intraspecific variability in stable isotope values. Moreover, recent studies showed that part of the intra-specific variability observed in fossil records could

be due to the co-occurrence of several different genotypes of a morphospecies (Darling et al., 2003).

Geochemical analyses of individual shells from deep sea cores have been conducted on a number of planktonic species from the tropical Atlantic, Pacific and Indian Oceans (Schiffelbein and Hills, 1984; Spero and Williams, 1990; Tang and Stott, 1993; Billups and Spero, 1995; Stott and Tang, 1996; Spero, 1998). The $\delta^{18}O$ range observed within a single interval near the core tops (CTs) is ~1.2–1.5‰ for species such as *Orbulina universa*, *Globigerinoides sacculifer* and *G. ruber* (pink and white variations). For thermocline dwellers such as *Neogloboquadrina dutertrei* and *Pulleniatina obliquiloculata*, the range can exceed 2‰.

An example of the intraspecific variability in core intervals can be seen in Fig. 1. Between 8 and 12 individual *G. sacculifer* and *G. ruber* (250–350 μm size range) were analyzed from two adjacent Holocene intervals and two adjacent glacial intervals in eastern equatorial Pacific core TR163-19 (2°16′N, 90°57′W; 2348 m; with 3.5 cm/kyr sedimentation rate). $\delta^{13}C$ vs. $\delta^{18}O$ plots of the combined data (Fig. 1A and B) show the typical intraspecific variability that is observed in such data sets. When plotted with the original oxygen isotope stratigraphies published for this core (Fig. 1C) (Lea et al., 2000; Spero et al., 2003), it is easy to see the variability that makes up each isotope datum in an oxygen isotope stratigraphic record.

In the early years of mass spectrometry, each oxygen isotope measurement required large numbers of shells to generate a single value. Although the precision of the measurements was not as good as measurements made on more modern instruments, the use of a large number of shells produced a true average value for each interval in a core. With the development of sensitive isotope ratio mass spectrometers in the late 1980s, some researchers began to utilize fewer fossil shells per measurement, with reports of as few as 3–5 *G. ruber* per interval being used to generate oxygen isotope records (Bassinot et al., 1994). Given the level of intraspecific variability that has been observed in core intervals, the use of fewer shells in an analysis implies that the value obtained will not reflect the true average of a fossil population.

Schiffelbein and Hills (1984) explored this problem and estimated that the isotopic variance within a population, σ^2, combined with the mass spectrometer precision, σ_m^2 will define the number of shells, n, needed to obtain an average interval value with a predefined precision, σ_T^2 for that value, where $\sigma_T^2 = \sigma_m^2 + \sigma^2/n$.

For the data presented in Fig. 1, $\sigma = 0.3 - 0.8$‰, and assuming a machine precision of 0.06‰ for $\delta^{18}O$, the use of 20 shells in each interval would give a precision around the average value of 0.09–0.19‰ for both species. Clearly, the quality of the oxygen isotope data published in the literature is as much a function of

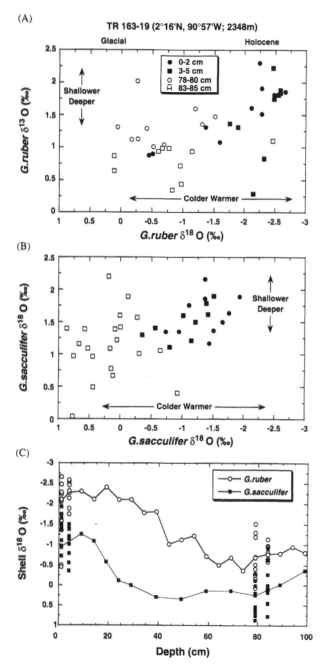

Fig. 1. Individually analyzed specimens of *G. ruber* and *G. sacculifer* from the 250 to 350 μm sieve fraction in TR163-19. Specimens were collected from two Holocene and two glacial intervals as determined by AMS dating. (A) Plot of individual *G. ruber* $\delta^{13}C$ and $\delta^{18}O$; (B) Plot of individual *G. sacculifer* $\delta^{13}C$ and $\delta^{18}O$; (C) Individual $\delta^{18}O$ data plotted with previously published downcore stratigraphy (Lea et al., 2000; Spero et al., 2003).

careful cleaning and analysis as it is of selecting the number of shells for an analysis. We recommend that as many shells/analysis as possible be used for down-core stratigraphic analyses. The number of shells/analysis should be especially large in low sedimentation cores, since the problem worsens with bioturbation.

Note that the standard deviation on the $\delta^{18}O$ average value of *n* measurements performed within the Late Holocene (LH) time slice reflects a combination of the intra-specific variability and climatic variability over the last 4 ky. Over 90% of the MARGO LH data have standard deviations $\sigma < 0.3$‰.

3.2. The effect of foraminifera size fraction

Previous studies have examined the influence of size fraction on the isotopic composition of living and fossil foraminifera (e.g., Berger et al., 1978; Bouvier-Soumagnac and Duplessy, 1985; Spero and Lea, 1996; Bemis et al., 1998; Peeters et al., 2002). The size dependency of $\delta^{18}O$ can be different for living and fossil foraminifera due to the compounding influence of dissolution on fossil assemblages (Bouvier-Soumagnac and Duplessy, 1985). Here, we briefly discuss the case of fossil foraminifera in the context of the MARGO data set. The impact of fossil foraminifera size on measured $\delta^{18}O$ values appears to depend on the species, size fraction and region. For instance, analyses of the small size fractions of *G. ruber* from the Mediterranean Sea (Fig. 2A), show that the $\delta^{18}O$ of the 200–250 μm fraction is up to 1‰ more positive than the $\delta^{18}O$ of the 250–315 μm fraction. In contrast, measurements on four discrete *G. ruber* size fractions between 300 and 500 μm from Equatorial Indian Ocean core ODP 714A show no size dependency on $\delta^{18}O$ (Fig. 2B). *G. sacculifer* $\delta^{18}O$ is influenced by both the fraction selected and water column hydrography. For instance, in the eastern equatorial Pacific (Spero et al., 2003), smaller *G. sacculifer* (250–350 μm) are 1‰ more positive than $\delta^{18}O$ in larger shells (>650 μm). This difference appears to be a function of the steep thermocline in this region combined with observations that suggest that light levels may control the final size of *G. sacculifer* through symbiont activity (Bé et al., 1982; Spero and Lea, 1993). However, recent modelling studies by Wolf-Gladrow et al. (1999) suggest that variations in respiration, calcifcation rate and symbiont photosynthesis can lead to considerable modifcations of the pH profile in the vicinity of the foraminiferal shell. Since these parameters might also vary with shell geometry and size, they could explain as much as 30–40% of the changes in $\delta^{18}O$ with size. The remainder should be related to environmental temperature and salinity differences. In contrast, no significant difference in $\delta^{18}O$ was found among 11 size fractions between 375 and 825 μm, in Caribbean cores V19-19 and V28-122 (Oppo and Fairbanks, 1989). The differences observed between smaller and larger shells for *G. ruber* and *G. sacculifer* could also be linked to the juvenile state of the smaller individuals. Analyses on 125–250 μm and >250 μm fractions of *N. pachyderma* left, *N. pachyderma* right in the North Atlantic do not reveal significant trends (Johannessen, 1992).

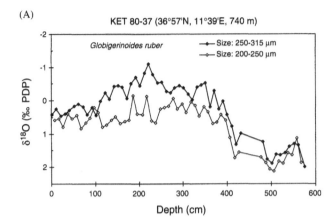

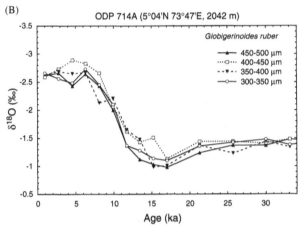

Fig. 2. (A) *G. ruber* δ^{18}O records of Mediterranean core KET80-37 for the 200–250 µm (seven shells samples) and 250–315 µm (four shells samples) size fractions. All samples have been cleaned by sonicating in methanol and vacuum roasted at 400 °C for 45 min (LSCE data). (B) *G. ruber* δ^{18}O record from equatorial Indian Ocean core ODP 714A. All analytical points are from 30 shells/sample cleaned by sonicating in methanol (Mielke, 2001).

In the present study, sorting the MARGO data base by size did not reveal a clear dependency of the δ^{18}O value on the analyzed size fraction. For each species, we drew separate plots of SST vs. the difference between foraminifera and water δ^{18}O (as in Fig. 7) for each size fraction and graphically verified that the data distribution does not depend on foraminifera size.

3.3. Preparation techniques

Sediment samples are first sorted using > 150 µm sieves in order to separate the foraminifera to be analyzed from the sediment fine fraction. Before measuring the oxygen and carbon isotopic composition of the foraminiferal calcite, it is necessary to clean the shells to remove organic matter and other contaminants which could remain after sieving. Cleaning may involve several steps, depending on the laboratory. It is thus important to evaluate the impact of the cleaning procedure on the final isotopic value, and to verify that

measurements made in different laboratories can be directly compared and discussed. Standard procedures may include the following cleaning steps:

(1) Sieving is usually done with deionized water. The sediment can however be soaked in H_2O_2 10% for 30 min before sieving to help desegregate the sediment matrix and remove part of the organic matter.
(2) In some cases, the selected foraminifera shells are crushed before immersing in methanol for cleaning.
(3) The foraminifera shells selected for mass spectrometer measurements can be washed in a methanol ultrasonic bath for a period of 5–30 s depending on the fragilility of the shells. Sonication removes clays and other materials that may be embedded within the shell aperatures or pores, or on the surface of the shells.
(4) Finally, some laboratories perform vacuum roasting at 375–400 °C for 30–45 min or more just prior to analysis to eliminate impurities that may be present (Duplessy, 1978).

Sensitivity studies were performed at LSCE (Gif) in order to test whether H_2O_2 treatment before sieving had any impact on the measured isotopic composition. A small, statistically insignificant, oxygen isotopic impoverishment of the H_2O_2 treated samples was observed on benthic (-0.11 ± 0.13‰) and planktonic (-0.11 ± 0.23‰) foraminifera records. Further systematic tests on larger data sets are however necessary before we can conclude that there is minimal effect of this type of treatment on the measured δ^{18}O.

Crushing the foraminifera shells has been included as part of the sample pretreatment by several groups in order to remove adhering calcite contaminants like coccoliths that could be located inside the shell chambers. Subsequently, a large number (e.g., 30 shells) of individual shells are pooled for each analysis to obtain average interval values and reduce the intra-species variability. A sensitivity study on two Indian Ocean cores performed at LSCE showed that crushed foraminifera do not have the same isotopic composition as whole foraminifera (Fig. 3): large fragments of *G. ruber* were found to be 0.2 ± 0.19‰ lighter in δ^{18}O and 0.19 ± 0.18‰ lighter in δ^{13}C than whole shells. Furthermore, δ^{18}O of large fragments were also 0.15 ± 0.08‰ lighter in δ^{18}O and 0.18 ± 0.16‰ lighter in δ^{13}C than the remaining powder (Fig. 3). There are several possible explanations for these observations: (1) contaminants (such as coccoliths) of higher δ^{18}O and δ^{13}C may be lining the inside of the shells. In that case, the isotopic composition of the large fragments would correspond to the foraminifera isotopic composition, whereas the remaining powder would be contaminated by the coccoliths originally present inside the shells.

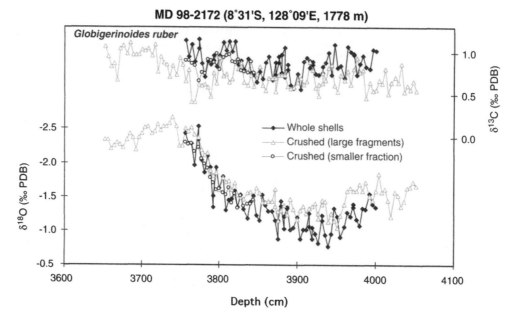

Fig. 3. $\delta^{18}O$ and $\delta^{13}C$ measurements on crushed and whole *G. ruber* shells in core MD98-2172 from the Banda Sea. All samples (crushed and uncrushed) have been cleaned by sonicating in methanol, prior to crushing, and vacuum roasted at 400 °C for 45 min, whole shells samples comprised five shells (LSCE data).

However, measurements of coccoliths exhibit lower $\delta^{18}O$ values than would be expected from foraminifera shells (Dudley et al., 1980), which is not consistent with this hypothesis. (2) One could object that as reaction rate of carbonate dissolution in phosphoric acid depends on carbonate grain size, differences in size could have an impact on the $\delta^{18}O$ value if the reaction is not complete. However, it has been shown that for reactions taking place at temperatures above 50 °C there is no systematic variation of $\delta^{18}O$ with grain size over the size interval 45–180 μm (Shackleton, 1974). As both the whole foraminifera and large fragments used in the present study are comprised in this size interval, this effect cannot explain the observed isotopic shift. (3) Alternatively, experimental evidence suggests that foraminifera shells do not have a homogeneous isotopic composition. For instance, measurements on single chambers of cultured *G. bulloides* showed that calcite $\delta^{18}O$ increases from inner to outer chambers (Spero and Lea, 1996; Bemis et al., 1998). Moreover, several studies showed that an additional layer of higher $\delta^{18}O$ calcite is added during gametogenesis in deeper waters before the shell reaches the sea floor (Bé, 1980; Duplessy et al., 1981b). The large fragments and the remaining powder could thus have different isotopic compositions.

A more detailed experimental study is needed to quantify the effect of crushing and other pre-analysis treatments on all the planktonic species commonly used in paleoceanographic studies. Regardless of the cleaning methodology employed, we recommend that a large

number of complete shells be used in each sample. For Finnigan mass spectrometers, it may be necessary to lower the mass spectrometer source emission to prevent amplifier saturation. This leads to a larger signal to noise ratio and does not induce any unknown bias.

Roasting was originally introduced to suppress organic impurities (Duplessy, 1978). Many laboratories do not roast the samples any longer. There are contradictory evidences concerning the potential effect of roasting on measured isotopic values (Duplessy, 1978; Sarkar et al., 1990). In some cases, a significant difference has been found between the isotopic composition of roasted vs. unroasted foraminifera or carbonate standards, whereas in other cases, no difference has been observed. For instance, roasted vs. unroasted experiments with split *O. universa* displayed no statistical difference in either oxygen or carbon isotopic composition. However, roasting helps reduce the variability in measured $\delta^{18}O$, presumably due to adsorption of water or other volatiles onto the samples (Spero, unpublished data).

It is difficult to evaluate the maximum bias that could result from the use of different preparation procedures. Assuming that the effect of crushing does indeed reflect a more efficient removal of the organic matter, roasting and crushing should induce isotopic shifts in the same direction that should thus not be additive. Adopting this optimistic point of view, we estimate that differences in sample preparation techniques among laboratories introduce an uncertainty on the measured isotopic value of the order of 0.2‰.

3.4. Mass spectrometry

Mass spectrometers measure the difference in isotopic composition between the CO_2 gas produced by complete acidification of the carbonate sample and a reference gas. Standards are routinely measured in addition to samples, in order to verify that the machine is stable as well as to be able to convert the measured sample over reference gas ratio into the V-PDB nomenclature. A first assessment of the error on the measurements is given by the external reproducibility of the carbonate standards, i.e., the standard deviation obtained on the standards. Standard deviations of 0.05–0.06‰ can easily be reached on Micromass/GV Prism, Optima and Iso-primes, or Finnigan MAT251, MAT252 and Delta +, equipped with a Kiel device. However, the isotopic values of standards measured on MAT251 and 252 are clearly dependent on the amount of gas. The bias is larger when the sample size is small and when its isotopic composition is very different from that of the reference gas. This is due to a decrease in the isotopic difference between the reference and sample gas because of mixing in the source (Ostermann and Curry, 2000). Ostermann and Curry (2000) established a correction of this bias using measurements on standards of very different isotopic compositions encompassing the entire range of isotopic compositions of glacial and interglacial benthic and planktonic foraminifera. A correction of this effect is now applied in most laboratories involved in paleoceanographic research. Were this bias not corrected, a systematic shift would appear between the isotopic measurements on a given species in a given region with respect to the values provided by the laboratories that apply the correction. Unlike the Finnigan 251 or 252 mass spectrometers, the Micromass/GV Prism, Optima and Isoprines are generally linear across a large range of values.

In MARGO database, analyses of carbonates with Micromass/GV instruments are typically processed using either an Isocarb common acid bath autocarbonate device or a Multiprep individual sample analysis system at 90 °C. Comparable results are obtained using either preparation system with no observable memory effect in the common acid bath system if samples are collected for a period of 14 min. We have no information on potential memory effects in the common acid bath system using shorter collection times.

Scanning the entire MARGO data set, we identified only one case of systematic offsets compared to values measured by the other laboratories. We did not include these data in the final tabulation. Note that some laboratories have retroactively applied a correction to all the measurements performed on Finnigan MAT251 and 252, but it is likely that, in other laboratories, old measurements have not been corrected for this bias. In those cases, it is not possible to detect systematic biases

and this induces an additional source of uncertainty on the $\delta^{18}O$ data.

We conclude that the error on planktonic $\delta^{18}O$ values resulting from the fact that the MARGO data set is the collection of measurements made over the last two decades, in more than 15 different laboratories, on different mass spectrometers, is of about 0.2‰.

4. Interpretation of planktonic $\delta^{18}O$

The $\delta^{18}O$ of fossil planktonic foraminifer is influenced by factors changing the $\delta^{18}O$ of living foraminifera, factors associated with processes at the end of the life cycle, with sedimentation, and by post-depositional effects. It is essential to assess the potential of fossil foraminifera to record surface water properties, since all reconstructions of past climatic conditions using planktonic foraminifera $\delta^{18}O$ are necessarily based on specimens from marine sediments.

4.1. Environmental factors influencing the $\delta^{18}O$ of living planktonic foraminifera

4.1.1. Seasonality

A number of studies have demonstrated that both $\delta^{18}O$ and shell flux to the sea floor vary seasonally (e.g. Williams et al., 1981; Thunnel et al., 1999). As an example, Fig. 4 shows the fluxes of several planktonic species at sediment trap station CB (off Mauritania, Cape Blanc) over the period from 1988 to 1991, together with the oxygen isotope composition of G. ruber. Seasonal variations in the oxygen isotope ratio of G. ruber are a direct function of surface water temperature as predicted by the temperature vs. $\delta^{18}O$ relationships derived from plankton tows and pumped samples (Mulitza et al., 2003a). These data show that the surface water temperature signal is transported into the sediment without a significant time lag.

Different species fluxes peak at different times of the year with an apparent succession of abundance peaks through the year. The oxygen isotope composition of a foraminiferal population in the sediment is the flux-weighted mean of all oxygen isotope values. The environmental conditions recorded by the oxygen isotope composition of fossil foraminifera are therefore distorted towards the peak season and highly productive years. In Fig. 4G we calculated the theoretical mean flux-weighted, SST recorded in each of the individual species. Apparently, G. ruber and G. sacculifer record a summer signal at the position of CB, while O. universa records annual mean conditions and G. bulloides and G. inflata are shifted towards the winter. This evidence suggests that the $\delta^{18}O$ of different species might be a useful recorder of the seasonal temperature contrast. However, although some attempts have been made to

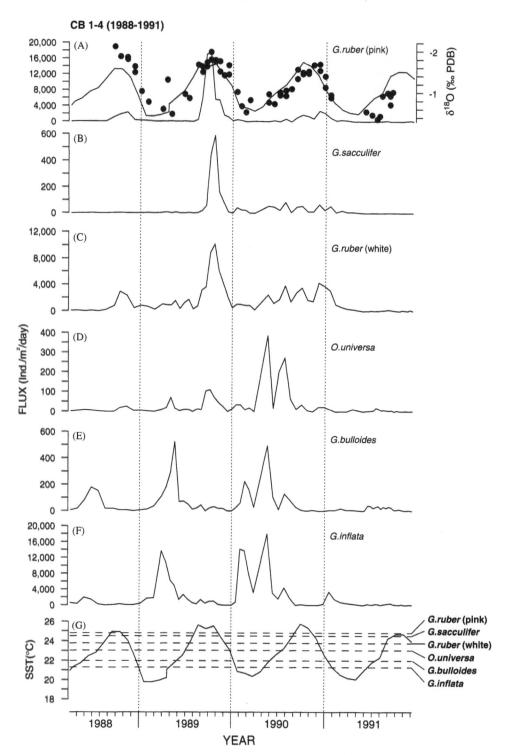

Fig. 4. (A)–(F) Seasonal succession of foraminiferal fluxes at sediment trap station CB off Cape Blanc (Mauritania) for *G. ruber* pink, *G. sacculifer*, *G. ruber* white, *O. universa*, *G. bulloides*, and *G. inflata*, from top to bottom. (A) Seasonal variations of the oxygen isotope composition of *G. ruber* pink (black dots, Fischer et al., 1999). Predicted $\delta^{18}O$ of foraminiferal calcite (upper curve) has been calculated using the regression equation on plankton tows data from Mulitza et al. (2003a) and measured SST given in (G). (G) Seasonal variations in measured SST. Dashed lines represent the flux weighted mean temperature of the different species. Redrawn from Mulitza et al. (2003b).

reconstruct seasonality from oxygen isotope values of planktonic foraminifera (e.g., Deuser, 1978; Mulitza et al., 1998), a quantification of seasonality is difficult

because most winter species are not surface-dwelling foraminifera and because the seasonal flux is not constant through time.

4.1.2. Vertical migration

Investigations from plankton tow data have shown that planktonic foraminifera live vertically dispersed in the upper water column and that the habitat depth range has a significant influence on the oxygen isotope composition of each species (e.g. Fairbanks and Wiebe, 1980; Bouvier-Soumagnac and Duplessy, 1985; Mortyn and Charles, 2003). Many planktonic foraminiferal species deposit their shells at the chlorophyll maximum zone (Fairbanks et al., 1982; Kohfeld et al., 1996; Mortyn and Charles, 2003), which is the depth of the major food resource exploited by planktonic foraminifera. However, the isotope data of primarily shallow-dwelling species indicate that calcification takes place in depth zones that are significantly narrower than implied by the overall vertical distribution of these species (Fairbanks and Wiebe, 1980). As an example, Fig. 5 shows the vertical distributions of *G. sacculifer and G. ruber* (white) at the plankton tow station GeoB 1402 located in the eastern equatorial Atlantic, together with the oxygen isotopic compositions of the species from the different towing intervals. The isotope values indicate calcification within the photic zone, where both species have their highest abundance. It is also visible that the

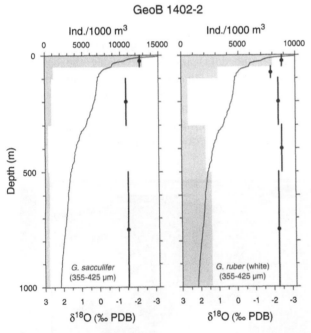

Fig. 5. Vertical distribution (shaded bars), oxygen isotope composition of *G. sacculifer* and *G. ruber* (white) (dots), and predicted $\delta^{18}O$ of foraminiferal calcite (solid line) at the plankton tow station GeoB 1402-2 in the eastern equatorial Atlantic. Predicted $\delta^{18}O$ of calcite has been calculated from least squares regression given in Mulitza et al. (2003a), a regional $\delta^{18}O$:salinity relationship and salinity and temperature measurements made with a CTD attached to the towing cable. Vertical error bars indicate the tow interval. $\delta^{18}O$ values of calcite are plotted at the midpoint of the tow interval. Data are from Kemle-von Mücke and Oberhänsli (1999). Redrawn from Mulitza et al. (2003b).

$\delta^{18}O$ values of *G. sacculifer* increase within the seasonal thermocline, while the $\delta^{18}O$ of *G. ruber* is relatively constant throughout the water column. This implies a much smaller and shallower depth range of calcification for *G. ruber* (white) than for *G. sacculifer* at this site.

The difference in calcification depth has been used to reconstruct the stratification of past oceans from $\delta^{18}O$ differences of several species (e.g., Mulitza et al., 1997; Rühlemann et al., 2001; Spero et al., 2003). This approach, however, is complicated by the fact that foraminifera are not restricted to a certain depth level, but can change their habitat depth. For example, Sautter and Thunell (1991) have shown that both *N. dutertrei* and *N. pachyderma* follow isothermal ranges, migrating to shallower waters during upwelling and subsequently descending after upwelling.

4.1.3. Temperature:$\delta^{18}O$ relationship for living planktonic foraminifera

Several studies have shown that $\delta^{18}O$ values of *O. universa* collected in plankton tows agree with data obtained from cultured individuals (Bouvier-Soumagnac and Duplessy, 1985; Bemis et al., 1998). More recently, this has also been shown for *G. bulloides* and *G. sacculifer* when cultured at surface water pH (Bemis et al., 1998; Mulitza et al., 2003a). Moreover, Mulitza et al. (2003a) argue that the temperature vs. $\delta^{18}O$ relationship for living *G. ruber* collected from the water column is not statistically different from the equations for living *G. sacculifer* and *G. bulloides*, whereas the equation derived for living *N. pachyderma* has a significantly lower slope and intercept, in agreement with the equilibrium equation for inorganic calcite. Importantly, all these studies consistently show that the $\delta^{18}O$ of living foraminifera species is much lighter than predicted by equilibrium equations derived for inorganic calcite (O'Neil, 1969; Kim and O'Neil, 1997). A lighter $\delta^{18}O$ value cannot be explained by vertical migrations, nor by seasonality as these observations are based on in situ temperature and $\delta^{18}O$ values. The most likely explanation of this discrepancy lies in the sensitivity of isotopic fractionation to pH conditions within the water column or calcifying microenvironment (Spero et al., 1997). For instance, water column data demonstrate that the pH of the upper 50 m of the modern ocean can vary from ~7.6 in upwelling regions of the eastern equatorial Pacific (Millero et al., 1998) to >8.25 in the tropical Atlantic (Bates et al., 1996). During glacial times, surface pH may have been substantially higher (Broecker and Peng, 1982; Sanyal et al., 1995; Lea et al., 1999). Microelectrode experiments with living foraminifera have demonstrated that the pH within the calcifying microenvironment of symbiotic and non-symbiotic foraminifera can range from 7.9 to 8.8 (Rink et al., 1998). Because calcite $\delta^{18}O$ decreases by ~1.1‰ per unit pH

C. Waelbroeck et al. / Quaternary Science Reviews 24 (2005) 853–868

increase (Zeebe, 1999), a decrease in foraminifera $\delta^{18}O$ of 0.5‰ could be easily explained by the pH effect.

4.2. $\delta^{18}O$ of fossil planktonic foraminifera

4.2.1. Post-depositional effects

Foraminifera from sediments contain the integrated signal of all processes discussed above. Moreover, two processes may alter the isotopic signal of the mean foraminiferal population in the sediment after deposition: bioturbation and calcite dissolution. The main effect of bioturbation on marine sediment records is a truncation of the glacial/interglacial amplitude because of mixing of glacial and interglacial shells. Bioturbation in cores with low sedimentation rates would thus lead to abnormally heavy Holocene $\delta^{18}O$. In the present study, we discarded $\delta^{18}O$ values heavier than the regression value $+1$ root mean square error for cores with sedimentation rates lower than 5 cm/ky. Dissolution can also alter the isotopic composition of foraminiferal shells after deposition. The effect of dissolution depends on the homogeneity of the foraminiferal shells. Many foraminifera add different kinds of calcite during their life cycle. For example G. sacculifer adds a thick calcite crust usually in deeper and colder waters (Duplessy, 1981b). If the initial shell is preferentially dissolved, the $\delta^{18}O$ of the remainder is shifted towards heavier values. In the case of the MARGO data set, plotting measured calcite $\delta^{18}O$ with respect to core water depth did not reveal any clear pattern.

4.2.2. Temperature:$\delta^{18}O$ relationship for planktonic foraminifera from recent sediments

Numerous studies have compared $\delta^{18}O$ of core tops and living planktonic foraminifera (e.g., Duplessy et al.,

1981b; Bemis et al., 1998). However, previous studies were based on limited regional data sets. MARGO data set allows us to examine the temperature:$\delta^{18}O$ relationship of planktonic foraminifera on a wide geographical scale, based on a large number of high quality data checked for internal consistency, from recent sediment with thorough age control. Fig. 6 shows the geographical distribution and values of MARGO planktonic foraminifera $\delta^{18}O$ data.

Fig. 7A–D show annual mean SST vs. the difference between foraminifera and water $\delta^{18}O$ ($\delta^{18}O_w$) for G. bulloides, G. ruber (white and pink), G. sacculifer, N. pachyderma left, and N. pachyderma right from the MARGO core top (CT) and Late Holocene (LH) samples. These five species were chosen because their geographical distribution covers the entire present-day temperature range, because they are all among the shallowest dwelling species, and because they are the most frequently used species for paleoceanographic investigations. For each species, we grouped LH samples and CT samples with good chronostratigraphic control (i.e., chronozone level of 1–4), and computed the corresponding linear regression, except for G. ruber pink for which we had too few samples of chronozone level 1–4. The plots also show the CT data for which there is no chronological control.

$\delta^{18}O_w$ was computed by applying regional linear regressions between salinity and $\delta^{18}O_w$ based on GEOSECS data (Paul et al., 1999). In the Arctic Ocean, some of our data points are outside the calibration range of the linear regression (annual mean salinity < 29.5 psu). At those sites, $\delta^{18}O_w$ is taken from the interpolated surface water $\delta^{18}O$ field of Schmidt et al. (1999). Mean annual temperature and salinity at 10 m depth were retrieved from the World Ocean Atlas

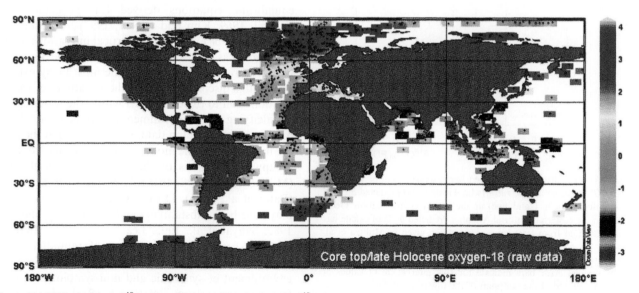

Fig. 6. MARGO planktonic $\delta^{18}O$ LH and CT data. The plot includes $\delta^{18}O$ data of G. sacculifer, G. ruber white and pink, G. bulloides, N. pachyderma left, and N. pachyderma right in ‰ PDB.

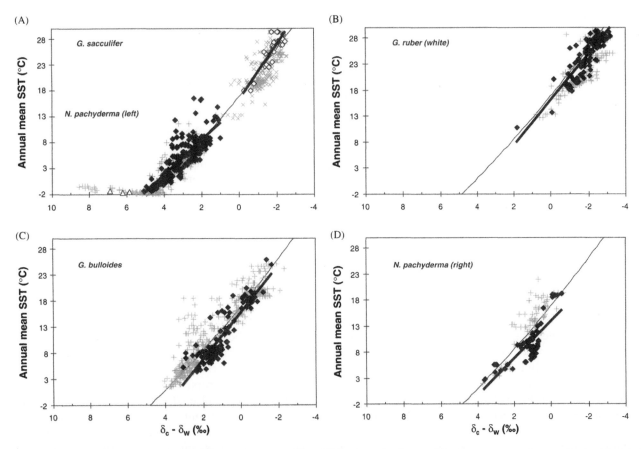

Fig. 7. Annual mean SST vs. foraminifera $\delta^{18}O$–surface water $\delta^{18}O$ ($\delta^{18}O_w$) for: (A) *G. sacculifer* (gray crosses and open diamonds) and *N. pachyderma left* (gray plus signs and filled diamonds), (B) *G. ruber*, (C) *G. bulloides*, and (D) *N. pachyderma right*. Diamonds denote LH and CT samples with chronozone level 1–4, gray crosses and plus signs denote undated CTs. Black thin line represents the temperature equation for inorganic calcite; bold black lines are linear regressions through LH and well dated CT samples. For *G. sacculifer*, 12 out of the 46 samples with chronozone level 1–4 were not taken into account in the computation of the regression because their sedimentation rates are inferior to 3 cm/ky, likely resulting in bioturbation biases. For *N. pachyderma* left, three data points with chronozone level 1–4 below 0 °C (empty triangles symbols) were not taken into account in the computation of the regression because of the very large uncertainties on $\delta^{18}O_w$ in high latitudes. $\delta^{18}O_w$ was computed by applying regional linear regressions between salinity and $\delta^{18}O_w$ based on GEOSECS data (see text).

1998 (Conkright et al., 1998), following the recommendations of the MARGO project. The uncertainty in $\delta^{18}O_w$ is extremely high in certain regions, such as regions with strong hydrological gradients, coastal regions in the vicinity of large estuaries, or high latitudes. In these regions, the uncertainty in $\delta^{18}O_w$ can reach 0.5–1‰, so that part of the dispersion in Fig. 7 is unfortunately not interpretable. We will thus restrict the following discussion to the general patterns.

Fig. 7A–D illustrate that the data dispersion is much larger for undated CT than for LH and level 1–4 CT samples. Many of the undated *G. bulloides* and *N. pachyderma* right CT samples are clearly isotopically heavier than the well-dated samples (Fig. 7C and D). This could result from mixing with glacial sediment due to bioturbation in cores with low sedimentation rates, or from loss of the Holocene sediment during coring, core handling, etc. However, for *G. ruber* and *G. sacculifer*, undated CT data exhibit values that are lighter than the

well dated samples (Fig. 7A and B). Plotting warm season instead of annual mean SST, we were able to verify that this pattern is not simply due to the fact that these species preferentially live during the warm months. These lighter values could thus reflect warmer conditions or enhanced precipitation that characterized the Early Holocene in certain regions. Finally, the large spread of undated *N. pachyderma* left CT data below 0 °C is largely imputable to the very large uncertainties on $\delta^{18}O_w$ in high latitudes.

The regressions obtained for recent sediment foraminifera exhibit increasing slopes from lower to higher temperatures, in agreement with the equilibrium equation for inorganic calcite. T-tests indicate that the slope obtained for *N. pachyderma* left data is significantly lower than those for *G. bulloides*, *G. ruber* white and *G. sacculifer* data at the 90% confidence level. Similarly, the slope of the regression on *N. pachyderma* right data is also significantly lower than the slope for *G. sacculifer* data.

These regressions can be compared to the species-specific regressions obtained on plankton tows and pumped foraminifera by Mulitza et al. (2003a) and laboratory-derived relationships in Bemis et al. (1998) and Spero et al. (2003). For all species, the calcite of fossil planktonic foraminifera is enriched in ^{18}O with respect to that of living foraminifera. We find a shift of 0.2 to 0.8‰ in δ^{18}O between living and fossil foraminifera (Table 1). This shift towards higher δ^{18}O values of the foraminifera calcite found in the sediment with respect to the foraminifera sampled in surface waters could reflect calcite precipitation in deeper waters during ontogenesis or gametogenesis. This is illustrated in Fig. 5 for *G. sacculifer* and has been observed in other studies on *G. sacculifer* and *G. ruber* collected in plankton tows and recent sediment (Duplessy, 1981b) and in cultures of *G. sacculifer* and *O. universa* (Caron et al., 1990).

Maps of the difference between the δ^{18}O from recent sediments and equilibrium calcite δ^{18}O computed using regressions established on living foraminifera indicate that this difference is of a regional nature (Fig. 8) and increases in areas where surface stratification is high, as in low latitudes (Mulitza et al., 1997) and in regions of low surface salinity (e.g., Gulf of Guinea). This pattern suggests that the regressions derived from sediment foraminifera reflect calcification at or within the thermocline. On the other hand, the differences are usually smallest at the lower temperature limit. For example both *G. sacculifer* and *G. ruber* display the smallest deviation from equilibrium calcite in the North Atlantic and in the upwelling regions off NW-Africa. This might either indicate a shallower depth habitat, less stratification or a weighing of the flux towards the warmer season (Mulitza et al., 1998).

The general pattern of the differences between "equilibrium calcite" and fossil foraminifera δ^{18}O demonstrates that the relationship between "equilibrium calcite" δ^{18}O and the δ^{18}O of foraminiferal populations recorded in the sediment is complex and depends on local hydrography. From our study, it seems unlikely that a transfer function in form of a simple linear regression equation is the appropriate tool to correct for these distortions in paleoceanographic reconstructions.

5. Conclusion

The large discrepancy between δ^{18}O values of living and recent fossil foraminifera highlights the difficulty in reconstructing past surface conditions based on fossil planktonic foraminifera δ^{18}O. Indeed, as stratification of surface waters and seasonality do not remain constant in time, the difference between δ^{18}O of fossil and living foraminifera will likely not be constant in time either. This might make temperature:δ^{18}O relationships established on fossil foraminifera from recent sediment less reliable tools for paleoceanographic reconstructions. Hence, it is necessary to gain more information on the nature and the magnitude of the environmental sensitivity of the main foraminifera species during their entire life cycle before developing models which enable us to predict seasonal flux, abundance and δ^{18}O composition of a foraminiferal populations in the sediment. Although laboratory cultures have contributed much to the understanding of "vital effects" of foraminifera, there is still a need for combined laboratory and field studies to derive information about the influence of the environment on foraminiferal isotopic composition. Eventually, all these improvements will enable global circulation models to mechanistically simulate the oxygen isotopic composition of foraminifera calcite. The oxygen isotope composition of foraminifera will then provide a set of constraints to test coupled climate models under boundary conditions different from the present climate state. In the meantime, a good approach to interpret fossil planktonic foraminifera δ^{18}O is to study relative changes in δ^{18}O with respect to recent sediment values. These changes can be decomposed into three components: the impact of temperature changes on the oxygen fractionation between foraminifera calcite and water, the change in water δ^{18}O related to sea level changes (global ice volume signal), and the change in local water δ^{18}O due to regional hydrological variation. A cautious interpretation is then necessary, knowing that changes in planktonic foraminiferal seasonality, depth habitat or water pH might translate into apparent changes in temperature or local water δ^{18}O.

Table 1
Comparison of ($\delta^{18}O_{calcite}-\delta^{18}O_{water}$) for living and fossil foraminifera

	Common temperature range covered by measurements			Living foraminifera[a] middle ($\delta_c-\delta_w$) (‰)	Fossil foraminifera middle ($\delta_c-\delta_w$) (‰)	Average shift (‰)
	Lower limit (°C)	Upper limit (°C)	Middle (°C)			
G. sacculifer	17.51	29.1	23.31	−1.93	−1.35	0.58
G. ruber white	15.27	30	22.64	−1.90	−1.50	0.40
G. bulloides	4.14	23.05	13.60	0.22	0.46	0.24
N. pachyderma left	−1.34	11.69	5.18	2.12	2.95	0.83

[a]Multza et al., 2003a.

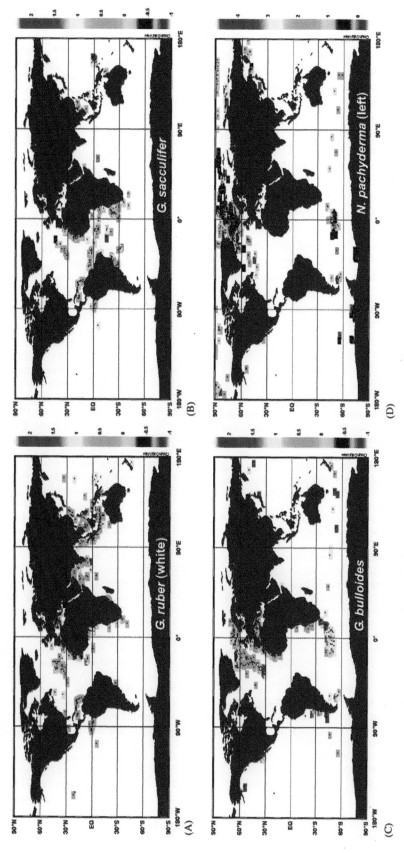

Fig. 8. (A–D) Maps of the difference between the $\delta^{18}O$ from recent sediments and equilibrium calcite $\delta^{18}O$ computed using species-specific regressions established on living foraminifera by Mulitza et al. (2003a) and the same annual mean SST and water $\delta^{18}O$ values as in Fig. 7.

Acknowledgements

We thank J.C. Duplessy and L. Labeyrie for fruitful discussions, as well as P. De Deckker, M. Kucera and an anonymous reviewer for constructive comments on the first version of the manuscript. We are grateful to S. Jung and M.T. Chen for contributing unpublished data. We thank D. Lea, M. Weinelt, V. Meyer-Stumborg, M. Kienast and L. Senneset for assistance in compiling data. We acknowledge B. Lecoat and J. Tessier for running isotopic measurements in Gif. The Marion-Dufresne cores have been collected under the responsibility of the IPEV oceanic team directed by Y. Balut. This work has been supported by the CNRS/INSU, CEA, IPEV, NERC and NSF, as well as by the French Programme National de la Dynamique du Climat. S.M. was supported by the Deutsche Forschungsgemeinschaft (DFG Research Center Ocean Margins) and the Bundesministerium für Bildung und Forschung (DEKLIM E, grant 01 LD 0019). H.J.S. also thanks the Hanse Institute for Advanced Study, Delmenhorst Germany for support. This is contribution no 0990 of the LSCE.

References

Bassinot, F., Labeyrie, L., Vincent, E., Quidelleur, X., Shackleton, N., Lancelot, Y., 1994. The astronomical theory of climate and the age of the Brunhes–Matuyama magnetic reversal. Earth and Planetary Science Letters 126, 91–108.

Bates, N.R., Michaels, A.F., Knap, A.H., 1996. Seasonal and interannual variability of oceanic carbon dioxide species at the US JGOFS Bermuda Atlantic Time-series Study (BATS) site. Deep-Sea Research II 43 (2–3), 347–383.

Bé, A.W.H., 1980. Gametogenic calcification in a spinose planktonic foraminifer, Globigerinoides sacculifer (Brady). Marine Micropalaeontology 5, 283–310.

Bé, A.W.H., Spero, H.J., Anderson, O.R., 1982. Effects of symbiont elimination and reinfection on the life processes of the planktonic foraminifer Globigerinoides sacculifer. Marine Biology 70, 73–86.

Bemis, B.E., Spero, H.J., Bijma, J., Lea, D.W., 1998. Reevaluation of the oxygen isotopic composition of planktonic foraminifera: experimental results and revised paleotemperature equations. Paleoceanography 13 (2), 150–160.

Berger, W.H., Killingley, J.S., Vincent, E., 1978. Stable isotopes in deep-sea carbonates: Box core ERDC-92, west equatorial Pacific. Oceanologica Acta 1, 203–216.

Bijma, J., Erez, J., Hemleben, C., 1990. Lunar and semi-lunar reproductive cycles in some spinose planktonic foraminifers. Journal of Foraminiferal Research 20, 117–127.

Billups, K., Schrag, D.P., 2000. Surface ocean density gradients during the Last Glacial Maximum. Paleoceanography 15 (1), 110–123.

Billups, K., Spero, H.J., 1995. Relationship between shell size, thickness and stable isotopes in individual planktonic foraminifera from two equatorial Atlantic cores. Journal of Foraminiferal Research 25, 24–37.

Bouvier-Soumagnac, Y., Duplessy, J.C., 1985. Carbon and oxygen isotopic composition of planktonic foraminifera from laboratory culture, plankton tows and recent sediment: implications for the reconstruction of paleoclimatic conditions and of the global carbon cycle. Journal of Foraminiferal Research 15 (4), 302–320.

Boyle, E.A., 1984. Sampling statistic limitations on benthic foraminifera chemical and isotopic data. Marine Geology 58, 213–224.

Broecker, W.S., Peng, T.-H., 1982. Tracers in the Sea. Eldigio Press, Palisades, New York.

Caron, D.A., Anderson, O.R., Lindsey, J.L., Faber, W.W.J., Lim, E.L., 1990. Effects of gametogenesis on test structure and dissolution of some spinose planktonic foraminifera and implications for test preservation. Marine Micropalaeontology 16, 93–116.

Charles, C.D., Fairbanks, R.G., 1992. Evidence from Southern Ocean sediments for the effect of North Atlantic deep water flux on climate. Nature 355, 416–419.

Conkright, M.E., Levitus, S., O'Brien, T., Boyer, T.P., Stephens, C., Johnson, D., Baranova, O., Antonov, J., Gelfeld, R., Rochester, J., Forgy, C., 1998. World Ocean Database 1998, CD-ROM Data Set Documentation Version 2.0, Ocean Climate Laboratory National Oceanographic Data Center, National Oceanographic Data Center Internal Report 14.

Coplen, T.B., 1996. Editorial: more uncertainty than necessary. Paleoceanography 11 (4), 369–370.

Craig, H., Gordon, L.I., 1965. Deuterium and oxygen 18 variations in the ocean and the marine atmosphere. In: Tongiorgi, E. (Ed.), Stable Isotopes in Oceanographic Studies and Paleotemperatures. Consiglio nazionale delle ricerche, Spoleto, pp. 9–122.

Darling, K.F., Kucera, M., Wade, C.M., von Langen, P., Pak, D., 2003. Seasonal distribution of genetic types of planktonic foraminifer morphospecies in the Santa Barbara Channel and its paleoceanographic implications. Paleoceanography 18 (2), 1032.

Dudley, W.C., Duplessy, J.C., Blackwelder, P.L., Brand, L.E., Guillard, R.R.L., 1980. Coccoliths in Pleistocene-Holocene nannofossil assemblages. Nature 285, 222–223.

Duplessy, J.C., 1978. Isotope studies. In: Gribbin, J. (Ed.), Climatic Changes. Cambridge University Press, Cambridge, pp. 46–67.

Duplessy, J.C., Bé, A.W.H., Blanc, P.L., 1981a. Oxygen and carbon isotopic composition and biogeographic distribution of planktonic foraminifera in the Indian Ocean. Palaeogeography, Palaeoclimatology, Palaeoecology 33, 9–46.

Duplessy, J.-C., Blanc, P.L., Bé, A.W., 1981b. Oxygen-18 enrichment of planktonic foraminifera due to gametogenic calcification below the euphotic zone. Science 213, 1247–1250.

Duplessy, J.-C., Labeyrie, L., Juillet-Leclerc, A., Maitre, F., Duprat, J., Sarnthein, M., 1991. Surface salinity reconstruction of the North Atlantic Ocean during the last glacial maximum. Oceanology Acta 14 (4), 311–324.

Epstein, S., Mayeda, T., 1953. Variation of ^{18}O content of waters from natural sources. Geochimica and Cosmochimica Acta 4, 213–224.

Fairbanks, R.G., Wiebe, P.H., 1980. Foraminifera and chlorophyll maxima: vertical distribution, seasonal succession, and paleoceanographic significance. Science 209, 1524–1526.

Fairbanks, R.G., Sverdlove, M., Free, R., Wiebe, P.H., Bé, A.W.H., 1982. Vertical distribution and isotopic fractionation of living planktonic foraminifera from the Panama Basin. Nature 298, 841–844.

Fischer, G., Kalberer, M., Donner, B., Wefer, G., 1999. Stable isotopes of pteropod shells as recorders of sub-surface water conditions: Comparison to the record of G. ruber and to measured values. In: Fischer, G., Wefer, G. (Eds.), Use of Proxies in Paleoceanography. Springer, Berlin, pp. 191–206.

Hoefs, J., 1987. Stable Isotope Geochemistry. Springer, Berlin.

Hut, G., 1987. Stable Isotope Reference Samples for Geochemical and Hydrological Investigations. Consultant Group Meeting IAEA, Vienna 16–18 September 1985, Report to the Director General, International Atomic Energy Agency, Vienna.

Imbrie, J., Hays, J.D., Martinson, D.G., McIntyre, A., Mix, A.C., Morley, J.J., Pisias, N.G., Prell, W.L., Shackleton, N.J., 1984. The orbital theory of Pleistocene climate: support from a revised chronology of the marine $\delta^{18}O$ record. In: Berger, A.,

Imbrie, J., Hays, J., Kukla, G., Saltzman, B. (Eds.), Milankovitch and Climate. Reidel Publishing Company, pp. 269–305.

Johannessen, T., 1992. Stable isotopes as climatic indicators in ocean and lake sediments. Ph.D. Thesis, University of Bergen.

Kemle-von Mücke, S., Oberhänsli, H., 1999. The distribution of living planktic foraminifera in relation to southeast Atlantic oceanography. In: Fischer, G., Wefer, G. (Eds.), Use of Proxies in Paleoceanography: Examples from the South Atlantic. Springer, Berlin, pp. 91–115.

Kim, S.-T., O'Neil, J.R., 1997. Equilibrium and nonequilibrium oxygen isotope effects in synthetic carbonates. Geochimica et Cosmochimica Acta 61 (16), 3461–3475.

Kohfeld, K.E., Fairbanks, R.G., Smith, S.L., Walsh, I.D., 1996. Neogloboquadrina pachyderma (sinistral coiling) as paleoceanographic tracers in polar oceans: evidence from northeast water polynia plankton tows, sediment traps, and surface sediments. Paleoceanography 11 (6), 679–699.

Lea, D.W., Bijma, J., Spero, H.J., Archer, D., 1999. Implications of a carbonate ion effect on shell carbon and oxygen isotopes for glacial ocean conditions. In: Fischer, G., Wefer, G. (Eds.), Use of Proxies in Paleoceanography—Examples from the South Atlantic. Springer, Berlin, pp. 513–522.

Lea, D.W., Pak, D.K., Spero, H.J., 2000. Climate impact of late Quaternary equatorial Pacific sea surface temperature variations. Science 289, 1719–1724.

Mashiotta, T.A., Lea, D.W., Spero, H.J., 1999. Glacial-interglacial changes in Subantarctic sea surface temperature and $\delta^{18}O$ using foraminiferal Mg. Earth and Planetary Science Letters 170, 417–432.

McCrea, J.M., 1950. On the isotopic chemistry of carbonates and a paleotemperature scale. The Journal of Chemical Physics 18 (6), 849–857.

Mielke, K.M., 2001. Reconstructing surface carbonate chemistry and temperature in paleoceans: Geochemical results from laboratory experiments and the fossil record. Unpublished M.Sc. Thesis, University California Davis.

Millero, F.J., Yao, W., Lee, K., Zhang, J.-Z., Campbell, D.M., 1998. Carbonate system in the waters near the Galapagos Islands. Deep-Sea Research II 45, 1115–1134.

Mortyn, P.G., Charles, C.D., 2003. Planktonic foraminiferal depth habitat and $\delta^{18}O$ calibrations: Plankton tow results from the Atlantic sector of the Southern Ocean. Paleoceanography 18 (2) 15-1-15-14.

Mulitza, S., Dürkoop, A., Hale, W., Wefer, G., Niebler, H.S., 1997. Planktonic foraminifera as recorders of past surface-water stratification. Geology 25 (4), 335–338.

Mulitza, S., Wolff, T., Pätzold, J., Hale, W., Wefer, G., 1998. Temperature sensitivity of planktic foraminifera and its influence on the oxygen isotope record. Marine Micropalaeontology 33, 223–240.

Mulitza, S., Boltovskoy, D., Donner, B., Meggers, H., Paul, A., Wefer, G., 2003a. Temperature: $\delta^{18}O$ relationships of planktic foraminifera collected from surface waters. Palaeogeography, Palaeoclimatology, Palaeoecology 202, 143–152.

Mulitza, S., Donner, B., Fischer, G., Paul, A., Pätzold, J., Rühlemann, C., Segl, M., 2003b. The South Atlantic Oxygen-isotope record of planktic foraminifera. In: Wefer, G., Mulitza, S., Ratmeyer, V. (Eds.), The South Atlantic in the Late Quaternary: Reconstruction of Material Budgets and Current Systems. Springer, Berlin, pp. 121–142.

O'Neil, J.R., Clayton, R.N., Mayeda, T.K., 1969. Oxygen isotope fractionation in divalent metal carbonates. Journal of Chemical Physics 51, 5547–5558.

Oppo, D.W., Fairbanks, R.G., 1989. Carbon isotope composition of tropical surface water during the past 22,000 years. Paleoceanography 4, 333–351.

Ostermann, D.R., Curry, W.B., 2000. Calibration of stable isotopic data: An enriched $\delta^{18}O$ standard used for source gas mixing detection and correction. Paleoceanography 15 (3), 353–360.

Paul, A., Mulitza, S., Pätzold, J., Wolff, T., 1999. Simulation of oxygen isotopes in a global ocean model. In: Fischer, G., Wefer, G. (Eds.), Use of Proxies in Paleoceanography. Springer, Berlin, pp. 655–686.

Peeters et al., 2002. The effect of upwelling on the distribution and stable isotope composition of G. bulloides and G. ruber in modern surface waters of the NW Arabian Sea. Global and Planetary Change 34, 269–291.

Rink, S., Kühl, M., Bijma, J., Spero, H.J., 1998. Microsensor studies of photosynthesis and respiration in the symbiotic foraminifer Orbulina universa. Marine Biology 131, 583–595.

Rohling, E.J., Cooke, S., 1999. Stable Oxygen and Carbon Isotope Ratios in Foraminiferal Carbonate in Gupta, Modern Foraminifera, Kluwer Academic, Dordrecht, The Netherlands, pp. 239–258.

Rühlemann, C., Diekmann, B., Mulitza, S., Frank, M., 2001. Late Quaternary changes of western equatorial Atlantic surface circulation and Amazon lowland climate recorded in Ceara Rise deep-sea sediments. Paleoceanography 16 (3), 293–305.

Sanyal, A., Hemming, N.G., Hanson, G.N., Broecker, W.S., 1995. Evidence for a higher pH in the glacial ocean from boron isotopes in foraminifera. Nature 373, 234–236.

Sarkar, A., Ramesh, R., Bhattacharya, S.K., 1990. Effect of sample pretreatment and size fraction on the $\delta^{18}O$ and $\delta^{13}C$ values of foraminifera in Arabian Sea sediments. Terra Nova 2, 489–493.

Sautter, L.R., Thunell, R.C., 1989. Seasonal succession of planktonic foraminifera: results from a four-year time-series sediment trap experiment in the northeast Pacific. Journal of Foraminiferal Research 19, 253–267.

Sautter, L.R., Thunell, R.C., 1991. Planktonic foraminiferal response to upwelling and seasonal hydrographic conditions: Sediment trap results from San Pedro Basin Southern California Bight. Journal Foraminiferal Research 21, 347–363.

Schiffelbein, P., Hills, S., 1984. Direct assessment of stable isotope variability in planktonic foraminifera populations. Palaeogeography, Palaeoclimatology, Palaeoecology 48, 197–213.

Schmidt, G.A., Mulitza, S., 2002. Global calibration of ecological models for planktic foraminifera from coretop carbonate oxygen-18. Marine Micropalaeontology 44, 125–140.

Schmidt, G.A., Bigg, G.R., Rohling, E.J., 1999. Global Seawater Oxygen-18 Database, http://www.giss.nasa.gov/data/o18data/

Shackleton, N.J., 1974. Attainment of isotopic equilibrium between ocean water and benthonic foraminifera genus Uvigerina: isotopic changes in the ocean during the last glacial. Les méthodes quantitatives d'étude des variations du climat au cours du Pleistocène, Gif-sur-Yvette. Colloque international du CNRS 219, 203–210.

Shackleton, N.J., Opdyke, N.D., 1973. Oxygen isotope and paleomagnetic stratigraphy of equatorial Pacific core V 28-238: Oxygen isotope temperatures and ice volumes on a 10^5 year scale. Quaternary Research 3, 39–55.

Spero, H.J., 1998. Life history and stable isotope geochemistry of planktonic foraminifera. In: Norris, R.D., Corfield, R.M. (Eds.), Isotope Paleobiology and Paleoecology Paleontological Society Papers. The Paleontological Society, pp. 7–36.

Spero, H.J., Lea, D.W., 1993. Intraspecific stable isotope variability in the planktic foraminifera Globigerinoides sacculifer: Results from laboratory experiments. Marine Micropalaeontology 22, 221–234.

Spero, H.J., Lea, D.W., 1996. Experimental determination of stable isotope variability in G bulloides: implications for paleoceanographic reconstructions. Marine Micropalaeontology 28, 231–246.

Spero, H.J., Williams, D.F., 1990. Evidence for seasonal low-salinity surface waters in the Gulf of Mexico over the last 16,000 years. Paleoceanography 5, 963–975.

Spero, H.J., Bijma, J., Lea, D.W., Bemis, B.E., 1997. Effect of seawater carbonate concentration on foraminiferal carbon and oxygen isotopes. Nature 390, 497–500.

Spero, H.J., Mielke, K.M., Kalve, E.M., Lea, D.W., Pak, D.K., 2003. Multispecies approach to reconstructing eastern equatorial Pacific thermocline hydrography during the past 360 kyr. Paleoceanography 18 (1) 22-1–22-15.

Stott, L.D., Tang, C.M., 1996. Reassessment of foraminiferal-based tropical sea surface $\delta^{18}O$ paleotemperatures. Paleoceanography 11, 37–56.

Tang, C.M., Stott, L.D., 1993. Seasonal salinity changes during Mediterranean sapropel deposition 9000 years BP: evidence from isotopic analyses of individual planktonic foraminifera. Paleoceanography 8, 473–493.

Thunnel, R., Tappa, E., Pride, C., Kincaid, E., 1999. Sea-surface temperature anomalies associated with the 1997–1998 El Nino recorded in the oxygen isotope composition of planktonic foraminifera. Geology 27, 843–846.

Urey, H.C., 1947. Thermodynamic properties of isotopic substances. Journal of Chemical Society 562–581.

Willams, D.F., Healy Williams, N., 1980. Oxygen isotopic-hydrographic relationships among recent planktonic foraminifera from the Indian Ocean. Nature 283, 848–852.

Williams, D.F., Bé, A.W., Fairbanks, R.G., 1981. Seasonal stable isotopic variations in living planktonic foraminifera from Bermuda plankton tows. Palaeogeography, Palaeoclimatology, Palaeoecology 33, 71–102.

Wolf-Gladrow, D.A., Bijma, J., Zeebe, R.E., 1999. Model simulation of the carbonate chemistry in the microenvironment of symbiont bearing foraminifera. Marine Chemistry 64, 181–198.

Zeebe, R.E., 1999. An explanation of the effect of seawater carbonate concentration on foraminiferal oxygen isotopes. Geochimica et Cosmochimica Acta 63, 2001–2007.

Quaternary Science Reviews 24 (2005) 869–896

Sea-surface temperature and sea ice distribution of the Southern Ocean at the EPILOG Last Glacial Maximum—a circum-Antarctic view based on siliceous microfossil records ☆

Rainer Gersonde[a],*, Xavier Crosta[b], Andrea Abelmann[a], Leanne Armand[c]

[a] *Alfred Wegener Institute for Polar and Marine Research, Bremerhaven, Germany*
[b] *DGO, UMR-CNRS 5805 EPOC, Université de Bordeaux I, Talence cedex, France*
[c] *School of Earth Sciences, University of Tasmania, Hobart, Tasmania, Australia*

Received 8 January 2004; accepted 10 July 2004

Abstract

Based on the quantitative study of diatoms and radiolarians, summer sea-surface temperature (SSST) and sea ice distribution were estimated from 122 sediment core localities in the Atlantic, Indian and Pacific sectors of the Southern Ocean to reconstruct the last glacial environment at the EPILOG (19.5–16.0 ka or 23 000–19 000 cal yr. B.P.) time-slice. The statistical methods applied include the Imbrie and Kipp Method, the Modern Analog Technique and the General Additive Model. Summer SSTs reveal greater surface-water cooling than reconstructed by CLIMAP (Geol. Soc. Am. Map Chart. Ser. MC-36 (1981) 1), reaching a maximum (4–5 °C) in the present Subantarctic Zone of the Atlantic and Indian sector. The reconstruction of maximum winter sea ice (WSI) extent is in accordance with CLIMAP, showing an expansion of the WSI field by around 100% compared to the present. Although only limited information is available, the data clearly show that CLIMAP strongly overestimated the glacial summer sea ice extent. As a result of the northward expansion of Antarctic cold waters by 5–10° in latitude and a relatively small displacement of the Subtropical Front, thermal gradients were steepened during the last glacial in the northern zone of the Southern Ocean. Such reconstruction may, however, be inapposite for the Pacific sector. The few data available indicate reduced cooling in the southern Pacific and give suggestion for a non-uniform cooling of the glacial Southern Ocean.

This study is part of MARGO, a multiproxy approach for the reconstruction of the glacial ocean surface.
© 2004 Elsevier Ltd. All rights reserved.

1. Introduction

About 25 years ago Hays et al. (1976) published their pioneering study on the Southern Ocean surface water temperature and sea ice extent during the Last Glacial Maximum (LGM), based on 34 cores from the Atlantic and the western Indian sectors of the Southern Ocean. Austral summer (February) and winter (August) sea-surface temperatures (SST) were estimated using the

Imbrie and Kipp (1971) transfer function technique, applying a radiolarian-based paleoecological equation from Lozano and Hays (1976). The standard error was 1.5 °C for the summer and 1.4 °C for the winter estimates. The LGM sea ice boundary was reconstructed by mapping the lithological boundary between diatom-rich and diatom-poor sediments. Because in most cores no continuous calcareous microfossil records were preserved allowing the establishment of an oxygen isotope record, the definition of the LGM level, set at 18 ka, was based on the abundance pattern of the radiolarian *Cycladophora davisiana*. The pattern was calibrated in four cores recovered from the Subantarctic and Subtropical Zone with the oxygen isotope stratigraphy obtained from three planktic and one benthic

☆ This paper is dedicated to the diatom paleoceanographer Dr. Jean-Jacques Pichon who died on September 9, 2003 in a tragic accident.

*Corresponding author. Tel.: +49-471-4831-203; fax: +49-471-4831-1923.

E-mail address: rgersonde@awi-bremerhaven.de (R. Gersonde).

foraminiferal records and a few ^{14}C measurements, all Holocene in age. The stratigraphic determination of the LGM and the SST reconstruction proposed by Hays et al. (1976) was used by Climate Long-range Investigation, Mapping, and Prediction (CLIMAP) (1976, 1981) to estimate circum-Antarctic SST and sea ice distribution as a part of the first global ocean LGM reconstruction. CLIMAP (1976, 1981) placed the austral winter and summer sea ice edge at the faunally identified 0 °C winter and summer isotherm, respectively. Later, Cooke and Hays (1982) presented a revised estimation of the LGM summer and winter sea ice (WSI) extent, considering additional parameters, such as changes in sedimentation rates and the quantification of ice-rafted debris. Burckle et al. (1982) and Burckle (1983) supported this approach, but proposed that the lithological boundary between silty diatomaceous clay and diatom ooze identifies the spring/summer sea ice limits. Alternative attempts to reconstruct past sea ice cover rely on the distribution of diatom sea ice indicators preserved in the sediment record as reviewed in Armand and Leventer (2003). Crosta et al. (1998a, b) were the first to use the Modern Analog Technique (MAT), established by Hutson (1980), for quantitative reconstruction of circum-Antarctic sea ice distribution (months/year) at the LGM levels defined by CLIMAP (1976, 1981). Most recently, Gersonde et al. (2003a) presented a new LGM reconstruction of the Atlantic and western Indian sectors of the Southern Ocean applying IKM or MAT on siliceous (diatoms, radiolarians) and calcareous (planktic foraminifers) microfossil assemblages for SST estimation, and diatom indicator species for the identification of the sea ice extent. This multi-proxy approach was part of "Glacial Atlantic Ocean Mapping" (GLAMAP-2000), an initiative for the reconstruction of the Atlantics SST and sea ice at well-defined LGM time slices (Sarnthein et al., 2003).

Here we present a new circum-Antarctic view of the Southern Oceans SST and sea ice fields during the LGM. This represents a "state-of-the-art" compilation of yet published and new data sets of SST and sea ice estimates from a total of 122 Southern Ocean sediment cores generated from the siliceous microfossil record (diatoms, radiolarians) (Fig. 1). Our compilation is part of the international "Multiproxy Approach for the Reconstruction of the Glacial Ocean Surface"(MARGO) initiative, started in 2002. The data have been assembled following the rules agreed upon by the MARGO scientific community (Kucera et al., 2004). This includes a well-defined quality control for the selection of used sample material and quality ranking of the obtained data. As suggested by MARGO, we follow the LGM time slice definition (19.5–16.0 ka, equal to 23 000–19 000 cal yr B.P.) proposed by the international "Environment Processes of the Ice Age: Land, Ocean, Glaciers" (EPILOG) working group (Mix et al., 2001).

Our Southern Ocean EPILOG-LGM (E-LGM) compilation represents a major step to describe and understand environmental conditions and processes in a part of the World Ocean that acts as a major player in global climate change through feedback mechanisms driven by changes in albedo, ocean/atmosphere exchange rates, physical parameters of ocean surface waters, water mass structure and formation, and biological productivity. Of crucial interest is the reconstruction of sea ice and its seasonal variability. Sea ice represents a fast reacting climate amplifier, causing enhanced variability during glacial intervals, as a result of its impact on water mass production and circulation, as well as air-sea gas and energy exchange (Stephens and Keeling, 2000; Keeling and Stephens, 2001). Sea ice also impacts the Earths albedo and it gears latitudinal thermal gradients and storminess and thus, involves the lofting of dust, micronutrient iron and sea salt into the atmosphere (Broecker, 2001). Amalgamated with the results obtained from other ocean basins within MARGO, the presented reconstructions of summer and winter conditions will help to provide information on climate end-member conditions at the global scale required to test climate models and to increase their fidelity to simulate future climate change. Our study also identifies current deficiencies in the methods used for reconstruction of past Southern Ocean conditions, as acknowledging existing gaps of information.

2. Material, methods, age determination and quality control

2.1. Sample preparation and counting

Preparation of sediment samples for light-microscopic investigations was done according to various techniques. Diatom samples collected during R.V. *Polarstern* cruises (PS indexed sample sites) were cleaned according to the method described by Gersonde and Zielinski (2000) for diatoms. All other diatom samples were treated using a method adapted from Schrader and Gersonde (1978) and Pichon et al. (1992a). Radiolarian samples have been cleaned according to the method described by Abelmann (1988) and Abelmann et al. (1999).

Preparation of permanent mounts for light microscopic investigation was completed according to Gersonde and Zielinski (2000) for diatom slides, and according to Abelmann et al. (1999) for radiolarian slides.

Diatom counts followed the conventions of Schrader and Gersonde (1978) and Laws (1983). A minimum of 300 diatom valves or radiolarian skeletons (in average around 400) was counted in each sample using high quality photomicroscopes (Leitz, Olympus, Zeiss) at a

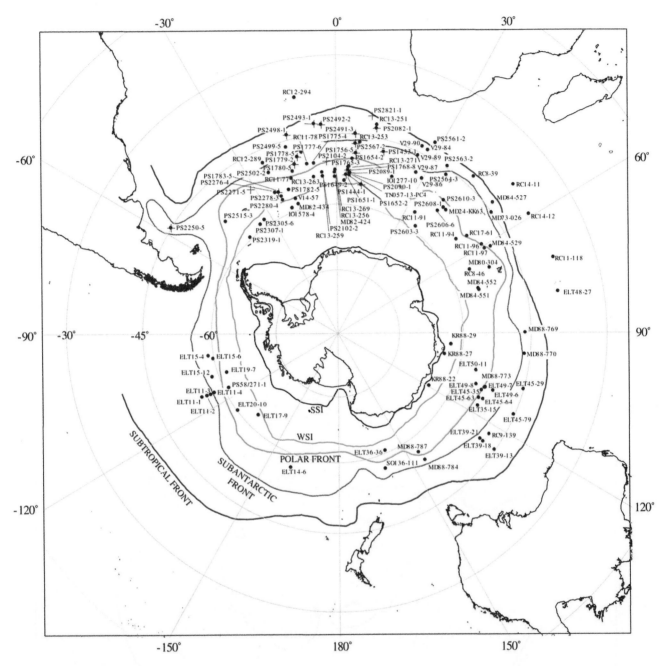

Fig. 1. Distribution of cores used for E-LGM reconstruction. Closed points indicate locations with diatom-based reconstruction, crosses indicate location with radiolarian-based reconstruction. For core location information see Table 3. Location of oceanic fronts according to Belkin and Gordon (1996). The sea ice distribution is from data of Comiso (2003), WSI indicates >15% September sea ice concentration average (1979–1999), summer sea ice (SSI) >15% February concentration average (1979–1999).

magnification of 1000 × for diatom counts and of 250 × or 400 × for radiolarian counts.

2.2. Reconstruction techniques and reference data sets

Statistical methods used for the estimation of SSTs and sea ice include the classical Imbrie and Kipp Method (IKM; Imbrie and Kipp, 1971), the MAT (Hutson, 1980) and the recently proposed Generalized Additive Model (GAM; Hastie and Tibshirani, 1990).

The basic assumption of these techniques is that the modern spatial variability of microfossil assemblages in surface sediment samples deposited at known environmental conditions serves as a proxy for past environmental variability documented down-core.

By means of factor analysis, the IKM resolves microfossil assemblages preserved in surface sediment samples. The resulting varimax factors are calibrated in terms of hydrographic parameters of the surface waters, such as temperature, by using a stepwise multiple regression analysis, which is then applied to down-core assemblages to estimate past hydrographic parameters. For further reading on the IKM, we refer to Imbrie and Kipp (1971), Maynard (1976), Jöreskog et al. (1976), Malmgren and Haq (1982), Le (1992), and Le and Shackleton (1994).

IKM was used for E-LGM SST reconstruction at 45 sites in the Atlantic and eastern Indian sector of the Southern Ocean (Gersonde et al., 2003a) and at one site in the eastern Pacific sector (Wittling and Gersonde, unpublished data) applying a regional diatom and/or radiolarian reference data set. According to sensitivity tests by Le (1992), regional data sets exhibit better

statistical results using the IKM than other techniques. For this reason we carefully selected 93 surface sediment samples from a total of 218 samples in the Atlantic and Indian Ocean sectors (Zielinski and Gersonde, 1997) (Fig. 2, Table 1). This diatom reference data set contains 29 species or species groups. Many of the selected surface samples have been recovered with a multicorer (MC) device, allowing undisturbed sampling of the sediment surface (Table 1). For SST estimates a logarithmic ranking of the diatom abundance data was applied in order to compensate the dominance of one diatom species, *Fragilariopsis kerguelensis*. The documentation of the diatom reference data set and the

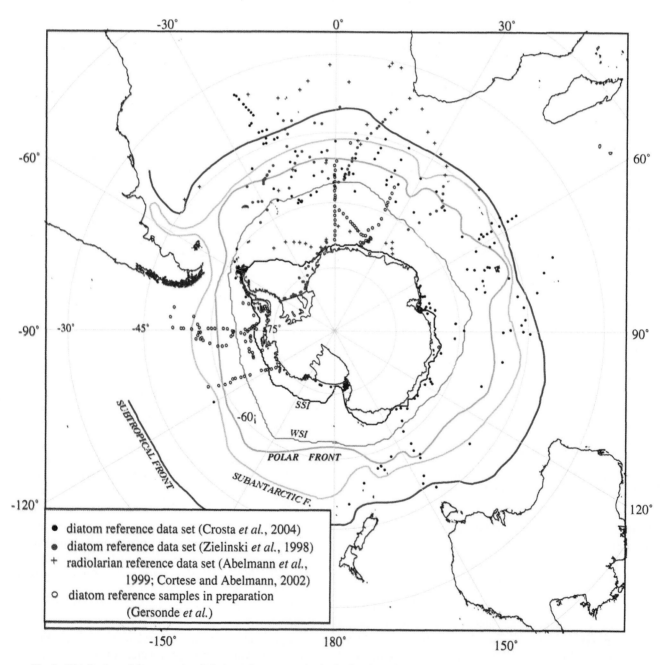

Fig. 2. Distribution of diatom and radiolarian reference samples in the Southern Ocean. For sample location information see Table 1.

Table 1
Compilation of reference surface sediment samples used for diatom (D) and radiolarian (R) based estimation of summer SST and sea-ice (see also Fig. 2 for map)

Core	Longitude	Latitude	Water depth (m)	Coring device	Sampling level	Strat. quality	Fossil group	Use	Ref.
AA93-7/105GR	−66.56	62.74	1882	BC	Top	4	D	MAT/GAM-SST/SI	1
AA93-7/106GR	−66.87	63.16	434	BC	Top	4	D	MAT/GAM-SST/SI	1
AA93-7/12GR	−68.7	77.51	707	BC	Top	4	D	MAT/GAM-SST/SI	1
AA93-7/13GR	−68.67	77.27	538	BC	Top	4	D	MAT/GAM-SST/SI	1
AA93-7/14GR	−68.91	76.9	700	BC	Top	4	D	MAT/GAM-SST/SI	1
AA93-7/15GR	−68.82	77.17	760	BC	Top	4	D	MAT/GAM-SST/SI	1
AA93-7/17GR	−68.78	76.8	798	BC	Top	4	D	MAT/GAM-SST/SI	1
AA93-7/18GR	−68.71	76.74	820	BC	Top	4	D	MAT/GAM-SST/SI	1
AA93-7/19GR	−68.65	76.72	775	BC	Top	4	D	MAT/GAM-SST/SI	1
AA93-7/21GR	−68.01	76.55	460	BC	Top	4	D	MAT/GAM-SST/SI	1
AA93-7/23GR	−67.35	76.59	318	BC	Top	4	D	MAT/GAM-SST/SI	1
AA93-7/24GR	−66.97	79.26	330	BC	Top	4	D	MAT/GAM-SST/SI	1
AA93-7/37GR	−68.96	75.19	775	BC	Top	4	D	MAT/GAM-SST/SI	1
AA93-7/38GR	−68.61	74.52	667	BC	Top	4	D	MAT/GAM-SST/SI	1
AA93-7/39GR	−68.55	74.42	775	BC	Top	4	D	MAT/GAM-SST/SI	1
AA93-7/41GR	−68.94	73.57	792	BC	Top	4	D	MAT/GAM-SST/SI	1
AA93-7/42GR	−68.18	75.87	695	BC	Top	4	D	MAT/GAM-SST/SI	1
AA93-7/43GR	−69.23	76.1	548	BC	Top	4	D	MAT/GAM-SST/SI	1
AA93-7/59GR	−68.41	72.01	509	BC	Top	4	D	MAT/GAM-SST/SI	1
AA93-7/60GR	−68.1	72.25	788	BC	Top	4	D	MAT/GAM-SST/SI	1
AA93-7/73GR	−66.56	69.4	1435	BC	Top	4	D	MAT/GAM-SST/SI	1
AA93-7/78GR	−67.51	68.2	460	BC	Top	4	D	MAT/GAM-SST/SI	1
AA93-7/9GR	−68.43	77.81	173	BC	Top	4	D	MAT/GAM-SST/SI	1
AA93-7/GR158	−68.92	76.62	700	BC	Top	4	D	MAT/GAM-SST/SI	1
DF86-119TC	−66.95	−69.86	600	TRIG	Top	4	D	MAT/GAM-SST/SI	1
DFBC83-10II	−76.95	166.33	878	BC	Top	4	D	MAT/GAM-SST/SI	1
DFBC83-19III	−77.3	−158.72	677	BC	Top	4	D	MAT/GAM-SST/SI	1
DFBC83-1II	−76.17	168.96	540	BC	Top	4	D	MAT/GAM-SST/SI	1
DFBC83-1III	−77.17	169.12	930	BC	Top	4	D	MAT/GAM-SST/SI	1
DFBC83-20II	−76.95	166.68	750	BC	Top	4	D	MAT/GAM-SST/SI	1
DFBC83-21II	−76.69	167.82	768	BC	Top	4	D	MAT/GAM-SST/SI	1
DFBC83-23II	−76.52	170.09	860	BC	Top	4	D	MAT/GAM-SST/SI	1
DFBC83-27II	−75.7	170.65	322	BC	Top	4	D	MAT/GAM-SST/SI	1
DFBC83-28II	−75.85	169.3	485	BC	Top	4	D	MAT/GAM-SST/SI	1
DFBC83-29II	−76.02	167.2	622	BC	Top	4	D	MAT/GAM-SST/SI	1
DFBC83-2II	−76.62	164.35	540	BC	Top	4	D	MAT/GAM-SST/SI	1
DFBC83-30II	−76.09	166.7	668	BC	Top	4	D	MAT/GAM-SST/SI	1
DFBC83-40II	−76.35	167.2	732	BC	Top	4	D	MAT/GAM-SST/SI	1
DFBC83-41III	−76.67	−164.02	516	BC	Top	4	D	MAT/GAM-SST/SI	1
DFBC83-42III	−76.63	−166.05	420	BC	Top	4	D	MAT/GAM-SST/SI	1
DFBC83-43III	−76.72	−176.32	541	BC	Top	4	D	MAT/GAM-SST/SI	1
DFBC83-5II	−76.5	166	640	BC	Top	4	D	MAT/GAM-SST/SI	1
DFBC83-6II	−77.5	165.8	823	BC	Top	4	D	MAT/GAM-SST/SI	1
DFBC83-7II	−77.35	165.88	880	BC	Top	4	D	MAT/GAM-SST/SI	1
DFBC83-8II	−77.17	165.8	871	BC	Top	4	D	MAT/GAM-SST/SI	1
DFBC83-9II	−77.09	166.32	915	BC	Top	4	D	MAT/GAM-SST/SI	1
ELT33-21	−56.54	−119.8	2240	TRIG	1–2	5	D	MAT/GAM-SST/SI	1
ELT36-33	−57.77	154.88	1877	TRIG	Top	4	D	MAT/GAM-SST/SI	1
ELT36-38	−56.47	161.76	2258	TRIG	3–4	5	D	MAT/GAM-SST/SI	1
GC33	−67.68	68.5	320	GRAV	Top	2	D	MAT/GAM-SST/SI	1
GC5	−67.05	69.09	376	GRAV	Top	2	D	MAT/GAM-SST/SI	1
GeoB2004-1	−30.87	14.34	2569	MC	0–0.5	4	R	IKM-SST	4
GeoB2007-1	−30.44	12.16	3906	BC	0–1	4	R	IKM-SST	4
GeoB2008-1	−31.1	11.72	4312	BC	0–1	4	R	IKM-SST	4
GeoB2016-3	−31.91	−1.3	3385	MC	0–0.5	4	R	IKM-SST	4
GeoB2018-1	−34.66	−6.56	4241	MC	0–0.5	4	R	IKM-SST	4
GeoB2019-2	−36.05	−8.77	3825	MC	0–0.5	4	R	IKM-SST	4
GeoB2021-4	−36.83	−14.4	3575	MC	0–0.5	4	R	IKM-SST	4
GeoB2022-3	−34.44	−20.89	4025	MC	0–0.5	4	R	IKM-SST	4
GeoB6402-9	−39.75	−22.76	3878	BC	Top	4	D	MAT/GAM-SST/SI	1
GeoB6403-4	−40.01	−23.36	4226	MC	Top	4	D	MAT/GAM-SST/SI	1

Table 1 (*continued*)

Core	Longitude	Latitude	Water depth (m)	Coring device	Sampling level	Strat. quality	Fossil group	Use	Ref.
GeoB6405-8	−42.01	−21.85	3862	MC	Top	4	D	MAT/GAM-SST/SI	1
GeoB6406-1	−42.01	−20.79	3514	MC	Top	4	D	MAT/GAM-SST/SI	1
GeoB6407-2	−42.04	−19.5	3384	MC	Top	4	D	MAT/GAM-SST/SI	1
GeoB6408-3	−43.61	−20.45	3797	BC	Top	4	D	MAT/GAM-SST/SI	1
GeoB6409-3	−44.51	−21.72	4269	BC	Top	4	D	MAT/GAM-SST/SI	1
GeoB6410-1	−44.52	−20.9	4038	MC	Top	4	D	MAT/GAM-SST/SI	1
GeoB6411-4	−44.37	−18.35	3893	BC	Top	4	D	MAT/GAM-SST/SI	1
GeoB6413-4	−44.21	−17.34	3768	BC	Top	4	D	MAT/GAM-SST/SI	1
GeoB6418-3	−38.43	−21.53	4126	BC	Top	4	D	MAT/GAM-SST/SI	1
GeoB6419-1	−37.78	−21.83	3568	BC	Top	4	D	MAT/GAM-SST/SI	1
GeoB6420-2	−37.12	−22.16	3998	BC	Top	4	D	MAT/GAM-SST/SI	1
GeoB6421-1	−36.45	−22.47	4216	BC	Top	4	D	MAT/GAM-SST/SI	1
GeoB6422-5	−35.71	−22.73	3972	BC	Top	4	D	MAT/GAM-SST/SI	1
GeoB6424-2	−34.61	−23.28	3820	BC	Top	4	D	MAT/GAM-SST/SI	1
GeoB6425-1	−33.83	−23.59	4352	BC	Top	4	D	MAT/GAM-SST/SI	1
GeoB6426-2	−33.5	−24.02	4381	MC	Top	4	D	MAT/GAM-SST/SI	1
IO1176-55	−53.39	6.66	2926	KULL	Top	4	D	MAT/GAM-SST/SI	1
IO1176-65	−57.21	8.21	5483	KULL	Top	4	D	MAT/GAM-SST/SI	1
IO1176-82	−49.52	13.2	4100	KULL	Top	4	D	MAT/GAM-SST/SI	1
IO1176-86	−48.03	13.82	4338	KULL	Top	4	D	MAT/GAM-SST/SI	1
IO1176-88	−46.95	14.31	5106	KULL	Top	4	D	MAT/GAM-SST/SI	1
IO1176-91	−44.94	15.05	4649	KULL	Top	4	D	MAT/GAM-SST/SI	1
IO1277-10	−52.02	20.47	2740	KULL	Top	4	D	MAT/GAM-SST/SI	1
IO1277-12	−54.01	19.79	3178	KULL	Top	4	D	MAT/GAM-SST/SI	1
IO1277-2	−45.03	22.45	4806	KULL	Top	4	D	MAT/GAM-SST/SI	1
IO1277-4	−47.98	21.58	3150	KULL	Top	4	D	MAT/GAM-SST/SI	1
IO1277-8	−50.54	20.89	4492	KULL	Top	4	D	MAT/GAM-SST/SI	1
IO1678-112	−48.15	−27.98	3250	KULL	Top	4	D	MAT/GAM-SST/SI	1
IO1678-64	−54.01	−24.19	4515	KULL	Top	4	D	MAT/GAM-SST/SI	1
IO1678-80	−47.95	−13.03	3120	KULL	Top	4	D	MAT/GAM-SST/SI	1
IO1678-84	−51.96	−14.42	3952	KULL	Top	4	D	MAT/GAM-SST/SI	1
IO1678-89	−57.06	−18.54	4285	KULL	Top	4	D	MAT/GAM-SST/SI	1
KR 88-04	−49.93	100.08	3350	BC	Top	4	D	MAT/GAM-SST/SI	1
KR87-06	−63.05	−63.05	630	BC	Top	4	D	MAT/GAM-SST/SI	1
KR87-07	−62.35	−57.96	2810	BC	Top	4	D	MAT/GAM-SST/SI	1
KR87-08	−60.92	−56.44	2150	BC	Top	4	D	MAT/GAM-SST/SI	1
KR87-10	−59.65	−51.27	2820	BC	Top	4	D	MAT/GAM-SST/SI	1
KR88-01	−46.69	79.48	2925	BC	Top	4	D	MAT/GAM-SST/SI	1
KR88-02	−45.75	82.94	3480	BC	Top	4	D	MAT/GAM-SST/SI	1
KR88-03	−46.07	90.11	3400	MC	Top	4	D	MAT/GAM-SST/SI	1
KR88-05	−52.95	109.92	3510	BC	Top	4	D	MAT/GAM-SST/SI	1
KR88-06	−49.02	128.78	3850	BC	Top	4	D	MAT/GAM-SST/SI	1
KR88-07	−47.15	145.79	2890	BC	Top	4	D	MAT/GAM-SST/SI	1
KR88-08	−49.26	148.8	3885	BC	Top	4	D	MAT/GAM-SST/SI	1
KR88-09	−50.59	147.16	4350	BC	Top	4	D	MAT/GAM-SST/SI	1
KR88-10	−54.19	144.8	2785	BC	Top	4	D	MAT/GAM-SST/SI	1
KR88-11	−54.92	144.07	2880	BC	Top	4	D	MAT/GAM-SST/SI	1
KR88-12	−56.4	145.29	3020	BC	Top	4	D	MAT/GAM-SST/SI	1
KR88-13	−57.95	144.58	3740	BC	Top	4	D	MAT/GAM-SST/SI	1
KR88-14	−61.23	144.44	4200	BC	Top	4	D	MAT/GAM-SST/SI	1
KR88-15	−63.31	141.93	3880	BC	Top	4	D	MAT/GAM-SST/SI	1
KR88-17	−66.2	140.5	180	BC	Top	4	D	MAT/GAM-SST/SI	1
KR88-18	−65.75	138.2	615	BC	Top	4	D	MAT/GAM-SST/SI	1
KR88-19	−64.57	135.63	2930	BC	Top	4	D	MAT/GAM-SST/SI	1
KR88-20	−64.94	129	1670	BC	Top	4	D	MAT/GAM-SST/SI	1
KR88-21	−64.82	126.72	2250	BC	Top	4	D	MAT/GAM-SST/SI	1
KR88-22	−64.67	119.5	3140	BC	Top	4	D	MAT/GAM-SST/SI	1
KR88-23	−63.3	117.26	3292	BC	Top	4	D	MAT/GAM-SST/SI	1
KR88-24	−63.75	116.75	2600	BC	Top	4	D	MAT/GAM-SST/SI	1
KR88-25	−64.3	115.7	2232	BC	Top	4	D	MAT/GAM-SST/SI	1
KR88-27	−63.65	101.15	1210	BC	Top	4	D	MAT/GAM-SST/SI	1
KR88-30	−61	93.2	4300	BC	Top	4	D	MAT/GAM-SST/SI	1
KR88-31	−59	89.6	4595	BC	Top	4	D	MAT/GAM-SST/SI	1

Table 1 (*continued*)

Core	Longitude	Latitude	Water depth (m)	Coring device	Sampling level	Strat. quality	Fossil group	Use	Ref.
KTB01	−49.12	57.02	1235	MC	Top	4	D	MAT/GAM-SST/SI	1
KTB08	−51.98	61.11	4710	MC	Top	4	D	MAT/GAM-SST/SI	1
KTB12	−49	57.98	4390	MC	Top	4	D	MAT/GAM-SST/SI	1
KTB14	−50	57.98	4610	MC	Top	4	D	MAT/GAM-SST/SI	1
KTB18	−48	57.98	4245	MC	Top	4	D	MAT/GAM-SST/SI	1
KTB20	−47	58.02	4550	MC	Top	4	D	MAT/GAM-SST/SI	1
KTB21	−45.96	55.98	4195	MC	Top	4	D	MAT/GAM-SST/SI	1
KTB22	−45.98	55.98	4260	MC	Top	4	D	MAT/GAM-SST/SI	1
KTB25	−45.02	57.94	4680	BC	Top	4	D	MAT/GAM-SST/SI	1
KTB26	−43.97	55.95	4527	MC	Top	4	D	MAT/GAM-SST/SI	1
KTB29	−43	58.02	4765	MC	Top	4	D	MAT/GAM-SST/SI	1
KTB31	−40.98	57.98	5077	MC	Top	4	D	MAT/GAM-SST/SI	1
KTB34	−41.98	58.02	4800	MC	Top	4	D	MAT/GAM-SST/SI	1
MD24-KK02	−54.22	3.52	1522	KULL	Top	4	D	MAT/GAM-SST/SI	1
MD24-KK32	−54.5	3.81	2020	KULL	Top	4	D	MAT/GAM-SST/SI	1
MD24-KK35	−53.11	19.41	2725	KULL	Top	4	D	MAT/GAM-SST/SI	1
MD24-KK37	−52.97	23.77	2905	KULL	Top	4	D	MAT/GAM-SST/SI	1
MD24-KK63	−51.93	42.88	2550	KULL	Top	4	D	MAT/GAM-SST/SI	1
MD80-301	−54	66.83	3750	KULL	Top	4	D	MAT/GAM-SST/SI	1
MD80-304	−51.07	67.73	1950	KULL	Top	4	D	MAT/GAM-SST/SI	1
MD82-422	−52.56	2.24	3750	KULL	Top	4	D	MAT/GAM-SST/SI	1
MD82-424	−54.09	−0.34	2350	KULL	Top	4	D	MAT/GAM-SST/SI	1
MD82-425	−55.58	−0.72	1940	KULL	Top	4	D	MAT/GAM-SST/SI	1
MD82-428	−57.32	−7.98	3750	KULL	Top	4	D	MAT/GAM-SST/SI	1
MD82-430	−57.87	−10.68	3863	KULL	Top	4	D	MAT/GAM-SST/SI	1
MD82-432	−58.64	−14.93	4150	KULL	Top	4	D	MAT/GAM-SST/SI	1
MD82-433	−58.88	−15.2	4750	KULL	Top	4	D	MAT/GAM-SST/SI	1
MD82-434	−58.87	−16.65	3640	KULL	Top	4	D	MAT/GAM-SST/SI	1
MD82-443	−58.78	−15.43	5650	KULL	Top	4	D	MAT/GAM-SST/SI	1
MD82-445	−58.3	−16.03	5750	KULL	Top	4	D	MAT/GAM-SST/SI	1
MD84-521	−50.14	6.79	4150	KULL	Top	4	D	MAT/GAM-SST/SI	1
MD84-529	−48.9	61.99	2600	KULL	Top	4	D	MAT/GAM-SST/SI	1
MD84-530	−66.11	73.98	2412	KULL	Top	4	D	MAT/GAM-SST/SI	1
MD84-531	−66.96	75.41	365	KULL	Top	4	D	MAT/GAM-SST/SI	1
MD84-532	−66.12	76.76	2700	KULL	Top	4	D	MAT/GAM-SST/SI	1
MD84-533	−65.13	78.35	3363	KULL	Top	4	D	MAT/GAM-SST/SI	1
MD84-540	−60.74	86.39	3964	KULL	Top	4	D	MAT/GAM-SST/SI	1
MD84-552	−54.92	73.83	1780	KULL	Top	4	D	MAT/GAM-SST/SI	1
MD84-557	−53.33	75.8	1080	KULL	Top	4	D	MAT/GAM-SST/SI	1
MD84-561	−53.09	71.61	1754	KULL	Top	4	D	MAT/GAM-SST/SI	1
MD84-562	−51.92	68.23	3553	KULL	Top	4	D	MAT/GAM-SST/SI	1
MD84-563	−50.71	68.15	1720	KULL	Top	4	D	MAT/GAM-SST/SI	1
MD84-569	−47.64	73.38	1720	KULL	Top	4	D	MAT/GAM-SST/SI	1
MDBX94-01	−42.5	79.42	2895	BC	Top	4	D	MAT/GAM-SST/SI	1
MDBX94-02	−45.58	86.52	3205	BC	Top	4	D	MAT/GAM-SST/SI	1
MDBX94-03	−46.47	88.05	3559	BC	Top	4	D	MAT/GAM-SST/SI	1
MDBX94-04	−50.37	90.27	3460	BC	Top	4	D	MAT/GAM-SST/SI	1
MDBX94-05	−48.8	89.53	4036	BC	Top	4	D	MAT/GAM-SST/SI	1
MDBX94-06	−44.65	90.69	3709	BC	Top	4	D	MAT/GAM-SST/SI	1
P1010	−77.33	−35	476	KULL	Top	4	D	MAT/GAM-SST/SI	1
PCDF82-1	−63.95	−56.36	430	KULL	Top	4	D	MAT/GAM-SST/SI	1
PCDF82-102	−64.31	−61.88	540	KULL	Top	4	D	MAT/GAM-SST/SI	1
PCDF82-167	−63.93	−56.61	448	KULL	Top	4	D	MAT/GAM-SST/SI	1
PCDF82-174	−64.17	−56.81	288	KULL	Top	4	D	MAT/GAM-SST/SI	1
PCDF82-197	−63.72	−57.23	750	KULL	Top	4	D	MAT/GAM-SST/SI	1
PCDF82-20	−64.23	−55.9	381	KULL	Top	4	D	MAT/GAM-SST/SI	1
PCDF82-34	−62.3	−57.62	1979	KULL	Top	4	D	MAT/GAM-SST/SI	1
PCDF82-35	−62.36	−57.37	1484	KULL	Top	4	D	MAT/GAM-SST/SI	1
PCDF82-47	−62.92	−58.4	723	KULL	Top	4	D	MAT/GAM-SST/SI	1
PCDF82-51	−63.72	−60.05	560	KULL	Top	4	D	MAT/GAM-SST/SI	1
PCDF82-60	−63.39	−59.57	673	KULL	Top	4	D	MAT/GAM-SST/SI	1
PCDF82-61	−63.28	−59.34	728	KULL	Top	4	D	MAT/GAM-SST/SI	1
PCDF82-69	−63	−59.63	916	KULL	Top	4	D	MAT/GAM-SST/SI	1

Table 1 (*continued*)

Core	Longitude	Latitude	Water depth (m)	Coring device	Sampling level	Strat. quality	Fossil group	Use	Ref.
PCDF82-71	−62.64	−59.54	1350	KULL	Top	4	D	MAT/GAM-SST/SI	1
PCDF82-93	−64.07	−61.33	690	KULL	Top	4	D	MAT/GAM-SST/SI	1
PS1195-2	−76.85	−50.49	257	BC	0–1	4	D	IKM-SST	2
PS1200-4	−76.53	−52.72	374	BC	0–1	4	D	IKM-SST	2
PS1208-1	−75.34	−58.79	628	BC	0–1	4	D	IKM-SST	2
PS1209-1	−75.54	−57.72	516	BC	0–1	4	D	IKM-SST	2
PS1214-1	−77.1	−48.61	241	BC	0–1	4	D	IKM-SST	2
PS1222-1	−75.86	−34.31	670	BC	0–1	4	D	IKM-SST	2
PS1223-1	−75.98	−33.55	754	BC	0–1	4	D	IKM-SST	2
PS1273-1	−75.16	−27.33	333	BC	0–1	4	D	IKM-SST	2
PS1277-1	−77.51	−43.19	447	BC	0–1	4	D	IKM-SST	2
PS1278-1	−77.54	−42.13	632	BC	0–1	4	D	IKM-SST	2
PS1366-2	−70.44	−8.42	380	BC	0–1	4	D	IKM-SST	2
PS1372-2	−72.21	−16.72	792	BC	0–1	4	D	IKM-SST	2
PS1374-2	−72.22	−16.93	1458	BC	0–1	4	D	IKM-SST	2
PS1380-1	−70.01	−9.97	2072	BC	0–1	4	R	IKM-SST	3
PS1384-1	−70.46	−9.62	704	BC	0–1	4	D	IKM-SST	2
PS1386-1	−68.33	−5.62	4405	BC	0–1	4	R	IKM-SST	3
PS1387-1	−68.73	−5.84	2435	BC	0–1	4	R	IKM-SST	3
PS1388-1	−69.03	−5.89	2521	BC	0–1	4	D	IKM-SST	2
PS1394-1	−70.08	−6.68	1948	BC	0–1	4	R	IKM-SST	3
PS1395-1	−70.22	−6.98	1489	BC	0–1	4	D	IKM-SST	2
PS1399-1	−76.82	−51.02	251	BC	0–1	4	D	IKM-SST	2
PS1400-4	−77.55	−36.4	1064	BC	0–1	4	D	IKM-SST	2
PS1401-2	−77.6	−35.9	691	BC	0–1	4	D	IKM-SST	2
PS1402-2	−77.48	−34.73	320	BC	0–1	4	D	IKM-SST	2
Ps1403-1	−76.89	−33.39	431	BC	0–1	4	D	IKM-SST	2
PS1407-1	−71.24	−13.57	421	BC	0–1	4	D	IKM-SST	2
PS1410-1	−71.19	−13.55	1511	BC	0–1	4	D	IKM-SST	2
PS1419-1	−74.67	−35.08	479	BC	0–1	4	D	IKM-SST	2
PS1424-1	−76.59	−49.78	286	BC	0–1	4	D	IKM-SST	2
PS1425-1	−70.35	−6.76	456	BC	0–1	4	D	IKM-SST	2
PS1427-1	−70.32	−6.84	612	BC	0–1	4	D	IKM-SST	2
PS1428-1	−70.28	−6.9	1165	BC	0–1	4	D	IKM-SST	2
PS1455-4	−65.42	1.83	2730	BC	0–1	4	R	IKM-SST	3
PS1472-4	−76.58	−30.54	258	BC	0–1	4	D	IKM-SST	2
PS1485-1	−72.56	−18.78	2075	BC	0–1	4	D	IKM-SST	2
PS1486-2	−73.4	−23.09	2572	BC	0–1	4	D	IKM-SST	2
PS1649-1	−54.91	3.29	2446	BC	0–1	4	D	IKM-SST	2
PS1651-2	−53.64	3.84	2089	BC	0–1	4	D	IKM-SST	2
PS1652-1	−53.67	5.08	1960	BC	0–1	4	D	IKM-SST	2
PS1654-1	−50.16	5.77	3763	BC	0–1	2	D	IKM-SST	2
PS1751-2	−44.5	10.48	4802	MC	0–0.5	4	D/R	IKM-SST	2,3
PS1752-5	−45.62	9.61	4553	MC	0–0.5	4	D/R	IKM-SST	2,3
PS1755-1	−47.79	7.1	4321	MC	0–0.5	4	D/R	IKM-SST	2,3
PS1759-1	−50.15	5.76	3793	MC	0–0.5	4	D/R	IKM-SST	2,3
PS1764-2	−50.87	5.71	3936	MC	0–0.5	4	D	IKM-SST	2
PS1765-1	−51.83	4.86	3812	MC	0–0.5	4	D/R	IKM-SST	2,3
PS1768-1	−52.59	4.45	3331	MC	0–0.5	1	D/R	IKM-SST	2,3
PS1771-4	−53.76	3.78	1811	MC	0–0.5	4	R	IKM-SST	3
PS1772-6	−55.46	1.17	4140	MC	0–0.5	4	D/R	IKM-SST	2,3
PS1773-2	−56.32	−0.48	3259	MC	0–0.5	4	D/R	IKM-SST	2,3
PS1774-1	−54.65	−2.87	2453	MC	0–0.5	4	D/R	IKM-SST	2,3
PS1775-5	−50.95	−7.5	2523	MC	0–0.5	4	D/R	IKM-SST	2,3
PS1776-6	−49.73	−8.77	3155	MC	0–0.5	4	D/R	IKM-SST	2,3
PS1777-7	−48.23	−11.03	2575	MC	0–0.5	4	D/R	IKM-SST	2,3
PS1778-1	−49.01	−12.7	3384	MC	0–0.5	4	D/R	IKM-SST	2,3
PS1779-3	−50.4	−14.08	3574	MC	0–0.5	4	D/R	IKM-SST	2,3
PS1780-1	−51.68	−15.27	4258	MC	0–0.5	4	R	IKM-SST	3
PS1782-6	−55.19	−18.6	5131	MC	0–0.5	4	D/R	IKM-SST	2,3
PS1783-2	−54.91	−22.72	3390	MC	0–0.5	4	R	IKM-SST	3
PS1786-2	−54.93	−31.74	5771	MC	0–0.5	1	D/R	IKM-SST	2,3
PS1794-2	−73.54	−25.91	3381	MC	0–0.5	4	D	IKM-SST	2

Table 1 (*continued*)

Core	Longitude	Latitude	Water depth (m)	Coring device	Sampling level	Strat. quality	Fossil group	Use	Ref.
PS1805-5	−66.19	35.31	4149	MC	0–0.5	4	R	IKM-SST	3
PS1813-3	−64.96	33.63	2225	MC	0–0.5	4	R	IKM-SST	3
PS1821-5	−67.07	37.48	4028	MC	0–0.5	4	R	IKM-SST	3
PS1823-1	−65.93	30.84	4442	MC	0–0.5	4	R	IKM-SST	3
PS1825-5	−66.33	8.89	4341	BC	0–1	4	R	IKM-SST	3
PS1831-5	−65.74	13.66	2354	BC	0–1	4	R	IKM-SST	3
PS1957-1	−65.67	−37.48	4727	MC	0–0.5	4	R	IKM-SST	3
PS1967-1	−65.96	−30.07	4847	MC	0–0.5	4	R	IKM-SST	3
PS1973-1	−66.89	−25.55	4841	MC	0–0.5	4	R	IKM-SST	3
PS1975-1	−67.51	−22.52	4893	MC	0–0.5	4	R	IKM-SST	3
PS1977-1	−68.28	−19.34	4838	MC	0–0.5	4	R	IKM-SST	3
PS1979-1	−69.37	−16.5	4735	MC	0–0.5	4	R	IKM-SST	3
PS2073-1	−39.59	14.57	4692	MC	0–0.5	4	R	IKM-SST	3
PS2076-1	−41.14	13.48	2086	MC	0–0.5	4	D/R	IKM-SST	2,3
PS2080-1	−41.72	13.05	5078	MC	0–0.5	4	D/R	IKM-SST	2,3
PS2081-1	−42.69	12.19	4794	MC	0–0.5	4	D/R	IKM-SST	2,3
PS2082-3	−43.22	11.76	4661	MC	0–0.5	4	D/R	IKM-SST	2,3
PS2083-2	−46.37	7.04	1955	MC	0–0.5	4	R	IKM-SST	3
PS2084-2	−47.02	7.96	1664	MC	0–0.5	4	R	IKM-SST	3
PS2087-1	−49.13	6.71	3451	MC	0–0.5	4	D/R	IKM-SST	2,3
PS2102-1	−53.08	−5	2388	MC	0–0.5	1	D	IKM-SST	2
PS2103-2	−51.33	−3.32	2947	MC	0–0.5	4	R	IKM-SST	3
PS2104-1	−50.74	−3.21	2592	MC	0–0.5	4	D/R	IKM-SST	2,3
PS2105-2	−48.69	−2.85	3618	MC	0–0.5	4	D/R	IKM-SST	2,3
PS2108-1	−39.84	1.03	4920	MC	0–0.5	4	D	IKM-SST	2
PS2109-3	−35	3.17	5041	MC	0–0.5	4	R	IKM-SST	3
PS2254-1	−43.97	−50.07	5341	MC	0–0.5	4	R	IKM-SST	4
PS2256-4	−44.51	−44.47	5111	MC	0–0.5	4	R	IKM-SST	4
PS2270-5	−50.88	−32.32	4273	MC	0–0.5	4	D	IKM-SST	2
PS2299-1	−57.51	−30.23	3375	MC	0–0.5	4	D	IKM-SST	2
PS2307-1	−59.06	−35.58	2527	MC	0–0.5	4	D	IKM-SST	2
PS2487-2	−35.83	18.11	2942	MC	0–0.5	4	R	IKM-SST	4
PS2488-1	−38.56	15.8	4888	MC	0–0.5	4	R	IKM-SST	4
PS2489-4	−42.89	8.98	3795	MC	0–0.5	4	R	IKM-SST	4
PS2491-4	−44.96	5.97	4323	MC	0–0.5	4	D/R	IKM-SST	2,4
PS2492-1	−43.18	−4.05	4197	MC	0–0.5	4	D/R	IKM-SST	2,4
PS2493-3	−42.89	−6.02	4174	MC	0–0.5	4	D/R	IKM-SST	2,4
PS2494-1	−41.69	−12.34	3324	MC	0–0.5	4	D/R	IKM-SST	2,4
PS2495-1	−41.29	−14.5	3135	MC	0–0.5	2	D/R	IKM-SST	2,4
PS2496-2	−42.99	−14.64	3518	MC	0–0.5	4	D/R	IKM-SST	2,4
PS2498-2	−44.15	−14.23	3782	MC	0–0.5	3	D/R	IKM-SST	2,4
PS2499-1	−46.51	−15.33	3176	MC	0–0.5	3	D	IKM-SST	2
PS2501-4	−49.4	−21.39	4043	MC	0–0.5	4	D	IKM-SST	2
PS2502-1	−50.25	−23.24	4462	MC	0–0.5	4	D	IKM-SST	2
PS2503-1	−50.75	−24.32	4473	MC	0–0.5	4	D	IKM-SST	2
PS2504-1	−50.84	−24.51	4765	MC	0–0.5	4	D	IKM-SST	2
PS2505-1	−51.19	−25.47	1864	MC	0–0.5	4	D	IKM-SST	2
PS2506-1	−51.41	−25.7	2990	MC	0–1	4	D	IKM-SST	2
PS2507-1	−51.37	−26.2	3275	MC	0–1	4	D	IKM-SST	2
PS2508-1	−51.67	−26.53	3394	MC	0–0.5	4	D	IKM-SST	2
PS2509-1	−52.07	−26.89	4454	MC	0–0.5	4	D	IKM-SST	2
PS2511-1	−55.33	−30.4	2888	MC	0–0.5	4	D	IKM-SST	2
PS2512-1	−54.4	−33.63	4803	MC	0–0.5	4	D	IKM-SST	2
PS2557-2	−36.92	21.83	3371	MC	0–0.5	4	R	IKM-SST	4
PS2560-3	−40.54	25.57	2641	MC	0–0.5	4	R	IKM-SST	4
PS2561-1	−41.86	28.55	4471	MC	0–0.5	4	R	IKM-SST	4
PS2562-1	−43.18	31.58	5193	MC	0–0.5	4	R	IKM-SST	4
PS2563-3	−44.55	34.78	3515	MC	0–0.5	4	R	IKM-SST	4
PS2564-2	−46.14	35.9	3035	MC	0–0.5	4	R	IKM-SST	4
PS2566-1	−48.25	37.49	4422	MC	0–0.5	4	D	IKM-SST	2
PS2602-3	−60.38	36.58	5293	MC	0–0.5	4	D	IKM-SST	2
PS2604-4	−57.6	38.59	5083	MC	0–0.5	4	D	IKM-SST	2
PS2605-1	−54.66	39.92	2996	MC	0–0.5	4	D	IKM-SST	2

Table 1 (continued)

Core	Longitude	Latitude	Water depth (m)	Coring device	Sampling level	Strat. quality	Fossil group	Use	Ref.
PS2606-1	−53.24	40.87	2552	MC	0–0.5	1	D	IKM-SST	2
PS2607-1	−51.89	41.52	2859	MC	0–0.5	4	D	IKM-SST	2
PS2609-2	−51.5	41.6	3116	MC	0–0.5	4	D	IKM-SST	2
PS2610-1	−50.68	40.12	3579	MC	0–0.5	4	D	IKM-SST	2
PS2611-3	−49.51	38.83	4261	MC	0–0.5˙	4	D	IKM-SST	2
RC11-118	−37.8	71.53	4354	KULL	Top	4	D	MAT/GAM-SST/SI	1
RC11-119	−40.3	74.57	3709	KULL	Top	4	D	MAT/GAM-SST/SI	1
RC11-77	−53.05	−16.45	4098	KULL	Top	4	D	MAT/GAM-SST/SI	1
RC11-79	−49	−4.6	3100	KULL	2-3	5	D	MAT/GAM-SST/SI	1
RC11-80	−46.75	−0.05	3656	KULL	2-3	5	D	MAT/GAM-SST/SI	1
RC11-90	−56.63	25.72	5334	KULL	2-3	5	D	MAT/GAM-SST/SI	1
RC11-91	−56.57	34.18	5150	KULL	3-4	5	D	MAT/GAM-SST/SI	1
RC11-95	−52.8	54.08	3150	KULL	2-3	5	D	MAT/GAM-SST/SI	1
RC11-98	−47.65	61.48	4650	KULL	Top	4	D	MAT/GAM-SST/SI	1
RC12-292	−39.69	−15.47	3541	KULL	Top	4	D	MAT/GAM-SST/SI	1
RC13-263	−53.81	−8.22	3389	KULL	Top	4	D	MAT/GAM-SST/SI	1
RC15-91	−49.92	−15.57	3775	KULL	3-4	5	D	MAT/GAM-SST/SI	1
RC8-40	−43.75	46.08	2250	KULL	Top	4	D	MAT/GAM-SST/SI	1
RC8-46	−55.33	65.47	2761	KULL	Top	4	D	MAT/GAM-SST/SI	1
SO136-111	−56.66	160.23	3910	KULL	Top	2	D	MAT/GAM-SST/SI	1
SO136-BX043	−50.15	174.67	956	BC	Top	4	D	MAT/GAM-SST/SI	1
SO136-BX068	−54.08	168.5	981	BC	Top	4	D	MAT/GAM-SST/SI	1
SO136-BX110	−56.69	160.25	3907	BC	Top	4	D	MAT/GAM-SST/SI	1
SO136-BX116	−55.66	159.42	4462	BC	Top	4	D	MAT/GAM-SST/SI	1
TNO57-13-PC4	−53.2	5.1	2851	KULL	Top	1	D	MAT/GAM-SST/SI	1
V14-53	−56.72	−24.52	7906	KULL	Top	4	D	MAT/GAM-SST/SI	1
V16-60	−49.99	36.76	4575	KULL	Top	4	D	MAT/GAM-SST/SI	1
V29-87	−49.57	30.02	4550	KULL	Top	4	D	MAT/GAM-SST/SI	1

Position is given in decimals. Coring devices include BC = box corer, GRAV = gravity corer, KULL = Kullenberg piston corer, MC = multicorer or minicorer, TRIG = trigger corer. Sampling level in cm below sea floor, Top indicates sampling of topmost sediment sequence. Stratigraphic quality according to MARGO-defined late Holocene chronozone quality level (1–5). Use of reference samples indicates statistical method (IKM, MAT, GAM) and reconstructed parameter sea surface temperature (SST) and sea ice (SI). Source of reference sample, (1) D204/31 reference set of Crosta et al. (2004), emended from Crosta et al. (1998a, b), (2) Zielinski et al., 1998, (3) Abelmann et al. (1999), (4) Cortese and Abelmann (2002).

Table 2
Summary of statistical methods and equations used for generation of SST and sea ice estimates presented in this paper

Fossil group	Method	Data set/equation	Summer SST see (°C)	Sea ice see (month/year)	Sea ice conc. see (%)	Reference
Diatoms	MAT	D204/31	0.85	0.53	4/5	Crosta et al. (2004)
Diatoms	MAT	D201/25	0.97	1.05		Armand et al. (unpublished data)
Diatoms	IKM	D93/29lg/3	0.66	—	—	Zielinski et al. (1998)
Radiolaria	IKM	R53/23/4	1.2	—	—	Abelmann et al. (1999)
Radiolaria	IKM	R73/24lg/4	1.16	—	—	Cortese and Abelmann (2002)

SEE: standard error of estimates of the used diatom and radiolarian reference data sets. SEE of sea ice concentration indicates error of summer (February)/winter (September) estimate. Data set and equation designations indicate fossil group (D = diatoms, R = radiolarians)/number of reference samples/number of taxa or taxa groups/number of IKM factors, lg indicates logarithmic conversion of species abundance data used to compensate the dominance of single taxa.

paleoceanographic equations (for a summary see Table 2) are presented in Zielinski et al. (1998).

The Southern Ocean radiolarian data set, originally presented by Abelmann et al. (1999), was augmented by Cortese and Abelmann (2002) by extension into the southern subtropical area. This increased the sample/taxa number from 53/23 to 73/24 and provides a reference data set covering a SST range from the Antarctic cold to the subtropical warm water regime in the Atlantic sector (Fig. 2, Table 2). In contrast to previous southern latitude radiolarian data sets, used by CLIMAP (1976, 1981), the new sets of data only consider surface-dwelling radiolarian taxa that show a clear relationship to the surface water distribution pattern. This provides unrestricted comparison of radiolarian and diatom-based SST estimates, both reflecting conditions in the euphotic ocean mixed layer.

The MAT of Hutson (1980) compares the floral assemblage from each down-core sample to a sub-set of modern floral core-top analogs. It calculates a dissimilarity coefficient, which measures the difference between the assemblage of the down-core sample and the assemblage of the analog. Calculation of the dissimilarity coefficient is based on the squared chord distance (Prell, 1985). The estimate is then a simple average of the modern values associated with the analogs chosen by the MAT, and is assumed to represent the climate at the core locality where the fossils of the down-core sample were produced. Down-core estimates are generally calculated on five analogs, except when the dissimilarity threshold of 0.25 is crossed. Below three analogs, the MAT program was set not to provide an estimate, and the fossil assemblage is considered to have no modern equivalent.

MAT was used for E-LGM SST and sea ice reconstruction at 73 locations from all sectors of the Southern Ocean based on a comprehensive diatom reference data set, including a total of 204 samples and 31 taxa (D204/31; Tables 1 and 2). The reference sample locations cover the Atlantic, Indian and westernmost Pacific sector of the Southern Ocean (Fig. 2). Although the samples have been recovered using a large variety of coring systems, ranging from MC to piston coring devices (e.g. Kullenberg corer), the majority are within the MARGO quality levels 1–4 (Kucera et al., 2004) ensuring that the surface sample is derived from a sediment interval not older than 4 ka (Table 1). The reference data set D204/31 has been developed by Crosta et al. (2004) from a database including 195 surface samples, originally prepared by Pichon et al. (1992a, b) and later revised by Crosta et al. (1998a, b). The latter revision considered (i) calibration of the diatom taxonomy approach among diatom paleoceanographers involved in the present study at AWI, Bordeaux and Hobard, (ii) the exclusion of artificially dissolved samples, and (iii) the composition of the considered diatom taxa.

The IKM and MAT calculations have been accomplished using the PaleoToolBox software package developed by Sieger et al. (1999).

Only recently Armand et al. (unpublished material) proposed the GAM (Hastie and Tibshirani, 1990) as a new statistical technique for the estimation of past environment, e.g. sea ice concentration. GAM is used to model a non-linear response between the factors derived from a Q-mode factor analysis and their mean response. Backward elimination procedures assist in selecting appropriate factors and quadratic terms for the estimation equations. Predictions are made from the final bootstrapped model at the 95% confidence level. Here we present GAM-derived E-LGM sea ice concentration estimates obtained from two cores located in the eastern Indian sector between Tasmania and Antarctica using a

diatom reference data set that includes 201 surface reference samples and 25 taxa (D201/25) presented by Armand et al. (unpublished data). D201/25 has been created based on the data compiled in the D204/31 Crosta data set. In D201/25, species in the genera *Rhizosolenia* and *Thalassionema* have been excluded, considering that the majority of these taxa are not related to the sea ice environment (Moreno-Ruiz and Licea, 1995; Zielinski and Gersonde, 1997; Crosta et al., 1998a; Armand and Zielinski, 2001).

The SSTs of the hydrographic reference data sets used for diatom and radiolarian-based estimations represent values measured at 10 m below sea surface and were retrieved from Olbers et al. (1992) and Conkright et al. (1998), representing data of the World Ocean Atlas (WOA) (Table 1). Extraction of data was partly achieved using the software available on MARGO web site (http://www.pangaea.de/Projects/MARGO). The temperature values are computed as the area-weighted average of the four temperature values surrounding the sample location. Given that the biogenic particle flux to the sea floor in the Southern Ocean is restricted to austral summer, also in areas unaffected by ice cover (Abelmann and Gersonde, 1991; Gersonde and Zielinski, 2000; Fischer et al., 2002), only summer (January–March average) SST have been estimated.

For MAT and GAM derived sea ice reconstruction, a 13.25-years series (1978–1991) of monthly sea ice concentration averages (Schweitzer, 1995) is employed as the data set of sea ice concentrations and annual duration of sea ice at the specified locations for all the core top samples. The monthly averaged sea ice data set contains information derived from the SMMR and SSM/I satellite instruments and allows the user to specify the locations of retrieval. The data for the Antarctic region uses the 'Total sea ice NASA TEAM algorithm' to compile the information of the CD data set. Thus, the data set here represents the time-averaged probability of finding sea ice at a given location and its corresponding typical monthly averaged sea ice concentration. Sea ice concentration is the percentage of a given area of ocean that is covered by sea ice; it represents the amount of sea ice versus open water (Zwally et al., 1983). The monthly average of modern sea ice concentration was extracted for every sample location of our modern data sets for February and September, these months being representative of the minimum and maximum seasonal extent within the annual sea ice cycle, respectively (Comiso, 2003). Sea ice duration in number of months per year at the core locations was calculated from the sea ice concentration data. For MAT-derived sea ice estimates based on the D204/31 data set of Crosta, a sea ice concentration > 40% was selected as a threshold to determine the presence or absence of sea ice during a month. Sea ice concentrations within the 40–50% isopleths correlate well with the compact ice edge location (Gloersen et al., 1992). For each month, the presence (noted 1) or absence (noted 0) was determined based on the concentration threshold, and the yearly sea ice duration was calculated by summing the monthly presence. When monthly sea ice concentration was between 30% and 40%, a 0.5-month duration is reported.

Under the GAM technique (Armand et al., unpublished material), the 15% sea ice concentration threshold was chosen to determine the presence or absence of monthly sea ice cover with respect to the limit of the unconsolidated outer sea ice edge as employed by the sea ice community (Gloersen et al., 1992).

We also include estimates of sea ice extent derived from the abundance pattern of sea ice indicator diatoms (SI-Ind.), as presented by Gersonde et al. (2003a) for the Atlantic sector of the Southern Ocean. The method, proposed by Gersonde and Zielinski (2000), has been developed from combined sediment trap, surface sediment and down-core studies in the Atlantic sector. It considers the relative abundance of the diatom species *Fragilariopsis curta* and *F. cylindrus* (combined into the *F. curta/cylindrus* group) higher than 3% of the total assemblage to represent a qualitative threshold between the average presence of winter (September) sea ice and year-round open waters documented from a 9 years sea ice observation time series (Naval Oceanography Command Detachment, 1985). The average WSI edge corresponds with a mean sea ice concentration of 50–80%. Sea ice indicator values between 3% and 1% of the total diatom assemblage are considered to monitor the maximum winter (September) sea ice extension (mean concentration $<20\%$). The proximity of the summer sea ice limit was deduced through the enhanced presence of *Fragilariopsis obliquecostata*, a taxon restricted to very cold waters ($<-1\,^{\circ}$C) (Zielinski and Gersonde, 1997). *F. obliquecostata* is relatively thickly silicified and thus insignificantly affected by opal dissolution. It thus remains a valuable tracer of sea ice cover even in conditions of low sedimentation and enhanced opal dissolution that are typical in areas close to the perennial ice edge (Gersonde and Zielinski, 2000).

Only recently, Curran et al. (2003) concluded that the Antarctic sea ice cover decreased by about 20% after 1950. This was based on a study of the methanesulfonic acid (MSA) records obtained from a coastal Antarctic ice core (Law Dome). Such findings would also imply the warming of the sea surface, therefore challenging the use of hydrographic and sea ice reference data sets for paleoenvironmental reconstruction, which rely on observations obtained during the past 25 years. In spite of this, the conclusions of Curran et al. (2003) require further support from other Antarctic ice core records, since the use of MSA as an indicator of sea ice extent has been questioned by other authors (Wolff, 2003). Future

paleoceanographic reconstruction studies should nonetheless keep in mind a possible mismatch between recent environment information and surface sediment reference data sets that may integrate environmental conditions over a period of more than the past 50 years.

2.3. Quality control of statistically derived estimates

For IKM derived SST estimates based on diatom and radiolarian assemblages, three estimate quality levels (EQL) have been defined using the communality value obtained for the down-core samples. The communality describes the amount of variance accounted for by the factors. Additionally, we considered the observation (Gersonde et al., 2003a) that diatom estimates obtained from Subantarctic and warmer core locations may be biased towards colder temperatures (for further details see Section 3). Consequently, EQLs of diatom estimates obtained from warmer water locations ($>9\,^{\circ}$C modern SSST, $>5\,^{\circ}$C E-LGM SSST), where both diatom and radiolarian estimates are available, have been downgraded to ensure that radiolarian estimates will be used preferentially at these locations. The observation of Gersonde et al. (2003a) indicates that other diatom estimates from Subantarctic and warmer sites that lack radiolarian estimates should also be treated with caution. In case of the occurrence of no-analog samples the EQL has been downgraded to 3 when no-analogs represented the majority of the samples ranging in the E-LGM time slice at the individual core locations.

Estimate quality level 1: communality >0.8. For diatom estimates only: SST difference between radiolarian and diatom based estimate $<1.5\,^{\circ}$C.

Estimate quality level 2: communality 0.7–0.8. For diatom estimates only: SST difference between radiolarian and diatom based estimate $<1.5\,^{\circ}$C.

Estimate quality level 3: communality <0.7. For diatom estimates only: SST difference between radiolarian and diatom based estimate $>1.5\,^{\circ}$C.

For MAT derived estimates of SST and sea ice extent based on the diatom record, three EQLs have been defined taking into account the dissimilarity index. This index indicates the distance between the down-core and the reference surface sediment samples (a zero value indicates down-core and reference samples are identical, whereas a value of 1 indicates total dissimilarity between the down-core and the reference sample).

Estimate quality level 1: dissimilarity <0.1

Estimate quality level 2: dissimilarity 0.1–0.2

Estimate quality level 3: dissimilarity 0.2–0.25

Samples with dissimilarity values above 0.25 indicate no-analog situations and have been discarded.

As the GAM technique employs a Q-mode factor analysis prior to its non-linear GAM regression, we are able to assign EQLs to the determined communalities undertaken in factor analysis as defined above. All determined IKM communalities for the E-LGM data upon which the GAM estimates are based fall into the EQL 1 definition, as they are greater than 0.8 in value (Armand et al., unpublished data).

2.4. Age assignment and stratigraphic quality levels

The establishment of accurate stratigraphic age models for late Pleistocene and Holocene sediments deposited south of the Subantarctic Zone (SAZ) is complicated by the scarcity or lack of biogenic carbonate, especially during glacial intervals, and hence by the lack of continuous benthic and planktic foraminiferal stable isotope records that can be correlated with the standard isotope stratigraphic records. In addition, the lack of sufficient foraminifers makes it difficult to obtain AMS ^{14}C datable samples from foraminiferal shell materials. As a consequence, most of the surface reference samples fall into the MARGO-defined Late Holocene Chronozone Quality Level (CQL) 4 (Table 1) (for definition of quality levels see Kucera et al., 2004). Of those, most have been taken with MC and box corer (BC) devices from areas not affected by winnowing or slumping, and thus represent undisturbed surface sediments. Only few samples from the D204/33 reference set prepared by Crosta have been taken up to a few centimeters below the sea floor surface. Such samples were placed in the Late Holocene CQL 5 (Table 1). However, the statistical treatment of all samples shows that the assemblages considered as reference of modern conditions are clearly related with the modern SST and sea ice distribution (see Crosta et al., 1998a, b; Zielinski et al., 1998; Abelmann et al., 1999; Cortese and Abelmann, 2002).

To identify the EPILOG time-slices as accurately as possible considerable effort has been made to calibrate the abundance fluctuations of siliceous microfossils, such as the radiolarian C. davisiana and the diatom Eucampia antarctica, with benthic and planktic oxygen isotope records and AMS ^{14}C measurements of organic carbon extracted from planktic foraminifers, or from the humic acid fraction in diatomaceous ooze samples that did not allow for extraction of sufficient amounts of foraminiferal carbonate (Gersonde et al., 2003a). Comparison of AMS ^{14}C dates obtained from organic carbon extracted from planktic foraminifers and the humic acid fraction of the bulk sediment (Bianchi and Gersonde, unpublished data) demonstrates the applicability of the humic acid fraction for ^{14}C dating of diatomaceous ooze from latest Pleistocene and Holocene sediment cores recovered in the Southern Ocean. This places all E-LGM values from PS cores presented by Gersonde et al. (2003a) in the MARGO-defined LGM Chronozone Level 3 or better (Tables 3 and 4; Fig. 3A). Ample stratigraphic accuracy has also been obtained for a number of R.V. Marion Dufresne (MD),

R.V. Robert Conrad (RC) and R.V. Sonne (SO) cores, based on AMS ^{14}C measurements and oxygen isotope records. All E-LGM values that have been obtained from a single sample taken at the core depth level defined to represent the LGM by CLIMAP (1976, 1981) have been downgraded to the LGM Chronozone Level 4. The CLIMAP (1976, 1981) age assignment of these cores has mainly been based on the abundance pattern of C. davisiana. In contrast, we only took into account such samples in the range of the C. davisiana abundance pattern that are assigned to the E-LGM based on AMS ^{14}C dating and isotope records. Lack of attention towards the single sample-based values would not permit the generation of a circumantarctic E-LGM reconstruction, but would provide a restricted reconstruction centered primarily in the Atlantic sector of the Southern Ocean (Fig. 3B).

2.5. Definition of average quality levels

To provide general quality information on each E-LGM value we combined both the stratigraphic and the estimate quality value. This resulted in the definition of two average quality levels (AQL). AQL 1 includes E-LGM values with an E-LGM CQL and an EQLs 1 or 2. All other combinations are in AQL 2 (Tables 3 and 4, Fig. 3C).

3. Results

Southern Ocean summer SST and WSI distribution at the E-LGM have been reconstructed at a total of 122 core locations (Fig. 2). This includes diatom-based reconstructions from 104 locations (Table 3) and radiolarian-based reconstructions from 19 locations (Table 4). Highest spatial coverage of investigated core locations has been obtained in the Atlantic—western Indian sector between 30°W and 45°E, and in the eastern Indian sector between 90 and 150°E. In the Pacific sector the coverage is poor, except in a narrow segment around 110°W. Most of the investigated E-LGM sections are from cores recovered between the WSI edge and the Subtropical Front. Only a few cores, located in the western Atlantic and in the eastern Indian sectors, have been collected from the seasonal sea ice covered zone of the Southern Ocean. The reasons for the small number of cores from this zone that could be considered for E-LGM reconstruction are (i) the widespread lack of well-preserved siliceous microfossil assemblages documenting LGM conditions in the present sea ice covered areas, and (ii) strongly reduced glacial sedimentation rates, which preclude accurate definition of the E-LGM level. In glacial sediment sequences deposited close to, or north of the modern Subtropical Front diatom assemblages are affected by

Table 3
Summary of averaged southern summer SST (SSST) and sea ice (SI) estimates from the Epilog Last Glacial Maximum (E-LGM) time slice derived from diatom assemblage information

Core	Lat.	Long.	Depth (m)	Modern SSST (°C)	E-LGM SSST (°C)	Delta LGM/mod. SSST (°C)	Modern SI pres. (m/yr)	E-LGM SI pres. (m/yr)	Delta LGM/mod. SI (m/yr)	Modern Feb. SI conc. (%)	E-LGM Feb. SI conc. (%)	Delta LGM/mod. Feb. SI conc. (%)	Modern Sept. SI conc. (%)	E-LGM Sept. SI conc. (%)	Delta LGM/mod. Sept. SI conc. (%)	E-LGM SI ind F.c + F.c (%)	E-LGM SI ind F.O. (%)	E-LGM SI pres.	Estimation method	E-LGM CQL	EQL	AQL	Ref.
ELT11-1	-54.91	-114.7	3475	6.37	5.7	-0.7	0	0	0	0	0	0	0	2.8	2.8	0	0	0	MAT	4	1	2	1
ELT11-2	-56.06	-115.06	3109	5.54	5.5	0.0	0	0	0	0	0	0	0	2.6	2.6	0	0	0	MAT	4	1	2	1
ELT11-3	-56.9	-115.24	4023	4.99	3.4	-1.6	0	0.4	0.4	0	0	0	0	12	12	0	0.3	1	MAT	4	1	2	1
ELT11-4	-57.83	-115.21	4773	4.46	4.2	-0.3	0	1.1	1.1	0	0	0	0	20	20	0	0	1	MAT	4	1	2	1
ELT14-6	-57.02	-160.09	4520	5.57	3.4	-2.2	0	1	1	0	0	0	0	22.4	22.4	0	0	1	MAT	4	1	2	1
ELT15-12	-58.68	-108.8	4572	4.87	4.0	-0.9	0	0	0	0	0	0	0	2.6	2.6	0	0	0	MAT	4	1	2	1
ELT15-4	-59.02	-99.76	4910	5.45	4.7	-0.7	0	0.4	0.4	0	0	0	0	2.6	2.6	0	0	1	MAT	4	1	2	1
ELT15-6	-59.97	-101.32	4517	4.63	4.8	0.2	0.5	2.8	2.3	0	0	0	0	9.2	9.2	0	0	1	MAT	4	1	2	1
ELT17-9	-63.08	-135.12	4848	1.16	0.9	-0.3	0	1.9	1.9	0	0	0	28	49.2	21.2	0	0	1	MAT	2	1	1	1
ELT19-7	-62.16	-109.09	5051	3.01	1.6	-1.4	0	0	0	0	0	0	27	34	7	0	0.9	1	MAT	4	1	2	1
ELT20-10	-60.22	-127.03	4471	3.68	2.7	-1.0	0	0	0	0	0	0	0	7	7	0	0	1	MAT	4	1	2	1
ELT35-15	-52.94	116.99	3764	7.49	3.9	-3.6	0	2.5	2.5	0	0	0	0	2.6	2.6	0	0	1	MAT	4	1	2	1
ELT36-36	-60.39	157.53	2816	2.79	1.5	-1.3	0	0	0	0	0	0	0	40	40	2.8	1.9	1	MAT	4	1	1	1
ELT39-13	-45.01	125.98	4535	12.30	5.0	-7.3	0	0	0	0	0	0	0	0	0	0	0	0	MAT	4	1	2	1
ELT39-18	-48.03	126.13	4612	10.56	4.3	-6.3	0	0	0	0	0	0	0	0	0	0	0	0	MAT	4	1	2	1
ELT39-21	-48.86	126.02	4078	9.93	3.5	-6.4	0	0	0	0	0	0	0	0	0	0	0.3	1	MAT	3	1	1	1
ELT45-29	-44.88	106.52	3819	10.39	5.6	-4.8	0	1.1	1.1	0	0	0	0	22.8	22.8	0	0	1	MAT	4	1	1	1
ELT45-35	-53.5	111.33	3840	4.51	2.9	-1.7	0	2	2	0	0	0	0	37.2	37.2	0	0	1	MAT	4	1	1	1
ELT45-63	-53.44	114.26	3917	5.04	1.5	-3.5	0	0.9	0.9	0	0	0	0	22.6	22.6	0	0.3	1	MAT	4	1	1	1
ELT45-64	-52.48	114.09	3822	5.37	1.7	-3.7	0	0	0	0	0	0	0	0	0	0.3	0.3	1	MAT	4	1	1	1
ELT45-79	-45.06	114.37	4097	10.58	5.2	-5.4	0	0	0	0	0	0	0	0	0	0.3	0.3	1	MAT	4	1	1	1
ELT48-27	-38.54	79.87	3277	17.07	16.3	-0.8	0	0	0	0	0	0	0	0	0	0	0	0	MAT	4	3	3	1
ELT49-6	-51.01	109.99	3325	5.44	1.6	-3.8	0	1.7	1.7	0	0	0	0	33.4	33.4	0.3	0.3	1	MAT	4	1	1	1
ELT49-7	-53.04	110.05	3590	4.45	1.4	-3.1	0	2.7	2.7	0	0	0	0	45.2	45.2	0.3	0.3	1	MAT	4	1	1	1
ELT49-8	-55.07	110.02	3692	3.89	1.1	-2.8	0	2.7	2.7	0	0	0	0	45.2	45.2	0	0	0	MAT	4	1	1	1
ELT50-11	-55.95	104.95	3921	3.30	1.7	-1.6	0	1.8	1.8	0	0	0	0	34.6	34.6	0.9	0.3	1	MAT	4	1	1	1
IO1277-10	-52.02	20.47	2740	2.19	1.4	-0.8	0	3.02	3.02	0	0	0	0	37.6	37.6	1.4	2.1	1	MAT	4	1	1	1
IO1578-4	-59.23	-19.73	4217	0.58	0.5	0.0	5	4.83	-0.17	0	1	1	64	62.33	-1.67	0.4	0.8	1	MAT	3	3	3	1
KR88-22	-64.67	119.51	3140	0.59	0.4	-0.2	7	7	0	0	1	1	73	81.4	8.4	1.3	8.2	1	MAT	3	1	1	1
KR88-27	-63.65	101.15	1210	0.03	0.4	0.4	8	7	-1	0	1	1	88	81.4	-6.6	2.2	9.3	1	MAT	3	1	1	1
KR88-29	-62.49	95.89	3790	0.87	3.0	-0.5	6	6.7	0.7	0	1	1	83	76.8	-6.2	1.9	3.4	1	MAT	1	1	1	1
MD24-KK63	-51.91	42.88	2550	3.32	0.9	-2.5	0	4.8	4.8	0	0.4	0.4	0	55.2	55.2	3.7	1.6	1	MAT	2	2	1	1
MD73-026	-44.98	53.28	3429	8.99	7.7	-1.3	0	0	0	0	0.8	0.8	0	0	0	0	0	0	MAT	2	2	1	1
MD80-304	-51.06	67.72	1950	4.29	2.1	-2.2	0	2	2	0	0.4	0.4	0	32	32	1.3	1.6	1	MAT	2	2	1	1
MD82-424	-54.09	-0.35	2350	1.43	0.6	-0.8	0	5.4	5.4	0	0.4	0.4	5	61.6	56.6	5.2	3.9	1	MAT	2	3	1	1
MD82-434	-58.85	51.32	3640	0.62	0.7	0.1	5	4.8	-0.2	0	0.8	0.8	64	62.2	-1.8	0	1.3	1	MAT	3	3	1	1
MD84-527	-43.49	51.32	3262	10.51	8.2	-2.3	0	0	0	0	1	1	0	0	0	0	0	0	MAT	4	1	1	1
MD84-529	-48.89	61.66	2600	5.64	3.0	-2.7	0	0	0	0	0	0	0	2.6	2.6	0	0	1	MAT	1	2	1	1
MD84-551	-55	73.26	1504	2.49	0.9	-1.6	0	4.3	4.3	0	0.4	0.4	0	49.6	49.6	2.2	0	1	MAT	1	1	1	1
MD84-552	-54.91	73.84	1780	2.42	1.1	-1.3	0	1.9	1.9	0	0.4	0.4	0	34.8	34.8	0.9	0	1	MAT	2	2	2	1
MD88-769	-46.06	90.09	3420	9.11	7.2	-1.9	0	0	0	0	0	0	0	0	0	0	0	0	MAT	2	2	1	1
MD88-770	-46.02	96.43	3290	8.63	5.7	-3.0	0	0	0	0	0	0	0	0	0	0	0	0	MAT	2	2	1	1
MD88-773	-52.9	109.85	2460	4.47	2.3	-2.2	0	1.37	1.37	0	0	0	0	23.72	23.72	0	0	1	MAT	2	1	1	2
MD88-784	-54.19	144.79	2800	5.60	3.5*	-2.1	0	1.8*/1.2**	1.8/1.2	0	0.1**	0.1	0	13.4**	13.4	2.2	0	1	MAT/GAM	2	1	1	2
MD88-787	-56.38	145.3	3020	3.75	2.33*	-1.4	0	0.8*/4.3**	0.8/4.3	0	0.2**	0.2	0	19**	19	2.5	0.6	1	MAT/GAM	2	1	1	2
PS1433-1	-47.54	15.36	4810	6.18	2.7	-3.5	0	0	0	0	0	0	0			1.23	0.6	1	IKM/SI-Ind.	2	2	2	3
PS1444-1	-55.37	9.98	4862	0.51	0.1	-0.4	0	0	0	0	0	0	22			4.11	0.69	1	IKM/SI-Ind.	2	2	1	3
PS1649-2	-54.91	3.31	2427	-0.04	-0.1	-0.1	1	0	0	1	0	0	3			7.68	1.29	1	IKM/SI-Ind.	2	3	2	3
PS1651-1	-53.63	3.86	2075	0.44	0.0	-0.4	0	0	0	0	0	0	3			8.4	2.4	1	IKM/SI-Ind.	1	2	1	3
PS1652-2	-53.66	5.10	1963	0.24	-0.5	-0.8	0	0	0	0	0	0	0			18.29	1.1	1	IKM/SI-Ind.	2	2	1	3
PS1654-2	-50.16	5.72	3744	4.56	1.4	-3.1	0	0	0	0	0	0	0			2.1	0.3	1	IKM/SI-Ind.	2	2	1	3
PS1756-5	-48.90	6.71	3828	5.03	1.6	-3.4	0	0	0	0	0	0	0			1.84	0.2	1	IKM/SI-Ind.	1	2	2	3
PS1765-3	-51.83	4.81	3760	2.93	0.8	-2.1	0	0	0	0	0	0	0			4.68	0.42	1	IKM/SI-Ind.	2	2	2	3
PS1768-8	-52.59	4.48	3299	1.47	0.5	-0.9	0	0	0	0	0	0	0			7.85	0.6	1	IKM/SI-Ind.	1	2	2	3
PS1775-4	-50.95	-7.51	2507	2.03	0.4	-1.7	0	0	0	0	0	0	0			8.24	0.4	1	IKM/SI-Ind.	2	2	2	3
PS1777-6	-48.23	-11.04	2577	4.92	2.0	-3.0	0	0	0	0	0	0	0			1.56	0.1	1	IKM/SI-Ind.	2	2	1	3

Sample	Lat	Lon	Depth	Va	Vb	Vc	MAT SST	MAT err	C8	C7	Method	Cat1	Cat2	Cat3	Cat4
PS1778-5	-49.01	-12.70	3407	4.81	1.7	-3.1			1.29	0.1	IKM/SI-Ind.	3	1	2	2
PS1779-2	-50.40	-14.08	3570	3.97	1.2	-2.7			2.45	0.1	IKM/SI-Ind.	3	1	2	2
PS1780-5	-51.70	-15.30	4280	2.86	1.1	-1.7			3.71	0	IKM/SI-Ind.	3	1	2	2
PS1782-5	-55.19	-18.61	5160	1.16	1.4	0.2			3.65	0.4	IKM/SI-Ind.	3	2	3	3
PS1783-5	-54.91	-22.71	3394	0.85	0.4	-0.4			0.34	0.1	IKM/SI-Ind.	3	1	2	2
PS2082-1	-43.22	11.74	4610	11.35	4.8	-6.6			10.18	0.25	IKM/SI-Ind.	3	1	2	2
PS2089-1	-53.19	5.33	2615	1.36	0.2	-1.2			5.76	0.8	IKM/SI-Ind.	3	1	2	2
PS2090-1	-53.18	5.13	2819	1.36	0.4	-0.9			12	0.4	IKM/SI-Ind.	3	1	2	1
PS2102-2	-53.07	-4.99	2390	0.55	0.0	-0.5			0.22	0	IKM/SI-Ind.	3	1	2	2
PS2250-5	-45.10	-57.95	3181	12.10	3.0	-9.1			3.36	0	IKM/SI-Ind.	3	1	2	2
PS2276-4	-54.64	-23.95	4383	0.89	0.5	-0.4			2.76	0.1	IKM/SI-Ind.	3	1	2	2
PS2278-3	-55.97	-22.22	4418	0.66	0.6	0.0			2.65	0.1	IKM/SI-Ind.	3	1	3	2
PS2280-4	-56.84	-22.32	4750	0.22	0.8	0.6			7.9	0.3	IKM/SI-Ind.	3	2	2	2
PS2305-6	-58.72	-33.04	3243	1.40	0.2	-1.2			7	0	IKM/SI-Ind.	3	1	2	3
PS2307-1	-59.05	-35.61	2532	0.58	0.0	-0.6			10.21	1.4	IKM/SI-Ind.	3	1	2	2
PS2319-1	-59.79	-42.68	4323	0.82	0.5	-0.3			0.76	0	IKM/SI-Ind.	3	1	3	3
PS2491-3	-44.96	5.97	4324	9.32	3.2	-6.1			0.3	0	IKM/SI-Ind.	3	1	2	2
PS2492-2	-43.17	-4.06	4207	11.42	4.1	-7.3			0.41	0	IKM/SI-Ind.	3	1	3	3
PS2493-1	-42.88	-6.02	4153	11.42	4.0	-7.4			0.28	0	IKM/SI-Ind.	3	1	3	3
PS2498-1	-44.15	-14.23	3783	11.11	4.5	-6.6			0.67	0.2	IKM/SI-Ind.	3	1	2	2
PS2499-5	-46.51	-15.33	3175	5.54	2.7	-2.8			3.16	0	IKM/SI-Ind.	3	2	2	1
PS2502-2	-50.25	-23.24	4461	4.37	1.2	-3.1			7.16	0.7	IKM/SI-Ind.	3	1	2	2
PS2515-3	-53.55	-45.29	3467	4.35	0.8	-3.5			0.84	0.5	IKM/SI-Ind.	3	2	3	3
PS2561-2	-41.86	28.54	4465	16.40	10.9	-5.5			0.28	0.8	IKM/SI-Ind.	3	1	2	3
PS2563-2	-44.56	34.79	3514	9.38	4.1	-5.3			0.33	0.24	IKM/SI-Ind.	3	1	3	3
PS2564-3	-46.14	35.90	3034	9.03	4.3	-4.7			0.48	0	IKM/SI-Ind.	3	1	3	2
PS2567-2	-46.94	6.26	4102	6.90	3.9	-3.0			3.66	0.7	IKM/SI-Ind.	3	1	2	1
PS2603-3	-53.23	37.63	5289	1.51	0.9	-0.6			3.64	0.5	IKM/SI-Ind.	3	1	2	2
PS2606-6	-58.99	40.80	2545	3.09	0.3	-2.6			5	0.8	IKM/SI-Ind.	3	1	3	3
PS2608-1	-51.88	41.65	2787	3.74	0.1	-3.0			5.09	0.24	IKM/SI-Ind.	3	1	3	2
PS58/271-1	-61.24	-116.05	3593	2.95	0.6	-1.3			1.18	0	IKM/SI-Ind.	4	1	3	2
RC11-118	-37.8	71.53	5214	17.15	16.3	-0.9	0	0			MAT	1	2	2	4
RC11-77	-53.05	-16.45	4354	2.46	0.9	-1.6	52	4.1			MAT	1	2	2	4
RC11-78	-50.87	-9.87	3115	2.53	0.7	-1.8	61	4.6			MAT	1	1	2	4
RC11-91	-56.57	34.18	5373	2.19	1.8	-0.4	45	2.6			MAT	1	2	1	1
RC11-94	-54.48	53.05	4303	2.88	0.8	-2.1	58	3.5			MAT	1	1	2	4
RC11-96	-50.47	59.58	4839	4.71	1.5	-3.3	45.2	2.7			MAT	1	2	3	4
RC11-97	-50.32	61.2	4638	4.65	2.2	-2.4	27.7	1.45			MAT	1	2	2	4
RC12-289	-47.9	-23.7	4484	7.08	2.1	-4.9	21.2	1.9			MAT	1	2	3	4
RC12-294	-42.52	-10.1	3308	17.92	17.3	-0.6	0	0			MAT	1	2	2	4
RC13-251	-37.27	11.67	4341	11.92	10.0	-1.9	0	0			MAT	1	2	3	4
RC13-253	-46.6	7.63	2494	1.80	4.1	-3.2	2.8	0			MAT	1	2	2	4
RC13-256	-53.18	-0.35	2525	7.23	0.9	-0.9	44	3.8			MAT	1	2	2	4
RC13-259	-53.88	-4.93	2677	1.61	0.8	-0.8	48.4	3.8			MAT	1	2	2	4
RC13-263	-53.8	-8.22	3389	1.48	0.8	-0.7	55.8	4.6			MAT	1	2	2	4
RC13-269	-52.63	-0.13	2591	2.12	1.9	-0.2	36.5	2.65			MAT	1	2	2	4
RC13-271	-51.98	4.52	3634	2.97	2.3	-0.6	15.8	0.9			MAT	1	2	2	3
RC14-11	-38	51.18	3268	19.01	14.0	-5.0	0	0			MAT	1	2	3	2
RC14-12	-38.75	59.3	5271	17.79	13.6	-4.2	0	0			MAT	1	2	2	3
RC17-61	-52.2	54.47	3947	4.06	1.9	-2.2	16.8	0.8			MAT	1	2	2	4
RC8-39	-42.88	42.35	2761	12.11	9.7	-2.5	0	0			MAT	1	2	3	4
RC8-46	-55.33	65.47	4330	3.07	0.9	-2.2	47	2.6			MAT	1	2	2	4
RC9-139	-47.77	123.1	4158	9.91	4.8	-5.2	0	0			MAT	1	2	2	4
SO136-111	-56.67	160.23	3912	5.54	2.2	-3.3	22.3	0.9			MAT	1	1	1	1
TN057-13-PC4	-53.17	5.11	2848	2.18	1.2	-1.0	48.23	3.93			MAT	1	2	3	4
V14-57	-57.57	-17.1	4978	1.02	0.9	-0.1	45.4	1.3			MAT	1	2	2	4
V29-84	-43.85	27.6	5451	11.57	7.7	-3.9	0	0			MAT	1	2	2	3
V29-86	-49.57	30.02	5614	4.67	4.2	-0.5	20	1.1			MAT	1	2	3	4
V29-87	-49.1	27.38	5314	4.82	3.0	-1.8	20	1.1			MAT	1	2	3	3
V29-89	-45.73	25.65	5945	8.10	6.9	-1.2	0	0			MAT	1	2	3	3
V29-90	-43.7	25.73	5148	11.84	10.1	-1.8	0	0			MAT	1	2	2	4

* refers to MAT derived estimate.
** refers to GAM derived estimate.

Table 4
Summary of southern summer SST (SSST) estimates derived from radiolarian-based transfer functions from the Epilog Last Glacial Maximum (E-LGM) time slice

Core	Lat.	Long.	Depth (m)	Modern SSST (°C)	E-LGM SSST (°C)	Delta LGM/mod. SSST (°C)	Estimation method	E-LGM CQL	EQL	AQL	Ref.
PS1433-1	−47.54	15.36	4810	6.18	5.2	−1.0	IKM	2	1	1	3
PS1444-1	−55.37	9.98	4862	0.51	1.1	0.6	IKM	2	2	1	3
PS1651-1	−53.63	3.86	2075	0.44	0.5	0.1	IKM	2	2	1	3
PS1756-5	−48.90	6.71	3828	5.03	3.8	−1.2	IKM	2	2	1	3
PS1768-8	−52.59	4.48	3299	1.47	0.9	−0.6	IKM	1	3	2	3
PS1778-5	−49.01	−12.70	3407	4.81	2.0	−2.8	IKM	2	2	1	3
PS1779-2	−50.40	−14.08	3570	3.97	0.8	−3.2	IKM	2	3	2	3
PS1783-5	−54.91	−22.71	3394	0.85	1.0	0.2	IKM	2	2	1	3
PS2082-1	−43.22	11.74	4610	11.35	6.6	−4.8	IKM	2	1	1	3
PS2089-1	−53.19	5.33	2615	1.36	0.5	−0.9	IKM	2	3	2	3
PS2104-2	−50.74	−3.23	2611	3.6	1.9	−1.7	IKM	2	2	1	3
PS2250-5	−45.10	−57.95	3181	12.1	2.4	−9.7	IKM	2	1	1	3
PS2271-5	−51.53	−31.35	3645	3.1	2.7	−0.4	IKM	2	2	1	3
PS2491-3	−44.96	5.97	4324	9.32	5.0	−4.3	IKM	3	2	2	3
PS2492-2	−43.17	−4.06	4207	11.42	6.5	−4.9	IKM	2	1	1	3
PS2493-1	−42.88	−6.02	4153	11.42	6.6	−4.8	IKM	2	2	1	3
PS2498-1	−44.15	−14.23	3783	11.11	6.2	−4.9	IKM	1	1	1	3
PS2567-2	−46.94	6.26	4102	6.9	3.7	−3.2	IKM	2	2	1	3
PS2821-1	−40.94	9.89	4575	15.3	12.1	−3.2	IKM	2	2	1	3

Modern SSST (°C at 10 m water depth) at core location according to Olbers et al. (1992) and Conkright et al. (1998) representing data of the World Ocean Atlas (WOA). Estimation method: Imbrie and Kipp Method (IKM). Quality levels include MARGO-defined E-LGM chronozone quality level (E-LGM CQL), estimate quality level (EQL) and average quality level (AQL). Estimates are from reference (3) Gersonde et al. (2003a).

dissolution. Where this biases the interpretation of the diatom records radiolarian assemblages are able to provide a far more useful paleoceanographic signal. Overall, the set of investigated E-LGM sections/samples is well suited for reconstruction of the glacial northward expansion of the Southern Ocean cold-water realm and WSI field.

Locations with highest E-LGM CQLs are concentrated in the Atlantic and western Indian sectors, while most of the E-LGM estimates in the central and eastern Indian and the Pacific sectors have been placed in the lowest E-LGM CQL (Fig. 3A). Nevertheless, EQLS are remarkably high (EQLs 1 and 2) in most of the investigated E-LGM core sections (Fig. 3B). Thus, the combination of both attribute levels substantiates that high quality information on E-LGM SSST and sea ice conditions are concentrated in the Atlantic sector and western Indian sector of the Southern Ocean (Fig. 3C). This result is due to the low stratigraphic control on many of the samples from the other sectors, mostly representing single samples from an interval defined by CLIMAP (1976, 1981) to represent the LGM.

3.1. E-LGM sea ice reconstruction

E-LGM sea ice reconstruction was based on MAT and GAM-derived estimates of sea ice annual duration (month/year) and concentration during winter (Septem-

ber) and summer (February). In the Atlantic sector this is combined with winter estimates based on the presence of sea ice indicator species (*F. curta* and *F. cylindrus*) presented in Gersonde et al. (2003a). Although MAT and WSI indicator-based reconstructions resulted in a coherent estimation of the WSI extent in the Atlantic sector, there is less conformity between the MAT derived sea ice estimates and the available data on WSI indicators in the Pacific and Indian sectors.

Maximum E-LGM WSI with a concentration > 15% extended in the Atlantic and Indian sector close to 47°S, and in the Pacific sector as far north as 57°S. This extension penetrates into the modern Polar Front Zone (PFZ), between the Polar Front and the Subantarctic Front, and indicates a northward displacement of maximum WSI occurrence by 7–10° in latitude in the different sectors of the Southern Ocean (Fig. 4). WSI with concentrations greater than 40% extended as far north as the present Polar Front area. The available data indicate that the strongest expansion of the ice cover occurs in the Atlantic and western Indian sectors. This indicates that during the E-LGM the WSI field expanded to around $39 \times 10^6 \, km^2$, which presents a ca 100% increase compared to modern conditions ($19 \times 10^6 \, km^2$; Comiso, 2003).

Only limited information is available on the E-LGM summer sea ice extent. This is due to the fact that in areas covered by perennial ice no microfossil

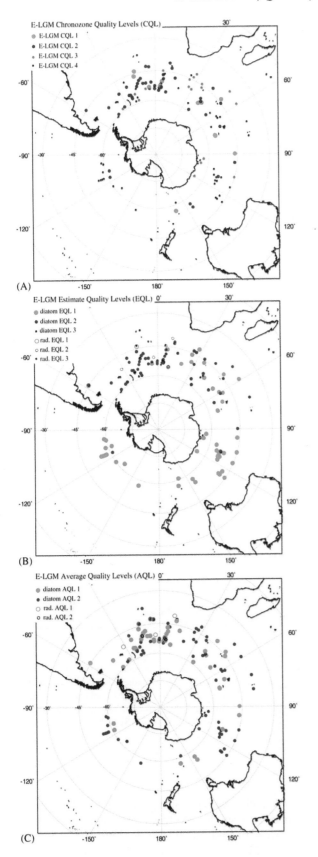

Fig. 3. Spatial distribution of quality level information assigned to E-LGM reconstruction. (A) Distribution of CQL assignments. (B) Distribution of EQL assignments. (C) Distribution of AQL assignments (for details see Tables 3 and 4).

assemblages are preserved in the sediment record that would allow paleoenvironmental reconstructions, such as SST and sea ice extent. Siliceous microfossil assemblages preserved in the sediment record are generally produced during the austral summer season in open water conditions. The siliceous microorganism flux from the sea ice itself is extremely low or absent, and is generally not discernible in the sediment record (Gersonde and Zielinski, 2000; Fischer et al., 1988, 2002). As a consequence, siliceous microfossil-based estimates of summer sea ice extent only represent a rough approximation of the nearby presence of summer sea ice or the irregular occurrence of summer sea ice. In the few cores recovered from the Atlantic sector, close to the present WSI edge, and three Indian sector cores located in the seasonal sea ice covered zone, we found diatom-derived indications of the sporadic presence of summer sea ice during the E-LGM (Fig. 4). This signal is based on the concomitant occurrence of the cold-water indicator *F. obliquecostata* (>1% of diatom assemblage) and a MAT estimation that indicates presence of summer (February) sea ice (sea ice concentration >0%) (Table 3). Diatom assemblages at two locations in the Indian sector north of the seasonal sea ice covered zone also signal sporadic presence of summer sea ice during the E-LGM. However, these observations are isolated and diatom assemblages from surrounding locations do not support the presence of sporadic occurrence of summer sea ice at these latitudes.

3.2. E-LGM summer SST reconstruction

Although diatom-based SST estimates have been obtained using different statistical methods and reference data sets in the Atlantic and western Indian sectors, the resulting values show a coherent picture of glacial conditions. This suggests that the different methodical approaches produce sound results when based on a uniform diatom taxonomical approach. Consistency also occurs between the diatom and radiolarian based results from core locations in the modern Antarctic Zone (AZ, south of the Polar Front) and PFZ (between the Polar Front and the Subantarctic Front) (Fig. 5). This supports the reliability of the resulting SST data obtained from both siliceous microfossil groups. However, in cores from the modern SAZ (between the Subantarctic Front and Subtropical Front), the radiolarian-based SST generally give values that are up to 2 °C above the diatom-based estimates (see Tables 3 and 4). The mismatch is interpreted to result in part from a bias of the diatom-based estimates towards colder values due to selective dissolution leading to a relative increase of the coarsely silicified colder-water diatom *F. kerguelensis* (Zielinski and Gersonde, 1997; Crosta et al., 1998a) and also from the shift in phytoplankton communities as a result of nutrient stress and varying

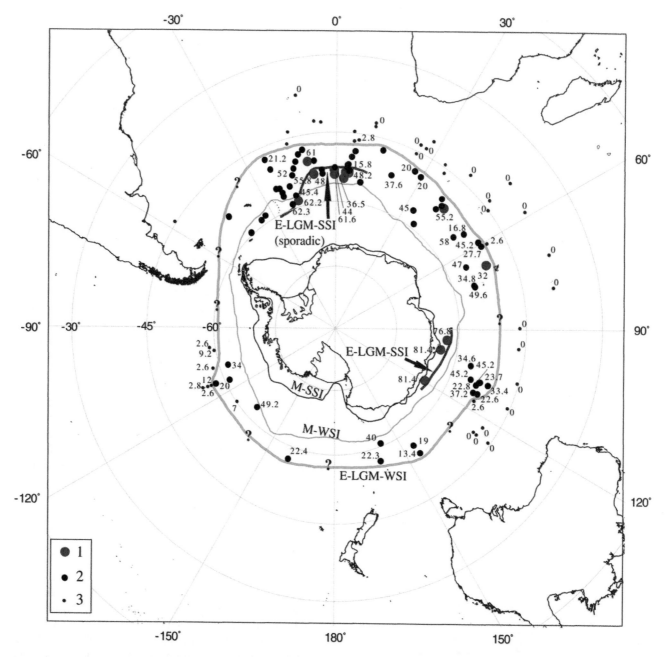

Fig. 4. Sea ice distribution at the E-LGM time slice. E-LGM-WSI indicates maximum extent of WSI (September concentration > 15%). Modern winter sea ice (M-WSI) shows extent of > 15% September sea ice concentration according to Comiso (2003). Values indicate estimated winter (September) sea ice concentration in percent derived with MAT and GAM. Signature legend: (1) concomitant occurrence of cold-water indicator *F. obliquecostata* (> 1% of diatom assemblage) and summer sea ice (February concentration > 0%) interpreted to represent sporadic occurrence of E-LGM summer sea ice; (2) presence of WSI (September concentration > 15%, diatom WSI indicators > 1%); (3) no WSI (September concentration < 15%, diatom WSI indicators < 1%). For data compilation see Table 3.

physical conditions. Such observations indicate that diatom estimates from Subantarctic and warmer locations should be treated with caution.

The E-LGM summer SSTs obtained at locations in the modern AZ generally display values below 1 °C in the Atlantic sector, and below 2 °C in the Indian and Pacific sectors of the Southern Ocean (Fig. 5), and thus are up to 3 °C colder than modern SST values (Fig. 6).

The cooler SSTs in the Atlantic sector can be related to the Weddell Sea cold-water gyre, which at present extends between the Antarctic Peninsula and around 20°E (western Enderby Basin) to about 60–55 °S (Olbers et al., 1992), and is closely related to the oceanic frontal system named the Scotia Front (Belkin and Gordon, 1996) (Fig. 7A). During the E-LGM, this cold-water gyre expands further to the east by approximately 10°

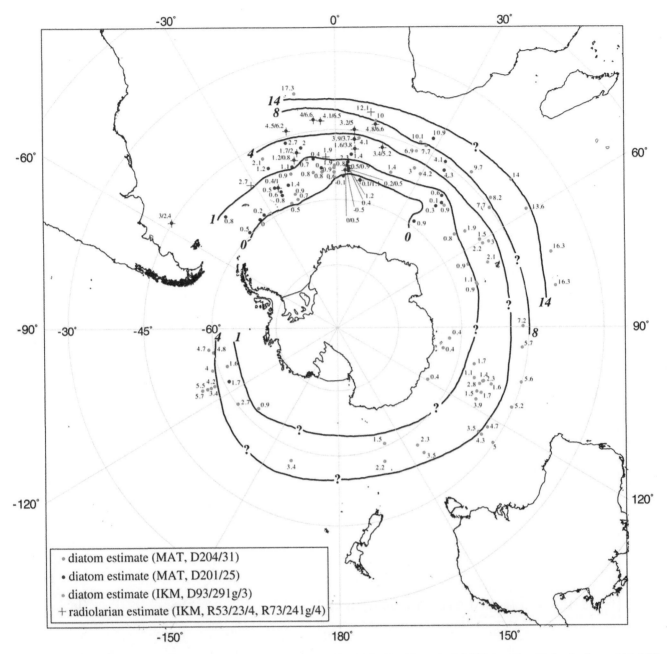

Fig. 5. Estimated austral summer SST (°C) and averaged summer sea surface isotherms (°C) at the E-LGM time slice. Surface isotherms 4 °C, 8 °C and 14 °C stand for average locations of E-LGM PF, Subantarctic Front and Subtropical Front. In the case of multiple SST estimates at any one location, only the highest quality estimate (see Tables 3 and 4) was considered for definition of the isotherm location. At locations with both, diatom and radiolarian estimates, values are labeled diatom SST/radiolarian SST. For data compilation see Tables 3 and 4.

longitude (Figs. 5 and 7A), indicating enhanced formation of cold surface water in the Weddell Sea area. In the modern PFZ, E-LGM summer SSTs range between 1 and 4 °C, and thus were approximately 3–4 °C colder than present. The strongest cooling of summer surface waters during the E-LGM is recorded in sediments collected from the modern SAZ, where the SSTs generally decreased by 4–6 °C (Fig. 6). Highest anomalies, reaching values around 7 °C, come from diatom-based estimates in the Atlantic and Indian sector. However, comparison with radiolarian-based values shows that the E-LGM cooling in the modern SAZ did generally not exceed 5 °C and that the diatom-based estimates may be biased towards colder values by selective species dissolution, hence resulting in higher anomalies.

In the Pacific sector, only a few cores document the E-LGM temperature regime in the northern zone of the

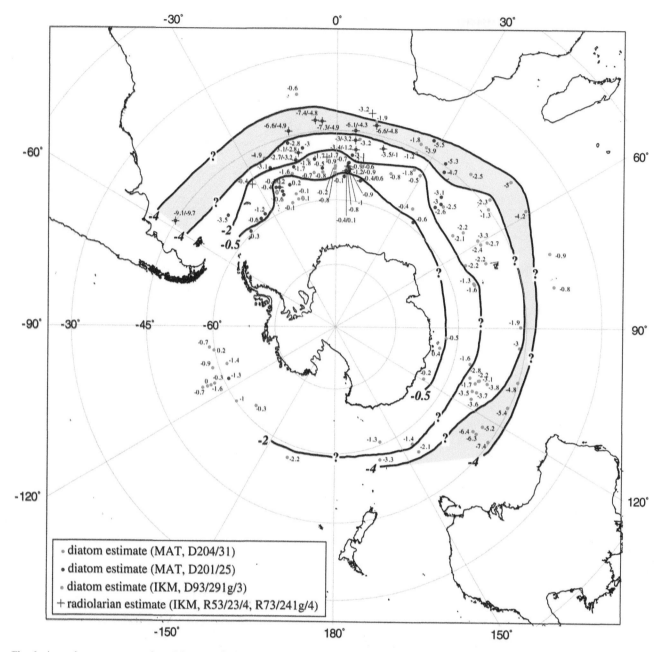

Fig. 6. Austral summer sea surface (°C) anomaly (modern/E-LGM) and averaged anomaly isotherms (°C). Area with anomalies > 4 °C is shaded blue. In cases with more than one SST anomaly available at any one location, only anomalies derived from highest quality estimate (see Tables 3 and 4) was considered for definition of an anomaly isotherm location. At locations with anomalies derived from both, diatom and radiolarian estimates, values are labeled diatom derived anomaly/radiolarian derived anomaly. For data compilation see Tables 3 and 4.

Southern Ocean. These cores are located around the present Subantarctic Front (Fig. 1). In contrast to the E-LGM anomalies obtained from the Atlantic and Indian sectors, we do not observe significant SST change from these Pacific locations (Fig. 6). The few cores located north of the present Subtropical Front (Fig. 1) indicate only a minor decrease in E-LGM summer SST, which suggests that the present southern subtropical realm was not strongly affected by cooling during the E-LGM.

4. Discussion

Comparison of the E-LGM WSI extent obtained from diatom-based studies with previous estimates relying on the sediment facies distribution (CLIMAP 1976, 1981; Cooke and Hays, 1982) yield a rather consistent pattern of LGM sea ice maximum extent (Fig. 8A). In the Atlantic and Indian sectors the WSI expanded by ca 10° in latitude during the E-LGM. In

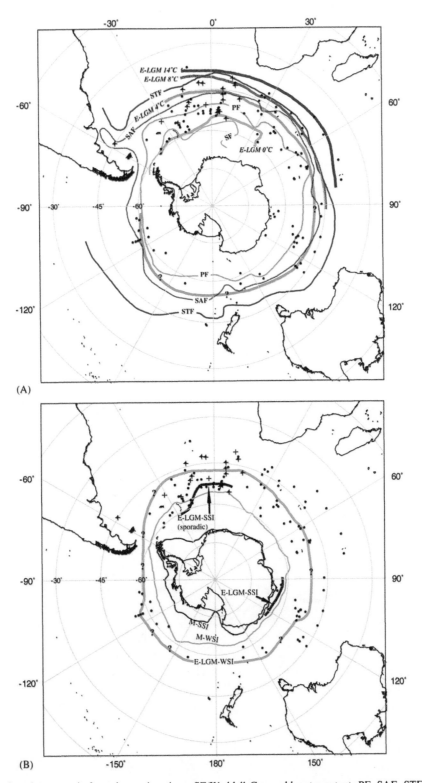

Fig. 7. (A) Comparison of modern oceanic frontal zone locations, SF/Weddell Gyre cold water extent, PF, SAF, STF, according to Belkin and Gordon (1996), with average E-LGM summer sea surface isotherms (0 °C isotherm approximates E-LGM Weddell Gyre cold water extent, 4 °C approximates E-LGM Polar Front, 8 °C approximates E-LGM Subantarctic Front, 14 °C approximates E-LGM Subtropical Front). (B) Comparison of modern sea ice edge, M-WSI, summer sea ice (M-SSI), data from Comiso (2003) (see also Fig. 4 legend), with E-LGM sea ice (E-LGM-WSI, E-LGM-SSI sporadic occurrence of summer sea ice).

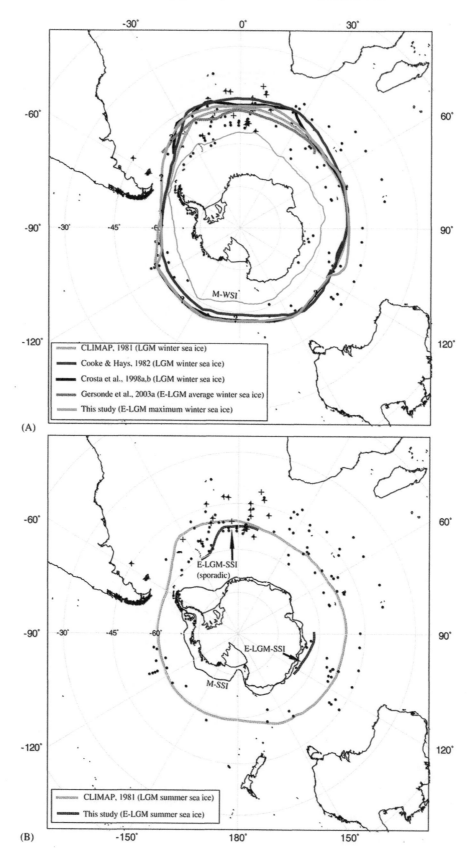

Fig. 8. (A) Comparison of LGM WSI edge reconstructions presented by different studies. (B) Comparison of LGM summer sea ice edge reconstructions from CLIMAP (1981) and this study.

these sectors the E-LGM sea ice edge approaches the area around 47°S, which is in the modern PFZ, close to the Subantarctic Front. Although the AQL of the sea ice estimates for the Indian sector is generally low, due to low-E-LGM CQLs (Fig. 3A), the obtained reconstruction yields a consistent pattern among all sample locations, supporting the reliability of the reconstruction. The scarce data from the Pacific sector point to smaller expansions of the E-LGM WSI in this sector. To substantiate the Pacific sector sea ice distribution, more data points are urgently needed. In the Atlantic sector the estimation of the average maximum WSI edge derived from the abundance pattern of the diatom sea ice indicators *F. curta* and *F. cylindrus* matches well the MAT-derived 15% sea ice concentration limit, indicating the edge of dense sea ice cover (Fig. 4). In the Pacific and Indian sectors, there is generally less conformity between the MAT derived sea ice estimates and the available data on sea ice indicators used for WSI reconstruction. The approach to reconstruct past sea ice by means of winter percent concentration recently proposed by Crosta et al. (2004) and Armand et al. (unpublished observations) and using MAT or GAM, respectively, significantly improves our ability of past circum-Antarctic WSI reconstruction.

Major uncertainties concern the reconstruction of summer sea ice extent. Here we interpret the concomitant occurrence of the cold-water indicator *F. obliquecostata* ($>1\%$ of diatom assemblage) and MAT-derived summer sea ice concentration $>0\%$ to indicate the "sporadic occurrence" of summer sea ice. In the eastern Atlantic sector of the Southern Ocean such sporadic summer sea ice occurrences have extended during the E-LGM as far north as 52°S, in close relation with an expanded Weddell Sea cold-water gyre (Figs. 7A and B). Nevertheless, as it has already been outlined by Crosta et al. (1998a, b) and Gersonde et al. (2003a), our results definitely rule out a strongly expanded E-LGM summer sea ice extent as proposed by CLIMAP (1981), reaching 50–52°S in the Atlantic and Indian sectors, and around 60°S in the Pacific sector of the Southern Ocean (Fig. 8B). Perennial LGM sea ice cover as suggested by CLIMAP (1981) would not allow the production and preservation of siliceous microfossil assemblages documented in E-LGM intervals obtained from cores located between 50°S and 64°S. Crosta et al. (1998a, b) speculate that the location of the LGM summer sea ice edge was similar to its modern position in the Indian sector of the Southern Ocean. This is documented in E-LGM intervals obtained from three KR88 cores recovered close to the East Antarctic coast. From the Pacific and Atlantic sector no cores located proximal to the Antarctic continent are available that would provide first-order documentation of the E-LGM summer sea ice extent. Widespread opal dissolution in the Weddell Sea (Schlüter et al., 1998) precludes reconstruction of the sea

ice edge and of temperatures based on the siliceous microfossil record in a large area of the southernmost Atlantic sector. However, the obtained summer SST and the indication of a patchy northward expansion of the E-LGM summer sea ice field point to a larger than present summer sea ice extent in the Weddell Sea area at the E-LGM time slice, as already outlined by Gersonde et al. (2003a). The relatively small expansion of the E-LGM summer sea ice extent and a strong expansion of the WSI field resulted in increased seasonality of the E-LGM sea ice compared with present conditions. It can be speculated that the seasonal changes of the sea ice field that at present range between $4 \times 10^6\,\mathrm{km}^2$ (summer) and $19 \times 10^6\,\mathrm{km}^2$ (winter) (Comiso, 2003) changed to a range between ca 5–$6 \times 10^6\,\mathrm{km}^2$ (E-LGM summer) and $39 \times 10^6\,\mathrm{km}^2$ (E-LGM winter).

Such enhanced seasonal sea ice production would have a strong impact on the production of Southern Ocean cold deep water via brine rejection and the velocity pattern of the Antarctic Circumpolar Current (ACC). Extremely cold Southern Ocean deep water, close to the freezing point, has been reported from the LGM (Duplessy et al., 2002; Mackensen et al., 2001). Keeling and Stephens (2001) hypothesize that Antarctic sea ice expansion affects not only the amplification of climate variability, but it also steers thermohaline overturning, due to the associated changes in the oceans salinity structure (Shin et al., 2003).

There is evidence that during the E-LGM time slice, which coincides with a maximum sea-level-low stand at ca 135 m below present (Yokoyama et al., 2000), climate conditions of the Southern Ocean were not the coldest of the last glacial. Summer sea ice indicators show distinct northward expansion of the summer sea ice field during the pre-E-LGM period between 30 000 and 25 000 cal yr. B.P., reaching 55°S in the eastern Atlantic sector (Gersonde et al., 2003a). Stronger Southern Ocean cooling during this period is also indicated by SST records from the Atlantic Subantarctic sector (Gersonde et al., 2003b) as well as SST and sea ice records from the eastern Indian sector (Armand and Leventer, 2003; Crosta et al., 2004; Armand et al., unpublished data) (Fig. 9). Such patterns point to significant changes in sea ice seasonality and production during the last glacial in the Southern Ocean. This observation calls for the need of accurate dating of any time-slice assigned to represent the LGM, to prevent data and varying environmental conditions from different periods being averaged into reconstructions of the LGM environment. The recent approach to reconstructing past Antarctic sea ice production based on the flux rates of sea salt recorded in ice cores (Wolff et al., 2003) may represent an additional method for sea ice reconstruction at high-resolution. Distinctly higher glacial sea salt flux rates observed in the Dome C record may be indicative for stronger sea ice seasonality and

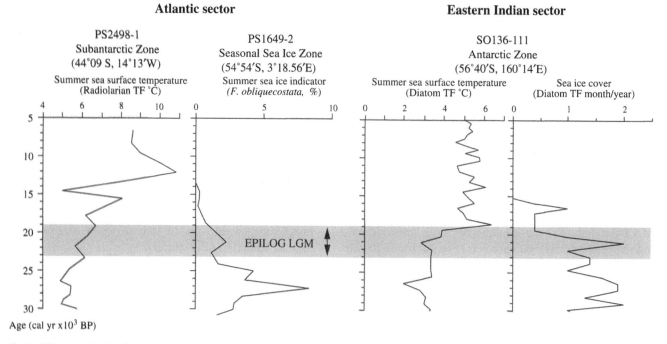

Fig. 9. Siliceous microfossil derived summer sea surface temperature and sea ice estimates for the past 30 000–5000 calendar years (BP), showing pre-EPILOG LGM maximum cooling in different sectors and latitudes of the Southern Ocean. Radiolarian transfer function (TF) derived SST record in Core PS2498-1 from Gersonde et al. (2003b), summer sea ice indicator record in Core PS1649-2 from Gersonde et al. (2003a), diatom SST and sea ice record in Core SO136-111 from Crosta et al. (2004).

related sea ice production during glacial periods. However, the record presented by Wolff et al. (2003) does not show significant variability in glacial sea ice seasonality as it can be deduced from the marine record.

Comparison of the E-LGM summer SSTs with those presented by CLIMAP (1981) shows that both reconstructions result in strongest LGM cooling being located in the northern Southern Ocean, with temperature decreases of more than 4 °C. While CLIMAP (1981) presents a more patchy areal distribution of maximum cooling, our compilation of E-LGM summer SSTs indicates a more continuous zone of enhanced cooling, and thus generally colder summer sea-surface conditions than estimated by CLIMAP (1981). We also remark that CLIMAP (1981) placed the belt of strongest cooling around 50°S in the Atlantic and the Indian sector, while our reconstruction points to a more southerly located zone of maximum cooling (Fig. 6). The rather small number of observations available from the Pacific Southern Ocean does not provide an overall picture of the E-LGM surface water temperature pattern from this sector. The few data available point to a reduced E-LGM cooling in the southern Pacific, a pattern also presented by CLIMAP (1981). In the event that additional data from the Pacific sector confirm a non-uniform E-LGM cooling of the Southern Ocean surface waters, major questions on the mechanisms responsible for such differentiation will arise. Possible mechanisms may be related to the configuration of the glacial Ross Ice Shelf that may have a major impact on the

generation of last glacial cold surface waters in the Pacific sector of the Southern Ocean.

Reconstruction of summer SST and sea ice clearly demonstrates that there was an expansion of the Southern Ocean cold water realm. Assuming that the modern relationship between summer SST and the location of the oceanic fronts can be applied to the E-LGM, the Polar Front in the Atlantic, Indian and Pacific sectors would have shifted to the North during the E-LGM by around 4°, 5–10°, and 2–3° in latitude, respectively, compared to their present location. In the Atlantic and Indian sector, the Subantarctic Front would have shifted by around 4–5° and 4–10° in latitude, respectively. The Subtropical Front displacement would have been minor, by around 2–3° and 5° in latitude in the Atlantic and Indian sector, indicating a compression of the SAZ during the E-LGM (Fig. 7A). The resolution of our data set makes it impossible to trace the location of past fronts based on the spatial mapping of surface water gradients, and thus we cannot contribute much to the ongoing debate if the oceanographic fronts were displaced during glacial cooling periods along with the surface isotherms. Our microfossil analyses indicate a northward migration of the planktic communities from which we inferred the glacial cooling. Stable isotope measurements ($\delta^{18}O$) on planktic and benthic foraminifers are interpreted to show similar Holocene and glacial oceanic circulation, arguing against frontal migration (Matsumoto et al., 2001). The latter authors attributed such robust ACC circulation pattern to factors such as bottom topography and land–sea configuration, both exerting strong controls over the large-scale ocean circulation. Hydrographic sections and satellite-derived data on frontal patterns, frontal locations and their spatial and temporal variability outline the complexity of these oceanographic features (Belkin and Gordon, 1996; Moore et al., 1999). In areas with distinct bottom topography, the mean path of fronts is closely linked to topographic structures, while over deep ocean basins fronts may meander over a wide latitudinal range extending over 5–7° in latitude (Moore et al., 1999). Comparison of different observations such as compiled by Belkin and Gordon (1996) and Moore et al. (1999) indicate that fronts may jump from one to another topographic structure, over more than 5° in latitude, a mechanism that could also have occurred following climate related changes in Southern Ocean hydrography. Assuming that the location of the E-LGM fronts was not significantly changed compared to present, they were represented by cooler isotherms, and the Polar Front was seasonally south of the sea ice edge.

The location and oceanographic pattern of the ACC zones in the Pacific sector has major implications on the functionality of the Drake Passage "cold water route" (CWR; Rintoul, 1991) that regulates Pacific–Atlantic water mass and heat exchange. Together with the Agulhas "warm water route" (Gordon, 1986), the CWR represents the most important return-flow portal in global thermohaline circulation, having crucial importance on global climate development (Drijfhout et al., 1996). Gersonde et al. (2003a) speculated that a strong E-LGM cooling in the area east of the southern Argentine coast (Fig. 6) might indicate the reduced import of surface waters with temperature properties of the present PFZ via the CWR. They relate this reduction to a northward displacement of the Southern Ocean zones that caused truncation of the northern warmer waters passing the Drake Passage at present. A resulting deflection of Southern Ocean waters along the South American coast in the southeast Pacific is corroborated by foraminiferal studies in the eastern tropical Pacific showing that the cooling in the LGM Equatorial Pacific was related to an increased northward advection of Southern Ocean waters (Feldberg and Mix, 2002, 2003).

As a consequence of northward expansion of Southern Ocean cold waters and of minor changes of E-LGM summer SST in the present southern subtropical realm observed in the Atlantic (Gersonde et al., 2003a; Niebler et al., 2003) and the Indian sectors, the surface water temperature gradients steepened during the E-LGM around the Subtropical Front and the SAZ compared to modern conditions (Fig. 7A). Such steepening of hydrographic gradients should have had an impact on the velocity of zonal water transport in the northern realm of the E-LGM ACC, as indicated by sedimento-

logical and isotope geochemical studies showing intensified glacial deep water mass transport in the Indian sector and in the western Pacific (Dézileau et al., 2000; Hall et al., 2001). The gradient change should have also affected atmospheric circulation, e.g. a northward-shift of the westerly winds, as proposed by Sigman and Boyle (2000) in a glacial Southern Ocean model. Such northward displacement of the wind field may result in the displacement of the Polar Front, as postulated from a simulation with an simple, one-layer Southern Ocean model forced by westerlies, northward shifted by 5° in latitude and increased in strength (Klinck and Smith, 1993).

The reconstruction of sea ice and summer SST at the E-LGM time slice documents distinct changes in physical parameters that potentially enhance glacial CO_2 draw down in the Southern Ocean. The Southern Ocean cold-water sphere expansion (between 5° and 10° in latitude) and the enlargement of the sea ice field would have increased Southern Ocean carbon uptake capacity (Bakker et al., 1997) and reduced air–sea exchange of CO_2 (Stephens and Keeling, 2000; Morales Maqueda and Rahmstorf, 2002), respectively.

5. Summary and outlook

Based on the quantitative study of diatoms and radiolarians, we estimated summer SST and sea ice distribution at 122 sediment core localities in the Atlantic, Indian and Pacific sector of the Southern Ocean to reconstruct the last glacial environment at the EPILOG (19.5–16.0 ka, equal to 23 000–19 000 cal yr B. P.) time-slice (E-LGM time slice). The applied statistical methods include the Imbrie and Kipp Method, the MAT and the General Additive Model. Age assignment of the samples selected to represent the E-LGM time slice and the obtained estimates have been ranked according to defined quality levels. Highest AQLs concentrate in the Atlantic and western Indian sector. Although the AQL of the estimates is generally low in the other sectors due to low stratigraphic control, the obtained reconstruction yields a rather consistent pattern supporting the reliability of the reconstructions.

Even though diatom-based SST estimates have been derived from different statistical methods and reference data sets, they result in a coherent SST pattern. This is also true for diatom and radiolarian-based SST estimates from the Antarctic and PFZ locations in the Atlantic sector. However, at locations from Subantarctic and warmer areas, diatom-based estimates may be biased towards colder values due to selective species dissolution.

The obtained E-LGM reconstructions can be summarized as follows:

Maximum E-LGM WSI (concentration > 15%) extended in the Atlantic and Indian sector close to 47°S, and in the Pacific sector as far north as 57°S. This reflects an E-LGM northward displacement by 7–10° in latitude in the various Southern Ocean sectors and converts to a ca 100% increase of the sea ice field $(39 \times 10^6 \, \text{km}^2)$ compared with modern conditions.

E-LGM summer sea ice extent information is rather limited. In the Indian sector the extent was close to the modern summer sea ice distribution. In the Atlantic, second-order information points to more expanded than present summer sea ice coverage that is restricted to the Weddell Sea area. There is indication for sporadic summer sea ice occurrence reaching as far north as 52°S in the eastern Atlantic sector. Estimates from the southwest Pacific region clearly indicate that no summer sea ice cover was present in the current PFZ (Armand et al., unpublished data).

The distinct enlargement of the E-LGM WSI field and indications of only minor summer-sea ice expansion in two sectors of the Southern Ocean, lend support to the theory of increased sea ice seasonality compared with present day.

The reconstruction of maximum WSI extent is broadly in accordance with CLIMAP (1981). The data, however, clearly show that CLIMAP (1981) strongly overestimated the glacial summer sea ice extent.

E-LGM Summer SST indicate a northward displacement of the Antarctic cold waters between 5° and 10° in latitude in the Atlantic and Indian sectors. Strongest cooling occurs in the present SAZ, reaching values between 4 and 6 °C. As a result of northward expansion of Antarctic cold waters and a relatively small displacement of the averaged Subtropical Front, thermal gradients were steepened during the last glacial in the northern zone of the Southern Ocean. This may, however, be inapplicable to the Pacific sector. The few data points available indicate reduced cooling the southern Pacific, and give hint to a non-uniform cooling of the glacial Southern Ocean.

Generally, the summer SSTs reveal greater surface-water cooling than those reconstructed by CLIMAP (1981).

Despite the progress in the dating, the use of statistical reconstruction methods and the establishment of surface sediment, hydrographic and sea ice reference data sets, as well as the collection of cores, there are still deficiencies that reduce our ability of accurate LGM reconstruction of the entire Southern Ocean. This includes the lack of appropriate surface sediment sample sets and sediment cores especially in the Pacific sector, but to some extent also in the Indian sector. Some improvement will be available in the near future from the establishment of surface sediment references for the central and eastern Pacific sector (Gersonde et al., unpublished data) (Fig. 2). Improved methods for a more accurate reconstruction of the glacial summer sea ice field are urgently needed. Such methods may be based on a combination of paleo sea ice reconstructions and paleo sea ice modeling, using SST prescription from paleoceanographic reconstructions. This should allow better determination of the distribution of the perennial ice cover as well as the sea ice seasonality, both of which impact sea–air gas exchange, biological productivity, atmospheric circulation and water-mass formation. Considering the strong value of radiolarian assemblages for generation of SST in the northern zone of the Southern Ocean, the number of radiolarian-based LGM reconstructions should be increased considerably together with the enlargement of reference data sets and expansion into the Indian and Pacific sectors. Such studies may also improve our knowledge on the processes related to water mass and heat exchange via the Drake Passage cold-water route. A major effort is also required to better describe past salinity changes at the Southern Oceans surface that are of major importance for the understanding of the Southern Ocean hydrography and its role in past thermohaline circulation changes (Stocker, 2003). A new promising tool to generate estimates of surface water salinity changes is the determination of the isotopic composition of oxygen in marine diatoms, as proposed by Shemesh et al. (1995, 2002). Last, but not least, AMS ^{14}C dating methods need further improvement to obtain more accurate dating of sediments dominated by siliceous microfossils.

Acknowledgment

We thank Nalan Koc and an anonymous reviewer, as well as Giuseppe Cortese for review of the paper and useful comments. Discussions among the MARGO scientific community during two MARGO workshops were also fruitful. Hans-Werner Schenke and Martin Klenke helped with graphic areal estimates. Oscar Romero provided surface sediment samples from the South Atlantic that have been included to the D204/31 data set prepared by Xavier Crosta. The generation of this paper has largely profited from the remarkable effort of the MARGO steering group and editors Michal Kucera, Ralph Schneider, Claire Waelbroeck and Mara Weinelt.

Electronic versions of data shown in this paper are available at http://www.pangaea.de/Projects/MARGO and will be updated according to future progress in the acquisition of E-LGM paleoceanographic data for the Southern Ocean.

References

Abelmann, A., 1988. Freeze-drying simplifies the preparation of microfossils. Micropaleontology 34, 361.

Abelmann, A., Gersonde, R., 1991. Biosiliceous particle flux in the Southern Ocean. Marine Chemistry 35, 503–536.

Abelmann, A., Brathauer, U., Gersonde, R., Sieger, R., Zielinski, U., 1999. A radiolarian-based transfer function for the estimation of summer sea-surface temperatures in the Southern Ocean (Atlantic sector). Paleoceanography 14 (3), 410–421.

Armand, L.K., Leventer, A., 2003. Paleo sea ice distribution—reconstruction and paleoclimatic significance. In: Thomas, D.N., Diekmann, G.S. (Eds.), Sea Ice an Introduction to its Physics, Chemistry, Biology and Geology. Blackwell, Oxford, pp. 333–372.

Armand, L.K., Zielinski, U., 2001. Diatom species of the genus *Rhizosolenia* from Southern Ocean sediments: Distribution and taxonomic notes. Diatom Research 16, 259–294.

Bakker, D.C.E., de Baar, H.J.W., Bathmann, U.V., 1997. Changes of carbon dioxide in surface waters during spring in the Southern Ocean. Deep-Sea Research II 44, 91–127.

Belkin, I.M., Gordon, A.L., 1996. Southern Ocean fronts from the Greenwich meridian to Tasmania. Journal Geophysical Research 101 (C2), 3675–3696.

Broecker, W.S., 2001. The big climate amplifier ocean circulation-sea ice-storminess-dustiness-albedo. In: Seidov, D., Haupt, B.J., Maslin, M. (Eds.), The Oceans and Rapid Climate Change: Past, Present, Future. Geophysical Monograph 126, AGU, Washington, DC, pp. 53–56.

Burckle, L.H., 1983. Diatom dissolution patterns in sediments of the Southern Ocean. Geological Society of American Journal 15, 536–537.

Burckle, L.H., Robinson, D., Cooke, D., 1982. Reappraisal of sea-ice distribution in Atlantic and Pacific sectors of the Southern Ocean at 18 000 yr BP. Nature 299, 435–437.

Climate: Long-Range Investigation, Mapping, and Prediction (CLI-MAP) Project Members, 1976. The surface of the Ice Age Earth. Science 191, 1131–1137.

Climate: Long-Range Investigation, Mapping, and Prediction (CLI-MAP) Project Members, 1981. Seasonal reconstructions of the Earth's surface at the Last Glacial Maximum. Geological Society of American Map Chart Series MC-36, 1–18.

Comiso, J.C., 2003. Large-scale characteristics and variability of the global sea ice cover. In: Thomas, D.N., Diekmann, G.S. (Eds.), Sea Ice an Introduction to its Physics. Chemistry, Biology and Geology, Blackwell, Oxford, pp. 112–142.

Conkright, M., Levitus, S., O'Brien, T., Boyer, T., J. Antonov, J., Stephens, C., 1998. World Ocean Atlas 1998 CD-ROM Data Set Documentation. National Oceanographic Data Center (NODC) Internal Report, Silver Spring, Maryland, pp. 16.

Cooke, D.W., Hays J.D., 1982. Estimates of Antarctic ocean seasonal ice-cover during glacial intervals. In: Craddock, C. (Ed.), Antarctic Geoscience, IUGS, Ser. B, No. 4, pp. 1017–1025.

Cortese, G., Abelmann, A., 2002. Radiolarian-based paleotemperatures during the last 160 kyr at ODP Site 1089 (Southern Ocean, Atlantic Sector). Palaeogeography, Palaeoclimatology, Palaeoecology 82, 259–286.

Crosta, X., Pichon, J.-J., Burckle, L.H., 1998a. Application of the modern analog technique to marine Antarctic diatoms: reconstruction of maximum sea-ice extent at the Last Glacial Maximum. Paleoceanography 13 (3), 284–297.

Crosta, X., Pichon, J.-J., Burckle, L.H., 1998b. Reappraisal of Antarctic seasonal sea ice at the Last Glacial Maximum. Geophysical Research Letters 25 (14), 2703–2706.

Crosta, X., Sturm, A., Armand, L., Pichon, J.-J., 2004. Late Quaternary sea ice history in the Indian sector of the Southern Ocean as record by diatom assemblages. Marine Micropaleontology 50, 209–223.

Curran, M.A.J., van Ommen, T.D., Morgan, V.I., Philips, K.L., Palmer, A.S., 2003. Ice core evidence for Antarctic sea ice decline since the 1950s. Science 302, 1203–1206.

Dézileau, L., Bareille, G., Reiss, J.L., Lemoine, F., 2000. Evidence for strong sediment redistribution by bottom currents along the southeast Indian Ridge. Deep-Sea Research I 47, 1899–1936.

Drijfhout, S.E., Maier-Reimer, E., Mikolajewicz, U., 1996. Tracing the conveyor belt in the Hamburg large scale geostrophic ocean general circulation model. Journal of Geophysical Research 101, 22563–22575.

Duplessy, J.-C., Labeyrie, L., Waelbroeck, C., 2002. Constraints on the ocean oxygen isotopic enrichment between the Last Glacial Maximum and the Holocene. Paleoceanographic implications. Quaternary Science Reviews 21, 315–330.

Feldberg, M.J., Mix, A.C., 2002. Sea-surface temperature estimates in the Southeast Pacific based on planktonic foraminiferal species: modern calibration and Last Glacial Maximum. Marine Micropaleontology 44, 1–29.

Feldberg, M.J., Mix, A.C., 2003. Planktonic foraminifera, sea surface temperatures, and mechanisms of oceanic change in the Peru and south equatorial currents, 0–150 ka BP. Paleoceanography 18 (1), 1016.

Fischer, G., Fütterer, D., Gersonde, R., Honjo, S., Ostermann, D., Wefer, G., 1988. Seasonal variability of particle flux in the Weddell Sea and its relation to ice-cover. Nature 335, 426–428.

Fischer, G., Gersonde, R., Wefer, G., 2002. Organic carbon, biogenic silica and diatom fluxes in the Northern Seasonal Ice Zone in the Polar Front Region in the Southern Ocean (Atlantic Sector): Interannual variation and changes in composition. Deep-Sea Research II 49, 1721–1745.

Gersonde, R., Zielinski, U., 2000. The reconstruction of late Quaternary Antarctic sea-ice distribution—The use of diatoms as a proxy for sea-ice. Palaeogeography, Palaeoclimatology, Palaeoecology 162, 263–286.

Gersonde, R., Abelmann, A., Brathauer, U., Becquey, S., Bianchi, C., Cortese, G., Grobe, H., Kuhn, G., Niebler, H.-S., Segl, M., Sieger, R., Zielinski, U., Fütterer, D.K., 2003a. Last glacial sea-surface temperatures and sea-ice extent in the Southern Ocean (Atlantic-Indian sector)—A multiproxy approach. Paleoceanography 18 (3), 1061 doi:1029/2002PA000809.

Gersonde, R., Abelmann, A., Cortese, G., Becquey, S., Bianchi, C., Brathauer, U., Niebler, H.-S., Zielinski, U., Pätzold, J., 2003b. The late Pleistocene South Atlantic and Southern Ocean Surface—A Summary of time-slice and time-series studies. In: Wefer, G., Mulitza, S., Ratmeyer, V. (Eds.), The South Atlantic in the Late Quaternary: Reconstruction of Material Budgets and Current systems. Springer, Berlin, Heidelberg, pp. 499–529.

Gloersen, P., Campbell, W.J., Cavalieri, D.J., Comiso, J.C., Parkinson, C.L., Zwally, H.J., 1992. Arctic and Antarctic sea ice, 1978-1987. Satellite Passive Microwave Observations and Analysis. NASA Special Publication 511, 289 pp., National Aeronautics and Space Administration, Washington, DC.

Gordon, A.L., 1986. Interocean exchange of thermocline water. Journal of Geophysical Research 91, 5037–5046.

Hall, I.R., McCave, N., Shackleton, N.J., Weedon, G., Harris, S.E., 2001. Intensified deep pacific inflow and ventilation in Pleistocene glacial times. Nature 412, 809–812.

Hastie, T., Tibshirani, R., 1990. Generalized Additive Models. Chapman & Hall, London, 335pp.

Hays, J.D., Lozano, J.A., Shackleton, N., Irving, G., 1976. Reconstruction of the Atlantic and western Indian Ocean sectors of the 18 000 B.P. Antarctic Ocean. Geological Society of American Memories 145, 337–372.

Hutson, W.H., 1980. The Agulhas Current during the late Pleistocene: analysis of modern faunal analogs. Science 207, 64–66.

Imbrie, J., Kipp, N.G., 1971. A new micropaleontological method for Quantitative Paleoclimatology: Application to a late Pleistocene Caribbean Core. In: Turekian, K.K. (Ed.), The Late Cenozoic

Glacial Ages. Yale University Press, New Haven, Conn., pp. 71–181.

Jöreskog, K.G., Klovan, J.E., Reyment, R.A., 1976. Geological Factor Analysis, Elsevier Science. New York, 178pp.

Keeling, R.F., Stephens, B.B., 2001. Antarctic sea ice and the control of Pleistocene climate instability. Paleoceanography 16, 112–131.

Klinck, J.M., Smith, D.A., 1993. Effect of wind changes during the Last Glacial Maximum on the circulation in the Southern Ocean. Paleoceanography 8, 427–433.

Kucera, M., Rosell-Melé, A., Schneider, R., Waelbroeck, C., Weinelt, M., 2004. Multiproxy Approach for the Reconstruction of the Glacial Ocean surface (MARGO). Quaternary Science Review, this issue, doi:10.1016/j.quascirev.2004.07.017.

Laws, R.A., 1983. Preparing strewn slides for quantitative microscopical analysis: a test using calibrated microspheres. Micropaleontology 24, 60–65.

Le, J., 1992. Palaeotemperature estimation methods: sensitivity test on two western equatorial Pacific cores. Quaternary Science Review 11, 801–820.

Le, J., Shackleton, N.J., 1994. Reconstructing paleoenvironment by transfer function. Model evaluation with simulated data. Marine Micropaleontology 24, 187–199.

Lozano, J.A., Hays, J.D., 1976. Relationship of radiolarian assemblages to sediment types and physical oceanography in the Atlantic and western Indian oceanic sectors of the Antarctic Ocean. In: Cline, R.M., Hays, J.D. (Eds.), Investigations of Late Quaternary Paleoceanography and Paleoclimatology. Geological Society of American Memoirs 145, pp. 303–336.

Mackensen, A., Rudolph, M., Kuhn, G., 2001. Late Pleistocene deep water circulation in the subantarctic eastern Atlantic. Global and Planetary Change 30, 197–229.

Malmgren, B.A., Haq, B.U., 1982. Assessment of quantitative techniques in paleobiogeography. Marine Micropaleontology 7, 213–236.

Matsumoto, K., Lynch-Stieglitz, J., Anderson, R.F., 2001. Similar glacial and Holocene Southern Ocean hydrography. Paleoceanography 16, 445–454.

Maynard, N.G., 1976. Relationship between diatoms in surface sediments of the Atlantic Ocean and the biology and physical oceanography of overlying waters. Paleobiology 2, 99–121.

Mix, A.C., Bard, E., Schneider, R., 2001. Environmental processes of the ice age: land, oceans, glaciers (EPILOG). Quaternary Science Review 20, 627–658.

Moore, K.J., Abbott, M.R., Richman, J.G., 1999. Location and dynamics of the Antarctic Polar Front from satellite sea surface temperature data. Journal Geophysical Research 104 (C2), 3059–3073.

Morales Maqueda, M.A., Rahmstorf, S., 2002. Did Antarctic sea-ice expansion cause glacial CO_2 decline? Geophysical Research Letters 29, 111–113.

Moreno-Ruiz, J.L., Licea, S., 1995. Observations on the valve morphology of Thalassionema nitzschioides (Grunow) Hustedt. In: Marino, D., Montresor, M. (Eds.), Proceedings of the 13th International Diatom Symposium. Biopress, Bristol, pp. 393–413.

Naval Oceanography Command Detachment, 1985. In: Sea-ice Climatic Atlas Vol. 1. Antarctic National Space Technology. Lab, Ashville, 131pp.

Niebler, H.-S., Arz, W., Donner, B., Mulitza, S., Pätzold, J., Wefer, G., 2003. Sea-surface temperatures in the equatorial and South Atlantic Ocean during the Last Glacial Maximum (23–19 ka). Paleoceanography 18 (3), 1069.

Olbers, D., Gouretski, V., Seiß, G., Schröter, J., 1992. Hydrographic Atlas of the Southern Ocean. Alfred Wegener Institute, Bremerhaven, pp. 82.

Pichon, J.-J., Labeyrie, L.D., Bareille, G., Labracherie, M., Duprat, J., Jouzel, J., 1992a. Surface water temperature changes in the high latitudes of the Southern Hemisphere over the Last Glacial–Interglacial cycle. Paleoceanography 7 (3), 289–318.

Pichon, J.-J., Bareille, G., Labracherie, M., Labeyrie, L.D., 1992b. Quantification of the biogenic silica dissolution in the Southern Ocean. Quaternary Research 37, 361–378.

Prell, W.L., 1985. The stability of low-latitude sea-surface temperatures: An evaluation of the CLIMAP reconstruction with emphasis on the positive SST anomalies. Rep. TR 025, pp. 1-60, US Department of Energy, Washington, DC.

Rintoul, S.R., 1991. South Atlantic interbasin exchange. Journal of Geophysical Research 97, 5493–5550.

Sarnthein, M., Gersonde, R., Niebler, S., Pflaumann, U., Spielhagen, R., Thiede, J., Wefer, G., Weinelt, M., 2003. Overview of Glacial Atlantic Ocean Mapping (GLAMAP 2000). Paleoceanography 18 (2), 1030 doi:1029/2002PA000769.

Schlüter, M., Rutgers van der Loeff, M.M., Holby, O., Kuhn, G., 1998. Silica cycle in surface sediments of the South Atlantic. Deep-Sea Research I 45, 1085–1109.

Schrader, H.J., Gersonde, R., 1978. Diatoms and Silicoflagellates. Micropaleontological counting methods and techniques—An exercise on an eight meters section of the lower Pliocene of Capo Rossello. In: Zachariasse, W.J., Riedel, W.R., Sanfilippo, A., Schmidt, R.R., Brolsma, M.J., Schrader, H.J., Gersonde, R., Drooger, M.M., Broekman, J.A. (Eds.), Utrecht Micropaleontological Bulletin 17, 129–176.

Schweitzer, P.N., 1995. Monthly averaged polar sea ice concentration. US Geological Survey Digital Data Series: Virginia. CD, Ed. 1. DDS-27.

Shemesh, A., Burckle, L.H., Hays, J.D., 1995. Late Pleistocene oxygen isotope records of biogenic silica from the Atlantic sector of the Southern Ocean. Paleoceanography 10, 179–196.

Shemesh, A., Hodell, D., Crosta, X., Kanfoush, S., Charles, C., Guilderson, T., 2002. Sequence of events during the last deglaciation in Southern Ocean sediments and Antarctic ice cores. Paleoceanography 17 (4), 1056.

Shin, S.I., Liu, Z., Otto-Bliesner, B.L., Kutzbach, J.E., Vavrus, S., 2003. Southern Ocean control of the glacial North Atlantic thermohaline circulation. Geophysical Research Letters 30 (2), 1096.

Sieger, R., Gersonde, R., Zielinski, U., 1999. New software package available for quantitative paleoenvironmental reconstructions. EOS 80 (19), 223.

Sigman, D.M., Boyle, E.A., 2000. Glacial/interglacial variations in atmosphere carbon dioxide. Nature 407, 859–869.

Stephens, B.B., Keeling, R.F., 2000. The influence of Antarctic sea ice on glacial–interglacial CO_2 variations. Nature 404, 171–174.

Stocker, T.F., 2003. South dials North. Nature 424, 496–499.

Wolff, E.W., 2003. Whiter Antarctic sea ice? Science 302, 1164.

Wolff, E.W., Rankin, A.M., Röthlisberger, R., 2003. An ice core indicator of Antarctic sea ice productivity? Geophysical Research Letters 30, 2158.

Yokoyama, Y., Lambeck, K., de Dekker, P., Johnson, P., Fifield, K., 2000. Timing for the maximum of the Last Glacial constrained by lowest sea-level observations. Nature 406, 713–716.

Zielinski, U., Gersonde, R., 1997. Diatom distribution in Southern Ocean surface sediments: Implications for paleoenvironmental reconstructions. Palaeogeography, Palaeoclimatology, Palaeoecology 129, 213–250.

Zielinski, U., Gersonde, R., Sieger, R., Fütterer, D.K., 1998. Quaternary surface water temperature estimations—calibration of diatom transfer functions for the Southern Ocean. Paleoceanography 13, 365–383.

Zwally, H.J., Comiso, J.C., Parkinson, C.L., Campbell, W.J., Carsey, F.D., Gloersen, P., 1983. Antarctic Sea Ice 1973–1976: Satellite Passive Microwave Observations. NASA, Washington, DC, Report SP-459.

ELSEVIER

Quaternary Science Reviews 24 (2005) 897–924

QSR

Reconstruction of sea-surface conditions at middle to high latitudes of the Northern Hemisphere during the Last Glacial Maximum (LGM) based on dinoflagellate cyst assemblages

A. de Vernal[a],*, F. Eynaud[b], M. Henry[a], C. Hillaire-Marcel[a], L. Londeix[b], S. Mangin[b], J. Matthiessen[c], F. Marret[d], T. Radi[a], A. Rochon[e], S. Solignac[a], J.-L. Turon[b]

[a]*GEOTOP, Université du Québec à Montréal, P.O. Box 8888, Montréal, Qué., Canada H3C 3P8*
[b]*Département de Géologie et Océanographie, UMR 5805 CNRS, Université Bordeaux I, Avenue des Facultés, 33405 Talence Cedex, France*
[c]*Alfred Wegener Institute for Polar and Marine Research, P.O. Box 120161, D27515 Bremerhaven, Germany*
[d]*School of Ocean Sciences, University of Wales Bangor, Menai Bridge LL59 5EY, UK*
[e]*Institut des Sciences de la mer de Rimouski (ISMER), Université du Québec à Rimouski, 310, allée des Ursulines, Rimouski, Qué., Canada G5L 3A1*

Received 21 November 2003; accepted 30 June 2004

Abstract

A new calibration database of census counts of organic-walled dinoflagellate cyst (dinocyst) assemblages has been developed from the analyses of surface sediment samples collected at middle to high latitudes of the Northern Hemisphere after standardisation of taxonomy and laboratory procedures. The database comprises 940 reference data points from the North Atlantic, Arctic and North Pacific oceans and their adjacent seas, including the Mediterranean Sea, as well as epicontinental environments such as the Estuary and Gulf of St. Lawrence, the Bering Sea and the Hudson Bay. The relative abundance of taxa was analysed to describe the distribution of assemblages. The best analogue technique was used for the reconstruction of Last Glacial Maximum (LGM) sea-surface temperature and salinity during summer and winter, in addition to sea-ice cover extent, at sites from the North Atlantic ($n = 63$), Mediterranean Sea ($n = 1$) and eastern North Pacific ($n = 1$). Three of the North Atlantic cores, from the continental margin of eastern Canada, revealed a barren LGM interval, probably because of quasi-permanent sea ice. Six other cores from the Greenland and Norwegian seas were excluded from the compilation because of too sparse assemblages and poor analogue situation. At the remaining sites ($n = 54$), relatively close modern analogues were found for most LGM samples, which allowed reconstructions. The new LGM results are consistent with previous reconstructions based on dinocyst data, which show much cooler conditions than at present along the continental margins of Canada and Europe, but sharp gradients of increasing temperature offshore. The results also suggest low salinity and larger than present contrasts in seasonal temperatures with colder winters and more extensive sea-ice cover, whereas relatively warm conditions may have prevailed offshore in summer. From these data, we hypothesise low thermal inertia in a shallow and low-density surface water layer.
© 2004 Elsevier Ltd. All rights reserved.

1. Introduction

The earliest reconstructions of the Last Glacial Maximum (LGM) ocean published by CLIMAP (1981) constituted a major breakthrough in paleoceanography and paleoclimatology. These reconstructions of summer and winter sea-surface temperatures (SSTs) were principally established from transfer functions based on multiple regression techniques and planktonic foraminifer data (Imbrie and Kipp, 1971). Since this pioneer work, many methodological approaches have been developed for the reconstruction of past climatic parameters based on an array of biological indicators, notably pollen grains, diatoms, dinoflagellate cysts, radiolarians, planktonic foraminifera, ostracods, and coccoliths. Various data treatment techniques were also

*Corresponding author. Tel.: +1-514-987-3000x8599; fax: +1-514-987-3635.

E-mail address: devernal.anne@uqam.ca (A. de Vernal).

0277-3791/$ - see front matter © 2004 Elsevier Ltd. All rights reserved.
doi:10.1016/j.quascirev.2004.06.014

developed or adapted to the analyses of the diverse micropaleontological populations. They mainly include techniques using the degree of similarity between fossil and modern assemblages (e.g., Guiot, 1990; Pflaumann et al., 1996; Waelbroeck et al., 1998), and the artificial neural network techniques (e.g., Malmgren and Nordlund, 1997; Weinelt et al., 2003). In addition to the above-mentioned approaches based on the analyses of microfossil populations, biogeochemical analyses of organic compounds, such as alkenones produced by coccolithophorids, or the measurement of trace elements, such as Mg/Ca or Sr/Ca in biogenic calcite, yielded insights into past temperatures in the water column (e.g., Rosell-Melé, 1998; Lea et al., 1999; Nürnberg et al., 2000).

Many of these recently developed methods have been applied to re-evaluate the sea-surface conditions which prevailed during the LGM. In addition to the CLIMAP (1981) scenario, there are now many LGM data sets available on regional scales. For example, at the scale of the northern North Atlantic, there are data sets based on planktonic foraminifera (Weinelt et al., 1996; Pflaumann et al., 2003; Sarnthein et al., 2003), dinoflagellate cysts (de Vernal et al., 2000, 2002), and alkenone biomarkers (Rosell-Melé, 1997; Rosell-Melé and Comes, 1999; Rosell-Melé et al., 2004). Comparison of the paleoceanographical data sets has revealed significant discrepancies, notably in terms of paleotemperature estimates.

With the aim to compare and eventually to reconcile paleoceanographical reconstructions based on different proxies, an intercalibration exercise has been undertaken within the frame of the Multiproxy Approach for the Reconstruction of the Glacial Ocean (MARGO) Project. The first step was to adopt a common hydrography for the calibration of the temperature vs. proxy relationships, in order to avoid any bias that can be related to initial oceanographical data inputs. The "standardised" hydrography that has been selected for the present MARGO exercise is the 1998 version of the World Ocean Atlas produced by the National Oceanographic Data Center (NODC). In the present paper, we are thus reporting on (i) the updated modern database of dinoflagellate cyst assemblages, (ii) the results from calibration exercises with the standardised hydrography (summer and winter SSTs) and other key parameters such as salinity and sea-ice cover, (iii) the sea-surface condition reconstructions for the LGM interval defined by Environmental Processes of the Ice age: Land, Oceans, and Glaciers (EPILOG) criteria as the interval of maximum continental ice volume during the last glaciation, which spanned from ca. 23 to 19 kyr before present (Schneider et al., 2000; Mix et al., 2001).

Data presented here are representative of middle to high latitudes of the Northern Hemisphere. The reference dinocyst database for the hemisphere includes 940 sites from the North Atlantic, North Pacific and Arctic oceans,

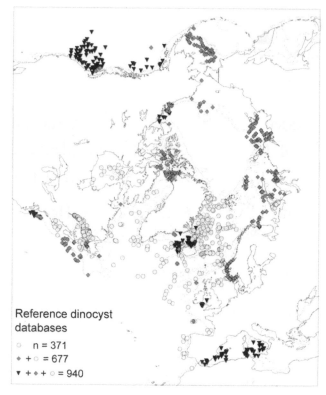

Reference dinocyst databases

○ n = 371
◆ + ○ = 677
▼ + ◆ + ○ = 940

Fig. 1. Location of surface sediment samples used to establish the updated "n = 940" reference dinocyst database, which was developed after the "n = 371" database (de Vernal et al., 1997; Rochon et al., 1999) and the "n = 677" database (de Vernal et al., 2001), and includes the regional data sets from the northeastern North Pacific (Radi and de Vernal, 2004) and the Mediterranean Sea (Mangin, 2002), notably. The isobaths correspond to 1000 and 200 m of water depth.

and their adjacent seas (see Fig. 1). This database constitutes an update of the "n = 371" (cf. Rochon et al., 1999) and "n = 677" (cf. de Vernal et al., 2001) databases. The update notably includes additional sites from the North Atlantic (Marret and Scourse, 2003; Marret et al., 2004), the Mediterranean Sea (Mangin, 2002), and the North Pacific (Radi and de Vernal, 2004). This database was used here to produce an update of LGM reconstructions of SST, salinity and sea-ice cover, which were published previously for a number of sites from the northern North Atlantic (cf. de Vernal et al., 2000). Four additional LGM sites are included in the present compilation. Two are from the northern North Atlantic, one from the Mediterranean Sea, and one from the Gulf of Alaska in the northeastern Pacific.

2. Methodology for sea-surface reconstructions

2.1. Dinoflagellate cyst data

2.1.1. The ecology of dinoflagellates and their cysts

Dinoflagellates occur in most aquatic environments and constitute one of the main primary producers in

marine environments, together with diatoms and coccolithophorids. Living dinoflagellates are not fossilisable.
However, during their life cycle, after the fusion of the
gametes for sexual reproduction, some taxa produce
highly resistant organic-walled cysts protecting the
diploid cells for a dormancy period of variable length
(e.g., Wall and Dale, 1968; Dale, 1983). The organic-
walled cysts of dinoflagellates (or dinocysts) thus
represent only a fragmentary picture of the original
dinoflagellate populations (e.g., Dale, 1976; Head,
1996). Amongst dinoflagellates producing cysts currently recovered in geological samples, there are mainly
species belonging to the orders of Gonyaulacales,
Peridiniales and Gymnodiniales. Gonyaulacales are
autotrophic whereas Peridiniales and Gymnodiniales
may have heterotrophic or mixotrophic behaviour (e.g.,
Gaines and Elbrächer, 1987; Taylor and Pollingher,
1987). These taxa that belong to phytoplankton or
microzooplankton develop and bloom in surface waters.
They are usually recovered together, from plankton
samples collected on a routine basis in the upper 50 m
(e.g., Dodge and Harland, 1991) or 100 m (e.g., Raine et
al., 2002) of the water column. Their living depth is
relatively shallow since the autotrophic taxa are
dependant upon light penetration, and because the
habitat of the heterotrophic species appears to be closely
coupled to diatoms on which they feed and/or to the
maximum chlorophyll zone (e.g., Gaines and Elbrächer,
1987). Moreover, despite their ability to move vertically
with their flagella, dinoflagellates generally inhabit a
relatively thin and shallow surface layer, especially in
stratified marine environments, because they cannot
migrate across the pycnocline that constitutes an
important physical barrier (cf. Levandowsky and
Kaneta, 1987).

Planktonic dinoflagellates in the North Atlantic show
distribution patterns of species in surface water closely
related to salinity and temperature, which are controlled
by current patterns (e.g., Dodge and Harland, 1991;
Dodge, 1994; Raine et al., 2002). Nearshore assemblages
can also be distinguished from oceanic assemblages,
with regard to the species diversity and taxa dominance.
The biogeographical distributions of cyst-forming dinoflagellates in surface waters and that of dinocysts in
sediments are generally consistent with each other,
notably with respect to their onshore–offshore patterns
and latitudinal gradients (Dodge and Harland, 1991;
Dodge, 1994). However, the correspondence between
observations of motile and cyst assemblages is not
perfect, probably due to the fact that the motile
dinoflagellates in the plankton assemblages correspond
to an instantaneous time interval, whereas the cysts in
surface sediments may represent several years or decades
of sedimentary fluxes.

The distribution of dinocysts in sediments has been
relatively well documented and has contributed to our

understanding of the average sea-surface conditions that
determine the distribution pattern and abundances of
the taxa. Since the early works of Wall et al. (1977),
Harland (1983), and Turon (1984), the relative abundance of dinocyst taxa is known to follow distribution
patterns closely related to the temperature gradients and
to show distinct neritic, outer neritic and oceanic
assemblages. During the last two decades, many studies
have contributed to the description of the dinocyst
distribution on the sea floor. These illustrate qualita-
tively or quantitatively the relationships between dinocyst assemblages and sea-surface parameters including
temperature and salinity, sea-ice cover, productivity,
upwelling and eutrophication (for reviews, see e.g.,
Dale, 1996; Mudie et al., 2001; Marret and Zonneveld,
2003).

2.1.2. The establishment of the modern dinocyst database

To develop the reference database, we have analysed
surface sediment samples that were mostly collected
from box cores or gravity cores. Although samples were
taken from the uppermost 1 or 2 cm in the sedimentary
column, they may represent the last 10^1–10^3 years
depending upon sediment accumulation rates, and
biological mixing intensity and depth in sediment. More
information on sampling or subsampling, laboratory
procedures, the nature of palynological assemblages in
general, and the abundance, preservation, and species
diversity of dinocyst assemblages can be found in
original publications of the regional data sets available
for the northern Baffin Bay (Hamel et al., 2002), the
Canadian Arctic (Mudie and Rochon, 2001), the
Russian Arctic, including the Laptev Sea (Kunz-
Pirrung, 1998, 2001) and the Barents Sea (Voronina et
al., 2001), the Arctic Ocean as a whole (de Vernal et al.,
2001), the Labrador Sea (Rochon and de Vernal, 1994)
and northwest North Atlantic (de Vernal et al., 1994),
the northeast North Atlantic (Rochon et al., 1999), the
Norwegian and Greenland Seas (Matthiessen, 1995), the
Celtic Sea (Marret and Scourse, 2003), the Norwegian
Coast (Grøsfjeld and Harland, 2001), the Icelandic Sea
(Marret et al., 2004), the Estuary and Gulf of St.
Lawrence in eastern Canada (de Vernal and Giroux,
1991), the Bering Sea (Radi et al., 2001), the northeastern North Pacific (Radi and de Vernal, 2004), and
the Mediterranean Sea (Mangin, 2002).

Although it is derived from a number of regional data
sets, the $n = 940$ database is internally consistent with
respect to laboratory procedures and taxonomy. This
database actually results from a collective endeavour
that started about 15 years ago.

With regard to sample preparation, the standardised
protocol consists of repeated HCl and HF treatments of
the > 10 μm fraction (for details, see de Vernal et al.,
1999; Rochon et al., 1999). This protocol avoids
treatment with oxidant agents because the organic cyst

Table 1
List of dinocyst taxa in the n = 940 database

Taxa name	Code	Notes
Cyst of cf. *Scrippsiella trifida*	Alex	
Achomosphaera spp.	Acho	
Ataxiodinium choane	Atax	
Bitectatodinium tepikiense	Btep	
Impagidinium aculeatum	Iacu	
Impagidinium pallidum	Ipal	
Impagidinium paradoxum	Ipar	
Impagidinium patulum	Ipat	
Impagidinium sphaericum	Isph	
Impagidinium strialatum	Istr	
Impagidinium plicatum	Ipli	
Impagidinium velorum	Ivel	
Impagidinium japonicum	Ijap	
Impagidinium spp.	Ispp	
Lingulodinium machaerophorum	Lmac	
Nematosphaeropsis labyrinthus	Nlab	
Operculodinium centrocarpum sensu Wall & Dale 1966	Ocen	
O. centrocarpum sensu Wall & Dale 1966—short processes	Ocss	Grouped with *O. centrocarpum* sensu Wall & Dale 1966
Operculodinium centrocarpum—Arctic morphotype	Oarc	Grouped with *O. centrocarpum* sensu Wall & Dale 1966
Operculodinium israelianum	Oisr	
Operculodinium cf. *janduchenei*	Ojan	
Operculodinium centrocarpum—morphotype *cezare*	Ocez	Grouped with *O. centrocarpum* sensu Wall & Dale 1966
Polysphaeridium zoharyi	Pzoh	
Pyxidinopsis reticulata	Pret	
Spiniferites septentrionalis	Ssep	Grouped with *Achomosphaera* spp.
Spiniferites alaskum	Sala	
Spiniferites membranaceus	Smem	
Spiniferites delicatus	Sdel	
Spiniferites elongatus	Selo	
Spiniferites ramosus	Sram	
Spiniferites belerius	Sbel	Grouped with *S. membranaceus*
Spiniferites bentorii	Sben	
Spiniferites bulloideus	Sbul	Grouped with *S. ramosus*
Spiniferites frigidus	Sfri	Grouped with *S. elongatus*
Spiniferites lazus	Slaz	
Spiniferites mirabilis-hyperacanthus	Smir	
Spiniferites ramosus type *granosus*	Sgra	
Spiniferites pachydermus	Spac	
Spiniferites spp.	Sspp	
Tectatodinium pellitum	Tpel	
Cyst of *Pentapharsodinium dalei*	Pdal	
Islandinium minutum	Imin	
Islandinium? cesare	Imic	
Echinidinium cf. *karaense*	Espp	
Brigantedinium spp.	Bspp	
Brigantedinium cariacoense	Bcar	Grouped with *Brigantedinium* spp.
Brigantedinium simplex	Bsim	Grouped with *Brigantedinium* spp.
Dubridinium spp.	Dubr	
Protoperidinioids	Peri	
Lejeunecysta sabrina	Lsab	
Lejeunecysta oliva	Loli	
Lejeunecysta spp.	Lspp	
Selenopemphix nephroides	Snep	
Xandarodinium xanthum	Xand	
Selenopemphix quanta	Squa	
Cyst of *Protoperidinium nudum*	Pnud	
Protoperidinium stellatum	Pste	
Trinovantedinium applanatum	Tapp	
Trinovantedinium variabile	Tvar	
Votadinium calvum	Vcal	
Votadinium spinosum	Vspi	
Cyst of *Protoperidinium americanum*	Pame	

Table 1 (continued)

Taxa name	Code	Notes
Quinquecuspis concreta	Qcon	
Cyst of *Polykrikos schwartzii*	Psch	
Cyst of *Polykrikos* spp.—Arctic morphotype	Parc	
Cyst of *Polykrikos kofoidii*	Pkof	
Cyst of *Polykrikos* spp.—quadrangular morphotype	Pqua	Grouped with cyst of *Polykrikos* spp. Arctic morphotype
Echinidinium granulatum	Egra	
Gymnodinium catenatum	Gcat	
Gymnodinium nolleri	Gnol	

wall of some taxa can be altered by oxidation (cf. Marret, 1993). With the exception of a few studies suggesting in situ oxidation of the organic wall of protoperidinian cysts in sediment (cf. Zonneveld et al., 2001), dinoflagellate cysts are usually considered to be extremely resistant since they are composed of refractory organic matter called dinosporin (wax-like hydrocarbon; Kokinos et al., 1998). Their preservation is not affected by dissolution processes that result in alteration of siliceous or calcareous microfossils.

The taxonomy of dinoflagellates in the water column and that of their cysts preserved in sediment are mostly independent, because they reflect distinct stages in the dinoflagellate life cycle, i.e., a vegetative stage and a cyst stage following the sexual reproduction. The taxonomy of organic-walled dinoflagellate cysts is based on the morphology of the fossil remains. The taxonomy we are using for routine identification was developed after several workshops to ensure standardisation within the database. The nomenclature of dinocyst taxa used here conforms to Head (1996), Rochon et al. (1999), Head et al. (2001, 2005), Radi et al. (2001), de Vernal et al. (2001), Mangin (2002), and Radi and de Vernal (2004). A complete list of taxa used for statistical treatment and the application of the best analogue technique appears in Table 1. Counts of taxa in the 940 spectra of the reference database are reported following this taxa list (see GEOTOP site, www.geotop.uqam.ca/; see also MARGO data on the PANGAEA site, www.pangaea.de).

2.1.3. The dinocyst distribution in the calibration database

The overall dinocyst database, including 940 spectra and 60 taxa, has been submitted to multivariate analyses. Canonical correspondence analyses were performed using the CANOCO software of Ter Braak and Smilauer (1998) after logarithmic transformation (ln) of the relative frequency of taxa. Such a transformation is important to discriminate dinocyst assemblages in relation with environmental parameters because the dominant taxa are often opportunistic and ubiquitous, whereas accompanying taxa often show affinities for a narrow range of given hydrographical

parameters, such as salinity or temperature. The first and second axes, respectively, account for 14.9% and 12.2% of the total variance. Their geographical distribution and the weighting of the 60 taxa according to the two axes are shown in Fig. 2b. The spatial distribution of the values for the first axis reveals a latitudinal pattern, whereas the scores of the second axis show a nearshore to oceanic trend. Canonical correspondence analysis and cross-correlations of the axes with environmental parameters indicate that the assemblage distribution is predominantly controlled by SST and sea-ice cover extent, and that salinity also exerts a determinant role. The correspondence analysis results are consistent with those obtained from principal component analyses of samples from the North Atlantic and Arctic oceans (Rochon et al., 1999; de Vernal et al., 2001), which also demonstrated the dominant effect of the temperature and salinity on the dinocyst distribution. However, at regional scales, parameters other than temperature and salinity may determine dinocyst assemblages. Such is the case of the northeast Pacific margins, where productivity as estimated from satellite imagery (Antoine et al., 1996) seems to be the parameter that is most closely related to the spatial distribution of dinocysts in this area (cf. Radi and de Vernal, 2004).

2.2. Hydrographic data

Following the MARGO recommendation for hydrographical data standardisation, we have used the seasonal means of surface temperature at 10 m of water depth compiled from the 1998 version of the World Ocean Atlas (cf. National Oceanographic Data Center (NODC), 1994). However, at many stations, these data were not available. In such cases, we used seasonal means extracted from regional data sets, when possible. Alternatively, we have used the extrapolated fields of data available from the 1994 version of the World Ocean Atlas (cf. NODC, 1994), as developed for the $n = 677$ database (cf. de Vernal et al., 2001). Fig. 3 illustrates the location of sites with available NODC (1998) data, and the location of the sites where we had to use other sources of hydrographic data. These sites are principally

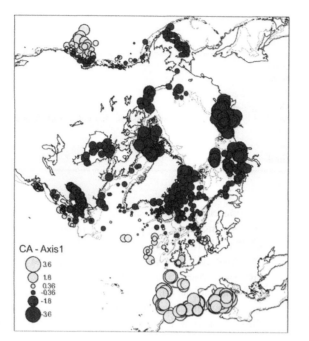

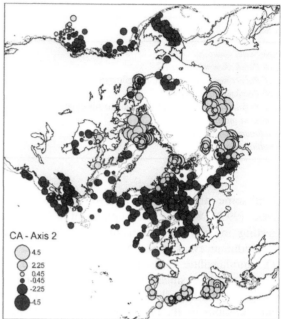

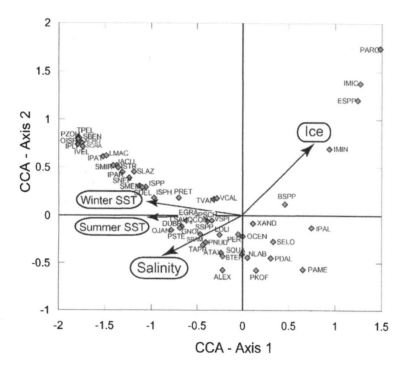

Correlation matrix		
	Axis 1	Axis 2
Winter SST	-0.891	0.0896
Summer SST	-0.881	-0.0235
Salinity	-0.7035	-0.2984
Ice	0.6754	0.5051

Fig. 2. Geographical distribution of the first two principal axes as defined from canonical correspondence analyses (axes 1 and 2 in the upper left and upper right diagrams, respectively), loading of the 60 dinocyst taxa in the 940 spectra of the reference database according to axes 1 and 2 (lower left diagram) and cross-correlation matrix between the axes and hydrographical parameters (lower right). Note that analyses were performed on logarithmic (ln) values of the relative frequency of taxa expressed in per mil, using the CANOCO software of Ter Braak and Smilauer (1998).

located in the Arctic or along continental margins, where instrumental data coverage is limited and where hydrographic conditions can be extremely variable from one year to another.

In addition to summer and winter mean SSTs, we have compiled the summer sea-surface salinity from NODC (1994). Note that salinity is reported as practical salinity units throughout the manuscript. We also

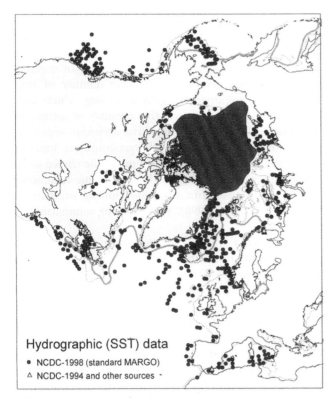

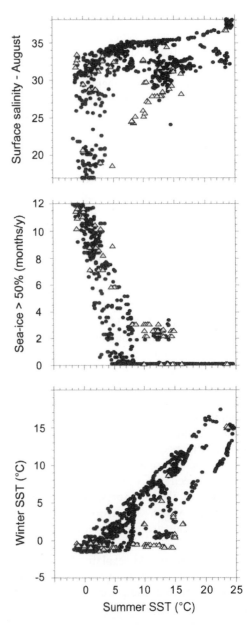

Fig. 3. Map showing the location of the 940 surface sediment samples in the calibration database. Different symbols illustrate the source of SST data. The black circles correspond to sites where hydrographic data from NODC (1998) following the standard defined for MARGO are available. The open triangles correspond to sites where NODC (1998) data were not available, and where compilations were made from the regional databases or from extrapolated values provided by NODC (1994). Details on the sources of these other data can be found in de Vernal et al. (2001). The minimum (grey zone) and maximum (grey line) limits of sea-ice cover are defined from a compilation of several sources.

Fig. 4. Graph showing the relationships of summer temperature vs. winter temperature, sea-ice cover, and summer salinity in the $n = 940$ database. As in Fig. 3, the black circles correspond to sites where seasonal temperature data from NODC (1998) are available, and the open triangles correspond to sites where other sources of data had to be used.

compiled the seasonal duration of sea-ice cover with concentration greater than 50%, as expressed in number of months per year after the 1953–1990 data set provided by the National Climate Data Centre in Boulder. We have limited the database to areas with salinity higher than 17 because instrumental data are sparse and show very large dispersal of values at nearshore and estuarine sites of lower salinities. The summer and winter SSTs, the summer salinity and the sea-ice cover extent at each station of the $n = 940$ database, which we use for the application of the best analogue technique, are archived on the websites of GEOTOP and PANGAEA.

The relationships between the summer and winter SSTs, the summer salinity and the sea-ice cover extent in the $n = 940$ database are illustrated in Fig. 4. They show the combinations of summer temperature vs. winter temperature, or sea-ice cover or salinity, and illustrate the range of hydrographical conditions we may thus reconstruct from dinocyst assemblages. It is of note that the range of salinity covered by the dinocyst database is

much larger than those of other micropaleontological tracers, notably planktonic foraminifera, which are much more stenohaline and representative of marine environments with salinity usually higher than 33 (Bé and Tolderlund, 1971). Diatoms that form assemblages characterised by large species diversity in a wide range of salinities and include sea-ice taxa would have been very useful as complementary indicators of sea-surface conditions. However, to date no attempt was made to set transfer functions for quantitative reconstruction of salinity and sea ice in the North Atlantic. Moreover, in the sediments of the last glacial episode, diatoms are not abundant and their assemblages suffer from poor preservation of the opal silica (Koç et al., 1993; Lapointe, 2000).

The hydrographical parameters we have used for reconstructions based on dinocysts are considered to be the most important determinants of the distribution of assemblages. In temperate marine environments, dinoflagellates generally develop during the warmest part of the year, following the diatom bloom, and their maximum growth rate occurs when close to optimal temperature establishes (Taylor and Pollingher, 1987). The summer SST or the maximum SST is thus the parameter exerting the most influential role on the distribution of dinoflagellate population in the upper water column, as reflected by the cyst populations in sediment traps (cf. Godhe et al., 2001) or in sediments (see Fig. 2). Salinity is also a very important parameter controlling the distribution of assemblages since the range of salinity tolerance varies among species, with euryhaline taxa being abundant in nearshore and estuarine environments as seen in living populations (Taylor and Pollingher, 1987) and cyst assemblages in sediments (e.g., de Vernal and Giroux, 1991). In addition to temperature and salinity, the annual cycle of temperature or seasonality most probably exerts a determinant control on the life cycle of dinoflagellates, notably on the respective duration of vegetative vs. encysted stages. The seasonality can be expressed as the difference between the warmest and the coolest temperatures. It can also be expressed as the length of the season during which autotrophic or heterotrophic metabolic activities are interrupted, because of limited light due to sea-ice cover or to reduced primary production. This would explain how the seasonal duration of sea-ice cover is one of the parameters that can be reconstructed using dinoflagellate cyst assemblages.

2.3. The approach for quantitative reconstructions

Different approaches for estimating past sea-surface conditions based on dinocyst assemblages have been tested, including canonical regressions, several variants of the best analogue technique (de Vernal et al., 1994, 1997, 2001; Rochon et al., 1999), and the artificial neural network technique (Peyron and de Vernal, 2001). Validation tests revealed that the best analogue and the artificial neural network techniques may yield similarly accurate results (cf. Peyron and de Vernal, 2001). Nevertheless, we have decided to use here the best analogue technique because it requires the least manipulation and transformation of data. The database, which covers three oceans, several epicontinental seas, and includes 60 taxa, implies distinct strategies of preparation depending upon the technique to be applied. In the case of the best analogue technique, we can use the entire database, without any discrimination of taxa and sites. In the case of the artificial neural network technique, however, the definition of regional calibration data sets with a reduced number of taxa would be a requirement. This is a step which may eventually help to constrain the accuracy of estimates, but which also relies on subjective decisions regarding the ultimate list of taxa and the geographical limits of the regional databases. Thus, we made the choice to be conservative by applying the best analogue technique, following the procedure adapted from the software of Guiot and Goeury (1996), which can be summarised as follows:

Prior to data analyses for the search of analogues, a few transformations are made. The abundance of taxa relative to the sum of dinocysts is calculated in per thousand instead of percentages in order to deal with whole numbers and to avoid decimals for further ln-transformation. One (1) is added to the frequency of each taxon in order to deal with values greater than zero. Another minor transformation consists of adjusting the frequency data ranging between 2 and 5 to the value of 5 in order to make a better discrimination between absence ($=1$) and presence (>5). This transformation is further justified because of the count limit, which is as low as 100 or 200 specimens in some instances. The zero elements are thus replaced by a value lower than the precision with which data were produced (cf. Kucera and Malmgren, 1998). After these transformations, a distance (d) between the spectrum to be analysed (t) and the spectra in the reference database (i) is calculated based on the difference in relative frequency (f) for each taxon ($j = 1$–60) as follows:

$$d = \sum_{j=1}^{n} [\ln f_{ij} - \ln f_{tj}]^2.$$

For estimating hydrographical conditions, we have used the five best analogues, which are the five modern samples with the lowest "d" values. The estimate for the "most probable" hydrographical values is obtained by calculating an average of the values for the five best analogues, weighted inversely by the distance. This most probable estimate is included within an interval corresponding to lower and upper limits, which are defined from the variances of the values below and above the most probable estimates, respectively. This technique leads to the calculation of a confidence interval that is not necessarily symmetric around the most probable estimate.

2.4. Validation of the approach

The degree of accuracy of reconstructions can be evaluated based on the estimations of the modern winter and summer temperatures, sea-ice cover and summer sea-surface salinity, which were made based on the calibration data excluding the spectrum to analyse (leaving-one-out technique; see Fig. 5). The linearity of the relationship with a slope close to one, and the coefficients of correlation between estimates and observations provide a first indication of the reliability of the approach. The degree of accuracy of the reconstruction is constrained by the standard deviation of the difference between estimates and observations. Values of ± 1.2 and $\pm 1.7\,^{\circ}$C have been calculated for the winter and summer SSTs, respectively, ± 1.7 for the salinity,

and ± 1.3 months/year for the sea-ice cover. On the whole, the degree of accuracy of estimates is of the same magnitude as the standard deviations around the mean for modern SST, salinity or sea-ice cover values collected instrumentally during the last decades (see also Rochon et al., 1999; de Vernal and Hillaire-Marcel, 2000). The degree of accuracy is better in open oceanic regions characterised by salinity higher than 33, and shows a larger spread of data in continental margin areas, estuaries, and ice marginal zones that are marked by highly variable hydrographical conditions on annual, decadal to centennial time scales. In the case of the

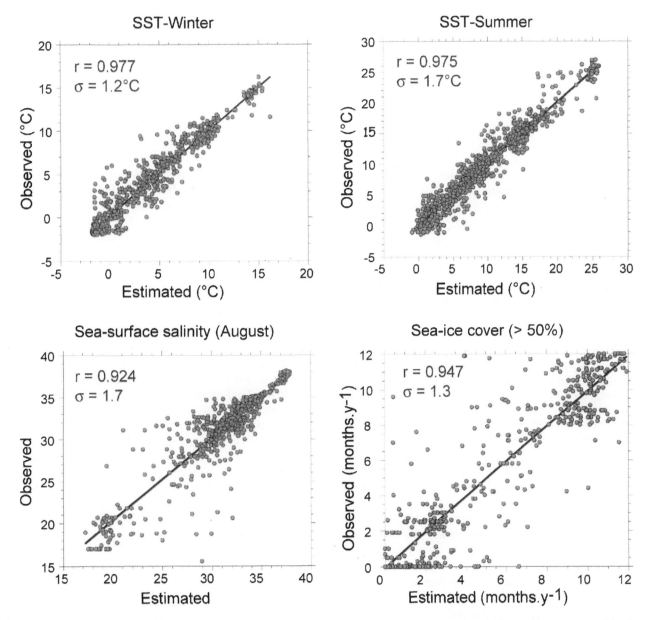

Fig. 5. Results of the validation test for the reconstruction of SST, salinity and sea-ice cover. The x-axis shows hydrographic averages resulting from instrumental observations, and y-axis shows estimates from the dinocyst data after the procedure described in the text. The coefficients of correlation (r) and the standard deviation (σ) of the difference between reconstruction and observation (i.e., the equivalent of the Root Mean Square Error of Prediction) provide the degree of accuracy of estimates. These accuracy indicators were calculated for all data points (n = 940) although the prediction error clearly depends upon the geographical domain considered.

Canadian and Russian Arctic, there is a particularly large error for salinity, and to a lesser extent for temperature, which can be explained by the high variability of these parameters and by the lack of accuracy of instrumental data (e.g., Mudie and Rochon, 2001). We estimate that about half of the spread of estimated vs. observed values could be attributed to inaccurate hydrographical measurements.

2.5. Definition of reliability indices

All methods developed for quantitative reconstructions of hydrographic parameters based on microfossil assemblages have intrinsic uncertainties due to the accuracy of the calibration databases themselves. Another source of uncertainty derives from the assumption that the present relationships between hydrographical parameters and microfossil assemblages were identical in the past. When dealing with past intervals such as the LGM, this assumption is debatable because conditions of biological production were different than at present. Therefore, the reliability of reconstructions is a question that has to be addressed.

In order to define a reliability index, we have used the degree of similarity between microssil spectra from LGM and modern based on the distance "d" as described above. From the calibration database, a threshold value of acceptable distance has been set on probabilistic grounds (i.e., a Monte-Carlo approach) for identification of a non-analogue or poor-analogue situation. In the case of the Northern Hemisphere $n = 940$ database, the distance between pairs randomly taken in the database averages 130.87 with a standard deviation of 56.46. The average minus standard deviation gives a threshold distance (here, 74.39) below which we consider the similarity to be significant. On these grounds, we defined a reliability index according to three categories (cf. Fig. 8):

(1) Good analogue situation when the distance is between 0 (perfect analogue) and half of the threshold value (37).
(2) Acceptable analogue situation when the distance is between half of the threshold value and the threshold (37–74).
(3) Poor analogue situation when the distance of the closest analogue is higher than the calculated threshold (>74).

The reliability index should be further constrained by the concentration of dinocysts, which depends on productivity and cyst fluxes, and sediment accumulation rates. When productivity and fluxes are low, reworking will have an increased influence on the assemblages and, therefore, on the reconstructed sea-surface conditions. Here, we have used a threshold value of 100 cysts/cm³ to define critically low concentration. Taking into account sedimentation rates of 10 cm/kyr, this concentration value corresponds to a flux of the order of 1 cyst/cm²/ year. For comparison, such a flux is lower than that of the modern Labrador Sea by one order of magnitude (Hillaire-Marcel et al., 1994), but is similar to the one presently recorded in the Baffin Bay basin (Rochon and de Vernal, 1994). In Table 2 and Fig. 8, "X" signs indicate which sites are characterised by critically low concentrations, below the threshold value of 100 cysts/cm³.

3. The LGM sea-surface conditions based on dinocyst data

3.1. The coring sites

A total of 65 cores have been analysed for their palynological content (see Fig. 6, Table 2) in order to reconstruct LGM conditions. Most of the cores are from the northern North Atlantic and adjacent subpolar seas: Labrador Sea and Baffin Bay, Irminger Basin, Norwegian and Greenland seas. One core from the Gulf of Alaska in northeastern North Pacific and one core from the western Mediterranean were also analysed.

The LGM time slice (~23,000–19,000 cal. years BP) has been defined following the recommendation made at the first EPILOG Workshop (cf. Schneider et al., 2000; Mix et al., 2001). In most of the cores, the LGM is defined with a good level of accuracy. It is based on radiocarbon dates, lithostratigraphical boundaries provided by the Heinrich layers H1 and H2, magnetic susceptibility or paleointensity correlations, and/or $\delta^{18}O$ stratigraphies. References about the stratigraphy of most cores we refer to can be found in de Vernal et al. (2000) or on the GEOTOP and PANGAEA websites. Additional information about the stratigraphy of the Mediterranean Sea core ODP161-976c is provided by von Grafenstein et al. (1999) and Combourieu-Nebout et al. (2002) and the stratigraphy of northeastern Pacific core PAR87-A10 can be found in de Vernal and Pedersen (1997). Stratigraphical data about the North Atlantic cores MD95-2002, MD95-2009 and MD95-2010 can be found in Zaragosi et al. (2001) and Eynaud et al. (2002), and about core MD99-2254 in Solignac et al. (2004).

3.2. The LGM dinocyst assemblages and their modern analogues

In many cores, the sediments of the LGM contain sparse palynological assemblages, with very low dinocyst concentrations. A few cores from Baffin Bay, and from the margins of Labrador and Greenland have revealed barren samples, with cyst concentration lower

Table 2
Summarised information about the cores analysed and the sea-surface conditions—estimates based on dinocyst data

Cores	Latitude	Longitude	n	nt	Annual SST Mean	Annual SST St. dev.	Summer SST Mean	Summer SST St. dev.	Winter SST Mean	Winter SST St. dev.	Salinity Mean	Salinity St. dev.	Ice cover Mean	Ice cover St. dev.	d—first analogue Mean	d—first analogue St. dev.	d—fifth analogue Mean	d—fifth analogue St. dev.	Dinocyst concentration	Reliability index
HU-76-029-033	71.33	-64.27	2	5	-0.2	0.1	1.9	0.2	-1.9	0.1	29.7	3.6	9.3	0.9	27.0	31.9	34.0	22.2	28	1X
M17045	52.43	-16.39	2	2	7.1	1.1	10.7	0.8	4.5	1.0	34.2	0.1	0.2	0.3	40.3	1.8	51.9	3.5	1361	2
M17724	76.00	8.33	4	6	6.7	2.4	12.4	3.3	1.7	1.9	31.1	0.5	1.5	2.7	47.8	14.9	59.1	16.8	72	2X
M23041	68.68	0.23	0	1	—	—	—	—	—	—	—	—	—	—	—	—	—	—	21	—
M23071	67.08	2.90	4	4	7.9	1.4	13.3	0.4	4.1	2.1	33.0	1.0	0.4	0.5	52.6	11.1	65.1	10.4	155	2
M23074	66.66	4.91	5	5	4.5	4.1	9.7	4.6	0.2	1.6	31.5	0.7	3.7	3.3	35.4	10.6	44.9	13.8	354	1
M23259	72.03	9.27	11	11	6.8	1.5	13.0	1.1	1.8	1.7	31.5	0.7	1.3	1.1	35.5	12.6	43.4	14.1	270	—
M23294	72.36	-10.36	2	3	8.2	0.9	11.8	1.9	5.5	0.5	34.2	0.1	0.0	0.0	55.0	13.0	72.2	17.3	38	2X
M23519	64.83	-29.56	2	4	6.9	0.2	10.2	0.5	4.7	0.1	33.9	0.7	1.5	1.3	20.5	11.2	37.1	3.1	66	1X
*MD95-2002	47.45	-8.53	8	19	7.5	1.2	12.9	1.3	3.5	1.9	32.4	1.2	0.2	0.4	36.0	10.1	41.5	11.5	11068	1
MD95-2009	62.74	-4.00	5	8	6.7	3.0	12.4	4.1	1.5	2.9	31.4	1.4	1.3	2.5	33.8	8.2	41.8	9.0	162	1
MD95-2010	66.68	4.57	25	35	4.3	1.7	9.9	3.1	-0.2	0.8	31.6	0.7	3.0	1.9	27.1	9.5	34.8	10.6	1483	1
POS0006	69.20	-16.81	0	2	—	—	—	—	—	—	—	—	—	—	—	—	—	—	26	—
POS0020	67.98	-18.53	7	7	8.4	1.4	13.0	1.4	5.0	2.0	33.2	1.1	0.4	0.8	42.6	12.7	50.8	13.0	334	2
PS1842-6a	69.45	-16.52	8	11	7.4	1.5	12.9	1.0	3.3	2.2	31.7	1.3	1.0	0.9	49.0	22.9	58.0	21.3	142	2
PS1919-2	74.98	-11.92	0	10	—	—	—	—	—	—	—	—	—	—	—	—	—	—	56	—
PS1927-2	72.48	-17.12	0	7	—	—	—	—	—	—	—	—	—	—	—	—	—	—	38	—
PS1951-1	68.83	-20.81	0	7	—	—	—	—	—	—	—	—	—	—	—	—	—	—	26	
SU8118	37.78	-10.19	9	13	10.7	7.3	13.0	8.5	8.9	6.2	34.6	1.8	2.8	4.8	86.7	20.7	110.9	23.7	169	3
SU9016	58.22	-45.16	7	7	4.4	3.2	7.7	4.6	2.2	2.6	33.2	1.4	3.0	3.9	55.5	30.7	66.2	27.5	92	2X
SU9019	59.53	-39.47	7	8	6.7	1.0	9.5	1.0	4.6	1.0	34.3	0.6	0.7	1.0	48.5	24.6	61.8	28.4	65	2X
SU9024	62.67	-37.38	4	7	6.7	1.0	10.1	1.0	4.3	1.1	34.0	1.2	0.3	0.4	51.6	27.1	64.0	30.6	140	2X
SU9032	61.79	-22.43	7	7	6.0	1.2	10.2	1.9	3.1	1.5	33.3	1.3	1.0	1.0	41.3	19.0	50.8	22.8	230	1
SU9033	60.57	-22.08	5	5	7.0	1.6	11.9	2.3	3.6	2.1	33.0	1.8	0.9	1.0	32.0	13.6	44.5	12.6	1021	2
SU9044	50.02	-17.10	2	2	8.8	1.1	11.7	1.7	6.6	0.6	34.7	0.6	0.0	0.0	60.2	0.1	71.9	7.5	7092	1
MD95-2033	44.66	-55.62	22	22	7.1	1.7	13.1	1.4	2.3	1.7	31.4	0.5	0.7	1.0	23.0	4.6	30.1	4.9	1527	2
NA87-22	55.51	-14.71	6	6	6.0	1.2	8.9	2.1	3.7	0.8	34.3	0.7	1.7	1.6	40.9	9.9	57.5	4.9		—
PS1730-2	70.12	-17.70	Barren	17	Barren		Barren		Barren		Barren		Barren		Barren		Barren		<10	—
SU8147	44.90	-3.31	17	17	7.6	1.5	13.3	1.5	3.2	2.1	32.2	1.0	0.6	0.9	48.0	11.1	57.7	10.0	10555	2
SU9039	52.57	-21.94	1	1	6.0	—	8.8	—	4.0	—	35.0	—	0.0	—	68.6	—	80.7	—		—
*MD99-2254	56.80	-30.66	5	5	6.7	1.0	9.6	1.2	4.6	0.8	34.6	0.6	0.6	1.0	37.2	13.6	53.1	9.1	43	2
*PAR87-A10	54.36	-148.47	3	3	3.3	3.1	9.6	1.2	-1.5	0.5	32.2	2.1	0.6	1.0	37.2	13.6	53.1	9.1	162	2
*ODP161-976c	36.21	-4.31	7	7	11.3	2.6	16.4	3.1	7.1	2.6	33.1	1.0	0.1	0.3	73.9	7.6	82.6	8.3	7041	2
HU-77-027-013	68.45	-63.53	Barren	7	Barren		Barren		Barren		Barren		Barren		Barren		Barren		<10	—
HU-84-030-003	53.32	-45.26	0	3	—	—	—	—	—	—	—	—	—	—	—	—	—	—	37	1
HU-84-030-021	58.37	-57.50	13	13	-0.3	1.7	1.2	2.6	-1.2	1.0	31.1	1.7	9.1	2.9	25.6	17.8	33.2	19.1	241	—
HU-85-027-016	70.51	-64.52	Barren	Barren	Barren		Barren		Barren		Barren		Barren		Barren		Barren		<10	—
HU-86-034-040	42.63	-63.10	5	5	6.4	1.1	12.6	0.5	1.7	0.9	31.6	0.7	1.7	0.6	20.5	7.2	27.2	5.8	2580	1
HU-87-033-007	65.40	-57.42	2	9	-0.7	1.1	0.1	1.3	-1.2	0.9	30.9	1.0	10.5	0.5	35.4	4.3	51.7	5.0	19	1X
HU-87-033-008	62.64	-53.88	7	7	0.0	2.4	1.6	4.1	-1.5	0.5	31.5	0.6	9.3	3.1	25.6	9.2	36.0	10.6	492	1
HU-87-033-009	62.51	-59.44	6	8	0.1	1.2	1.4	1.6	-0.8	0.6	31.1	0.7	9.6	1.1	25.7	16.3	41.5	20.7	85	2X
HU-90-013-012	58.92	-47.12	2	11	9.8	0.3	14.4	1.4	5.9	1.6	32.8	0.4	0.2	0.2	44.2	9.7	68.0	6.5	117	2
HU-90-013-013	58.21	-48.37	8	8	7.1	2.8	11.5	3.2	3.7	3.1	32.8	1.7	1.0	1.7	38.1	17.9	49.7	19.9	189	2
HU-90-015-017	42.78	-61.65	7	7	5.4	2.5	10.7	3.5	1.4	1.9	31.5	0.6	2.6	2.0	31.7	8.9	38.1	12.1	2277	1
HU-91-020-013	41.83	-62.33	6	6	5.9	3.7	10.9	5.2	1.4	2.4	30.1	1.5	2.5	3.7	25.8	5.4	35.0	9.1	1314	1
HU-91-045-025	55.03	-52.13	0	8	—	—	—	—	—	—	—	—	—	—	—	—	—	—	<10	—
HU-91-045-044	59.36	-43.45	4	5	-1.5	0.1	-0.8	0.1	-1.7	0.2	31.4	0.4	11.2	0.1	34.4	4.1	56.5	9.9	34	1X
HU-91-045-052	59.49	-39.30	7	9	6.3	2.2	10.2	2.5	3.6	2.5	33.0	2.0	1.7	2.5	49.2	9.6	60.1	5.3	55	2X

Table 2 (*continued*)

Cores	Latitude	Longitude	n	nt	Annual SST		Summer SST		Winter SST		Salinity		Ice cover		d—first analogue		d—fifth analogue		Dinocyst concentration	Reliability index
					Mean	St. dev.	Mean	St. dev.	Mean	St. dev.	Mean	St. dev.	Mean	St. dev.	Mean	St. dev.	Mean	St. dev.		
HU-91-045-058	59.84	−33.57	10	10	8.5	3.1	12.4	3.4	5.4	2.9	33.2	1.4	0.6	1.3	54.6	9.5	66.6	9.7	75	2X
HU-91-045-064	59.67	−30.57	6	6	7.7	1.8	11.1	2.4	5.3	1.2	34.0	0.6	0.4	0.7	56.6	11.0	68.7	10.9	65	2X
HU-91-045-072a	58.94	−28.74	9	9	5.2	2.8	7.9	3.8	3.4	2.2	33.6	1.0	2.9	2.9	49.4	20.9	66.1	22.7	70	2X
HU-91-045-074a	55.74	−30.22	5	5	7.8	2.4	11.7	2.7	4.9	1.8	34.2	0.6	0.2	0.3	60.2	9.3	71.1	10.0	76	2X
HU-91-045-080a	53.07	−33.52	10	10	8.5	2.3	12.9	2.8	5.2	2.5	33.2	1.4	0.5	0.9	65.5	16.8	75.3	20.6	104	2
HU-91-045-082a	52.86	−35.53	5	10	9.9	2.8	15.4	2.9	5.6	2.8	33.0	1.4	0.4	0.7	65.2	17.2	77.5	19.2	47	2X
HU-91-045-085	53.97	−38.64	3	3	11.0	3.0	14.8	2.9	8.1	2.8	33.5	0.7	0.2	0.3	62.5	8.9	81.5	4.1	134	2
HU-91-045-091a	53.33	−45.26	14	14	8.5	4.1	11.7	4.3	6.1	3.8	33.8	1.3	1.1	1.8	75.4	21.1	91.6	20.4	38	3X
HU-91-045-094a,c	50.20	−45.68	27	27	0.5	2.2	2.5	3.5	−0.7	1.2	31.0	1.8	8.4	3.1	30.3	19.2	39.6	20.5	1526	1
AII-94-PC3	62.09	−16.62	2	2	4.7	2.4	7.7	2.8	2.6	2.1	31.8	1.4	4.1	2.4	24.5	9.7	34.2	6.0	415	1
D89-BOFS-5K	50.69	−21.86	6	6	9.8	3.8	14.3	4.1	6.2	3.6	33.8	1.6	0.1	0.2	61.5	18.5	77.4	22.4	195	2
HM52-43	64.25	0.73	3	3	6.8	1.4	12.8	2.3	2.2	1.9	31.8	1.0	0.8	1.4	44.5	5.2	55.5	15.5	1143	2
HM71-19	69.49	−9.53	1	1	8.8	—	13.8	—	5.4	—	33.6	—	0.0	—	65.6	—	80.3	—	152	2
HM80-30	71.80	1.61	2	2	8.7	0.1	13.3	0.3	5.5	0.4	34.0	0.2	0.0	0.0	62.0	0.9	78.2	1.2	146	2
HM94-13	71.63	−1.63	1	1	3.3	—	6.4	—	1.1	—	31.9	—	5.3	—	63.6	—	69.3	—	238	2
HM94-25	75.60	1.30	1	1	7.1	—	10.9	—	4.1	—	33.0	—	2.2	—	63.3	—	76.7	—	37	2X
HM94-34	73.78	−2.54	3	3	8.7	1.6	13.0	1.7	5.5	1.7	33.6	1.2	0.7	1.2	71.8	22.0	88.8	30.6	202	2
M17730	72.05	7.31	1	1	9.2	—	13.6	—	5.8	—	33.9	—	0.0	—	89.1	—	101.4	—		3

All cores analysed for their LGM palynological content are listed in the table. The core location is also illustrated in Fig. 6. LGM data for the cores marked with an asterisk are reported in this manuscript for the first time. For all other cores, the reconstructions are updated from those published previously by de Vernal et al. (2000). The columns "*n*" and "*nt*" refer respectively to the number of spectra used for LGM reconstruction and to the total number of samples analysed in the LGM interval. The difference between the two numbers corresponds to the spectra discarded because of too low cyst counts, or no analogue situation. The mean and standard deviation (std. dev.) for each parameter are calculated on the basis of the most probable estimate of the "*n*" spectra retained for reconstructions. The mean distances of the first and fifth analogues also refer to the "*n*" spectra used for reconstructions. When available, the concentration of dinocysts is expressed as the number of cysts per cubic centimetre. The distance, which permits evaluation of the similarity of the modern analogues, is used to define a reliability index (1 = good; 2 = moderate; 3 = poor), which is further constrained by the concentration of dinocysts when possible (X for critically low concentration).

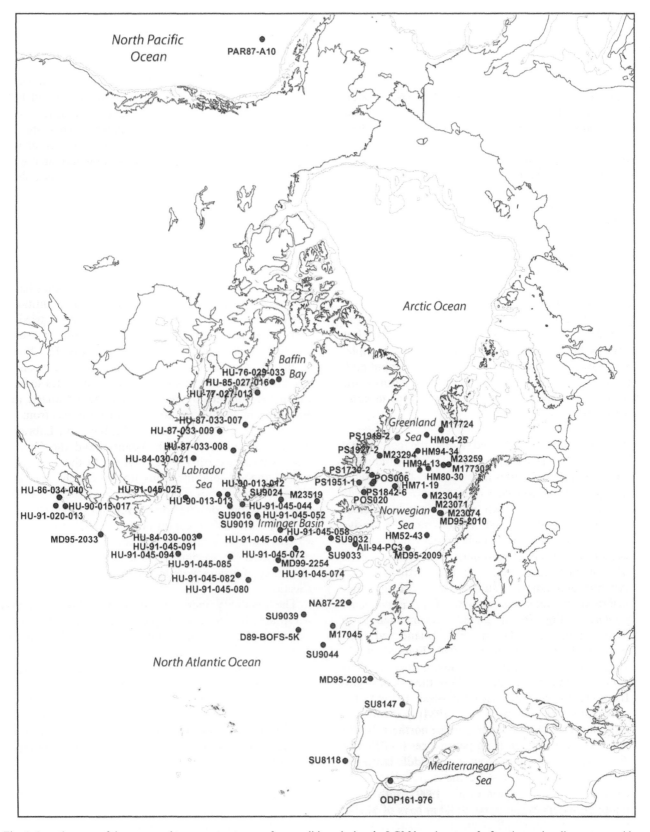

Fig. 6. Location map of the cores used to reconstruct sea-surface conditions during the LGM based on transfer functions using dinocyst assemblages (see Table 1 for core coordinates).

than 10 cysts/cm^3 (see Table 2), which we have interpreted as the result of limited biogenic production because of permanent or quasi-permanent sea-ice cover (see also de Vernal et al., 2000). In other cores, some samples within the LGM interval yielded low dinocyst concentrations (<40 cysts/cm^3), and their spectra were discarded (see Table 2 and Fig. 8; detailed counts and raw data tables are archived in the PANGAEA database and at GEOTOP).

As a general feature, the LGM samples from the northern North Atlantic contain dinocyst assemblages characterised by low concentrations, generally ranging between 10^1 and 10^3 cysts/cm^3, with higher values recorded along the margins of southeastern Canada and off western Europe and Scandinavia (Table 2). In the calibration database established from surface sediment samples (Rochon et al., 1999), the abundance of dinocysts is also at a maximum along the margins off southeastern Canada, Western Europe and Scandinavia. However, the concentrations are higher by one order of magnitude than those of the LGM samples. The low dinocysts concentrations in the LGM sediment samples indicate low productivity, due to low nutrient input and/ or harsh conditions. Beyond these broad features, the dinocyst assemblages show some peculiarities in comparison with the modern ones:

(1) The assemblages recovered along the continental margins of northeastern Canada and Scandinavia show a major southward shift of taxa usually associated with sea-ice cover, notably *Islandinium minutum* (Fig. 7a). Such assemblages have close modern analogues, and reveal more extensive seasonal sea ice, and much colder than present conditions especially in winter.

(2) The offshore assemblages of the northern North Atlantic and adjacent subpolar seas are all characterised by high percentages of *Bitectatodinium tepikiense* (Fig. 7b), as already documented from many studies (e.g., Turon, 1984; Duane and Harland, 1990; Graham et al., 1990; Eynaud et al., 2002). In surface sediment samples, this taxon is common but rarely exceeds 10% of the assemblages. Its modern occurrence has been associated with the cool temperate domain (Turon, 1984) and with the subpolar–temperate boundary in the northern North Atlantic (Dale, 1996). High percentages ($>10\%$) of *B. tepikiense* have been reported at middle latitudes, in the North Sea and along the margins of southeastern Canada (Rochon et al., 1999), with maximum abundances (up to 60–80%) in bays of Maine and Nova Scotia (Wall et al., 1977; Mudie, 1992). The modern distribution of *B. tepikiense* indicates a tolerance to a wide range of salinities and temperatures in winter, and a preference for summer temperatures greater than 10 °C. Its maximum occurrence in coastal bays of southern Nova Scotia (Mudie, 1992) suggests special affinities for stratified surface waters characterised by large seasonal amplitudes of temperature from winter to summer (up to 15 °C) and low salinity (30–32‰). Therefore, the LGM dinocyst assemblages recovered offshore in the northern North Atlantic demonstrate very different conditions than at present. They show a relatively high degree of dissimilarity when compared to modern spectra, and the closest modern analogues for these assemblages are located in nearshore environments of the cool temperate domain.

3.3. The reliability of sea-surface condition estimates for the LGM

Beyond intrinsic limitations of any approaches based on the use of microfossil assemblages for quantitative paleoceanographical reconstructions, we have tried to clarify the reliability of estimates from dinocyst data using the indices defined in Section 2.5 (see Fig. 8 and Table 2). The reliability index based on the distance reveals good analogue situations for most sites located along the continental margins of eastern Canada and Scandinavia, in addition to a few offshore sites from the Iceland Basin, Irminger Sea, Baffin Bay, and Labrador Sea in the northern North Atlantic, and the Gulf of Alaska in the North Pacific (Fig. 8). At these sites, the LGM dinocyst concentrations are moderately high, with the exception of sites from the Irminger Basin and Baffin Bay. Therefore, despite some limitations, the reliability of LGM sea-surface condition estimates for the southeastern Canadian margins, Labrador Sea, Iceland Basin and eastern Norwegian Sea is reasonably high, within the range of accuracy defined by the validation exercise (Section 2.4).

The reliability index based on the distance shows acceptable but weak analogue situations in many cores of the northern North Atlantic, and the Greenland and Irminger seas. The weak analogue situation is notably due to the high frequency of *B. tepikiense* in LGM assemblages, which have no close equivalent in offshore areas of the modern database. At the sites from the Greenland and Irminger seas, the situation is particularly critical in view of the low dinocyst concentrations. In these areas, the confidence level of reconstructions is therefore lower.

3.4. Results

3.4.1. Sea-surface temperatures

The SST estimates based on dinocyst assemblages reveal LGM conditions that differ significantly from the modern situation with regard to the geographical distribution pattern of temperatures (Fig. 9). While

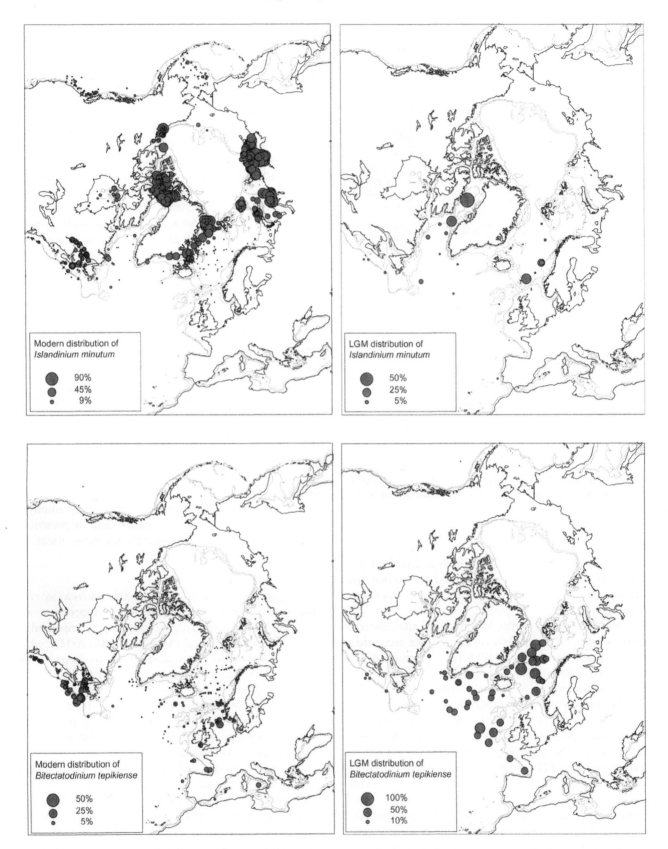

Fig. 7. Geographical distribution patterns of two characteristic dinocyst taxa in the modern (left diagrams) and LGM (right diagrams) databases. (a) Percentages of *I. minutum*; and (b) percentages of *B. tepikiense*.

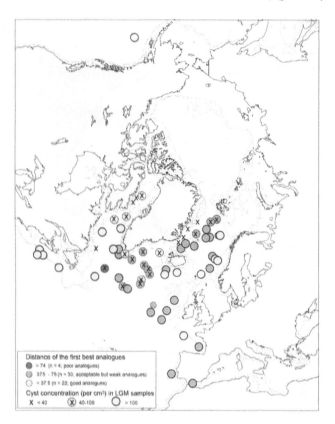

Fig. 8. Schematic illustration of the reliability of estimated sea-surface conditions for the LGM based on dinocyst data: the mean distance of the closest modern analogues permits the definition of a reliability index that is further constrained from the dinocyst concentrations (Table 2). Pale circles represent sites of relatively reliable LGM estimates and dark circles marked with "X" correspond to sites of less reliable LGM estimates (see text, Section 3.3).

the reconstructed LGM seasonal contrast of temperature in surface waters is larger than at present.

3.4.2. Sea-surface salinities

A particular feature of sea-surface condition estimates in the northern North Atlantic during the LGM is the low salinity, below 35, even at the most oceanic sites (see Fig. 10), which is much lower than at present. Very low salinities, ranging from 30 to 32, are reconstructed along the northeast margins of North America and off Scandinavia. The particularly low salinity recorded in surface waters of the Labrador Sea corresponds to areas also marked by extensive sea-ice cover. In such cases, the dilution in surface water can be associated with summer melting of sea ice. Such is not the case along the eastern Norwegian and southeastern Canadian margins, where the estimated average duration of seasonal sea ice during the LGM was restricted to a few weeks per year (cf. Fig. 10). The distribution pattern of sea-surface salinity estimates may indicate significant dilution resulting from meltwater discharges along the margins of the Laurentide and Fennoscandian ice sheets, which reached their maximum extent at that time.

In the offshore domain of the northern North Atlantic, the reconstructed sea-surface salinity ranges up to 35. This is lower by about one unit as compared to the present, suggesting that the dilution in surface waters was significant, beyond the degree of uncertainties of the reconstructions. The negative LGM anomaly of salinity in the northern North Atlantic is even more significant when taking into account that the salinity of the global LGM Ocean was higher than at present by approximately one (e.g., Broecker and Peng, 1982).

3.4.3. Sea-ice cover extent

In polar and subpolar environments, sea ice is a parameter closely interrelated with temperature and salinity. The freezing of sea water is accompanied by brine formation (e.g., Gascard et al., 2002), which usually sinks and may contribute to deep mixing, whereas the summer melt results in a low-salinity buoyant upper water layer. Thus, sea ice exerts a primary control on the thermohaline structure of the upper water mass and develops seasonally in areas that are often characterised by a strong stratification in the water column, at least during summer.

The LGM sea-ice distribution shows very important gradients in the northern North Atlantic, with an extensive cover in Baffin Bay and along the eastern continental margin of Canada and Greenland. At these locations, the proximity of the ice sheet margins, which were grounded on the shelves, may have fostered dense sea-ice formation: meltwater discharge from the base of the ice sheets together with iceberg flux no doubt resulted in the existence of a low saline and cold surface

some regions show negative anomalies (i.e., LGM minus present) as large as 10 °C, others are characterised by insignificant difference or even positive anomalies (Fig. 9; Table 3).

In the northwest North Atlantic, off the eastern margin of Canada, very cold conditions are recorded, both in summer and winter. Offshore, a sharp gradient of increasing temperatures is reconstructed, especially for the summer (Fig. 9). Over mid-latitudes, summer SSTs ranging up to 19 °C reveal relatively mild conditions, but still significantly cooler than the modern ones at most sites. In the subpolar basins of the Irminger, Greenland and Norwegian seas, however, LGM reconstructed summer SSTs are warmer than present (Fig. 9). The estimated SSTs in winter are less extreme, showing colder to cooler conditions than at present at most locations. The only exception concerns the Greenland Sea where warmer than present conditions are reconstructed. This is a peculiar, but apparently consistent feature.

On the whole, the anomalies in SSTs are more negative in winter than in summer (Fig. 9). Therefore,

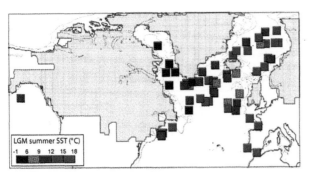
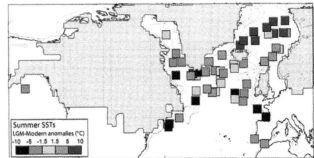
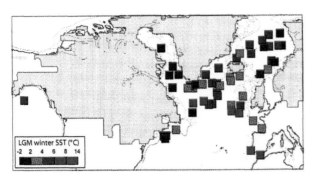
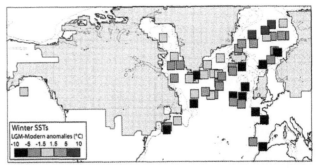

Fig. 9. Maps showing LGM SST estimates in summer (upper left diagram) and winter (bottom left diagram) and the LGM vs. modern SST anomalies in summer (upper right diagram) and winter (bottom right diagram). Note that anomalies within the ±1.5 °C range are not considered to be significant taking into account the accuracy of reconstruction and that of modern hydrographical averages (see text, Section 3.4). The continental ice limits are delimited after Peltier (1994).

water layer leading to seasonal freezing and pack ice development as in modern circum-Antarctic seas.

In some of the cores, the palynological analyses reveal barren or close to barren assemblages, which we associate to a close to nil productivity due to permanent or quasi-permanent multiyear sea ice, as it is the case in areas of the Arctic Ocean with permanent pack ice (cf. Rochon et al., 1999). Close to barren assemblages are recorded in Baffin Bay, on the slope off Labrador, and along the eastern Greenland margins (see Fig. 8). Analyses of nearby sequences (from Baffin Bay, and Labrador Sea, notably) support an interpretation of quasi-perennial sea ice, with extensive cover of more than 9 months/year. On these grounds, we may tentatively draw the probable limit of quasi-permanent sea ice during the LGM (see dashed gray line in Fig. 10, upper left), which seems to have been close to the limit of the continental shelf off eastern Canada and Greenland.

LGM data also indicate that the North Atlantic was characterised by a zone with dense sea-ice cover that was relatively narrow and confined to the eastern continental margins. Offshore, in subpolar seas, seasonal sea ice spanning up to a few weeks or a few months per year is reconstructed. The heterogeneity in estimates from one site to another in the Irminger Basin or the Greenland-Norwegian seas can be attributed to the extreme variability of the sea-ice parameter both in time and space (Comiso, 2002). In the eastern sector of the North Atlantic south of about 50°N (see the dashed pink line in

Fig. 10, upper left), however, the data are not equivocal and show that ice-free conditions prevailed throughout the year on an average basis.

3.5. Comparison with previous LGM estimates based on dinocysts

As mentioned before, this manuscript presents an update of the LGM reconstructions published by de Vernal et al. in 2000. Most primary data are the same, with the exception of a few additional sites or additional spectra for some cores (Table 2). However, there are some differences in the databases used for reconstructions, as summarised below.

(1) Here, we used seasonal averages of SSTs at 10 m depth from NODC (1998) as prescribed for the MARGO intercomparison exercise, instead of monthly averages compiled at 0 m depth, mainly on the basis of the data comprised in the 1994 version of the NODC Atlas. The differences from the temperatures compiled as described above are relatively low on the average but show a rather large dispersal as illustrated in Fig. 11. The August SSTs at 0 m in the NODC-1994 data set are slightly higher than the summer SSTs at 10 m in the NODC-1998 data set, with an average difference of 1.0 ± 1.1 °C. Inversely, the February SSTs at 0 m in the NODC-1994 data set are slightly lower than the winter

Table 3
Anomalies of sea-surface conditions between the LGM and the modern

Cores	Longitude	Latitude	Annual SSTs			Winter SSTs			Summer SSTs			Salinity			Sea ice (months/year)		
			LGM	Modern	Δ	LGM	Modern	Δ	LGM	Modern	Δ	LGM	Modern	Δ	LGM	Modern	Δ
HU-76-029-033	−64.27	71.33	−0.2	−0.2	0.0	−1.9	−1.6	−0.3	1.9	2.0	−0.2	29.7	29.0	0.6	9.3	9.3	0.0
M17045	−16.39	52.43	7.1	12.5	−5.4	4.5	10.6	−6.2	10.7	15.0	−4.4	34.3	35.5	−1.2	0.2	0	0.2
M17724	8.33	76	6.7	3.1	3.6	1.7	1.7	0.0	12.4	5.5	6.9	31.1	34.9	−3.8	1.5	0	1.5
M23071	2.9	67.08	7.9	7.9	0.0	4.1	6.2	−2.1	13.3	10.5	2.8	33.0	35.1	−2.1	0.4	0	0.4
M23074	4.91	66.66	4.5	8.2	−3.7	0.2	6.6	−6.5	9.7	10.9	−1.2	31.5	34.9	−3.4	3.7	0	3.7
M23259	9.27	72.03	6.8	5.6	1.1	1.8	4.2	−2.4	13.0	8.0	5.0	31.5	35.1	−3.7	1.3	0	1.3
M23294	−10.36	72.36	8.2	0.4	7.8	5.5	−0.9	6.3	11.8	3.1	8.8	34.2	32.1	2.1	0.0	4.5	−4.5
M23519	−29.56	64.83	6.9	5.7	1.2	4.7	4.7	0.0	10.2	7.2	3.0	33.9	35.1	−1.2	1.5	0	1.5
MD95-2002	−8.53	47.45	6.3	13.9	−7.5	2.1	11.4	−9.3	11.8	17.2	−5.4	32.5	35.5	−3.1	1.5	0	1.5
MD95-2009	−4	62.74	6.3	7.9	−1.6	0.9	6.1	−5.2	12.5	10.3	2.2	31.4	35.1	−3.8	1.5	0	1.5
MD95-2010	4.57	66.68	4.6	8.2	−3.6	0.0	6.6	−6.6	10.3	10.8	−0.5	31.6	35.1	−3.5	2.7	0	2.7
POS0020	−18.53	67.98	8.4	2.1	6.3	5.0	0.7	4.3	13.0	4.4	8.6	33.2	33.6	−0.4	0.4	0.8	−0.4
PS1842-6a	−16.52	69.45	7.4	1.1	6.4	3.3	0.0	3.3	12.9	3.3	9.6	31.7	33.7	−2.0	1.0	4.4	−3.4
SU8118	−10.19	37.78	10.7	17.7	−6.9	8.9	15.4	−6.5	13.0	20.0	−7.0	34.6	36.1	−1.5	2.8	0	2.8
SU9016	−45.16	58.22	4.4	4.8	−0.4	2.2	3.4	−1.2	7.7	7.2	0.5	33.2	34.6	−1.4	3.0	0	3.0
SU9019	−39.47	59.53	6.7	5.8	0.9	4.6	4.3	0.4	9.5	8.0	1.4	34.3	34.8	−0.5	0.7	0	0.7
SU9024	−37.38	62.67	6.7	5.6	1.1	4.3	4.4	−0.1	10.1	7.9	2.3	34.0	34.9	−0.9	0.3	0	0.3
SU9032	−22.43	61.79	6.0	9.1	−3.1	3.1	7.5	−4.4	10.2	11.3	−1.1	33.3	35.2	−1.9	1.0	0	1.0
SU9033	−22.08	60.57	7.0	9.4	−2.4	3.6	7.9	−4.4	11.9	11.5	0.5	33.0	35.1	−2.2	0.9	0	0.9
SU9044	−17.1	50.02	8.9	13.3	−4.5	6.6	11.2	−4.6	11.7	16.0	−4.4	34.7	35.5	−0.8	0.0	0	0.0
MD95-2033	−55.62	44.66	7.1	−1.4	8.6	2.3	−1.0	3.3	13.1	−1.5	14.6	31.4	32.1	−0.7	0.7	0	0.7
NA87-22	−14.71	55.51	6.0	11.4	−5.4	3.7	9.8	−6.1	8.9	13.8	−4.9	34.3	35.4	−1.1	1.7	0	1.7
SU8147	−3.31	44.9	7.6	15.2	−7.6	3.2	11.8	−8.5	13.3	19.4	−6.1	32.2	35.0	−2.9	0.6	0	0.6
SU9039	−21.94	52.57	6.0	12.0	−6.0	4.0	10.2	−6.2	8.8	14.5	−5.7	35.0	35.3	−0.4	0.0	0	0.0
MD99-2254	−30.66	56.8	6.7	8.6	−1.9	4.6	6.8	−2.2	9.6	10.9	−1.3	34.6	35.1	−0.5	0.6	0	0.6
PAR87-A10	−148.47	54.36	3.3	7.1	−3.8	4.6	4.2	0.5	9.6	11.3	−1.7	32.2	32.7	−0.4	0.6	0	0.6
ODP161-976	−4.31	36.21	11.3	18.0	−6.7	7.1	15.3	−8.3	16.4	21.4	−5.0	33.1	36.5	−3.4	0.1	0	0.14
HU-84-030-021	−57.5	58.37	−0.3	3.4	−3.7	−1.2	2.4	−3.6	1.2	6.2	−5.0	31.1	34.4	−3.3	9.1	0	9.1
HU-86-034-040	−63.1	42.63	6.4	10.5	−4.1	1.7	5.1	−3.4	12.7	17.0	−4.4	31.6	31.7	−0.1	1.7	0	1.7
HU-87-033-007	−57.42	65.4	−0.7	1.1	−1.8	−1.2	−0.1	−1.0	0.1	3.6	−3.5	30.9	32.6	−1.7	10.5	4	6.5
HU-87-033-008	−53.88	62.64	0.0	2.5	−2.6	−1.5	1.2	−2.7	1.6	5.2	−3.6	31.5	33.7	−2.2	9.3	0.5	8.8
HU-87-033-009	−59.44	62.51	0.1	2.1	−2.0	−0.8	1.7	−2.5	1.4	4.9	−3.5	31.1	33.8	−2.6	9.6	2.9	6.7
HU-90-013-012	−47.12	58.92	9.8	4.4	5.4	5.9	3.0	2.9	14.4	6.7	7.7	32.8	34.7	−1.8	0.2	0	0.2
HU-90-013-013	−48.37	58.21	7.1	4.7	2.4	3.7	3.2	0.6	11.5	7.3	4.2	32.9	34.6	−1.7	1.0	0	1.0
HU-90-015-017	−61.65	42.78	5.4	11.1	−5.7	1.4	5.8	−4.4	10.7	17.8	−7.1	31.5	32.0	−0.5	2.6	0	2.6
HU-91-020-013	−62.33	41.83	5.9	13.7	−7.8	1.4	8.8	−7.4	10.9	19.7	−8.8	30.1	33.7	−3.6	2.5	0	2.5
HU-91-045-044	−43.45	59.36	−1.5	4.9	−6.4	−1.7	3.4	−5.1	−0.8	6.9	−7.6	31.4	32.8	−1.4	11.2	2.5	8.7
HU-91-045-052	−39.3	59.49	6.4	5.9	0.5	3.6	4.3	−0.7	10.2	8.1	2.1	33.0	34.7	−1.7	1.7	0	1.7
HU-91-045-058	−33.57	59.84	8.5	7.3	1.2	5.4	5.7	−0.3	12.4	9.7	2.7	33.2	34.9	−1.7	0.7	0	0.7
HU-91-045-064	−30.57	59.67	7.8	8.0	−0.3	5.3	6.5	−1.2	11.1	10.2	0.9	34.0	35.0	−1.0	0.4	0	0.4
HU-91-045-072a	−28.74	58.94	5.2	8.6	−3.4	3.4	7.1	−3.7	7.9	10.6	−2.8	33.6	35.0	−1.4	2.9	0	2.9
HU-91-045-074a	−30.22	55.74	7.8	9.0	−1.2	4.9	7.2	−2.3	11.7	11.4	0.3	34.2	35.0	−0.8	0.2	0	0.2
HU-91-045-080a	−33.52	53.07	8.5	9.2	−0.6	5.2	7.0	−1.8	13.0	12.1	0.9	33.2	34.7	−1.5	0.5	0	0.5
HU-91-045-082a	−35.53	52.86	9.9	8.8	1.2	5.6	6.5	−0.9	15.4	11.7	3.7	33.0	34.6	−1.7	0.4	0	0.4
HU-91-045-085	−38.64	53.97	11.0	7.6	3.4	8.1	5.3	2.8	14.8	10.5	4.4	33.5	34.7	−1.1	0.2	0	0.2
HU-91-045-091a	−45.26	53.33	8.5	7.6	0.9	6.2	5.5	0.7	11.7	10.6	1.1	33.8	34.6	−0.8	1.1	0	1.1
HU-91-045-094a,c	−45.68	50.2	0.5	7.9	−7.5	−0.7	5.3	−6.0	2.5	11.5	−9.1	31.0	33.8	−2.9	8.4	0	8.4
AII-94-PC3	−16.62	62.09	4.7	9.3	−4.6	2.6	7.8	−5.2	7.7	11.3	−3.7	31.8	35.2	−3.4	4.1	0	4.1
D89-BOFS-5K	−21.86	50.69	9.8	12.8	−3.0	6.2	10.9	−4.7	14.4	15.4	−1.0	33.8	35.4	−1.6	0.1	0	0.1
HM52-43	0.73	64.25	6.8	8.2	−1.4	2.2	6.3	−4.1	12.9	11.0	1.9	31.8	35.2	−3.4	0.8	0	0.8
HM71-19	−9.53	69.49	8.8	2.5	6.3	5.4	0.7	4.7	13.8	5.6	8.2	33.6	34.7	−1.1	0.0	0	0.0
HM80-30	1.61	71.8	8.7	3.9	4.8	5.5	2.1	3.5	13.3	6.9	6.4	34.0	35.0	−1.1	0.0	0	0.0
HM94-13	−1.63	71.63	3.3	3.2	0.1	1.1	1.4	−0.3	6.5	6.4	0.1	31.9	34.7	−2.8	5.3	0	5.3
HM94-25	1.3	75.6	7.1	0.8	6.4	4.1	−1.0	5.1	10.9	3.9	6.9	33.0	34.4	−1.3	2.2	0.9	1.3
HM94-34	−2.54	73.78	8.7	0.6	8.0	5.5	−1.3	6.8	13.0	4.1	8.9	33.6	33.8	−0.2	0.7	2.1	−1.4
M17730	7.31	72.05	9.2	5.2	4.1	5.8	3.6	2.2	13.6	7.7	5.9	33.9	35.1	−1.2	0.0	0	0.0

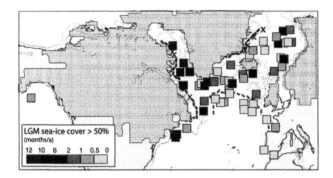
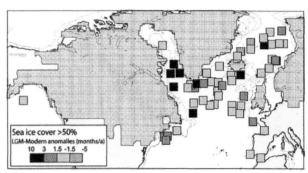
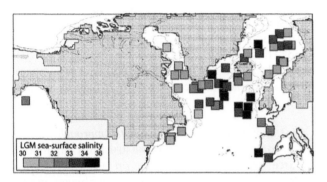
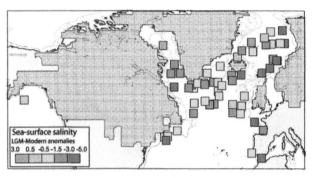

Fig. 10. Maps showing LGM sea-surface salinity estimates in summer (bottom left diagram), the seasonal extent in months per year of sea-ice cover with concentration greater than 50% (upper left diagram), the LGM vs. modern sea-surface salinity anomalies in summer (bottom right diagram) and the LGM vs. modern sea-ice cover extent (upper right diagram). The continental ice limits are delimited after Peltier (1994). In the upper left diagram, the dashed gray and pink lines would correspond to the southern limits of quasi-permanent pack-ice and extreme winter sea-ice cover, respectively.

SSTs at 10 m in the NODC-1998 data set, by $-0.2 \pm 0.9\,°C$. The largest differences concern the Arctic where there is limited information, and where the accuracy of hydrographic data is low.

(2) The reference dinocyst database includes 940 stations from three oceans (Arctic, Pacific, Atlantic) and 60 taxa, instead of 371 stations from one ocean (Atlantic) and 25 taxa (cf. Rochon et al., 1999). The updated database is representative of a wider range of environmental and hydrographic conditions, in both the Arctic and temperate domains.

(3) In addition to these differences with respect to databases used for the reconstructions, the procedures of data treatment were not exactly the same. In the case of the 2000 compilation, we have used the software provided by Guiot (1990) and we made estimates based on a set of 10 analogues, whereas we are now using the software 3PBase of Guiot and Goeury (1996) and we calculate them based on a set of five analogues. Tests of reproducibility have shown, however, that the two procedures yield almost identical results.

The LGM reconstructions of SSTs presented here (Table 2) are very similar to the ones which were published by de Vernal et al. in 2000 (see Table 4). On average, for the 50 sites used in both LGM compilations, the difference between summer and August SST reconstruction is $0.61 \pm 2.15\,°C$ and the average difference between winter and February SSTs was $0.64 \pm 1.1\,°C$. Such discrepancies are not significant given the differences in the two temperature databases and the calculated error of prediction. Similarly, the average differences in estimated salinity and sea ice are -0.23 ± 0.66 and -0.07 ± 1.14 months/year, respectively. Such differences are not significant either, given the range of accuracy of estimates. We are thus led to conclude that both sets of reconstruction are consistent and that the expansion of the reference database, from 371 to 940 stations, has a limited effect on estimating sea-surface conditions of the LGM in the northern North Atlantic.

4. Discussion

4.1. Uncertainties

4.1.1. Significance of anomalies

The reconstruction of hydrographical parameters based on microfossil assemblages implies a number of assumptions. One concerns the correspondence between

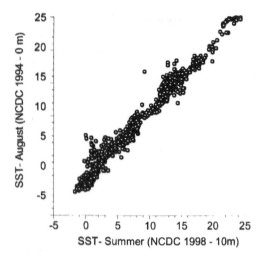

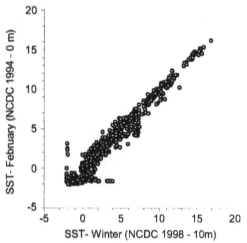

Fig. 11. Graphs showing the differences of SSTs in the two hydrographic databases (World Ocean Atlas versions of 1998 and 1994), which were used to estimate LGM sea-surface conditions in the present compilation and in the one published previously by de Vernal et al. (2000).

the "modern" assemblages recovered in surface sediment samples and the reference hydrographical data, which we assume to be contemporaneous. The interval represented by the microfossil assemblages may range from 10 to 1000 years, whereas mean value of hydrographic data collected over the last decades provide an average that is not necessarily accurate or representative of maximum productivity years. This is a problem especially when dealing with nearshore and circum-Arctic environments, where measurements are rare and where salinity, sea ice or temperature can be extremely variable in space and time. The degree of uncertainty or the inaccuracy of hydrographical averages can be illustrated by the comparison of salinity and temperature fields produced by NODC (1994) and Bedford Institute of Oceanography (BIO) (2003) following the method of analyses of Tang and Wang (1996). The comparison shows that there is a basic agreement in

open ocean, whereas significant differences are being recorded for the shelf and coastal ocean. As an example, the sharp front along the shelf edge of eastern Canada (Labrador Shelf and Grand Banks) clearly depicted in the gridded data from BIO (2003) is absent in the NODC Atlas. This is particularly critical in the case of the dinocyst database, which includes an important proportion of data points from epicontinental and nearshore areas. An illustration of the uncertainty concerning the hydrographical averages can also be found in the mapping of the standard deviations (one sigma) around the temperature average. The sigma value revealed to be very large, up to 4 °C, along transitional zones such as those marked by sea-ice limits or the polar front in the North Atlantic (cf. e.g., Isemer and Hasse, 1985). Actually, the standard deviation around the average for instrumental data is comparable to the accuracy of reconstruction defined by validation exercises.

4.1.2. Weaknesses of the actualistic approach

Beyond the accuracy of reconstructions, an important source of uncertainty relies on the fact that the relationships between microfossil assemblages and SSTs are not unequivocal and may have changed through time, because of changes in the structure of water masses or productivity (cf. e.g., Fairbanks and Wiebe, 1980; Faul et al., 2000; de Vernal et al., 2002). In the case of dinocyst assemblages, there are clear relationships with the distribution of seasonal temperature, salinity and sea ice. However, the dinocyst distribution is also dependent upon other parameters, such as the trophic structure of planktonic populations (e.g., Devillers and de Vernal, 2000; Radi and de Vernal, 2004). During the LGM, lower dinocyst fluxes than at present characterised the northern North Atlantic. This suggests that nutrient distribution and productivity were different, which may introduce a bias when making quantitative reconstructions of SST or salinity.

Another source of uncertainty lies in the fact that the reconstructed LGM sea-surface conditions are not well represented in the modern hydrographic database. The dinocyst database is representative of a particularly wide range of sea-surface conditions as compared to other biogenic tracers, such as foraminifera, coccoliths or alkenones, which show relationships with temperature, but within a narrow salinity spectra and almost exclusively in ice-free areas. The LGM sea-surface conditions in most of the northern North Atlantic apparently belong to a domain characterised by seasonal sea-ice cover and relatively low salinity. In such a context, dinocysts appear to be much more sensitive indicators than many other microfossils that are rather representative of open ocean conditions. Therefore, dinocysts are likely to provide more adequate estimates than stenohaline micro-organisms. However,

Table 4
Difference between estimates of LGM sea-surface conditions presented here (black, normal characters) and those published in 2000 by de Vernal et al. (italic characters)

Cores	Latitude	Longitude	n	n	Annual SSTs Mean	St. dev	Annual SSTs Mean	Δ-mean	Summer SSTs Mean	St. dev	August SSTs Mean	St. dev	Δ-mean	Winter SSTs Mean	St. dev	February SSTs Mean	St. dev	Δ-mean	Salinity Mean	St. dev	Salinity mean Mean	St. dev	Δ-mean	Sea ice Mean	St. dev	Sea ice Mean	St. dev	Δ-mean	Distance min mean
M17045	52.43	-16.39	2	2	7.1	1.1	6.2	**0.9**	10.7	0.8	9.2	0.3	**1.5**	4.5	1.0	3.2	0.8	**1.3**	34.2	0.1	34.8	0.1	**-0.6**	0.2	0.3	0.0	0.0	**0.2**	40.3
M17724	76.00	8.33	4	5	6.7	2.4	7.4	**-0.7**	12.4	3.3	13.3	3.1	**-0.9**	1.7	1.9	1.5	1.0	**0.2**	31.1	0.5	32.3	0.8	**-1.2**	1.5	2.7	1.1	1.9	**0.4**	47.8
M23071	67.08	2.90	4	4	7.9	1.4	7.6	**0.3**	13.3	0.4	13.7	1.3	**-0.4**	4.1	2.1	1.4	2.1	**2.7**	33.0	1.0	32.1	1.3	**0.9**	0.4	0.5	0.8	0.6	**-0.4**	52.6
M23074	66.66	4.91	5	4	6.9	0.3	6.7	**0.2**	9.7	4.6	13.8	1.3	**-4.1**	0.2	1.6	-0.5	1.6	**0.7**	31.5	0.7	30.8	0.7	**0.7**	3.7	3.3	1.8	3.3	**1.9**	35.4
M23259	72.03	9.27	11	13	6.8	1.5	7.5	**-0.7**	13.0	1.1	13.8	2.8	**-0.8**	1.8	1.7	1.2	1.4	**0.6**	31.6	0.7	31.6	0.7	**-0.1**	1.3	1.1	1.6	1.1	**-0.3**	35.5
M23294	72.36	-10.36	2	3	8.2	0.9	6.6	**1.6**	11.8	1.9	9.3	1.3	**2.5**	5.5	0.5	3.8	0.6	**1.7**	34.2	0.1	34.0	0.7	**0.2**	0.0	0.0	2.2	2.0	**-2.2**	55.0
M23519	64.83	-29.56	2	4	6.9	0.2	8.2	**-1.3**	10.2	0.5	13.1	1.8	**-2.9**	4.7	0.1	3.3	2.0	**1.4**	33.9	0.7	33.3	1.1	**0.6**	1.5	1.3	0.7	1.1	**0.8**	20.5
MD95-2009	62.74	-4.00	2	4	6.7	3.0	8.3	**-1.6**	12.4	4.1	15.4	3.4	**-3.0**	1.5	2.9	-1.0	2.0	**0.4**	31.4	1.4	31.3	0.9	**0.1**	1.5	1.3	2.0	0.9	**-0.7**	33.8
MD95-2010	66.68	4.57	25	7	4.3	1.7	6.2	**-1.9**	9.9	3.1	9.9	3.1	**-3.5**	-0.2	0.8	-1.0	0.8	**0.8**	31.6	0.7	31.1	0.6	**0.5**	3.0	1.9	2.5	1.6	**-0.5**	27.1
POS0020	67.98	-18.53	7	7	8.4	1.4	8.9	**-0.5**	13.0	1.0	14.2	1.4	**-1.2**	5.0	2.0	3.6	2.4	**1.4**	33.2	1.1	32.9	1.5	**0.3**	0.4	0.9	0.7	0.8	**-0.3**	42.6
PS1842-6a	69.45	-16.52	8	12	7.4	1.5	8.9	**-1.5**	12.9	1.0	14.1	1.9	**-1.2**	3.3	2.2	3.7	2.6	**-0.4**	31.7	1.3	33.1	1.5	**-1.4**	1.0	0.9	0.7	0.6	**0.2**	49.0
SU8118	37.78	-10.19	3	4	14.3	3.2	16.5	**-2.2**	17.1	4.6	18.8	4.7	**-1.7**	12.1	0.7	14.2	1.2	**-2.1**	35.5	0.8	36.1	1.7	**-0.6**	0.2	0.3	0.1	0.1	**0.1**	101.5
SU9016	58.22	-45.16	7	16	6.7	1.0	8.2	**-0.7**	7.7	4.6	8.6	1.6	**-0.9**	4.6	2.6	5.0	2.0	**0.5**	33.2	1.4	33.2	1.7	**0.0**	3.0	2.8	2.8	3.3	**0.2**	55.5
SU9019	59.53	-39.47	7	7	6.7	1.0	5.9	**-1.5**	9.5	1.0	11.3	1.6	**-1.8**	4.3	1.1	2.7	1.5	**1.6**	34.3	0.6	33.7	0.8	**0.3**	0.7	1.0	0.3	1.4	**-0.4**	48.5
SU9024	62.67	-37.38	4	7	6.7	1.0	5.9	**0.9**	10.1	1.0	9.0	1.4	**1.1**	4.3	1.1	2.6	1.9	**0.5**	34.0	1.3	33.7	0.8	**0.1**	0.7	0.4	1.0	1.5	**-1.3**	51.6
SU9032	61.79	-22.43	7	7	6.0	1.6	7.4	**-1.4**	10.2	1.9	12.2	1.8	**-2.0**	3.1	1.5	2.6	1.9	**0.5**	33.3	1.3	33.2	1.3	**0.1**	1.0	1.0	1.0	0.6	**0.0**	41.3
SU9033	60.57	-22.08	5	5	7.0	1.6	8.7	**-1.7**	11.9	2.3	14.0	0.9	**-2.1**	3.6	2.1	3.4	1.8	**0.2**	33.0	1.8	32.9	0.6	**0.1**	0.9	0.9	1.0	0.6	**0.0**	32.0
SU9044	50.02	-17.10	2	2	8.8	1.1	7.7	**1.2**	11.7	1.7	10.4	0.0	**1.3**	6.6	0.6	4.9	0.1	**1.7**	34.7	0.6	35.0	0.5	**-0.3**	0.0	0.1	0.8	1.0	**-0.1**	60.2
MD95-2033	44.66	-55.62	22	21	7.1	1.7	8.7	**-1.6**	13.1	1.4	15.8	0.6	**-2.7**	2.3	1.7	1.6	1.4	**0.7**	31.4	0.5	31.4	0.4	**-0.1**	0.7	1.0	0.8	0.4	**-0.1**	23.0
NA87-22	55.51	-14.71	6	6	6.0	1.2	6.1	**-0.1**	8.9	2.1	9.3	2.1	**-0.4**	3.7	0.8	2.9	0.4	**0.8**	34.3	0.7	34.8	0.4	**-0.5**	1.7	1.6	1.2	1.5	**0.5**	40.9
SU8147	44.90	-3.31	17	8	7.6	1.5	7.8	**-0.2**	13.3	1.5	14.5	1.5	**-1.2**	3.2	2.1	1.0	0.6	**2.2**	32.2	1.0	32.0	0.8	**0.2**	0.6	0.9	0.2	0.1	**0.4**	48.0
SU9039	52.57	-21.94	1	1	6.0	1.5	8.3	**-2.3**	8.8	—	11.5	—	**-2.7**	4.0	—	5.1	—	**-1.1**	35.1	—	35.1	—	**-0.1**	0.0	0.0	0.0	0.0	**0.0**	68.6
HU-84-030-021	58.37	-57.50	13	14	-0.3	1.7	0.1	**-0.4**	1.2	2.6	1.5	2.0	**-0.3**	-1.2	1.0	-1.3	0.5	**0.1**	31.3	1.7	31.3	0.9	**-0.2**	9.1	2.9	9.2	2.1	**-0.1**	25.6
HU-86-034-040	42.63	-63.10	5	5	6.4	1.1	8.4	**-2.0**	12.6	0.5	15.7	0.8	**-3.1**	1.7	1.7	1.0	1.6	**0.7**	31.4	0.7	31.4	0.7	**0.2**	1.7	0.6	1.0	0.6	**0.7**	20.5
HU-87-033-007	65.40	-57.42	2	11	-0.7	1.1	0.4	**-1.1**	0.1	1.3	1.6	1.0	**-1.5**	-1.2	0.9	-0.8	0.5	**-0.4**	30.9	1.0	31.8	0.3	**-0.9**	10.5	0.5	9.3	1.2	**1.2**	35.4
HU-87-033-008	62.64	-53.88	7	8	0.0	2.4	-1.3	**1.3**	1.6	4.1	-0.9	0.1	**2.5**	-1.5	0.5	-1.7	0.0	**0.2**	31.5	0.6	30.6	0.6	**0.9**	9.3	3.1	11.5	0.3	**-2.2**	25.6
HU-87-033-009	62.51	-59.44	6	10	0.1	1.2	-1.3	**1.4**	1.4	1.6	-0.8	1.6	**2.2**	-0.8	0.6	-1.7	0.2	**0.9**	31.1	0.7	30.7	0.2	**0.4**	9.6	1.1	11.2	0.1	**-1.6**	25.7
HU-90-013-012	58.92	-47.12	2	11	9.8	2.8	8.6	**1.2**	14.4	1.4	11.6	3.3	**2.8**	5.9	1.6	5.5	1.2	**0.4**	32.8	0.4	32.6	1.0	**0.2**	0.2	0.2	1.8	1.9	**-1.6**	44.2
HU-90-013-013	58.21	-48.37	8	9	7.1	2.8	4.7	**2.4**	11.5	3.2	8.4	1.2	**3.1**	3.7	3.1	0.9	1.2	**2.8**	32.8	1.7	32.6	0.9	**0.2**	1.0	1.7	2.9	1.3	**-1.9**	38.1
HU-90-015-017	42.78	-61.65	7	7	5.4	2.5	8.3	**-2.9**	10.7	3.5	14.7	2.3	**-4.0**	1.4	1.9	1.9	0.7	**-0.5**	31.5	0.6	31.7	0.6	**-0.2**	2.6	2.0	2.0	1.2	**0.6**	31.7
HU-91-020-013	41.83	-62.33	6	4	5.9	3.7	6.8	**-0.9**	10.9	5.2	13.7	2.7	**-2.8**	1.4	2.4	-0.1	0.8	**1.5**	30.1	1.5	30.5	0.4	**-0.9**	2.5	3.7	2.3	1.5	**0.2**	25.8
HU-91-045-044	59.36	-43.45	4	4	-1.5	2.2	-0.6	**-0.9**	-0.8	0.1	-1.4	0.4	**-0.3**	-1.7	0.2	-1.4	0.4	**-0.3**	31.4	1.5	30.5	0.9	**0.9**	11.2	0.1	9.7	0.1	**1.5**	34.4
HU-91-045-052	59.49	-39.30	7	8	6.3	2.2	8.3	**-2.0**	10.2	2.5	12.7	2.7	**-2.5**	3.6	0.2	3.8	0.2	**-0.2**	33.0	2.0	33.7	0.8	**-0.7**	1.7	2.5	0.6	0.3	**1.1**	49.2
HU-91-045-058	59.84	-33.57	10	9	8.5	3.1	9.5	**-1.0**	12.4	3.4	13.3	2.0	**-0.9**	5.4	2.9	5.7	1.5	**-0.3**	33.2	1.4	34.2	0.6	**-1.0**	0.6	1.3	0.3	0.2	**0.3**	54.6
HU-91-045-064	59.67	-30.57	6	6	7.7	1.8	8.3	**-0.5**	11.1	2.4	11.2	1.6	**-0.1**	5.3	1.2	5.3	1.0	**0.0**	34.5	0.6	34.5	0.7	**-0.2**	0.6	0.7	0.5	1.5	**-0.2**	56.6
HU-91-045-072a	58.94	-28.74	9	5	5.2	2.8	7.6	**-2.4**	7.9	3.8	10.7	2.7	**-2.8**	3.4	2.2	4.5	1.0	**-1.1**	33.6	1.0	34.5	0.5	**-0.9**	2.9	2.9	1.8	1.2	**1.1**	49.4
HU-91-045-074a	55.74	-30.22	5	4	7.8	2.4	8.5	**-0.7**	11.7	2.7	12.7	1.6	**-1.0**	4.9	1.8	4.2	0.7	**0.7**	34.2	0.6	33.7	0.5	**0.5**	0.2	0.9	0.5	0.2	**-0.3**	60.2
HU-91-045-080a	53.07	-33.52	10	10	8.5	2.3	9.3	**-0.8**	12.9	2.8	13.8	1.1	**-0.9**	5.2	2.5	4.8	1.8	**0.4**	33.2	1.4	33.6	0.7	**-0.4**	0.4	0.9	1.3	1.5	**0.0**	65.5
HU-91-045-082a	52.86	-35.53	5	5	9.9	2.8	8.5	**1.4**	15.4	2.9	13.1	2.0	**2.3**	5.6	2.8	3.9	1.5	**1.7**	33.2	1.4	33.6	1.1	**-0.3**	0.4	0.7	1.3	2.0	**-0.9**	65.2
HU-91-045-085	53.97	-38.64	3	3	11.0	3.0	8.1	**2.9**	14.8	2.9	11.7	3.0	**3.1**	8.1	2.8	4.5	2.8	**3.6**	33.5	0.7	33.8	1.1	**-0.3**	0.2	0.3	1.6	2.0	**-1.4**	62.5
HU-91-045-094a,c	50.20	-45.68	27	20	0.5	2.2	0.4	**0.1**	2.5	3.5	2.0	2.5	**0.5**	-0.7	1.2	-1.2	1.2	**0.5**	31.0	1.8	31.2	0.8	**-0.4**	8.4	3.1	9.6	1.7	**-1.2**	30.3
AII-94-PC3	62.09	-16.62	2	3	4.7	2.4	7.9	**-3.2**	7.7	2.8	12.8	1.3	**-5.1**	2.6	2.1	3.0	0.7	**-0.4**	31.8	1.4	32.8	0.3	**-1.0**	4.1	2.4	0.7	0.5	**3.4**	24.5
D89-BOFS-5K	50.69	-21.86	6	6	9.8	3.8	9.6	**0.2**	14.3	4.1	13.5	2.6	**0.2**	6.2	3.6	5.7	2.3	**0.5**	33.8	1.6	34.1	0.9	**-0.3**	0.1	0.2	0.1	0.3	**0.0**	61.5
HM52-43	64.25	0.73	3	3	6.8	1.4	7.8	**-1.0**	12.8	2.3	15.3	0.2	**-2.5**	2.2	1.9	0.3	1.9	**1.9**	31.8	1.0	31.1	0.1	**0.7**	0.8	1.4	0.4	0.5	**0.4**	44.5
HM71-19	69.49	-9.53	1	1	8.8	—	10.4	**-1.6**	13.8	—	14.4	—	**-0.6**	5.4	—	6.4	—	**-1.0**	33.6	—	34.9	—	**-1.3**	0.0	0.0	0.0	0.0	**0.0**	65.6
HM80-30	71.80	1.61	2	2	8.7	0.1	9.7	**-1.0**	13.3	0.3	13.4	0.2	**-0.1**	5.5	0.4	6.0	0.1	**-0.5**	34.0	0.2	35.1	0.0	**-1.1**	0.0	0.0	0.0	0.0	**0.0**	62.0
HM94-13	71.63	-1.63	1	1	3.3	—	1.9	**1.4**	6.4	—	4.2	—	**2.2**	1.1	—	-0.5	—	**1.6**	31.9	—	31.4	—	**0.5**	5.3	—	8.5	—	**-3.2**	63.6
HM94-25	75.60	1.30	1	1	7.1	—	5.6	**1.6**	10.9	—	8.9	—	**2.0**	4.1	—	2.2	—	**1.9**	33.0	—	34.9	—	**-1.9**	2.2	—	0.0	—	**2.2**	63.3
HM94-34	73.78	-2.54	3	3	8.7	1.6	8.9	**-0.2**	13.0	1.7	12.1	0.8	**0.9**	5.5	1.7	5.8	—	**-0.3**	33.6	1.2	34.6	0.5	**-1.0**	0.7	1.2	1.5	1.3	**-0.8**	71.8
M17730	72.05	7.31	1	1	9.2	—	6.9	**2.3**	13.6	—	9.9	—	**3.7**	5.8	—	3.8	—	**2.0**	33.9	—	34.7	—	**-0.8**	0.0	—	0.0	—	**0.0**	89.1

The differences are shown by bold characters.

the sea-surface reconstructions based on dinocysts are comprised in a domain of the reference database that is characterised by a low density of data points, with respect to summer SSTs vs. winter SSTs or salinity (Fig. 12). The dinocyst-based reconstructions of the LGM in the northern North Atlantic have no perfect analogue known from the modern hydrography of the high northern latitudes. Nevertheless, the LGM data points occur in a central area of the hydrographic domains represented by the modern database and likely occur within the range of values we may reconstruct based on dinocyst assemblages. In other words, the degree of accuracy of LGM reconstructions and estimated values may be discussed, but the hydrographical domains involved are probably correct.

4.2. The most salient features of the LGM based on dinocyst data

In spite of the uncertainties examined above, there are consistent features in the reconstructions of LGM sea-surface conditions based on dinocyst assemblages that appear as robust as possible given the context of weak analogue situation. These features are summarised below.

(1) Sea-ice cover was much more extensive than at present, notably along the eastern Canadian margin. Offshore, in the subarctic basins of the North Atlantic, i.e., the Labrador Sea, the Irminger Basin, the Greenland and Norwegian seas, seasonal sea ice developed for a few weeks to a few months per year, whereas in the northeastern part of the North Atlantic, ice-free conditions prevailed throughout the year.

(2) Relatively low sea-surface salinity during the LGM characterised the northern North Atlantic and adjacent basins (Fig. 13). In areas marked by extensive sea-ice cover, low sea-surface salinities may have been related to the summer melting of sea ice. However, the particularly low salinities recorded along the southern Canadian and Scandinavian margins (31–32) likely reflect mixing with large meltwater supplies from surrounding ice sheets, which resulted in the dilution of surface waters offshore. In the central part of the North Atlantic, a negative salinity anomaly of about one unit is recorded as compared with the modern situation. During the LGM, the buoyancy of the low-density surface water layer over the northern North Atlantic was responsible for reduced vertical convection from the surface and the absence of deep or intermediate water formation at the corresponding latitudes (cf. Hillaire-Marcel et al., 2001a, b; de Vernal et al., 2002).

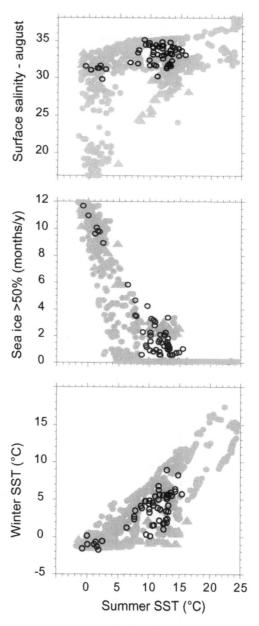

Fig. 12. Graph showing the hydrographical domains of the $n = 940$ calibration database and that of the LGM reconstructions. The light gray circles illustrate the distribution of summer temperature vs. winter temperature, sea-ice cover, and summer salinity for the 940 modern samples (same as Fig. 4) and the open circles illustrate the same distribution for the LGM cores. Note that the LGM reconstructions are included in domains characterised by a low density of modern data points.

(3) On the whole, winter conditions colder than at present and relatively mild summers resulted in significantly larger seasonal gradients of temperature in surface water masses (Fig. 13). The low thermal inertia of the shallow and buoyant upper water layer can be evoked to explain high energy uptake during the summer seasons, then characterised by an insolation roughly the same as at present, followed by rapid cooling during winters.

LGM vs. modern in mid-polar latitudes
of the Northern Hemisphere (n = 55)

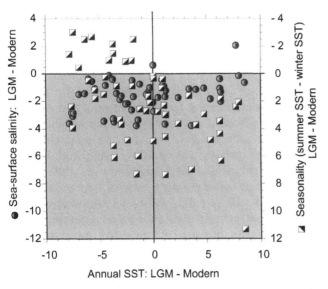

Fig. 13. Summary of the hydrographical anomalies between the LGM and the present.

(4) Reconstruction of the mean annual temperature shows values colder, cooler or equivalent to present conditions on an average basis during the LGM, with some exceptions, notably in the Greenland and Norwegian seas (Fig. 13; Table 3). LGM data from the Greenland Sea are not very robust because of low dinocyst concentrations and weak analogue situations (Fig. 8), but data from the Norwegian Sea appear reliable and suggest the existence of mild conditions at least episodically.

4.3. The northern North Atlantic LGM SST: contrasting pictures depending on the choice of proxies?

4.3.1. Contribution of dinocyst data

In the northern North Atlantic, the LGM paleoceanographic data from dinocysts are complementary to those from planktonic foraminifer assemblages (Weinelt et al., 1996; Pflaumann et al., 2003) and alkenones (Rosell-Melé and Comes, 1999; Rosell-Melé et al., 2004). First, they improve the geographical coverage of LGM reconstructions along the margins of south eastern Canada, the Labrador Sea and Baffin Bay, where other data are very rare. Second, they help to constrain the limits of sea ice and also provide quantitative information on the seasonal extent of the ice coverage (Fig. 10). Third, dinocysts permit the estimation of sea-surface salinity, which is an important hydrographic parameter of the North Atlantic Ocean with respect to thermohaline circulation. Finally, the

calibration database of dinocysts includes large numbers of samples from the Arctic seas and subarctic basins, unlike other proxies. Therefore, dinocyst assemblages should permit identification of analogues from Arctic and subarctic environments for the LGM interval. One of the important findings from this study is the lack of perfect modern analogues for LGM North Atlantic Ocean, even in the Arctic and subarctic seas.

4.3.2. Consistent features and discrepancies from the comparison with other proxies

Despite discrepancies, there are converging features in the LGM reconstructions based on the different proxies. One of these features concerns sea ice. All proxies suggest seasonally ice-free conditions in the northern North Atlantic and the Nordic seas, and the maximum extent of LGM sea-ice cover shows similar limits in the central North Atlantic estimated from dinocysts (pink dashed line in upper left diagram of Fig. 10) and indirectly estimated from planktonic foraminifera (cf. Pflaumann et al., 2003). Another convergent feature between dinocyst and planktonic foraminifer data is the gradient of LGM-modern SST that shows similar patterns, although the absolute values differ significantly.

The reconstructions of LGM SSTs based on dinocysts are not in contradiction with temperatures derived from alkenone data (Rosell-Melé and Comes, 1999; Rosell-Melé et al., 2004), but they show discrepancies with data from planktonic foraminifera well beyond the degrees of uncertainty of the respective approaches. On the whole, dinocyst data yield much warmer estimates than planktonic foraminifera, whatever the type of transfer function used (cf. CLIMAP, 1981; Pflaumann et al., 1996, 2003; Weinelt et al., 1996, 2003; Waelbroeck et al., 1998), but are compatible with those obtained from alkenones (cf. Rosell-Melé and Comes, 1999) or coccoliths (de Vernal et al., 2000). Therefore, we are tempted to distinguish two types of tracers. The first category includes dinocysts as well as coccoliths and their alkenones. Worth of mention here is the fact that the LGM coccolith assemblages of the northern North Atlantic are mostly dominated by *Emiliania huxleyi* (cf. de Vernal et al., 2000), one of the rare euryhaline taxa amongst coccolithophorids (e.g., Winter et al., 1994) that is also responsible for alkenone production. The second category of tracers consists of planktic foraminiferal assemblages that barely tolerate salinity below 32 or 33 (cf. Bé and Tolderlund, 1971). In the first case, i.e., that of coccoliths, alkenones and autotrophic dinocysts, the tracers relate to primary production in the photic zone of the upper surface water layer, whereas in the second case, they relate to heterotrophic production and characterise various depth habitats from the epipelagic to the mesopelagic domains. The two categories of tracers can thus be distinguished based on two patterns:

(1) their depth habitat and/or (2) their relative tolerance to low salinity. The discrepancy of temperature estimates obtained from the different types of proxies could thus simply reflect specificities in the vertical or temporal structure of the upper water masses.

Therefore, most LGM data from the various proxies, including microfossil assemblages, isotopic content of foraminifera or biomarkers, can be reconciled through an interpretation invoking either strong temperature-salinity gradients in the upper water column, with possible variations in the seasonal gradients of temperatures, or alternation of episodes of warm and dilute surface water with episodes of higher salinity but cold conditions. The existence of a sharp halo-thermocline separating a shallow and low saline mixed layer from a cold and dense mesopelagic water mass, could explain the co-occurrence of temperate autotrophic producers and polar zooplanktonic populations. As a matter of fact, sharp density gradients can be reconstructed for the LGM by combining information from dinocyst data and $\delta^{18}O$ in planktonic foraminifera (*Globigerina bulloides* and *Neogloboquadrina pachyderma* left coiled) recovered from different size fractions (Hillaire-Marcel et al., 2001a, b; de Vernal et al., 2002). The isotopic compositions of *N. pachyderma* shells of increasing size and density, which can be associated with gradational depths along the pycnocline, have been used to document the density gradient in the upper water column in the Labrador Sea (Hillaire-Marcel et al., 2001a) and the Arctic Ocean (Hillaire-Marcel et al., 2004). Using these data, together with $\delta^{18}O$ values in the epipelagic foraminifer *G. bulloides*, and estimated SSTs and salinities from dinocysts, it has been possible to propose the existence of vertical density gradients ranging from 1.5 to 3 density units (σ_θ) in the upper water masses of the northwestern North Atlantic during the LGM (de Vernal et al., 2002). However, it is not excluded that such gradients represent conditions occurring during distinct episodes.

4.3.3. The problem of the Nordic seas

Although the sea-surface conditions of the LGM reconstructed on the basis of dinocyst assemblages can be reconciled with other paleoceanographical data as outlined above, the case of the Nordic seas, and especially that of the Greenland Sea, remains problematic. The strongly positive LGM anomalies of temperatures obtained in the Greenland Sea based on dinocysts (de Vernal et al., 2000; this study) or alkenones (Rosell-Melé and Comes, 1999) are somewhat difficult to explain.

In the eastern part of the Nordic seas, along the Norwegian margins, the dinocyst assemblages are characterised by relatively high concentrations and large species diversity. The number of analysed samples, and the overall reliability of reconstructions based on dinocysts (Fig. 8 and Table 2) in the eastern Norwegian Sea suggest that relatively warm conditions existed in the Nordic seas during the LGM, at least episodically and or seasonally. The dinocyst assemblages and other paleoceanographical tracers show large-amplitude changes in oceanographical conditions which can be linked with the high-frequency climate oscillations during the last ice age as recorded in the isotopic record of Greenland ice cores (e.g., Rasmussen et al., 1996).

The reconstructions from the central and western parts of the Nordic seas, however should be considered with caution. In general, this area is characterised by low biogenic productivity and low sedimentation rates (cf. Sarnthein et al., 1995; Hebbeln et al., 1998). As a consequence, the sedimentary interval representative of the LGM is relatively thin. Moreover, the interpretation of the sparse micropaleontological assemblages is equivocal also because of possible reworking and biological mixing.

Two hypothesis, not exclusive to each other, are tentatively proposed to explain the overall records of the Nordic seas:

Hypothesis 1. The sparse assemblages that are observed in the LGM interval of the central and western Nordic seas might reflect highly variable conditions, from generally cold and quasi-perennial sea-ice cover (nil productivity) to episodically mild conditions (with some productivity), e.g., when large anticyclonic gyre developed and sea ice broke due to strong storms. At present, sharp fronts and extremely high interannual–interdecadal variability in sea-surface conditions and sea-ice cover are recorded in the central part of the Nordic seas, and we may hypothesise that a similar variability existed during the LGM. The existence of fronts controlling pressure gradient and storm tracks may have played an important role in the moisture supply to feed the northern ice sheet.

Hypothesis 2. The assemblages might reflect nil regional productivity due to quasi-permanent sea ice in the eastern and northern parts of the Nordic seas. In such a case, the very sparse assemblages of the Greenland Sea might be due to lateral advection of fine material (dinocysts, alkenone, coccoliths) through a subsurface current flowing from the south and penetrating into the central Nordic seas by subduction. A modern equivalent for such a situation could be found in the northern Barents Sea where the North Atlantic Drift surface water mass is subducting below the low saline Arctic waters.

5. Conclusions

The dinocyst database that has been developed from the analyses of surface sediment samples collected in middle- to high-latitude marine environments of the Northern Hemisphere covers a wide range of hydrographical conditions notably in the domain characterised by the presence of seasonal sea-ice cover. This database was used to reconstruct quantitatively the sea-surface conditions that prevailed during the LGM. It provides some clues about the paleoceanographical regime which prevailed over the northern North Atlantic when continental ice sheets reached their maximum extent over surrounding lands and continental shelves. The LGM reconstructions illustrate extensive sea-ice cover along the eastern Canadian margin, and some spreading of sea ice during winter, in subpolar basins and Nordic seas, whereas the northeast Atlantic south of 50°N remained free of sea ice. As a response to meltwater discharges from the surrounding ice sheets grounded on the shelves, relatively low salinity characterised surface waters, especially along the eastern Canadian and Scandinavian margins. A possible explanation would be the development of a buoyant and probably shallow mixed layer resulting in a low thermal inertia in the surface waters. This would explain the reconstructed large seasonal contrasts of temperatures with the cold winter and relatively mild summer. In this scenario, the sea-surface conditions reconstructed at the scale of the northern North Atlantic indicate strong stratification, unfavourable for vertical convection, and lead us to make a comparison with a large fjord-like system making the transition between the continental ice sheets and the ocean. In such a context, most bioindicators of "open ocean" conditions would have a limited sensitivity, which may explain the discrepancies between the different sets of LGM reconstructions based on different proxies. Another explanation would invoke alternation of episodes of low salinity and high summer SSTs in surface waters with episodes of higher salinity and lower temperature accounting for the reconstructed cold conditions based on foraminifera.

Acknowledgements

This study is a contribution to the Climate System, History and Dynamics (CSHD) project, supported by the National Science and Engineering Research Council (NSERC) of Canada, and to the international IMAGES program. Complementary support by the *Fonds québecois de Recherche sur la Nature et les Technologies* and by the Canadian Foundation for Climate and Atmospheric Sciences (project no. GR-240) is acknowledged. We are extremely grateful to many institutions for their most precious help in providing surface sediment samples, notably the Oregon State University, Lamont Doherty Earth Observatory, the Bedford Institute of Oceanography, the University of Kiel, GEOMAR, and the Alfred Wegener Institute. We also thank Karin Zonneveld, Barrie Dale and Michal Kucera for their critical and constructive review of the manuscript.

References

Antoine, D., André, J.M., Morel, A., 1996. Oceanic primary production. 2. Estimation at global scale from satellite (costal zone colour scanner) chlorophyll. Global Biogeochemical Cycles 10, 57–69.

Bé, A.W.H., Tolderlund, D.S., 1971. Distribution and ecology of living planktonic foraminifera in surface waters of the Atlantic and Indian oceans. In: Funnel, B.M., Riedel, W.R. (Eds.), The Micropaleontology of Oceans. Cambridge University Press, Cambridge, pp. 105–149.

Bedford Institute of Oceanography (BIO), 2003. A monthly gridded set of temperature and salinity for the northwest North Atlantic Ocean, CD-Rom data set.

Broecker, W.S., Peng, T.-H., 1982. Tracers in the Sea. Lamont–Doherty Geological Observatory. Columbia University, New York.

CLIMAP Project Members, 1981. Seasonal reconstructions of the earth's surface at the last glacial maximum. Geological Society of America Map and Chart Series MC-56.

Combourieu-Nebout, N., Turon, J.L., Zahn, R., Capotondi, L., Londeix, L., Pahnke, K., 2002. Enhanced aridity and atmospheric high-pressure stability over the western Mediterranean during the North Atlantic cold events of the past 50 kyrs. Geology 30, 863–866.

Comiso, J.C., 2002. Correlation and trend studies of the sea-ice cover and surface temperatures in the Arctic. Annals of Glaciology 34, 420–428.

Dale, B., 1976. Cyst formation, sedimentation, and preservation: factors affecting dinoflagellate assemblages in recent sediments from Trondheimsfjord, Norway. Review of Palaeobotany and Palynology 22, 39–60.

Dale, B., 1983. Dinoflagellate resting cysts: benthic plankton. In: Fryxell, G.A. (Ed.), Survival Strategies of Algae. Cambridge University Press, Cambridge, pp. 69–136.

Dale, B., 1996. Dinoflagellate cyst ecology: modeling and geological applications. In: Jansonius, J., McGregor, D.C. (Eds.), Palynology: Principles and Applications, vol. 3. American Association of Stratigraphic Palynologists Foundation, Dallas, TX, pp. 1249–1275.

de Vernal, A., Giroux, L., 1991. Distribution of organic-walled microfossils in recent sediments from the Estuary and Gulf of St. Lawrence: some aspects of the organic matter fluxes. Canadian Journal of Fisheries and Aquatic Sciences 113, 189–199.

de Vernal, A., Hillaire-Marcel, C., 2000. Sea-ice cover, sea-surface salinity and halo-thermocline structure of the northwest North Atlantic: modern versus full glacial conditions. Quaternary Science Reviews 19, 65–85.

de Vernal, A., Pedersen, T., 1997. Micropaleontology and palynology of core PAR 87 A-10: a 30,000 years record of paleoenvironmental changes in the Gulf of Alaska, northeast North Pacific. Paleoceanography 12, 821–830.

de Vernal, A., Turon, J.-L., Guiot, J., 1994. Dinoflagellate cyst distribution in high latitude environments and quantitative reconstruction of sea-surface temperature, salinity and seasonality. Canadian Journal of Earth Sciences 31, 48–62.

de Vernal, A., Henry, M., Bilodeau, G., 1999. Technique de préparation et d'analyse en micropaléontologie. Les Cahiers du GEOTOP, Université du Québec à Montréal, 3, Unpublished report.

de Vernal, A., Hillaire-Marcel, C., Turon, J.-L., Matthiessen, J., 2000. Reconstruction of sea-surface temperature, salinity, and sea-ice cover in the northern North Atlantic during the last glacial maximum based on dinocyst assemblages. Canadian Journal of Earth Sciences 37, 725–750.

de Vernal, A., Rochon, A., Turon, J.-L., Matthiessen, J., 1997. Organic-walled dinoflagellate cysts: palynological tracers of sea-surface conditions in middle to high latitude marine environments. GEOBIOS 30, 905–920.

de Vernal, A., Hillaire-Marcel, C., Peltier, W.R., Weaver, A.J., 2002. The structure of the upper water column in the northwest North Atlantic: modern vs. Last Glacial Maximum conditions. Paleoceanography 17, 1050.

de Vernal, A., Henry, M., Matthiessen, J., Mudie, P.j., Rochon, A., Boessenkool, K., Eynaud, F., Grøsfjeld, K., Guiot, J., Hamel, D., Harland, R., Head, M.j., Kunz-pirrung, M., Levac, E., Loucheur, V., Peyron, O., Pospelova, V., Radi, T., Turon, J.-L., Voronina, E., 2001. dinoflagellate cyst assemblages as tracers of sea-surface conditions in the northern North Atlantic, Arctic and sub-arctic seas: the new "n = 677" database and application for quantitative paleoceanographical reconstruction. Journal of Quaternary Science 16, 681–699.

Devillers, R., de Vernal, A., 2000. Distribution of dinocysts in surface sediments of the northern North Atlantic in relation with nutrients and productivity in surface waters. Marine Geology 166, 103–124.

Dodge, J.D., 1994. Biogeography of marine armoured dinoflagellates and dinocysts in the NE Atlantic and North Sea. Review of Palaeobotany and Palynology 84, 169–180.

Duane, A., Harland, R., 1990. Late Quaternary dinoflagellate cyst biostratigraphy for sediments of the Porcupine Basin, offshore western Ireland. Review of Palaeobotany and Palynology 63, 1–11.

Dodge, J.D., Harland, R., 1991. The distribution of planktonic dinoflagellates and their cysts in the eastern and northeastern Atlantic Ocean. New Phytologist 118, 593–603.

Eynaud, F., Turon, J.-L., Matthiessen, J., Kissel, C., Peypouquet, J.-P., de Vernal, A., Henry, M., 2002. Norwegian Sea surface palaeoenvironments of the Marine Isotopic Stage 3: the paradoxical response of dinoflagellate cysts. Journal of Quaternary Science 17, 349–359.

Fairbanks, R.G., Wiebe, P.H., 1980. Foraminifera and chlorophyll maximum: vertical distribution, seasonal succession, and paleoceanographic significance. Science 209, 1524–1526.

Faul, K., Ravelo, A.C., Delanay, M.L., 2000. Reconstructions of upwelling, productivity, and photic zone depth in the eastern equatorial Pacific ocean using planktonic foraminiferal stable isotopes and abundances. Journal of Foraminiferal Research 30, 110–125.

Gaines, G., Elbrächer, M., 1987. Heterotrophic nutrition. In: Taylor, F.J.R. (Ed.), The Biology of Dinoflagellates, Botanical Monograph 21. Blackwell Scientific, Oxford, pp. 224–281.

Gascard, J.-C., Watson, A.J., Messias, M.-J., Olsson, K.A., Johannessen, T., Simonsen, K., 2002. Long-lived vortices as a mode of deep ventilation in the Greenland Sea. Nature 416, 525–527.

Godhe, A., Norén, F., Kuylenstierna, M., Ekberg, C., Karlson, B., 2001. Relationship between planktonic dinoflagellate abundance, cysts recovered in sediment traps and environmental factors in the Gullmar Fjord, Sweden. Journal of Plankton Research 23, 923–938.

Graham, D.K., Harland, R., Gregory, D.M., Long, D., Morton, A.C., 1990. The biostratigraphy and chronostratigraphy of BGS Borehole 78/4, North Minch. Scottish Journal of Geology 26, 65–75.

Grøsfjeld, K., Harland, R., 2001. Distribution of modern dinoflagellate cysts from inshore areas along the coast of southern Norway. Journal of Quaternary Science 16, 651–660.

Guiot, J., 1990. Methods and programs of statistics for paleoclimatology and paleoecology. In: Guiot, J., Labeyrie, L. (Eds.), Quantification des changements climatiques: méthode et pro-

grammes. Institut National des Sciences de l'Univers (INSU-France), Monographie No 1, Paris.

Guiot, J., Goeury, C., 1996. PPPbase, a software for statistical analysis of paleoecological data. Dendrochronologia 14, 295–300.

Hamel, D., de Vernal, A., Gosselin, M., Hillaire-Marcel, C., 2002. Organic-walled microfossils and geochemical tracers: sedimentary indicators of productivity changes in the North Water and northern Baffin Bay (High Arctic) during the last centuries. Deep-Sea Research II 49, 5277–5295.

Harland, R., 1983. Distribution maps of recent dinoflagellate cysts in bottom sediments from the North Atlantic Ocean and adjacent seas. Palaeontology 26, 321–387.

Head, M.J., 1996. Modern dinoflagellate cysts and their biological affinities. In: Jansonius, J., McGregor, D.C. (Eds.), Palynology: Principles and Applications, vol. 3. American Association of Stratigraphic Palynologists Foundation, Dallas, TX, pp. 1197–1248.

Head, M.J., Harland, R., Matthiessen, J., 2001. Cold marine indicators of the late Quaternary: the new dinoflagellate cyst genus Islandinium and related morphotypes. Journal of Quaternary Science 16, 621–636.

Head, M.J., Lewis, J., de Vernal, A., The cyst of the calcareous dinoflagellate Scrippsiella trifida, and the fossil record of its organic wall. Journal of Palaeontology, in press.

Hebbeln, D., Henrich, R., Baumann, K.-H., 1998. Paleoceanography of the last Interglacial/Glacial cycle in the polar North Atlantic. Quaternary Science Reviews 17, 125–153.

Hillaire-Marcel, C., de Vernal, A., Lucotte, M., Mucci, A., Bilodeau, G., Rochon, A., Vallières, S., Wu, G., 1994. Productivité et flux de carbone dans la mer du labrador au cours des derniers 40.000 ans. Canadian Journal of Earth Sciences 31, 139–158.

Hillaire-Marcel, C., de Vernal, A., Candon, L., Bilodeau, G., Stoner, J., 2001a. Changes of potential density gradients in the northwestern North Atlantic during the last climatic cycle based on a multiproxy approach. In: Seidov, D., Maslin, M., Haupt, B. (Eds.), The Oceans and Rapid Climate Changes: Past, Present and Future. Geophysical Monograph Series, vol. 126. American Geophysical Union, Washington, DC, pp. 83–100.

Hillaire-Marcel, C., de Vernal, A., Bilodeau, G., Weaver, A., 2001b. Absence of deep water formation in the Labrador Sea during the last interglacial. Nature 410, 1073–1077.

Hillaire-Marcel, C., de Vernal, A., Polyak, L., Darby, D., 2004. Size dependent isotopic composition of planktic foraminifers from Chukchi Sea vs. NW Atlantic sediments—implications for the Holocene paleoceanography of the western Arctic. Quaternary Science Reviews 23, 245–260.

Imbrie, J., Kipp, N.G., 1971. A new micropaleontological method for quantitative paleoclimatology: application to a late Pleistocene Caribbean core. In: Turekian, K.K. (Ed.), The Late Cenozoic Glacial Ages. Yale University Press, New Haven, Conn., pp. 71–181.

Isemer, H.-J., Hasse, L., 1985. The Bunker Climate Atlas of the North Atlantic Ocean, vol. 1: Observations. Springer, Berlin, pp. 57–61.

Koç, N., Jansen, E., Haflidason, H., 1993. Paleoceanographic reconstruction of surface ocean conditions in the Greenland, Iceland, and Norwegian seas through the last 14,000 years based on diatoms. Quaternary Science Reviews 12, 115–140.

Kokinos, J.P., Eglinton, T.I., Goñi, M.A., Boon, J.J., Martoglio, P.A., Anderson, D.M., 1998. Characterisation of a highly resistant biomacromolecular material in the cell wall of a marine dinoflagellate resting cyst. Organic Geochemistry 28, 265–288.

Kucera, M., Malmgren, B.A., 1998. Logratio transformation of compositional data—a resolution of the constant sum constraint. Marine Micropaleontology 34, 117–120.

Kunz-Pirrung, M., 1998. Rekonstruktion der Oberflächenwassermassen der östlichen Laptevsee im Holozän anhand von aquatischen Palynomorphen. Berichte zur Polarforschung 281, 117.

Kunz-Pirrung, M., 2001. Dinoflagellate cyst assemblages in recent sediments of the Laptev Sea (Arctic Ocean) and their relation to hydrographic conditions. Journal of Quaternary Science 16, 637–650.

Lapointe, M., 2000. Late Quaternary paleohydrology of the Gulf of St. Lawrence (Québec, Canada) based on diatom analysis. Palaeogeography, Palaeoclimatology, and Palaeoecology 156, 261–276.

Lea, D.W., Mashiotta, T.A., Spero, H.J., 1999. Controls on magnesium and strontium uptake in planktonic foraminifera determined by live culturing. Geochimica et Cosmochimica Acta 63, 2369–2379.

Levandowsky, M., Kaneta, P., 1987. Behaviour in dinoflagellates. In: Taylor, F.J.R. (Ed.), The Biology of Dinoflagellates. Botanical Monograph 21. Blackwell Scientific, Oxford, pp. 360–397.

Malmgren, B., Nordlund, U., 1997. Application of artificial neural network to paleoceanographic data. Palaeogeography, Palaeoclimatology, and Palaeoecology 136, 359–373.

Mangin, S., 2002. Distribution actuelle des kystes de dinoflagellés en Méditerranée occidentale et application aux fonctions de transfert, vol. 1. Memoir of DEA, University of Bordeaux, 34pp.

Marret, F., 1993. Les effets de l'acétolyse sur les assemblages de kystes de dinoflagellés. Palynosciences 2, 267–272.

Marret, F., Scourse, J., 2003. Control of modern dinoflagellate cyst distribution in the Irish and Celtic Seas by seasonal stratification dynamics. Marine Micropaleontology 47, 101–116.

Marret, F., Zonneveld, K., 2003. Atlas of modern organic-walled dinoflagellate cyst distribution. Review of Palaeobotany and Palynology 125, 1–200.

Marret, F., Eiriksson, J., Knudsen, K.-L., Turon, J.-L., Scourse, J., 2004. Distribution of dinoflagellate cyst assemblages in surface sediments from the northern and western shelf of Iceland. Review of Palaeobotany and Palynology 128, 35–53.

Matthiessen, J., 1995. Distribution patterns of dinoflagellate cysts and other organic-walled microfossils in recent Norwegian–Greenland Sea sediments. Marine Micropaleontology 24, 307–334.

Mix, A.E., Bard, E., Schneider, R., 2001. Environmental processes of the ice age: land, ocean, glaciers (EPILOG). Quaternary Science Reviews 20, 627–657.

Mudie, P.J., 1992. Circum-arctic Quaternary and Neogene marine palynofloras: paleoecology and statistical analysis. In: Head, M.J., Wrenn, J.H. (Eds.), Neogene and Quaternary Dinoflagellate Cysts and Acritarchs. American Association of Stratigraphic Palynologists Foundation, College Station, TX, pp. 347–390.

Mudie, P.J., Rochon, A., 2001. Distribution of dinoflagellate cysts in the Canadian Arctic marine region. Journal of Quaternary Sciences 16, 603–620.

Mudie, P.J., Harland, R., Matthiessen, J., de Vernal, A., 2001. Dinoflagellate cysts and high latitude Quaternary paleoenvironmental reconstructions: an introduction. Journal of Quaternary Science 16, 595–602.

National Oceanographic Data Center (NODC), 1994. World Ocean Atlas. National Oceanic and Atmospheric Administration, CD-Rom data sets.

National Oceanographic Data Center (NODC), 1998. World Ocean Atlas. National Oceanic and Atmospheric Administration, http://www.nodc.noaa.gov/OC5/pr_woaf.html.

Nürnberg, D., Müller, A., Schneider, R.R., 2000. Paleo-sea surface temperature calculations in the equatorial east Atlantic from Mg/Ca ratios in planktic foraminifera: A comparison to sea surface temperature estimates from $U_{37}^{K'}$, oxygen isotopes, and foraminiferal transfer function. Paleoceanography 15, 124–134.

Peltier, W.R., 1994. Ice age paleotopography. Science 265, 195–201.

Peyron, O., de Vernal, A., 2001. Application of Artificial Neural Network (ANN) to high latitude dinocyst assemblages for the reconstruction of past sea-surface conditions in Arctic and sub-arctic seas. Journal of Quaternary Science 16, 699–711.

Pflaumann, U., Duprat, J., Pujol, C., Labeyrie, L., 1996. SIMMAX: a modern analog technique to deduce Atlantic sea surface temperatures from planktonic foraminifera in deep-sea sediments. Paleoceanography 11, 15–35.

Pflaumann, U., Sarnthein, M., Chapman, M., d'Abreu, L., Funnell, B., Huels, M., Kiefer, T., Maslin, M., Schulz, H., Swallow, J., van Kreveld, S., Vautravers, M., Vogelsang, E., Weinelt, M., 2003. Glacial North Atlantic: sea-surface conditions reconstructed by GLAMAP 2000. Paleoceanography 18, 1065.

Radi, T., de Vernal, A., 2004. Dinocyst distribution in surface sediments from the northeastern Pacific margin (40–60°N) in relation to hydrographic conditions, productivity and upwelling. Review of Paleobotany and Palynology 128, 169–193.

Radi, T., de Vernal, A., Peyron, O., 2001. Relationships between dinocyst assemblages in surface sediment and hydrographic conditions in the Bering and Chukchi seas. Journal of Quaternary Science 16, 667–680.

Raine, R., White, M., Dodge, J.D., 2002. The summer distribution of net plankton dinoflagellates and their relation to water movements in the NE Atlantic Ocean, west of Ireland. Journal of Plankton Research 24, 1131–1147.

Rasmussen, T.L., Thomsen, E., van Weering, T.C.E., Labeyrie, L., 1996. Rapid changes in surface and deep water conditions at the Faeroe Margin during the last 58,000 years. Paleoceanography 11, 757–771.

Rochon, A., de Vernal, A., 1994. Palynomorph distribution in recent sediments from the Labrador Sea. Canadian Journal of Earth Sciences 31, 115–127.

Rochon, A., de Vernal, A., Turon, J.-L., Matthiessen, J., Head, M.J., 1999. Distribution of dinoflagellate cyst assemblages in surface sediments from the North Atlantic Ocean and adjacent basins and quantitative reconstruction of sea-surface parameters. Special Contribution Series of the American Association of Stratigraphic Palynologists 35.

Rosell-Melé, A., 1997. Appraisal of CLIMAP temperature reconstruction in the NE Atlantic using alkenone proxies. Eos 78 (46), F28.

Rosell-Melé, A., 1998. Interhemispheric appraisal of the value of alkenone indices as temperature and salinity proxies in high latitude locations. Paleoceanography 13, 694–703.

Rosell-Melé, A., Comes, P., 1999. Evidence for a warm last glacial maximum in the Nordic Seas, or an example of shortcomings in $U_{37}^{K'}$ and U_{37}^{K} to estimate low sea surface temperature? Paleoceanography 13, 694–703.

Rosell-Melé, A., Bard, E., Emeis, K.C., Grieger, B., Hewitt, C., Müller, P.J., Schneider, R., 2004. Sea surface temperature anomalies in the oceans at the LGM estimated from the alkenone-$U_{37}^{K'}$ index: comparison with GCMs. Geophysical Research Letters 31, L03208.

Sarnthein, M., Gersonde, R., Niebler, S., Pflaumann, U., Spielhagen, R., Thiede, J., Wefer, G., Weinelt, M., 2003. Overview of Glacial Atlantic Ocean Mapping (GLAMAP 2000). Paleoceanography 18, 1071.

Sarnthein, M., Jansen, E., Weinelt, M., Arnold, M., Duplessy, J.C., Erlenkeuser, H., Flatøy, A., Johannessen, G., Johannessen, T., Jung, S., Koc, N., Labeyrie, L., Maslin, M., Pflaumann, U., Schulz, H., 1995. Variations in Atlantic surface ocean paleoceanography, 50°–80°N: A time-slice record of the last 30,000 years. Paleoceanography 10, 1063–1094.

Schneider, R., Bard, E., Mix, A., 2000. Last Ice Age global ocean and land surface temperature: the EPILOG initiative. PAGES Newsletter 8, 19–21.

Solignac, S., de Vernal, A., Hillaire-Marcel, C., 2004. Holocene sea-surface conditions in the North Atlantic—contrasted trends and regimes between the eastern and western sectors (Labrador Sea vs. Iceland Basin). Quaternary Science Reviews 23, 319–334.

Tang, C.L., Wang, C.K., 1996. A gridded data set of temperature and salinity for the northwest Atlantic Ocean. Canadian Data Report of Hydrography and Ocean Sciences no. 148, 45pp.

Taylor, F.J.R., Pollingher, U., 1987. Ecology of Dinoflagellates. In: Taylor, F.J.R. (Ed.), The Biology of Dinoflagellates. Botanical Monograph 21. Blackwell Scientific, Oxford, pp. 399–529.

Ter Braak, C.J.F., Smilauer, P., 1998. CANOCO reference manual and user's guide to CANOCO for Windows software for canonical community ordination (Version 4). Centre for Biometry, Wageningen, 351pp.

Turon, J.-L., 1984. Le palynoplancton dans l'environnement actuel de l'Atlantique nord-oriental. Évolution climatique et hydrologique depuis le dernier maximum glaciaire. Mémoire de L'Institut Géologique du Bassin d'Aquitaine 17, 1–313.

von Grafenstein, R., Zahn, R., Tiedemann, R., Murat, A., 1999. Planktonic ^{18}O records at Sites 976 and 977, Alboran Sea: Stratigraphy, forcing, and paleoceanographic implications. In: Zahn, R., et al. (Eds.), Proceedings of the Ocean Drilling Program, Scientific Results, vol. 161. Ocean Drilling Program, College Station, TX, pp. 469–479.

Voronina, E., Polyak, L., de Vernal, A., Peyron, O., 2001. Holocene variations of sea-surface conditions in the southeastern Barents Sea based on palynological data. Journal of Quaternary Science 16, 717–727.

Waelbroeck, C., Labeyrie, L., Duplessy, J.-C., Guiot, J., Labracherie, M., Leclaire, H., Duprat, J., 1998. Improving paleo-SST estimates based on planktonic fossil faunas. Paleoceanography 12, 272–283.

Wall, D., Dale, B., 1968. Modern dinoflagellate cysts and the evolution of the Peridiniales. Micropaleontology 14, 265–304.

Wall, D., Dale, B., Lohman, G.P., Smith, W.K., 1977. The environmental and climatic distribution of dinoflagellate cysts in the North and South Atlantic Oceans and adjacent seas. Marine Micropaleontology 2, 121–200.

Weinelt, M., Sarnthein, M., Pflaumann, U., Schulz, H., Jung, S., Erlenkeuser, H., 1996. Ice-free Nordic seas during the last glacial maximum? Potential sites of deep water formation. Paleoclimates 1, 283–309.

Weinelt, M., Vogelsang, E., Kucera, M., Pflaumann, U., Sarnthein, M., Voelker, A., Erlenkeuser, H., Malmgren, B.A., 2003. Variability of North Atlantic heat transfer during MIS 2. Paleoceanography, 18, 1071-10.

Winter, A., Jordan, R.W., Roth, P.H., 1994. Biogeography of living coccolithophores in ocean waters. In: Winter, A., Siesser, W.G. (Eds.), Coccolithophores. Cambridge University Press, Cambridge, pp. 161–177.

Zaragosi, S., Eynaud, F., Pujol, C., Auffret, G.A., Turon, J-L., Garlan, T., 2001. Initiation of the European deglaciation as recorded in the northwestern Bay of Biscay slope environments (Meriadzek Terrace and Trevelyan Escarpment): a multi-proxy approach. Earth and Planetary Science Letters 188, 493–507.

Zonneveld, K.A., Versteegh, G.H., de Lange, G.J., 2001. Paleoproductivity and postdepositional aerobic organic matter decay reflected by dinoflagellate cyst assemblages in Eastern Mediterranean S1 Sapropel. Marine Geology 172, 181–195.

Quaternary Science Reviews 24 (2005) 925–950

Planktonic foraminiferal assemblages preserved in surface sediments correspond to multiple environment variables

Ann E. Morey*, Alan C. Mix, Nicklas G. Pisias

College of Oceanic and Atmospheric Sciences, Oregon State University, Corvallis, OR 97331, USA

Received 1 July 2003; accepted 1 September 2003

Abstract

Here we investigate the relationships between modern planktonic foraminiferal species assemblages from Atlantic and Pacific core-top sediment samples and 35 water-column and preservation properties using Canonical Correspondence Analysis (CCA). CCA finds two faunal dimensions (axes) that are most highly correlated to the environmental variables, and describes each axis in terms of the best linear combination of environmental variables. CCA Axis 1 (30.4% of the faunal variance) is related primarily to mean annual sea-surface temperature (SST, $r = -0.96$). CCA Axis 2 (7.9% of the faunal variance) is related to environmental variability associated with an inverse relationship between SST and surface salinity, as well as pycnocline phosphate concentrations, the seasonal range in nitrate concentrations, water depth, and chlorophyll concentrations at the sea surface. Based on this clustering of nutrient and chlorophyll on Axis 2, we infer an ecological response to oceanic fertility. No evidence is found for a unique dissolution influence, suggesting that sea-floor carbonate ion concentration cannot be estimated reliably from planktonic foraminiferal assemblages.

Our results support the use of foraminiferal assemblages to estimate mean annual SST.

© 2004 Elsevier Ltd. All rights reserved.

1. Introduction

Planktonic foraminifera preserved in deep-sea sediments have been used to estimate sea-surface temperature (Imbrie and Kipp, 1971; Kipp, 1976), primary production (Mix, 1989), annual temperature as a function of mixed-layer depth in the tropical Atlantic (Ravelo et al., 1990), thermocline depth in the tropical Pacific (Andreasen and Ravelo, 1997) and carbonate-ion saturation at the seafloor (Anderson and Archer, 2002). Does a unique response to each of these variables exist in the faunal data or is estimation of multiple environmental properties from foraminiferal assemblages an exercise in wishful thinking? Many water column and preservation variables are highly correlated in the

modern ocean, making the identification of causal relationships difficult.

Early studies comparing foraminiferal assemblages from plankton tows to water column characteristics suggested that species distributions are most highly correlated to SST, salinity and nutrients (Bradshaw, 1959; Parker, 1960; Berger, 1969), or to integrated water mass characteristics (Berger, 1968). Seafloor dissolution, which alters the composition of the original assemblages by the selective removal of the most susceptible species, could partially obscure these primary relationships (Parker and Berger, 1971; Coulbourn et al., 1980) and therefore may influence paleoenvironmental interpretations based on relative species abundances (Thompson, 1981; Berger, 1970).

CLIMAP (1981) utilized assemblages of planktonic foraminifera (and other fossil groups) from surface sediments to calibrate paleotemperature transfer functions, based on the method of Imbrie and Kipp (1971), and used these equations to reconstruct SST fields for

---

*Corresponding author. Tel. +1-541-737-5214; fax: +1-541-737-2064.

E-mail address: morey@coas.oregonstate.edu (A.E. Morey).

0277-3791/$ - see front matter © 2004 Elsevier Ltd. All rights reserved.
doi:10.1016/j.quascirev.2003.09.011

the last glacial maximum (LGM) and other time intervals. In this method, the strong empirical relationship between modern foraminiferal fauna and SST could have predictive value even if species don't respond directly to SST. The only requirement is that the relationship between the true causal factors and SST remained constant through time. CLIMAP's methods have been questioned in light of conflicting evidence for tropical cooling during the LGM (Rind and Peteet, 1985; Guilderson et al., 1994; Thompson et al., 1995; Stute et al., 1995). Attempts to resolve this controversy included application of different statistical methods (Prell, 1985; Le and Shackleton, 1994; Pflaumann et al., 1996; Waelbroeck et al., 1998), avoidance or mitigation of 'no-analog' conditions in which past faunal variability exceeds that observed in modern samples (Hutson, 1977; Mix et al., 1999), consideration of selective preservation (e. g. Prell, 1985; Le and Shackleton, 1992; Miao et al., 1994; Thunell et al., 1994; Dittert et al., 1999), and estimation of environmental variables other than SST on faunal assemblages (Ravelo et al., 1990; Ortiz et al., 1995; Watkins et al., 1996; Watkins and Mix, 1998; Schiebel et al., 2002a). In spite of all of these efforts, it remains unknown which (or how many) environmental properties most strongly influence the distributions of species abundances, and how reliably these relationships can be used for paleo-environmental estimation.

Here we attempt to identify empirically the environmental properties in the upper ocean (which likely influence the ecology of living planktonic foraminifera), and at the sea floor (which likely control the preservation of sedimentary assemblages) that best describe variability of species observed in modern sea-floor sediment samples. We first determine how many properties can be resolved by identifying the number of statistically independent dimensions of the environmental and faunal data. We then employ CCA, a statistical method that has proven useful in ecology, to identify relationships between community structure and multiple environmental properties. Finally, we test the predictive value of these empirical relationships in Pleistocene samples.

2. Methods

2.1. Canonical correspondence analysis

Canonical Correspondence Analysis (ter Braak, 1986) identifies dominant relationships between community data and multiple environmental variables. CCA is closely related to reciprocal averaging ordination (RA, also known as correspondence analysis), and both are derived from weighted averaging (WA) ordination. Ordination refers to methods that order samples along

gradients, and includes the familiar method of Q-mode factor analysis (QFA).

CCA, RA and WA are called unimodal methods because they assume the response of species along environmental gradients is unimodal, with a single peak and two tails. Under this model, the environmental preference of each species is represented by its weighted average maximum abundance, which is equivalent to the position of the peak for a symmetric distribution along an environmental gradient.

Gradients are represented as axes, and samples are positioned relative to one another along these axes based on their faunal composition. Sample positions on these axes are called sample scores and species positions are called species optima. These properties are conceptually analogous to factor loadings and factor scores (respectively) in QFA. Both methods create a simplified model of the original dataset.

QFA and CCA differ in several ways. Based on the similarity of samples, QFA simplifies a faunal dataset into a few orthogonal factors representing compositional end-members. The relationship between these faunal factors and environmental properties is then determined post-analysis. In contrast, CCA reduces a faunal dataset into a few orthogonal gradients that best reflect the environmental properties included in the analysis. For example, QFA may identify five independent factors, three of which appear to have different temperature preferences along a temperature gradient, while the remaining two represent nutrient rich and nutrient poor environments. In our example, CCA would identify two independent dimensions of the faunal data, one most highly correlated to temperature and the other most highly correlated to nutrients.

Relationships between QFA faunal factors and environmental properties are identified one at a time. The appropriateness of the relationship is determined typically by the ability of the resulting equation to reconstruct the environmental property of interest in the calibration data set. In many cases, environmental properties (such as summer and winter temperature) are highly correlated to one another on large spatial scales, and this can yield solutions that are difficult to verify. In contrast to the QFA method, CCA identifies the faunal dimensions that are most highly correlated to multiple environmental variables included in the analysis, and interprets the dimensions in terms of linear combinations of environmental variables that are statistically independent of each other. CCA thus objectively identifies an optimal combination of environmental parameters that relate to (and may cause) species variations.

CCA is best understood as a modification of WA and RA. The faunal and environmental datasets are formatted so that each row represents a sample and each column represents a variable (either a species or an environmental variable).

WA positions samples or species on an environmental gradient represented by a single environmental variable chosen by the investigator. Species optima approximate the positions of maximum occurrence of each species along the gradient. To find species optima along the environmental gradient, WA simply weights (multiplies) the relative abundance of each species in each sample of the faunal data by the value of the environmental variable associated with that sample, and then averages all weighted abundances for each species (column-wise averaging). One can use the optima to find the sample scores with respect to the environmental gradient by weighting the relative abundance of each species by it's respective optimum, and then averaging all weighted abundances within a sample (row-wise averaging). This procedure can be performed multiple times, each time relating the faunal data to a different environmental variable. As with QFA, WA provides no specific guidance as to which environmental properties can be estimated most reliably.

RA follows the above procedure, but repeatedly alternates between calculating species optima and sample scores, until sample scores converge (i.e., no change occurs from iteration to iteration within a defined tolerance). This method of identifying eigenvectors (by column-wise, and then row-wise averaging iteratively until convergence) is called the Power Method (Gourlay and Watson, 1973). RA does not require any associated environmental data: one may assign arbitrary but unequal initial sample scores instead of environmental data to begin the averaging procedure. The arbitrary sample scores are used to calculate species optima, then the optima are used to calculate new sample scores, and so on; repeating until sample scores converge. The resulting gradient, however, is now called a latent environmental gradient and is an eigenvector of the faunal data. As with QFA and WA, sample positions along this RA gradient must be compared to environmental variables after the analysis to describe the gradient in environmental terms.

CCA follows the RA procedure, but adds a regression of sample scores on environmental variables within each iteration. The resulting regression coefficients (called canonical coefficients) within each iteration are used to estimate new sample scores which are then used to calculate new species optima. The regression of sample scores on environmental variables effectively steers the faunal data to maximize the fit between the faunal composition of samples and the environmental data. The canonical coefficients from the final regression define the best linear combination of environmental variables that describe the final sample positions along the resulting gradient, and provide an environmental description of this gradient.

Final sample scores can be represented in one of two ways within CCA. They can be calculated either from the final species optima (WA, or weighted averaging, scores), or they can be calculated from the canonical coefficients resulting from the final regression of WA scores on the environmental variables (LC, or linear combination, scores). If LC scores are chosen, the canonical coefficients define the axes as a linear combination of environmental variables, and the samples themselves have been fitted to the environmental variables. If WA scores are chosen, the canonical coefficients represent the best linear combination of environmental variables that describe the sample positions along each axis, but do not actually define the axis. In this case, species optima have been fitted to the environmental variables, but the samples have not. We have chosen to use the WA scores because the relative positions of samples can be inferred from their faunal compositions and the positions of species optima (unlike the LC scores), and one can calculate the positions of samples from unknown environments (i.e. fossil samples) by using the positions of modern optima.

We summarize this iterative weighted averaging technique here. Algorithm details are given in the Appendix.

(1) Choose arbitrary, but unequal, initial sample scores (known as trial scores).
(2) Calculate species scores by weighted averaging of the sample scores with sample scores as weights.
(3) Calculate new sample scores by weighted averaging of the species scores with species scores as weights (at convergence these are the WA scores).
(4) Obtain regression coefficients by weighted multiple regression of the sample scores on the environmental variables.
(5) Calculate new sample scores from the regression coefficients (at convergence these are the LC scores).
(6) Center and standardize the sample scores to a mean of zero and standard deviation of 1.
(7) Repeat from step 2, using the sample scores from step 6 in place of the previous sample scores. Stop when new sample scores do not change within a prescribed tolerance from the sample scores from the previous iteration. The square root of the dispersion (Appendix) of sample scores at convergence equals the eigenvalue.

Calculation of a second axis follows the same procedure, however an orthogonalization step (Appendix) is included within each iteration (after step 6) that removes from the trial sample scores the linear correlation with the first axis. Trial sample scores for additional axes have the linear correlation with the first axis and all other axes removed.

A well-known problem called the arch effect sometimes arises when using CCA and other unimodal methods. This mathematical artifact distorts sample positions on the second and higher axes. The arch effect occurs because the method assumes that all species distributions are symmetrical when in fact the maximum abundance of a species may occur at gradient extremes. Variance from truncated distributions folds onto the second (and higher) axes, and yields a geometric relationship to the lower axes (Jongman et al., 1995). To remove this effect, we detrend by second order polynomials after visually inspecting the results for an arch. Detrending by polynomials involves a step in the orthogonalization procedure of CCA to make the sample scores uncorrelated to the quadratic function of the next lower axis (Jongman et al., 1995). We choose to detrend even though it may potentially remove useful information to avoid describing variation that is an artifact of the method after a pronounced arch was detected by visual inspection of our preliminary results.

Environmental variables enter the analysis in order of the amount of variance they explain. Variables that are not statistically significant or are highly correlated to others already included in the model are excluded. Significance is determined by 199 Monte Carlo randomizations: CCA is performed on each of 199 randomized permutations of the original dataset; a variable is significant if the additional faunal variance explained is greater than that explained by 95% of the permutation tests. To avoid including variables that are highly correlated to one another, we reject variables that result in variance inflation factors (VIF; Appendix) >10. A very large VIF (>20) indicates a variable is almost perfectly correlated to other variables already included in the analysis.

We used the program CANOCO version 4.02 (ter Braak and Smilauer, 1998) to perform the CCA. The forward selection procedure is outlined below:

1. Perform CCA using the manual forward selection option on all faunal and environmental data. CANOCO lists environmental variables in order of how much variance each explains when CCA is run with each variable separately. We select the variable that explains the most variance (after testing for significance; $p < 0.05$) as the first variable.
2. CANOCO lists all remaining variables in order of how much variance each explains in addition to the amount explained earlier variables. We select the variable that explains the largest amount of variance from this list as the second variable after checking for significance. This choice is retained if the VIFs of all variables in the model are below 10 with this variable included. If the VIFs of any of the environmental variables already in the analysis are greater than 10, this choice is discarded and the variable explaining the next largest amount of variance is selected and evaluated similarly for inclusion.
3. Continue selecting variables as in step 2 until each of the remaining variables explains less than 1% of the remaining faunal variance.

2.2. Data

The data analyzed using CCA included foraminiferal assemblage data from surface sediment samples throughout the Atlantic and Pacific basins and corresponding environmental and preservation variables at the location of each sample. CCA results are applied to foraminiferal assemblage data from a tropical Pacific sediment core RC13-110, and LGM horizon samples. We have chosen to evaluate both oceans together to determine if the relationships between fauna and environment in each ocean are consistent between oceans. If consistent, relationships missing from one ocean (as an artifact of sampling) will be filled in by relationships in the other ocean, potentially increasing the range and accuracy of the identified relationships in sparsely sampled regions. If the relationships are not consistent in the two ocean basins, the environmental variables are most likely not causal, or there are additional confounding variables. All data are archived under the MARGO project heading at PANGAEA (http://www.pangaea.de/Projects/MARGO/).

2.2.1. Species data

We use the Brown University Foraminiferal Database (Prell et al., 1999) as the surface faunal dataset Species identifications in this database were made by one research group using the taxonomies of Parker (1962) and Bé (1977). These data were supplemented with samples from the OSU laboratory after taxonomic intercalibration with Brown University. These 992 samples are widely distributed throughout much of the Atlantic and Pacific Oceans (Fig. 1). Following Mix et al. (1999), we combine the pink and white varieties of *Globigerinoides ruber* into *G. ruber* (total), and *Globigerinoides sacculifer* with and without a terminal chamber into *G. sacculifer* (total). We include the *Neogloboquadrina pachyderma—Neogloboquadrina dutertrei* ("P-D") intergrade of Kipp (1976) with *N. dutertrei*, and combine *Globorotalia menardii*, *Globorotalia tumida* and *Globorotalia neoflexuosa* into *G. menardii-tumida* complex. We calculate relative abundances with closure around 28 species groups (Table 1).

The faunal data are transformed by ln(percent abundance + 1) prior to the analysis. Relative abundances of species within a sample are frequently lognormal in distribution (MacArthur, 1960). The log transform helps to prevent a few highly abundant species from dominating the analysis by increasing the contributions from less abundant species.

The addition of a constant to each value prior to transformation prevents taking the log of zero. To assess

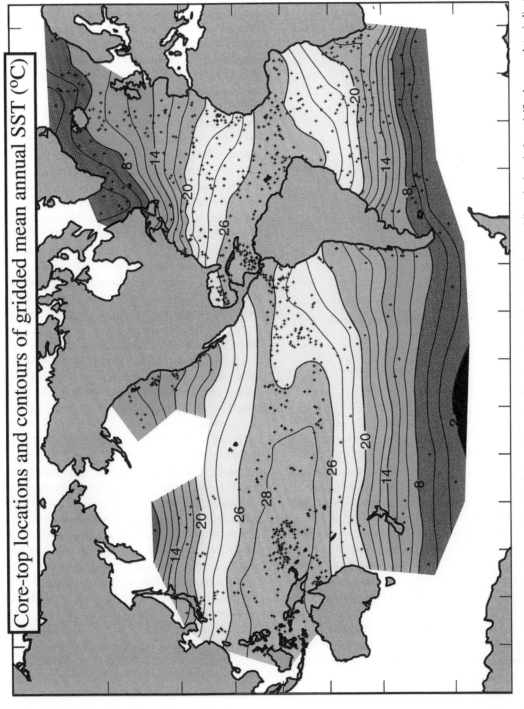

Core-top locations and contours of gridded mean annual SST (°C)

Fig. 1. Core locations and contours of gridded mean annual SST (°C). Locations of surface (core-top) sediment samples providing planktonic foraminiferal census data, indicated by a ' + ', used in this study to identify the relationship between modern species groups and environmental and preservation variables. These data are from the Brown University Foraminiferal Database (Prell et al., 1999) supplemented with data from Mix (1989), Feldberg and Mix (2002) and this paper.

Table 1
Species included in the analysis

Code	Name
UNIV	*Orbulina universa*
CONG	*Globigerinoides conglobatus*
RUBT	*Globigerinoides ruber* (total of white and pink varieties)
TENE	*Globigerinoides tenellus*
SACT	*Globigerinoides sacculifer* (total with and without final sac)
DEHI	*Sphaeroidinella dehiscens*
AEQU	*Globigerinella aequilateralis (G. siphonifera)*
CALI	*Globigerina calida*
BULL	*Globigerina bulloides*
FALC	*Globigerina falconensis*
DIGI	*Globigerina digitata*
RUBS	*Globigerina rubescens*
QUIN	*Turborotalita quinqueloba*
PACL	*Neogloboquadrina pachyderma* (left coiling)
PACR	*Neogloboquadrina pachyderma* (right coiling)
PDTR	*Neogloboquadrina dutertrei* (including *pachyderma-dutertrei* intergrade)
CGLM	*Globoquadrina conglomerata*
HEXA	*Globorotaloides hexagona*
OBLI	*Pulleniatina obliquiloculata*
INFL	*Globorotalia inflata*
TRUL	*Globorotalia truncatulinoides* (left coiling)
TRUR	*Globorotalia truncatulinoides* (right coiling)
CRAS	*Globorotalia crassaformis*
HIRS	*Globorotalia hirsuta*
SCIT	*Globorotalia scitula*
MTTL	*Globorotalia menardii, G. menardii flexuosa*, and *Globorotalia tumida*
GLUT	*Globigerinita glutinata*
THEY	*Globorotalia theyeri*

the choice of this arbitrary constant on statistical analyses, we compare our results using a constant of 1 to those using constants of 0.1, 0.5, 2.0 and 5.0. CCA eigenvalues are similar for all constants except for 5, which appears to increase the noise such that it interferes with the structure of the data. Also, the addition of constants from 0.1 to 2.0 does not influence the canonical coefficients for the first eigenvector and only slightly for the second eigenvector, while the addition of 5.0 greatly influences the canonical coefficients for both the first and second eigenvector. This suggests that our choice of one for the constant does not unduly influence the results.

Although we do not transform the environmental data prior to the analysis, CANOCO transforms the data within the analysis such that each variable has a mean of zero and standard deviation of one.

2.2.2. Environmental and preservation data

Physical and chemical water column variables for the sea-surface and the pycnocline were extracted from the World Ocean Atlas Database 1998 (Ocean Climate Laboratory, 1999), hereafter referred to as WOA98. Pycnocline depth was defined by the maximum in the rate of change in density with depth (averaged over 50 m). If more than one peak in the density gradient was found, the shallowest pycnocline was chosen. Annual averages, minima, maxima, and ranges (seasonal maximum minus minimum) of temperature, nutrients and other chemical and physical characteristics of the water are calculated at these levels. Primary production estimates (Behrenfeld et al., 2001), expressed as annual average and seasonal range at each location, were derived from satellite data (SeaWiFS; Sea-viewing Wide Field-of-view Sensor) collected between September 1997 and August 2000—an interval that includes both El Niño and La Niña conditions. We estimated carbon flux at the seafloor (annual average and range) from this primary production data using the relationship of carbon flux to water depth described by Suess (1980).

Carbonate ion concentration variables (Archer, 1996) are included as calcite dissolution proxies. Water depth was also included because it has been used to reflect the influence of water pressure on the solubility of calcite. We included sea-floor values of modeled carbonate ion concentration ($[CO_3^{2-}]$) and calcite saturation state ($\Delta[CO_3^{2-}]$); the difference between $[CO_3^{2-}]$ and calcite saturation. These carbonate concentration variables reflect the combined influence of water depth, acidification due to the respiration of organic matter, temperature, salinity, and alkalinity of the water at the location of each sample. We do not attempt to explicitly calculate local pore-water effects that may induce calcite dissolution above the lysocline, but if such effects are important, we expect that both calcite saturation and organic carbon fluxes will be selected by CCA analysis as important environmental variables.

Table 2 lists the 35 environmental and preservation variables that are included in our analysis, and their sources. The environmental variables were screened prior to analysis to remove redundant or problematic variables. Oxygen concentration at the sea surface was excluded from the analysis as it is primarily a function of temperature. Sea-surface and pycnocline density variables were excluded because they reflect the combined influence of temperature and salinity. Pycnocline nutrient range variables were not included in the analysis because some data are missing in winter for high-latitude samples.

2.3. Identifying the number of dimensions

Imbrie et al. (1973) state that the number of statistically independent environmental variables that can be estimated must be less than the number of statistically independent modes of the faunal data. This restriction on the number of environmental properties prevents the estimation of more dimensions than actually exist in the faunal data, and therefore is appropriate regardless of method, unless one has

Table 2
Environmental variables used in the analysis. All values are annual averages, except for range values, which are calculated from seasonal averages

Code	Variable	Source	Units
Sea-surface variables			
SST	Temperature	WOA98[a]	°C
SSTrg	Temperature range	WOA98[a]	°C
SSTmin	Minimum temperature	WOA98[a]	°C
SSTmax	Maximum temperature	WOA98[a]	°C
Salt	Salinity	WOA98[a]	psu
Saltrg	Salinity range	WOA98[a]	psu
Chlor	Chlorophyll	WOA98[a]	μM
Chlorrg	Chlorophyll range	WOA98[a]	μM
NO_3	Nitrate	WOA98[a]	μM
NO_3rg	Nitrate range	WOA98[a]	μM
PO_4	Phosphate	WOA98[a]	μM
PO_4rg	Phosphate range	WOA98[a]	μM
SiO_2	Silica	WOA98[a]	μM
SiO_2rg	Silica range	WOA98[a]	μM
PP	Primary production	Behrenfeld et al.[b]	g C/m^2/yr
PPrg	Primary production range	Behrenfeld et al.[b]	g C/m^2/yr
Sig(100–0)	Density contrast: 100 m density minus surface density	WOA98[c]	σ_t
Pycnocline variables			
PD	pycnocline depth	WOA98[c]	m
PDrg	pycnocline depth range	WOA98[c]	m
PT	Temperature	WOA98[c]	°C
PTrg	Temperature range	WOA98[c]	°C
PSalt	Salinity	WOA98[c]	psu
PSaltrg	Salinity range	WOA98[c]	psu
PdSig	Rate of change in density	WOA98[c]	σ_t
PdSigrg	Rate of change in density range	WOA98[c]	σ_t
PNO_3	Nitrate	WOA98[c]	μM
PPO_4	Phosphate	WOA98[c]	μM
$PSiO_2$	Silicate	WOA98[c]	μM
PO_2	Oxygen	WOA98[c]	μM
PO_2SAT	Oxygen percent saturation	WOA98[c]	percent saturation
Seafloor variables			
Depth	Water depth	NGDC[d]	m
$SFCO_3$	Carbonate conc. at seafloor	Archer[e]	μM
$SFdCO_3$	Carbonate conc. minus saturation	Archer[e]	
Cflux	Carbon flux	Behrenfeld et al.[b] data, and Suess[f] equation	gC/m^2/yr
Cfluxrg	Carbon flux range	Behrenfeld et al.[b] data, and Suess[f] equation	gC/m^2/yr

[a]Ocean Climate Laboratory, (1999).
[b]Behrenfeld et al. (2001).
[c]Calculated from WOA98.
[d]National Geological Data Center (http://www.ngdc.noaa.gov).
[e]Archer (1996).
[f]Suess (1980).

additional information. While not explicitly stated, this reflects the need to have at least two unimodal faunal variables (which could be Q-mode factor loadings) with different means with respect to a single environmental variable, to provide a unique solution to transfer function inversion (Imbrie et al., 1973; Imbrie and Kipp, 1971).

We investigate the dimensionality of our datasets using R-mode factor analysis, a method that identifies data structure through eigenanalysis of a cross-products matrix (in this case, the correlation matrix among variables) to determine how many significant independent modes of variability exist. We use two stopping criteria to evaluate how many eigenvectors should be retained. The broken-stick method identifies eigenvalues (the amount of variance explained by each eigenvector) as significant if larger than broken-stick eigenvalues (the amount of variance explained by random data) (Frontier, 1976). This method has been shown by Jackson (1993) to accurately reproduce the dimensions of simulated data matrices. We also apply the commonly used scree-plot method, which involves plotting the eigenvalues versus their rank order. The point at which eigenvalues deviate from a straight line drawn through the eigenvalues of higher rank indicates non-random variation. Cattell and Vogelmann (1977) argue that all eigenvalues to the left of this point, plus one to the right, should be retained.

An additional constraint is that we must limit our discussion to those dimensions that can be accurately identified by the statistical method employed. To do this, we calculate the gradient length, a measure of faunal turnovers along a gradient, for each axis identified by CCA. Gradient length is expressed in standard deviation units of species turnover (SD), and is calculated by dividing the range in sample scores by the average within-species standard deviation along the axis (ter Braak and Smilauer, 1998). Unimodal methods are typically considered appropriate to describe the variability along an axis if the gradient length is at least 2 SD.

3. Results

3.1. How many environmental dimensions can be identified from the faunal data?

R-mode factor analysis of the environmental data results in the eigenvalues shown in Table 3a. Three eigenvalues reflect non-random variation from the comparison of eigenvalues to broken-stick eigenvalues. The scree-plot method suggests that four eigenvalues should be retained. We conclude that three to four independent dimensions of non-random variation exist

Table 3
R-mode Factor Analysis results from the analysis of 35 environmental and preservation variables

(a) Eigenvalues

Axis	Eigenvalue	% of variance	Cum. % of var.	Broken-stick eigenvalue
1	10.832	30.949	30.949	4.147
2	5.743	16.407	47.356	3.147
3	3.902	11.148	58.505	2.647
4	2.172	6.207	64.711	2.313
5	1.894	5.410	70.122	2.063
6	1.544	4.411	74.533	1.863
7	1.298	3.709	78.241	1.697
8	1.122	3.204	81.446	1.554
9	1.061	3.031	84.477	1.429
10	0.796	2.273	86.750	1.318

(b) First four eigenvectors

Variable	Eig 1	Eig 2	Eig 3	Eig 4
Depth	0.0674	0.0349	−0.2193	−0.1666
SST	0.2900	−0.0512	0.0574	0.0853
SSTrg	−0.1497	0.2023	0.1220	−0.2471
SSTmin	0.2878	−0.0762	0.0338	0.1141
SSTmax	0.2859	−0.0155	0.0851	0.0433
Salt	0.0972	0.2795	−0.1943	0.1235
Saltrg	0.0941	−0.0729	0.3388	−0.1236
Chlor	−0.1815	0.1038	0.1787	0.2198
Chlorrg	−0.1299	0.1286	0.1688	0.1683
NO_3	−0.2301	−0.1593	−0.1115	−0.0353
NO_3rg	−0.2360	−0.0603	0.0159	0.0839
PO_4	−0.2252	−0.2012	−0.1048	−0.0438
PO_4rg	−0.1969	−0.0618	−0.0016	0.0579
SIO_2	−0.1834	−0.1502	0.0009	−0.2086
SIO_2rg	−0.1641	−0.0572	0.0142	−0.1326
PP	−0.1436	0.0751	0.1885	0.2340
PPrg	−0.2130	0.1018	0.1282	0.2070
$SFCO_3$	−0.0351	0.2636	−0.0229	0.1036
$SfdCO_3$	−0.0821	0.1895	0.1488	0.1916
Cflux	−0.0673	0.0102	0.1790	0.3147
Cfluxrg	−0.1385	0.0472	0.1900	0.3437
PD	0.1004	−0.1652	−0.2486	0.2193
PDrg	−0.1527	0.1208	−0.0341	0.1080
PT	0.2831	0.0246	0.0820	0.0220
PTrg	−0.0363	0.0836	0.2238	−0.1416
PSalt	0.1473	0.2417	−0.1602	0.1484
PSaltrg	0.0419	−0.0480	0.3294	−0.2261
PdSig	0.1401	−0.1877	0.2852	0.0141
PdSigrg	−0.0382	0.1205	0.3220	−0.2141
PNO_3	−0.1476	−0.3101	−0.0472	0.1366
PPO_4	−0.1429	−0.3364	−0.0429	0.1003
$PSIO_2$	−0.1176	−0.2927	0.0173	−0.0700
PO_2	−0.2420	0.1761	−0.0690	−0.2025
PO_2%SAT	−0.1017	0.3157	−0.0497	−0.2640
Sig(100-0)	0.1248	−0.1727	0.3085	−0.0347

Table 4
R-mode Factor Analysis results from the analysis of 28 faunal groups

(a) Eigenvalues

Axis	Eigenvalue	% of variance	Cum. % of var.	Broken-stick eigenvalue
1	6.814	24.337	24.337	3.927
2	5.257	18.775	43.111	2.927
3	2.339	8.354	51.466	2.427
4	1.883	6.726	58.191	2.094
5	1.400	5.002	63.193	1.844
6	1.114	3.979	67.172	1.644
7	0.932	3.328	70.500	1.477
8	0.887	3.169	73.669	1.334
9	0.806	2.878	76.547	1.209
10	0.710	2.534	79.081	1.098

(b) First three eigenvectors

Species group	Eig 1	Eig 2
UNIV	0.0939	−0.1492
CGLB	0.2144	−0.0592
RUBT	0.2920	−0.2201
TEN	0.1503	−0.2595
SACT	0.3311	−0.0365
DEHI	0.1121	0.1927
AEQU	0.3249	−0.0827
CALI	0.2289	−0.2292
BULL	−0.2190	−0.0839
FALC	−0.0665	−0.3408
DIGI	0.1333	−0.0354
RUBS	0.1807	−0.1903
QUIN	−0.2547	0.0174
PACL	−0.2884	0.0708
PACR	−0.3097	−0.0420
PDTR	0.0214	0.2349
CGLM	0.1131	0.1862
HEX	0.0558	0.1298
OBLI	0.2028	0.2108
INFL	−0.2061	−0.2204
TRUL	−0.1283	−0.2727
TRUR	0.0164	−0.2850
CRAS	0.0935	−0.0240
HIRS	−0.0868	−0.2811
SCIT	−0.0381	−0.2727
MTTL	0.2175	0.2302
GLUT	0.1673	−0.1593
THEY	0.0269	0.0722

within our environmental dataset. The first four eigenvectors are shown in Table 3b.

R-mode factor analysis of the faunal data yields the eigenvalues shown in Table 4a. Two eigenvalues pass the broken-stick eigenvalue criterion. The scree plot method, however, suggests that five eigenvalues may be non-random variation. Therefore at least two, and possibly up to five, independent dimensions exist within the faunal dataset. Considering the number of both environmental and faunal dimensions in our dataset, at least one and possibly as many as four dimensions may be interpreted. To be conservative, we will limit our discussion to two species-environment relationships in CCA. The first two eigenvectors are shown in Table 4b.

Table 5
Environmental variables selected for inclusion within CCA using the manual forward selection option in CANOCO (*F*-values and *p*-values calculated by CANOCO). Variables are listed in order of inclusion. Variance Explained refers to the percentage of total faunal variance explained by each environmental variable

	Variable name	Variance explained (%)	*F*-value	*p*-value
1	SST	29.75	420.0	<.005
2	PPO$_4$	7.72	122.1	<.005
3	PP	2.52	41.1	<.005
4	Salt	2.30	39.0	<.005
5	Depth	2.30	41.3	<.005
6	Sig(100-0)	1.56	27.9	<.005
7	Chlor	1.48	28.3	<.005
8	SSTrg	1.48	28.6	<.005
9	NO$_3$rg	1.17	22.9	<.005
10	SFCO$_3$	0.96	19.0	<.005

Table 6
Canonical (regression) coefficients scaled to standardized variables (based on the environmental variables centered and standardized to unit variance)

Name	Axis 1	Axis 2
SST	−0.7485	0.3127
Salt	−0.1558	−0.3030
NO$_3$rg	0.0178	0.2607
Depth	0.0105	0.2213
PPO$_4$	−0.0486	0.2064
Chlor	0.0175	0.1411
SFCO$_3$	0.0211	−0.0725
SSTrg	−0.0138	−0.0660
Sig(100-0)	−0.0063	0.0603
PPavg	0.0375	0.0531

3.2. Selection of environmental variables within CCA

Ten of the 36 variables in the environmental dataset were selected (Table 5) by manual forward selection within CCA. No environmental variables were rejected based on significance or VIF (based on retaining variables for which $p < 0.05$ and VIF < 10), however selection stopped when each variable explained less than 1% of the remaining faunal variance. At this point, many environmental variables explained a similar amount of the remaining variance, making selection of the next variable ambiguous.

Mean annual SST explained more of the faunal variance (29.8%) than any other variable. For comparison, the other temperature variables (all of which are significantly correlated to mean annual SST) include seasonal minimum surface temperature (29.4%), seasonal maximum surface temperature (28.5%), mean annual temperature at the pycnocline (27.7%), seasonal range in surface temperature (11.6%) and seasonal range in temperature at the pycnocline (11.1%). Water depth (2.4% of the faunal variance) and [CO$_3^{2-}$] (1.0% of the faunal variance) were selected, while Δ[CO$_3^{2-}$] was not, suggesting that a distinct dissolution signal is not identifiable in the combined Atlantic and Pacific faunal abundance data.

3.3. Species–environment relationships

CCA identifies four faunal axes that have significant ($p < 0.05$) relationships to the 10 environmental properties selected. These four axes explain 51% of the total faunal variance. We previously determined that one to four environmental properties could be estimated from our data. Because Axes 1 and 2 have gradient lengths greater than 2 (Axis 1 = 3.3 SD, Axis 2 = 2.2 SD), while

Axes 3 and 4 have gradient lengths less than 2 (Axis 3 = 1.0 SD, Axis 4 = 1.8 SD), we limit the interpretation of the CCA results to the first two axes and conclude that unimodal methods are not appropriate to describe the variability along Axes 3 and 4.

The two CCA axes we retained explain 38% of the total variance in the faunal dataset and 75% of the canonical variance (the percent of the faunal variance explained by all four axes). The relationship between environmental variables and the resulting axes can be inferred from the canonical coefficients (Table 6), which define the best linear combination of environmental variables that describe sample positions along each axis (Fig. 2).

The remaining 62% of the faunal variance is a result of noise (in both species abundance and environmental data), variability that is not explained by the environmental variables included, or non-linear relationships. We perform QFA on the faunal dataset to determine the amount of noise contained within the species percentage data alone. 60% of the total faunal variance is explained by five QFA factors (calculated from the coefficient of determination between the modeled species variability and the original faunal variability). The remaining 40% cannot be differentiated from noise. The two CCA axes, therefore explain approximately two-thirds (38/60) of the total faunal variance above the noise level. The 20% explained by the five QFA factors not captured by CCA is likely to be a result of non-linear relationships between sample scores along the axes and the environmental variables, environmental properties not included in the analysis, or noisy environmental data.

Percent fit is defined as the ratio of the modeled variance to the original variance for each species, multiplied by 100. Comparing the percent fit to the faunal variance explained by the two retained CCA axes (38%) quantifies each species utility as an environmental indicator (Table 7). The faunal variance is reproduced

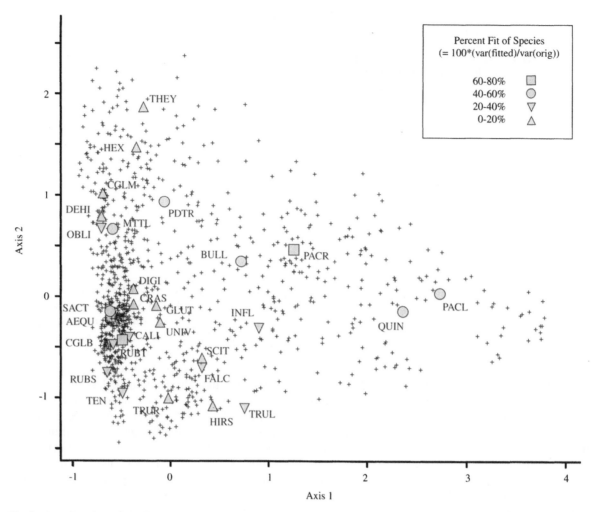

Fig. 2. Distribution of species optima (filled symbols) and samples (indicated by '+') within two dimensions of the foraminiferal data and associated environmental and preservation data. Each dimension represents a faunal gradient that CCA finds most highly correlated to a linear combination of environmental variables included in the analysis. The ability of the model to reconstruct the original variance of each species is indicated by symbols representing percent fit.

better than average if the percent fit of that species is greater than 38% and less than average if the percent fit is less than 38%. Species with high percent-fit values should be included in next-generation transfer functions, whereas species with low percent-fit values could be excluded without substantial loss of information.

3.3.1. CCA Axis 1

CCA Axis 1 explains 30.4% of the total faunal variance (59.4% of the canonical variance). Mean annual SST (canonical coefficient = −0.749, r = −0.96) best describes sample positions along this axis (Fig. 3a). Salinity also may explain some sample variability along Axis 1 (canonical coefficient = −0.156), but the contribution is much smaller than that of SST. Compared to SST and salinity, contributions from other environmental variables are negligible (canonical coefficients range between −0.05 and 0.04; Table 6).

Four species with obvious relationships to Axis 1 and no clear relationship to Axis 2 (Fig. 3b) are *Neogloboquadrina pachyderma* (left-coiling), *N. pachyderma* (right-coiling), *Globorotalia inflata*, and *Globigerina bulloides*. Living specimens of these taxa are most common outside the tropics, and typically reach greatest abundance in transitional to subarctic/subantarctic environments (Bé and Tolderlund, 1977), although *G. bulloides* is also found in tropical upwelling systems (Prell and Curry, 1981). *N. pachyderma* (left coiling) abundances are essentially zero at the negative extreme of Axis 1, but increase rapidly between 1.0 and 2.5 on Axis 1, and maintain very high (near 100%) levels at the positive extreme of Axis 1. *Turborotalita quinqueloba* has a similar relationship to Axis 1, however some samples with Axis 1 scores greater than 2.0 lack *T. quinqueloba*, suggesting that temperature may not be the only influence on this species. *G. inflata* and *N. pachyderma*

Table 7
Species optima and percent fit of each species for Axis 1 and Axis 2, and percent fit for both axes together. Species are listed in decreasing order of percent fit for both axes

Species	Species optima[a]		Percent fit		
	Axis 1	Axis 2	Axis 1	Axis 2	Axes 1 + 2
PACR	1.27	0.43	62.5	3.7	66.2
RUBT	−0.50	−0.40	49.0	15.9	64.9
AEQU	−0.62	−0.20	58.9	3.1	62.0
SACT	−0.62	−0.12	57.8	1.0	58.8
PACL	2.71	0.16	58.1	0.1	58.2
PDTR	−0.06	0.96	0.4	49.4	49.8
QUIN	2.37	−0.10	49.5	0.0	49.5
BULL	0.73	0.30	44.4	3.7	48.2
MTTL	−0.59	0.71	26.5	19.7	46.2
OBLI	−0.71	0.68	24.7	11.5	36.2
CALI	−0.43	−0.37	23.4	9.2	32.5
INFL	0.90	−0.32	29.9	1.9	31.8
RUBS	−0.66	−0.73	15.8	9.9	25.8
TEN	−0.49	−0.92	9.3	16.4	25.7
TRUL	0.73	−1.06	12.2	13.0	25.3
CGLB	−0.60	−0.43	19.6	5.0	24.6
FALC	0.32	−0.73	6.4	16.7	23.1
TRUR	−0.03	−1.01	0.0	18.9	19.0
HIRS	0.43	−1.13	4.0	14.3	18.3
DEHI	−0.70	0.82	7.9	5.5	13.4
CGLM	−0.68	1.04	6.0	7.1	13.2
SCIT	0.33	−0.72	3.4	8.4	11.8
HEX	−0.33	1.58	0.9	10.4	11.2
GLUT	−0.13	−0.14	4.7	2.7	7.4
DIGI	−0.38	0.05	7.2	0.1	7.3
THEY	−0.25	1.98	0.2	5.3	5.5
CRAS	−0.38	−0.03	3.1	0.0	3.1
UNIV	−0.11	−0.24	0.9	2.1	3.1

[a]To display both species optima and samples on the same diagram (Fig. 2), we used a scaling factor of $\alpha = .5$ in Eq. (5) (Appendix).

(right coiling) have unimodal distributions that peak near the middle of Axis 1 (between 0.5 and 2.0), but species abundances become more diffuse as scores along this axis increase. *G. bulloides* peaks in abundance near Axis 1 values of 2.0, but it also has a spike in abundance along Axis 1 from about −1.0 to −0.2.

The negative (warm) extreme of Axis 1 is dominated by tropical/subtropical species such as *Globigerinoides ruber* and *Globigerina aequilateralis* (Fig. 3b). These species vary substantially in abundance below a maximum at each point along this axis, however, suggesting that they are also distributed along CCA Axis 2. Of these species, *G. ruber* appears to be most highly correlated to Axis 1.

G. glutinata, like *G. bulloides,* has a multimodal distribution along Axis 1. The warm-water relationship is very similar to that of *G. aequilateralis*, with peak abundances occurring at the negative extreme of Axis 1. The second abundance maximum for *G. glutinata* occurs between values of 1 and 2 along Axis 1. This distribution

is reflected in the low percent fit of *G. glutinata* (7.4%), reflecting the inability of CCA to accurately reconstruct the variability of this species in our dataset. Multimodality may be a result of cryptic species (Kucera and Darling, 2002) that have been erroneously identified as a single species based on their similar morphology. Species with multimodal distributions along CCA Axis 1 should not be included in transfer functions of temperature.

3.3.2. CCA Axis 2

The second CCA axis explains 7.9% of the total variance in the faunal dataset (15.4% of the canonical variance). Several environmental variables contribute to the description of sample positions along Axis 2, as indicated by their canonical coefficients (Table 6). The contributions of mean annual SST (0.3127) and surface salinity (−0.3030) are greatest, although surface nitrate concentration range (0.2607), water depth (0.2213), pycnocline phosphate concentration (0.2064) and surface chlorophyll concentration (0.1411) also contribute to the explanation of sample positions along this axis. Of these variables, only salinity and pycnocline phosphate concentration are moderately ($r > 0.3$) correlated to sample positions along Axis 1 (Fig. 4a).

Several (mostly tropical and subtropical) species have significant relationships along Axis 2 (Fig. 4b). *Neogloboquadrina dutertrei* ($r = 0.78$) and *G. menardii–tumida* complex ($r = 0.52$) are positively correlated to this axis. Negative scores on Axis 2 are represented by high abundances of *G. ruber* ($r = −0.66$) and *Globigerinoides tenellus* ($r = −0.54$). Maximum abundances of *G. sacculifer* occur in the middle of this axis.

4. Discussion

4.1. CCA Axis 1

4.1.1. SST or SST and salinity?

SST is very highly correlated to sample positions along CCA Axis 1. Because the model also suggests that salinity describes some of the variability, we need to evaluate the viability of the relationship. We re-analyzed the data with SST*salinity included as a term with the environmental variables, to determine if SST and salinity interact together to influence species. The interaction of SST and salinity are selected over SST alone, but a comparison of the variance explained by each term suggests the improvement is very small (for SST*salinity, the variance explained is 0.408; for SST alone the variance explained is 0.401). Furthermore, the average change in sample positions along Axis 1 from the analysis with SST alone relative to the analysis with both SST and salinity (−0.0035) is very small compared to the total variability of sample positions (standard deviation of 0.2982 in Axis 1 units).

Culture experiments do not support the idea that salinity significantly influences foraminiferal species

distributions. For example, Bijma et al. (1990) cultured seven tropical and subtropical species (*G. sacculifer, G. ruber, G. conglobatus, G. aequilateralis, O. universa, N. dutertrei,* and *G. menardii*) to determine their biological

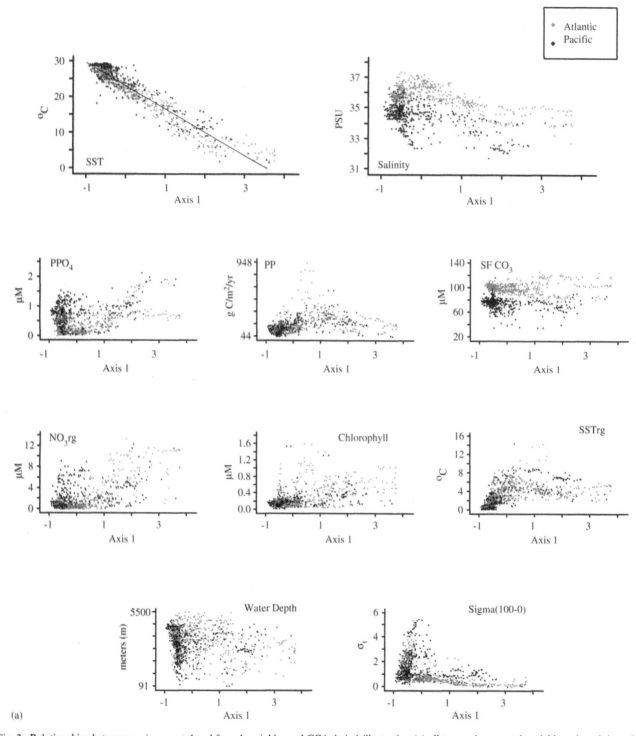

Fig. 3. Relationships between environmental and faunal variables and CCA Axis 1 illustrating (a) all ten environmental variables selected, in order of decreasing linear correlation (from left to right, top to bottom), and (b) species with strong relationships to Axis 1 (and highest values of percent fit). Examples of species that increase with higher Axis 1 sample scores (cold-water species) are on the left, those with maxima in the mid-range of Axis 1 are in the center, and those that increase with higher Axis 1 sample scores (warm-water species) are on the right. Gray dots indicate Atlantic samples, and black dots indicate Pacific samples.

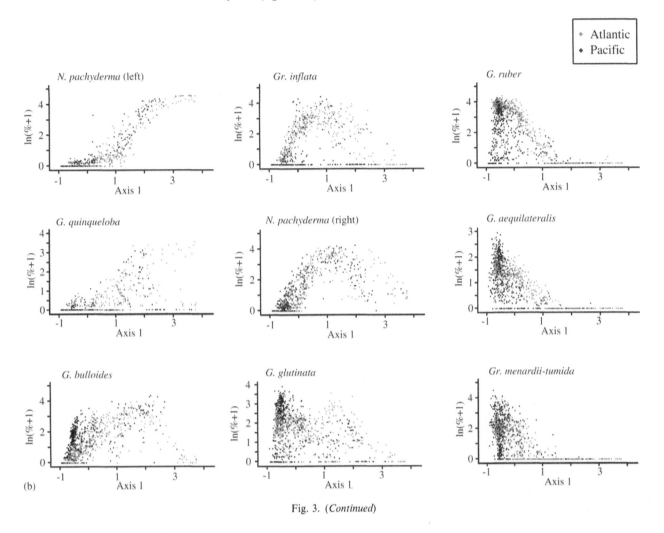

Fig. 3. (*Continued*)

response to a wide range of temperatures and salinities. The salinities tolerated by these species in culture exceeded the range found in the modern ocean, suggesting that real-world variations in salinity are not limiting. The temperatures tolerated by these species in culture, however, compares well with the geographic limits of the species in the real ocean, supporting the influence of temperature on foraminiferal biogeography. The lack of biological evidence supporting a causal relationship between salinity and foraminiferal species leads us to suspect that the weak relationship between salinity and CCA Axis 1 found here is an artifact of the partial correlation between salinity and temperature in the modern ocean.

To gain further insight into the question of salinity influences, we examine the residual variability from the prediction of SST from Axis 1, with respect to salinity. The correlation between salinity and Axis 1 SST residuals (Fig. 6) is essentially zero ($r^2 = 0.08$; slope $= 0.55$) overall, and outside the tropics in the Pacific ($r^2 = 0.04$; slope $= 0.54$) and Atlantic Oceans ($r^2 = 0.002$; slope $= -0.11$). The relationship of SST

residuals and salinity is slightly better, however, within the tropics (between 23 °N and 23 °S) in both the tropical Pacific ($r^2 = 0.37$; slope $= 1.45$) and Atlantic ($r^2 = 0.25$; slope $= 0.81$).

The relationship between SST residuals and salinity within the tropics may be an artifact because many environmental variables are moderately to highly correlated to one another in the tropics (from 23°N to 23°S). For example, variables most highly correlated to tropical sea-surface salinity in our dataset include sea-floor carbonate concentration ($r = 0.67$, possibly an artifact of sample distribution based on large-scale differences between the Atlantic and Pacific Oceans), rate of change in density at the pycnocline ($r = -0.65$), the density contrast between 100 meters and the surface ($r = -0.59$), silica concentration at the pycnocline ($r = -0.57$), the percentage of oxygen saturation at the pycnocline ($r = 0.45$), and the concentrations of nitrate ($r = 0.39$) and phosphate ($r = 0.37$) at the pycnocline. Most of these variables are also moderately correlated to the SST residuals with respect to Axis 1. Given this apparent intercorrelation of variables in the tropics, it may not be possible to accurately identify or verify secondary environmental influences from an analysis of globally distributed core-top samples.

4.1.2. The subtropical convergence zone: an ecotone?

A transition zone that separates tropical/subtropical species (such as *G. ruber* and *G. aequilateralis*) from species that are typically found outside the tropics appears between Axis 1 values of approximately 0 and 1 (Fig. 3b). Most tropical/subtropical species have a wide range in abundances along Axis 1 from approximately −1 to 0 (suggesting variability along CCA Axis 2), but decrease rapidly from 0 to 1 as extra-tropical species increase in abundance. This transition may be an ecotone, a transition zone separating two communities resulting from a physical boundary (Ricklefs, 1990). An example of an ecotone is the boundary between forests and grasslands. Most species comprising each community are unique to that community, but overlap at the transition between them. Ecotones are therefore gen-

erally more diverse than the communities they separate. This may explain why Rutherford et al. (1999) found modern foraminiferal assemblages from mid-latitudes to be more diverse than either low or high latitudes.

The center of the transition zone observed in the faunal data corresponds to sea-surface temperatures of about 18–20 °C, roughly coincident with the subtropical convergence zone (which is commonly identified by temperatures near 18 °C, Worthington, 1959) currently positioned at about 40°N and 40°S. The subtropical convergence zone separates tropical and subtropical waters, where a relatively strong thermocline exists throughout the year, from subpolar waters where winter cooling breaks through the thermocline and causes deep vertical mixing. The physical and chemical characteristics of these two regimes may result in a fundamental difference in foraminiferal communities.

The subtropical convergence zone has been previously identified from zooplankton assemblages. Howard and Prell (1992) found evidence of the subtropical

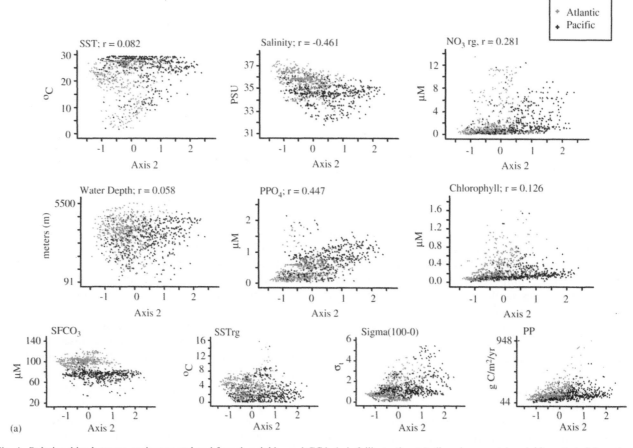

Fig. 4. Relationships between environmental and faunal variables and CCA Axis 2 illustrating (a) all environmental variables selected, in order of decreasing linear correlation to Axis 2 (from left to right, top to bottom), and (b) species with strong relationships to Axis 2 (and highest values of percent fit). Examples of species that correspond to more fertile environments increase where Axis 2 sample scores are high (right), whereas species associated with more oligotrophic environments increase where Axis 2 sample scores are low (left). Gray dots indicate Atlantic samples, and black dots indicate Pacific samples. Note that although SST, surface nitrate concentration range, water depth and chlorophyll have relatively large canonical coefficients, salinity and phosphate concentration at the pycnocline (PPO_4) have the most consistent relationships to Axis 2 sample positions in both oceans.

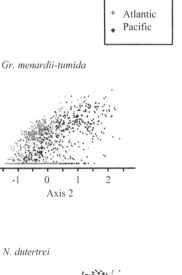

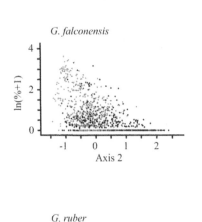

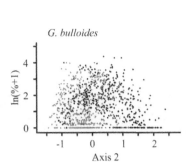

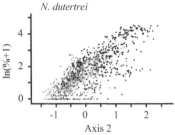

(b)

Fig. 4. (*Continued*)

convergence zone in planktonic foraminiferal assemblages from Indian Ocean sediment samples, and used this information to track the convergence zone through time. More recently, Schiebel et al. (2002a,b) reconstructed the late Quaternary variations in the Azores Front of the North Atlantic Ocean based on distributions of planktonic foraminifers, shelled gastropods and coccolithophorids (Schiebel et al., 2002a,b). The low latitude and high latitude communities separated by this transition zone may reflect a fundamental difference in the physical environmental requirements of the species that exist within them.

The presence of an ecotone in faunal assemblages may be important for paleoestimation studies because statistical methods cannot differentiate between variability due to the transition from one community to another across a physical boundary and continuous variation along an environmental gradient. If one community responds to different environmental properties than the other community, and the variability across the transition is larger than the faunal variability in response to environment within at least one of these communities, the true relationship between species and the environmental properties of interest may be obscured. Reanalyzing the data after careful partitioning with respect to a potential ecotone could validate or reject this influence, and confirm and/or refine the

relationships defined by the analysis of the entire dataset.

4.2. CCA Axis 2: environmental gradient, inter-ocean differences or dissolution?

4.2.1. Environmental contrasts

SST and salinity are most highly correlated to sample scores along both Axis 1 and Axis 2, but the environmental interpretation of Axis 2 is fundamentally different from that of Axis 1. Whereas Axis 1 predominantly reflects the latitudinal variability in species and is associated with the positive correlation between SST and salinity (compare Fig. 5a and Fig. 1), Axis 2 reflects species variability that occurs mostly within the tropics and subtropics (Fig. 5b), and is associated with a negative correlation between SST and salinity. Although SST has the highest canonical coefficient (0.3127), the overall correlation between SST and Axis 2 sample positions is low ($r = 0.082$). Other environmental variables associated with Axis 2 include nitrate, phosphate, and chlorophyll, suggesting that the apparent inverse relationship between SST and salinity in part reflects a fertility gradient between tropical upwelling areas and the oligotrophic subtropics.

Sample positions along Axis 2 primarily reflect a faunal gradient from high abundances of *G. ruber* to

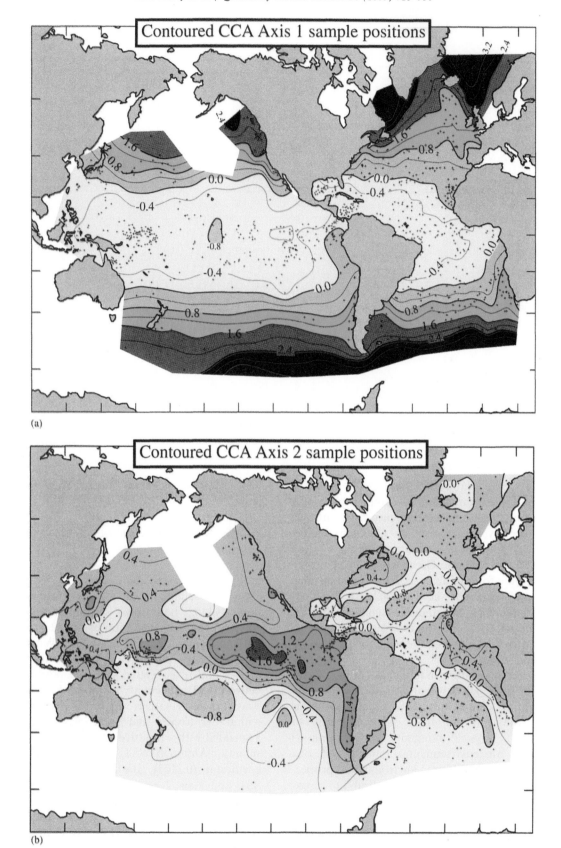

Fig. 5. Contour maps of modern (core-top) sample scores for CCA Axes. (a) CCA Axis 1 sample scores, which are highly correlated to SST, show a strong latitudinal trend as well as equator-ward deflection of latitudinal trends in the eastern boundary currents. (b) CCA Axis 2 sample scores vary primarily within the tropics and subtropics.

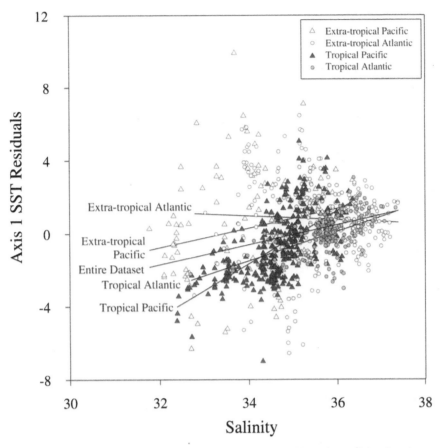

Fig. 6. The relationship of residuals from the regression of SST on Axis 1 sample scores to surface salinity. Samples are coded by ocean and by latitude (tropical or extra-tropical). There is no relationship between salinity and SST residuals outside the tropics, whereas the slopes of the tropical samples are likely different from 0. Because of the moderate to high correlations between salinity and other variables within the tropics, and culture studies that suggest salinity is not limiting to foraminiferal species (Bijma et al., 1990), the apparent link of SST residuals to salinity in the tropics may be spurious.

high abundances of *G. menardii-tumida* and *N. dutertrei*. *G. ruber* is a symbiont-bearing species that can survive in oligotrophic subtropical environments (characterized here by relatively high salinity, and relatively low nutrients and chlorophyll) where food is scarce (Watkins et al., 1996). In contrast, *N. dutertrei* (plus P-D intergrade) thrives in upwelling environments (characterized by relatively low salinity, and high nutrients and chlorophyll) that provide rich food sources (Berger, 1968).

4.2.2. Inter-ocean differences

The Atlantic and Pacific Oceans are different in terms of both fauna and environment, and these differences are reflected in the placement of species optima along Axis 2. Optima of species that are found in greater abundance in the Atlantic (such as *G. hirsuta* and *G. falconensis*) are positioned at the negative extreme of Axis 2, while those that are found in greater abundance in the Pacific (*N. dutertrei*, *Globorotaloides hexagona*, *Globigerina conglomerata*, and *P. obliquiloculata*) are

positioned at the positive extreme of this axis (Fig. 2). Sample scores along Axis 2 suggest these differences are likely a result of environmental properties that extend beyond the range found in the other ocean, as opposed to divergent adaptations of species to the same environmental properties (compare Figs. 4a and 4b). For example, salinity and sea floor $[CO_3^{2-}]$ are typically higher, and upper ocean nutrients are lower in the Atlantic than the Pacific. This finding supports the use of combined Atlantic and Pacific faunal data to develop transfer functions.

Sea-surface salinity and pycnocline phosphate concentration are negatively correlated to one another with respect to Axis 2, but no obvious relationship between salinity and Axis 2 occurs when each ocean is viewed separately (Fig. 7). Only samples from the extra-tropical north Atlantic reflect a correlation between salinity and Axis 2. These samples, however, do not show a clear relationship between *G. ruber* and Axis 2 (Fig. 7). We therefore suggest the relationship of salinity to Axis 2 is an artifact, just as we inferred it was for Axis 1. In

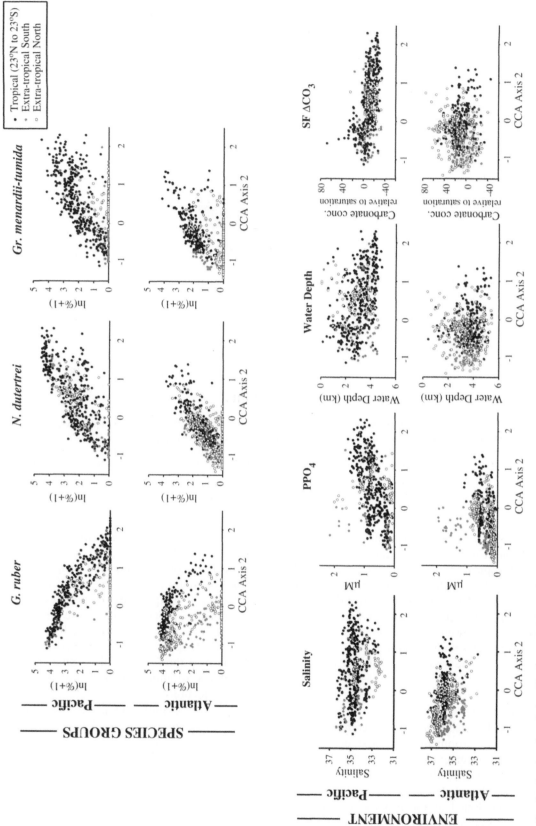

Fig. 7. Graphs showing the relationships between CCA Axis 2 and species (top) and environmental variables (bottom) highlighted by latitude and hemisphere. Black symbols represent tropical samples (23°N to 23°S), gray symbols represent extra-tropical samples from the southern hemisphere, and open circles represent extra-tropical samples from the northern hemisphere.

contrast, the relationship of pycnocline phosphate concentration to Axis 2 is consistent with the variability of species (especially *N. dutertrei*) along this axis in both oceans. The most robust relationship is between faunal variability and pycnocline nutrient concentration, a likely influence on food supply for these heterotrophs (Watkins et al., 1996; Watkins and Mix, 1998).

4.2.3. Selective dissolution

Sea-floor $[CO_3^{2-}]$ and water depth were selected over sea-floor $\Delta[CO_3^{2-}]$ (Table 5). The inclusion of water depth may have accounted for the portion of the faunal variability reflected by CCA Axis 2 that is correlated to the thermodynamic pressure influence on $\Delta[CO_3^{2-}]$. $\Delta[CO_3^{2-}]$ and water depth are correlated to each other ($r = -0.52$), however water depth cannot be interpreted strictly as a dissolution variable. Reanalysis of the data with $\Delta[CO_3^{2-}]$ in place of water depth does not change canonical coefficients for either axis.

Our CCA Axis 2 appears to reflect the same faunal variability that Berger (1968) attributed to dissolution within the low-latitude Pacific. Some species generally thought to be more susceptible to dissolution (*G. ruber*, *G. tenellus* and *G. rubescens*) are most abundant with negative scores along Axis 2, while some species generally thought to be dissolution resistant (*P. obliquiloculata*, *G. menardii-tumida* complex, *Globigerina conglomerata*) are most abundant with positive scores along Axis 2. However, other dissolution resistant species exhibit opposite trends, with negative scores along Axis 2 (*G. hirsuta* and left and right coiling *G. truncatulinoides*). A comparison of *G. menardii-tumida* relative abundance to our preservation variables (water depth and $\Delta[CO_3^{2-}]$) within each ocean separately supports this inference (Fig. 7). We find no evidence for an identifiable, unique response to carbonate under-saturation at the sea floor of the combined Atlantic and Pacific faunal data. This finding conflicts with suggestions by Berger (1968), Parker and Berger (1971), and Coulbourn et al. (1980), that selective dissolution, especially in the Pacific, alters faunal assemblages such that the primary environmental interpretations based on these assemblages are strongly biased.

Selective dissolution may influence assemblages in some instances, but CCA reveals that this effect is small relative to (or cannot be differentiated from) primary environmental influences in the upper ocean for the combined Atlantic and Pacific faunal data. It is therefore unlikely that paleo-estimation methods based on comparison to modern foraminiferal assemblages can be used to reconstruct the extent of dissolution, unless the primary environmental components are removed first. This finding calls into question recently published estimates of $\Delta[CO_3^{2-}]$ at the Last Glacial Maximum based on modern analogs of species assemblages (Anderson and Archer, 2002).

4.3. Paleoceanographic estimation

The strong relationship between sample positions along CCA Axis 1 and SST leads us to explore the use of this relationship to estimate paleotemperature from fossil assemblages. Although CCA has been used in paleolimnological studies to reconstruct lake chemistry through time (e.g. Fritz, 1990; Fritz et al., 1994), it has not been thoroughly explored as a tool to reconstruct oceanographic properties through time. The correlation between modern water column variables does not prevent the inversion of this relationship (see Birks, 1995) because of the overwhelming correlation between SST and CCA Axis 1. If there were an effect from the correlation between SST and salinity, the result would be to slightly underestimate temperature changes.

Positions of fossil samples with respect to CCA Axes can be determined either by including these samples as supplementary (samples are included, but do not influence, the analysis) or they can be calculated from the log-transformed abundances using the following calculations (equation numbers refer to equations in the Appendix):

(1) Use species optima in Table 7 to calculate weighted averages (Eq. (A.1)).
(2) Center and standardize these raw sample scores (Eq. (A.4)) by subtracting the centroid (Eq. (A.2)) and dividing by the square root of the dispersion (Eq. (A.3)).
(3) Multiply the sample scores from the previous step by the scaling factor calculated by Eq. (A.5).

To estimate SST from CCA sample scores one can apply the empirical relationship between SST and sample scores along CCA Axis 1 based on simple linear regression.

$$SST = -6.54(CCA\ Axis\ 1) + 23.3,$$
$$(r^2 = 0.91,\ Std.\ Err. = 2.0\,°C). \tag{1}$$

Alternatively, one can use the empirical relationship between SST and sample scores along both CCA Axes 1 and 2 based on multiple linear regression, however we found that the results of this more complex equation were not significantly different from the simpler equation based on CCA Axis 1, and we therefore restrict the following discussion to the results from the relationship between SST and CCA Axis 1. To test this CCA-transfer function approach we use the LGM dataset of CLIMAP (1981) supplemented with samples from Mix (1989), and the down-core data from the eastern tropical Pacific core, RC13-110 (Feldberg and Mix, 2003).

In general, SST is estimated well in core-top samples (Fig. 8a), although a systematic trend toward negative residuals (i.e., estimates that are cooler than modern

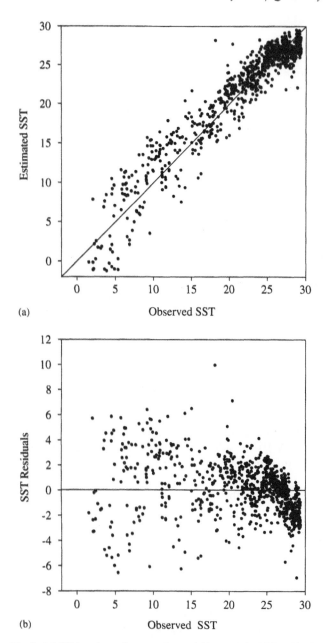

Fig. 8. (a) CCA-estimated versus observed SST, and (b) SST residuals (i.e., estimated minus observed SST) versus observed SST for the core-top calibration data set. The equation reproduces modern SSTs in an RMSE of 2.0 °C ($r^2 = .91$). A systematic trend of residuals occurs at temperatures >26 °C.

limit the ability of CCA methods to resolve the true relationships between faunas and temperatures in the warmest environments. These findings may help to explain why faunal transfer functions and modern analog methods both indicate relatively small temperature changes in the tropical warm pools, in apparent conflict with some geochemical data (Lea et al., 2000) and with evidence for advances of mountain glaciers in the tropics (Hostetler and Clark, 2000).

The range of residuals increases with decreasing temperatures, and is most pronounced for SSTs less than 6 °C. Samples from SSTs less than 6 °C with negative residuals are dominated by *N. pachyderma* (left-coiling) and lesser amounts of *T. quinqueloba*, and are primarily (but not exclusively) from the northern hemisphere. Samples from SSTs less than 6 °C with positive residuals are more diverse and are primarily (but not exclusively) from the southern hemisphere. Although *N. pachyderma* (left-coiling) remains most abundant, other species also occur in these samples including (in order of decreasing average abundance) *G. bulloides, G. inflata, N. pachyderma* (right-coiling), *N. glutinata T. quinqueloba,* and *G. truncatulinoides.* The positions of samples along CCA Axis 1 are heavily influenced by the presence of these species because their optima are positioned lower along this axis. In addition, two of these species (*G. bulloides* and *G. glutinata*) are multimodal on CCA Axis 1 and use of unrealistic optima for these species may bias the results. The cause of the high-latitude faunal differences at high latitudes between the northern and southern hemispheres is not known, although the high southern latitudes are higher in nutrients and are less productive than the high northern latitudes.

A map of estimated modern SST (Fig. 9a) suggests the CCA method and Eq. (1) reconstruct the general patterns of SST variability reasonably well, including the observed cool eastern boundary currents in both oceans, and the presence of a cool tongue in the eastern equatorial Pacific. Regions with positive (i.e., warmer than actual) SST residuals (Fig. 9b) tend to be in the south Pacific and the mid-latitudes of the Atlantic, while regions with negative residuals are located primarily in the warm pools of the western and eastern tropical Pacific and in the Labrador and Norwegian Seas.

Applying the CCA-based equation to LGM samples (Fig. 9c) yields substantial cooling at mid-to-high northern latitudes consistent with many other reconstructions based on foraminiferal transfer functions and modern analog methods (Trend-Staid and Prell, 2002; Pflaumann et al., 2003). The anomaly map of LGM minus modern SST (Fig. 9d) shows the magnitude of this cooling to be at least 6 °C in the north Pacific and >13 °C in the north Atlantic. The inferred magnitude of cooling at high southern latitudes was less than in the northern hemisphere, however this interpretation is based on only 6 Atlantic samples at latitudes greater than 40 °S.

values) occurs at temperatures >26°C (Fig. 8b). A similar effect occurs for warmest temperatures in many transfer functions based on foraminiferal species (Mix et al., 1999; Feldberg and Mix, 2002). This may simply be an effect of noisy data, which typically yields non-zero intercepts in standard regression analysis. Alternatively, foraminiferal assemblages have a lower sensitivity to temperature change than they do at colder temperatures. It may also reflect a possible artifact of WA methods, which assume two-sided Gaussian distributions of species abundances. Finally, the ecotone effect may

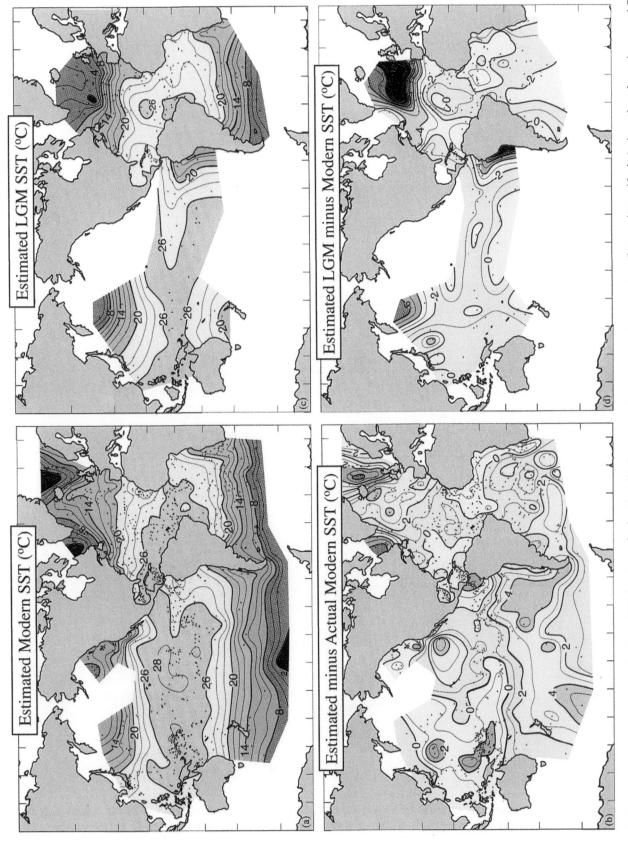

Fig. 9. Estimates of mean annual SST (°C) using a CCA-based transfer function. We have mapped (a) estimates of modern SST, (b) equation residuals (estimated modern minus actual SST), (c) estimated LGM SST, and (d) LGM temperature anomalies (estimated LGM SST minus estimated modern SST). The greatest LGM cooling occurred in the mid-to-high latitudes, especially the north Atlantic, and in the southeastern Pacific at low latitudes.

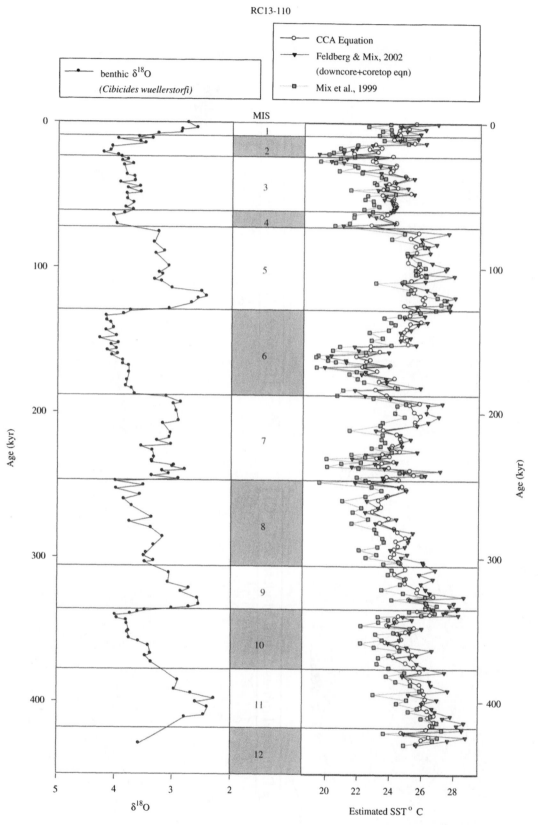

Fig. 10. (a) The oxygen isotope curve in core RC13-110 (0.1 °N, 95.7 °W, 3231 m depth) based on benthic foraminifera (*Cibicides wuellerstorfi*), from Mix (1989). SST estimates based on planktonic foraminiferal species assemblages in the core RC13-110. Gray squares refer to the downcore equation of Mix et al. (1999), inverted gray triangles refer to the downcore plus core-top equation of Feldberg and Mix (2002), and black dots represent SST estimates from this study. Time scales and Marine Isotope Stages are based on SPECMAP (Imbrie et al., 1984).

In the tropics, the CCA-based equation supports recent findings of cooling in the eastern tropical Pacific greater than that of CLIMAP (Mix et al., 1999; Feldberg and Mix, 2002; Trend-Staid and Prell, 2002). The CCA-based equation suggests that the eastern tropical Pacific cooled by 2–6 °C during the LGM and the eastern boundary current (at 20 °S) cooled by 8–9 °C. The tropical Atlantic was cooler by 2–3 °C (although individual points suggest cooling up to 5 °C), while the eastern boundary current cooled by <2 °C. This South Atlantic change is somewhat less than that reconstructed by Niebler et al. (2003), however this difference may in part reflect the location and smaller number of samples in our analysis of the South Atlantic.

The spatial pattern of SST anomalies also suggests cooling in the subtropical gyres ranging from 0–3 °C. This result supports the concept of relatively stable subtropical gyres at the LGM, but without the significant regional warming inferred by CLIMAP (1981).

We also applied Eq. (1) to sample scores calculated from down-core samples of RC13-110 (Fig. 10). The CCA-based temperature estimates vary over a lower total range of temperatures in this core than the tropical transfer function of Mix et al. (1999) or the eastern Pacific downcore plus core-top equation of Feldberg and Mix (2002). These differences may indicate either that the down-core factor analysis method of Mix et al. (1999) yielded a temperature equation too sensitive to faunal change, or that the magnitude of the estimates resulting from the CCA equation are dampened by the inclusion of high latitude samples in the CCA analysis (the ecotone effect mentioned earlier), or that the CCA equation, which considers only variability along Axis 1, is insensitive to faunal variability unique to the tropics (although estimates based on both Axes 1 and 2 are not significantly different from those based on Axis 1 alone).

A similar effect occurs for some other statistical transfer functions that include high-latitude samples, an effect that Mix et al. (1999) attributed to a lack of analogous samples in the modern (calibration) dataset. When applying a Q-mode factor model, such analog problems are diagnosed by low communalities (the sum of squares of factor loadings for each sample). The equivalent measure of fit for CCA results is based on the distance of samples from the centroid, and cannot be used as a diagnostic tool to recognize fossil samples with poor modern analogs. Nevertheless, the range of CCA sample scores for downcore samples in RC13-110 is within the range of sample scores of the modern dataset. Regional application of CCA to tropical and subtropical samples alone could help to discriminate between an ecotone effect and a no analog effect on the magnitude of resulting paleotemperature estimates.

5. Conclusions

The relationship of planktonic foraminiferal faunas to mean annual SST is stronger than to any of the 34 other environmental variables we considered. This finding strongly supports the use of fossil foraminiferal faunas to estimate paleotemperature. Secondary influences may include an ecotone effect, which tends to separate high-latitude faunas (in environments with seasonal erosion of the pycnocline), from low-latitude faunas (in environments with a permanent pycnocline). An apparent slight influence of salinity may be spurious, as an artifact of partial inter-correlation of temperature and salinity in the modern ocean. A second CCA dimension is dominated by faunal variability in the tropics and subtropics, which may represent a nutrient or fertility gradient.

We find no convincing evidence for species variability related to sea-floor carbonate ion concentrations or other proxies for partial dissolution. Similarly, we found no evidence for relationships to seasonal ranges of SST that are independent of mean annual SST. This analysis suggests that planktonic foraminiferal faunas cannot be used reliably to estimate seasonal temperatures or carbonate ion concentrations.

We used the relationship of CCA Axis 1 to mean annual (modern) SST to reconstruct modern, LGM and down-core mean-annual SST, however this method may underestimate temperature changes in some regions because its structure is dominated by large-scale (global and hemispheric) features, and may miss regional or local patterns of variability that do not fit the large-scale correlations among environmental variables that occur in the modern ocean.

Appendix

Mathematical descriptions of (1) the CCA algorithm, (2) centering and standardizing, (3) scaling of ordination scores and (4) orthogonalization procedures. Here 'sample scores' and 'species optima' replace 'site scores' and 'species scores' (the terminology used in the referenced publications) respectively to be consistent with this paper.

A.1. CCA algorithm (ter Braak, 1986; p. 1169)

Notation: The faunal data set (Y; each element represented as y_{ik}) contains i samples ($i = 1$ to n) and k species ($k = 1$ to m). The environmental data set (represented by Z, which includes a column of ones) contains environmental variables for the same i samples. λ is the eigenvalue (see step 7).

(1) Choose arbitrary, but unequal, initial sample scores (x_i).

(2) Calculate species optima (u_k) by weighted averaging of the sample scores:

$$\lambda u_k = \sum_{k=1}^{n} y_{ik} x_i / y_{+k}, \text{ where } y_{+k} = \sum_{k=1}^{n} y_{ik}.$$

(3) Calculate new sample scores by weighted averaging of the species optima (at convergence these are the WA scores):

$$x_i^* = \sum_{k=1}^{m} y_{ik} u_k / y_{i+}, \text{ where } y_{i+} = \sum_{k=1}^{m} y_{ik}. \tag{A.1}$$

(4) Obtain regression coefficients by weighted multiple regression of the sample scores on the environmental variables (**b** is a column vector of canonical coefficients, x* is a column vector of the new sample scores and **R** is a diagonal matrix of sample totals):

$$\boldsymbol{b} = (\boldsymbol{ZRZ})^{-1} \boldsymbol{ZR} x^*.$$

(5) Calculate new sample scores from the regression coefficients (at convergence these are the LC scores):

$$x = \boldsymbol{Zb}.$$

(6) Center and standardize the sample scores (see Section A.2 of this Appendix).

(7) Repeat from step 2, using the sample scores from step 6. Stop when the new sample scores do not change within a prescribed tolerance from the sample scores from the previous iteration (using the strict tolerance level of 10^{-13} suggested by Oksanen and Minchin, 1997). The square root of the dispersion of sample scores at convergence equals the eigenvalue.

(8) Additional axes may be extracted by repeating this procedure with the inclusion of an orthogonalization step (Section A.4 of this Appendix) to make trial sample scores of the current axis uncorrelated to sample scores of previous axes within each iteration after step 5.

A.2. Centering and standardizing procedure in CCA (Jongman et al., 1995, p. 100)

The following procedure is performed within each iteration of CCA after step 5 of the CCA Algorithm (Section 1 of this Appendix).

Step 1: Calculate the centroid, z, of sample scores (x_i)

$$z = \sum_{i=1}^{n} y_{+i} x / y_{++}. \tag{A.2}$$

Step 2: Calculate the dispersion of the sample scores

$$S^2 = \sum_{i=1}^{n} y_{+i} (x_i - z)^2 / y_{++}. \tag{A.3}$$

Step 3: Calculate

$$x_{i,new} = (x_{i,old} - z)/s \tag{A.4}$$

A.3. Scaling of ordination scores in CCA

Three options exist to scale ordination scores when using biplot scaling (the centering and standardizing in Section A.2 of this Appendix). These options are (1) scale ordination scores so that species optima approximate the chi-square distance between species, (2) scale ordination scores so that sample scores approximate the chi-square distance between samples, or as we have chosen, (3) to scale ordination scores such that the positions of species optima are placed so that they approximate the position of the species maximum relative to sample positions along a gradient. The third option is commonly referred to as a compromise between options 1 and 2. The constant used for rescaling using option 3 is

$$\lambda^{\alpha}, \tag{A.5}$$

where λ is the eigenvalue of the axis and $\alpha = 0.5$.

A.4. Orthogonalization procedure in CCA

The following procedure is performed within each CCA iteration after step 6 (Section A.1 of this Appendix). For detrended CCA by second order polynomials, the new sample scores are also made to be orthogonal to the square of the sample scores of previous axes in the same fashion.

Step 1: Denote the sample scores of the previous axis by f_i and the trial scores of the present axis by x_i.

Step 2: Calculate $v = \sum_{i=1}^{n} y_{+i} x_i f_i / y_{++}$, where $y_{+i} = \sum_{k=1}^{m} y_{ik}$ and $y_{++} = \sum_{i=1}^{n} y_{+i}$.

Step 3: Calculate $x_{i,\,new} = x_{i,\,old} - v f_i$. These are the new trial sample scores, which are now orthogonal to those of previous axes.

Step 4: Repeat Steps 1–3 for all previous axes.

References

Anderson, D.M., Archer, D., 2002. Glacial-interglacial stability of ocean pH inferred from foraminifer dissolution rates. Nature 416, 70–73.

Andreasen, D.J., Ravelo, A.C., 1997. Tropical Pacific Ocean thermocline depth reconstructions for the Last Glacial Maximum. Paleoceanography 12, 395–413.

Archer, D., 1996. An atlas of the distribution of calcium carbonate in deep sea sediments. Global Biogeochemical Cycles 10, 159–174.

Bé, A.W.H., 1977. An ecological, zoogeographic and taxonomic review of recent planktonic foraminifera. In: Ramsay, A.T.S. (Ed.), Oceanic Micropalaeontology. Academic Press, London, pp. 1–88.

Bé, A.W.H., Tolderlund, D.S., 1971. Distribution and ecology of living planktonic foraminifera in surface waters of the Atlantic and

Indian Oceans. In: Funnel, B.M., Riedel, W.R. (Eds.), Micropaleontology of the Oceans. Cambridge University Press, London, pp. 105–149.

Behrenfeld, M.J., Randerson, J.T., McClain, C.R., Feldman, G.C., Los, S.O., Tucker, C.J., Falkowski, P.G., Field, C.B., Frouin, R., Esaias, W.E., Kolber, D.D., Pollack, N.H., 2001. Biospheric production during an ENSO transition. Science 291, 2594–2598.

Berger, W.H., 1968. Planktonic foraminifera: Selective solution and paleoclimatic interpretation. Deep-Sea Research 15, 31–43.

Berger, W.H., 1969. Ecologic patterns of living planktonic foraminifera. Deep-Sea Research 16, 1–24.

Berger, W.H., 1970. Planktonic foraminifera: Selective solution and the lysocline. Marine Geology 8, 111–138.

Bijma, J.J., Faber, W.W., Hemleben, C., 1990. Temperature and salinity limits for growth and survival of some planktonic foraminifers in laboratory cultures. Journal of Foraminiferal Research 20, 95–116.

Birks, H.J.B., 1995. Quantitative Paleoenvironmental Reconstruction. In: Maddy, D., Brew, J.S. (Eds.), Statistical Modelling of Quaternary Science Data Technical Guide 5. Quaternary Research Association, Cambridge 271pp.

Bradshaw, J.S., 1959. Ecology of living planktonic foraminifera in the North and Equatorial Pacific Ocean. Cushman Foundation for Foraminiferal Research Contributions 10, 25–64.

Cattell, R.B., Vogelmann, S., 1977. A comprehensive trial of the scree and KG criteria for determining the number of factors. Multivariate Behavioral Research 12, 289–325.

CLIMAP, Climate: Long-Range Investigation, 1981. Mapping and Prediction Project Members. Seasonal reconstructions of the earth's surface at the Last Glacial Maximum. Geological Society of America Map Chart Series, MC- 36, 1–18.

Coulbourn, W.T., Parker, F.L., Berger, W.H., 1980. Faunal and solution patterns of planktonic foraminifera in surface sediments of the North Pacific. Marine Micropaleontology 5, 329–399.

Dittert, N., Baumann, K.-H., Bickert, T., Henrich, R., Huber, R., Kinkel, H., Meggers, H., 1999. Carbonate dissolution in the deep sea: Methods, quantification and paleoceanographic application. In: Fischer, G., Wefer, G. (Eds.), Use of Proxies in Paleoceanography: Examples from the South Atlantic. Springer-Verlag, Berlin, pp. 254–284.

Feldberg, M.J., Mix, A.C., 2002. Sea-surface temperature estimates in the Southeast Pacific based on planktonic foraminiferal species; modern calibration and Last Glacial Maximum. Marine Micropaleontology 44, 1–29.

Feldberg, M.J., Mix, A.C., 2003. Planktonic foraminifera, sea surface temperatures, and mechanisms of oceanic change in the Peru and south equatorial currents, 0-150 ka BP. Paleoceanography 18 (1), 1016 doi:10.1029/2001PA000740.

Fritz, S.C., 1990. Twentieth-century salinity and water-level fluctuations in Devils Lake, North Dakota: Test of a diatom-based transfer function. Limnology and Oceanography 35, 1771–1781.

Fritz, S.C., Engstrom, D.R., Haskell, B.J., 1994. "Little Ice Age" aridity in the North American Great Plains: a high-resolution reconstruction of salinity fluctuations from Devils Lake, North Dakota, USA. The Holocene 4, 69–73.

Frontier, S., 1976. Étude de la decroissance des valeurs propres dans une analyze en composantes principales: comparison avec le modèle de baton brisé. Journal of Experimental Marine Biology and Ecology 25, 67–75.

Gourlay, A.R., Watson, G.A., 1973. Computational methods for matrix eigen problems. Wiley, New York 130pp.

Guilderson, T.P., Fairbanks, R.G., Rubenstone, J.L., 1994. Tropical temperature variations since 20,000 years ago: Modulating interhemispheric climate change. Science 263, 663–665.

Hostetler, S., Clark, P.U., 2000. Tropical Climate at the Last Glacial Maximum Inferred from Glacier Mass-Balance Modeling. Science 290, 1747–1750.

Howard, W.R., Prell, W.L., 1992. Late Quaternary surface circulation of the Southern Indian Ocean and its relationship to orbital variations. Paleoceanography 7, 79–118.

Hutson, W.H., 1977. Transfer functions under no-analog conditions: Experiments with Indian Ocean planktonic foraminifera. Quaternary Research 8, 355–367.

Imbrie, J., Hays, J.D., Martinson, D.G., McIntyre, A., Mix, A.C., Morley, J.J., Pisias, N.G., Prell, W.L., Shackleton, N.J., 1984. The orbital theory of Pleistocene climate: Support from a revised chronology of the marine δ^{18}O record. In: Berger, A., et al. (Eds.), Milankovitch and Climate, vol. 1,. D. Reidel, Hingham Mass, pp. 269–305.

Imbrie, J., Kipp, N.G., 1971. A new micropaleontological method for quantitative paleoclimatology, Application to a late Pleistocene Caribbean core. In: Turekian, K.K. (Ed.), Late Cenozoic Glacial Ages. Yale Univ. Press, New Haven, Conn., pp. 71–182.

Imbrie, J., van Donk, J., Kipp, N.G., 1973. Paleoclimatic Investigation of a late Pleistocene Caribbean Deep-Sea Core: Comparison of Isotopic and Faunal Methods. Quaternary Research 3, 10–38.

Jackson, D.A., 1993. Stopping rules in principal components analysis: A comparison of heuristical and statistical approaches. Ecology 74, 2204–2214.

Jongman, R.H.G., ter Braak, C.J.F., van Tongeren, O.F.R., 1995. Data Analysis in Community and Landscape Ecology. Wageningen: Pudoc. Cambridge University Press, Cambridge.

Kipp, N.G., 1976. New Transfer Function for Estimating Past Sea-Surface Conditions from Sea-Bed Distribution of Planktonic Foraminiferal Assemblages in the North Atlantic. In: Cline, R.M., Hays, J.D. (Ed.), Investigation of Late Quaternary Paleoceanography and Paleoclimatology. Memoirs of the Geological Society of America, 145, pp. 3–41.

Kucera, M., Darling, K.F., 2002. Genetic diversity among modern planktonic foraminifer species: Its effect on paleoceanographic reconstructions. Philosphical Transactions of the Royal Society of London, Series A 360, 695–718.

Le, J., Shackleton, N.J., 1992. Carbonate dissolution fluctuations in the western equatorial Pacific during the late Quaternary. Paleoceanography 7, 21–42.

Le, J., Shackleton, N.J., 1994. Reconstructing paleoenvironment by transfer function: model evaluation by simulated data. Marine Micropaleontology 24, 187–199.

Lea, D.W., Pak, D.K., Spero, H.J., 2000. Climate impact of late Quaternary equatorial Pacific sea surface temperature variations. Science 289, 1719–1724.

MacArthur, R., 1960. On the relative abundance of species. The American Naturalist, XCIV 874, 25–36.

Miao, Q., Thunell, R.C., Anderson, D.M., 1994. Glacial-Holocene carbonate dissolution and sea surface temperatures in the South China and Sulu seas. Paleoceanography 9, 269–290.

Mix, A.C., 1989. Pleistocene paleoproductivity: Evidence from organic carbon and foraminiferal species. In: Berger, W.H., Smetacek, V.S., Wefer, G. (Eds.), Productivity of the Ocean: Present and Past. Wiley, New York, pp. 313–340.

Mix, A.C., Morey, A.E., Pisias, N.G., Hostetler, S.W., 1999. Foraminiferal faunal estimates of paleotemperature: circumventing the no-analog problem yields cool ice-age tropics. Paleoceanography 14, 350–359.

Niebler, H.-S., Arz, H.W., Donner, B., Mulitza, S., Pätzold, J., Wefer, G., 2003. Sea-surface temperatures in the equatorial and south Atlantic Ocean during the Last Glacial Maximum (23–19 ka). Paleoceanography 18 (3), 1069 doi:10.1029/2003PA000902.

Ocean Climate Laboratory, 1999. World Ocean Atlas 1998, Objective Analysis and Statistics (CD ROM data set), National Oceanic and Atmospheric Administration, Silver Spring, MD.

Oksanen, J., Minchin, P.R., 1997. Instability of ordination results under changes in input data order: explanations and remedies. Journal of Vegetation Science 8, 447–454.

Ortiz, J.D., Mix, A.C., Collier, R.W., 1995. Environmental control of living symbiotic and asymbiotic foraminifera of the California Current. Paleoceanography 10, 987–1009.

Parker, F.L., 1960. Living planktonic foraminifera from the equatorial and southeast Pacific. Tohoku University Science Reports Serial 2 (Geology) Special 4, 71–82.

Parker, F.L., 1962. Planktonic foraminiferal species in Pacific sediments. Micropaleontology 8, 219–254.

Parker, F.L., Berger, W.H., 1971. Faunal and solution patterns of planktonic foraminifera in surface sediments of the South Pacific. Deep-Sea Research 18, 73–107.

Pflaumann, U., Duprat, J., Pujol, C., Labracherie, L., 1996. SIMMAX: A modern analog technique to deduce Atlantic sea surface temperatures from planktonic foraminifera in deep-sea sediments. Paleoceanography 11, 15–35.

Pflaumann, U., Sarnthein, M., Chapman, M., D'Abreu, L., Funnell, B., Huels, M., Kiefer, T., Maslin, M., Schulz, H., Swallow, J., Van Kreveld, S., Vautravers, M., Vogelsang, E., Weinelt, M., 2003. The glacial North Atlantic: Sea-surface conditions reconstructed by GLAMAP-2000. Paleoceanography 18 (3), 1065 doi:10.1029/2002PA000774.

Prell, W.L., 1985. The stability of low-latitude sea-surface temperatures: An evaluation of the CLIMAP reconstruction with emphasis on the positive SST anomalies. Technical Report TR0-25, 60 pp., US Dept. of Energy, Washington, D.C.

Prell, W.L., Curry, W.B., 1981. Faunal and isotopic indices of monsoonal upwelling: Western Arabian Sea. Oceanologica Acta 4 (1), 91–98.

Prell, W.L., Martin, A., Cullen, J., Trend, M., 1999. The Brown University Foraminiferal Data Base. IGBP PAGES/World Data Center-A for Paleoclimatology, Data Contribution Series #1999-027, NOAA/NGDC Paleoclimatology Program, Boulder, CO, USA.

Ravelo, A.C., Fairbanks, R.G., Philander, S.G.H., 1990. Reconstructing tropical Atlantic hydrography using planktonic foraminifera and an ocean model. Paleoceanography 5, 409–431.

Ricklefs, R.E., 1990. Ecology, 3rd Edition. W. H. Freeman and Company, New York.

Rind, D., Peteet, D., 1985. Terrestrial conditions at the Last Glacial Maximum and CLIMAP sea-surface temperature estimates: Are they consistent. Quaternary Research 24, 1–22.

Rutherford, S., Hondt, S.D., Prell, W., 1999. Environmental controls on the geographic distribution of zooplankton diversity. Nature 400, 749–753.

Schiebel, R., Schmuker, G., Alves, M., Hemleben, Ch., 2002a. Tracking the Recent and Late Pleistocene Azores Front by the distribution of planktic foraminifers. Journal of Marine Systems 37, 213–227.

Schiebel, R., Waniek, J., Zeltner, A., Alves, M., 2002b. Impact of the Azores Front on the distribution of planktic foraminifers, shelled gastropods, and coccolithophorids, Deep-Sea Research II 49, 4035–4050.

Suess, E., 1980. Particulate organic carbon flux in the oceans – surface productivity and oxygen utilization. Nature 288, 260–263.

Stute, M., Forster, M., Frischkorn, H., Serejo, A., Clark, J.F., Schlosser, P., Broecker, W.S., Bonani, G., 1995. Cooling of tropical Brazil (5°C) during the Last Glacial Maximum. Science 269, 379–383.

ter Braak, C.J.F., 1986. Canonical correspondence analysis: a new eigenvector technique for multivariate direct gradient analysis. Ecology 67, 1167–1179.

ter Braak, C.J.F., Smilauer, P., 1998. CANOCO Reference Manual and User's Guide to Canoco for Windows: Software for Canonical Community Ordination (version 4). Microcomputer Power, Ithaca NY, USA 352 pp.

Thompson, P.R., 1981. Planktonic Foraminifera in the western north Pacific during the past 150,000 years: comparison of modern and fossil assemblages. Palaeogeography, Palaeoclimatology, Palaeoecology 35, 241–279.

Thompson, L.G., Mosley-Thompson, E., Davis, M.E., Lin, P.-N., Henderson, K.A., Cole-Dai, J., Bolzan, J.F., Liu, K.-B., 1995. Late glacial stage and Holocene tropical ice core records from Huascarán. Peru. Science 269, 46–50.

Thunell, R., Anderson, D., Gellar, D., Miao, Q., 1994. Sea surface temperature estimates for the tropical western Pacific during the last glaciation and their implications for the Pacific warm pool. Quaternary Research 41, 255–264.

Trend-Staid, M., Prell, W.L., 2002. Sea surface temperature at the Last Glacial Maximum: A reconstruction using the modern analog technique. Paleoceanography 17, 1–18.

Waelbroeck, C., Labeyrie, L., Duplessy, J.-C., Guiot, J., Labracherie, M., Leclaire, H., Duprat, J., 1998. Improving past sea surface temperature estimates based on planktonic fossil faunas. Paleoceanography 12, 272–283.

Watkins, J.M., Mix, A.C., 1998. Testing the effects of tropical temperature, productivity, and mixed-layer depth on foraminiferal transfer functions. Paleoceanography 13, 96–105.

Watkins, J.M., Mix, A.C., Wilson, J., 1996. Living planktic foraminifera: tracers of circulation and productivity regimes in the central equatorial Pacific. Deep-Sea Research II 43, 1257–1282.

Worthington, L.V., 1959. 18 °C water in the Sargasso Sea. Deep-Sea Research 5, 297–305.

ELSEVIER

Quaternary Science Reviews 24 (2005) 951–998

QSR

Reconstruction of sea-surface temperatures from assemblages of planktonic foraminifera: multi-technique approach based on geographically constrained calibration data sets and its application to glacial Atlantic and Pacific Oceans

Michal Kucera[a,*], Mara Weinelt[b], Thorsten Kiefer[c], Uwe Pflaumann[b], Angela Hayes[a,d], Martin Weinelt[e], Min-Te Chen[f], Alan C. Mix[g], Timothy T. Barrows[h], Elsa Cortijo[i], Josette Duprat[j], Steve Juggins[k], Claire Waelbroeck[i]

[a]*Department of Geology, Royal Holloway University of London, Egham, Surrey TW20 0EX, UK*
[b]*Institute for Geosciences, University of Kiel, Olshausen Strasse 40, D-24098 Kiel, Germany*
[c]*Department of Earth Sciences, University of Cambridge, Downing Street, Cambridge CB2 3EQ, UK*
[d]*Department of Geography, Mary Immaculate College, University of Limerick, Limerick, Ireland*
[e]*kk+w Digital Cartography, Lornsenstrasse 41, D-24105 Kiel, Germany*
[f]*Institute of Applied Geosciences, National Taiwan Ocean University, 2 Pei-Ning Road, Keelung 20224, Taiwan*
[g]*College of Oceanic and Atmospheric Sciences, Oregon State University, Corvallis, OR 97331, USA*
[h]*Department of Nuclear Physics, Australian National University, Canberra, ACT 0200, Australia*
[i]*Laboratoire des Sciences du Climat et de l' Environnement, Laboratoire mixte CNRS-CEA, Gif sur Yvette F-91198, France*
[j]*Dept. Géologie et Océanographie, Université de Bordeaux, CNRS-UMR, Avenue des Facultés, F-33405 Talence-Cedex, France*
[k]*School of Geography, Politics and Sociology, University of Newcastle, Newcastle upon Tyne NE1 7RU, UK*

Received 15 July 2004; accepted 19 July 2004

Abstract

We present a conceptual framework for a new approach to environmental calibration of planktonic foraminifer census counts. This approach is based on simultaneous application of a variety of transfer function techniques, which are trained on geographically constrained calibration data sets. It serves to minimise bias associated with the presence of cryptic species of planktonic foraminifera and provides an objective tool for assessing reliability of environmental estimates in fossil samples, allowing identification of adverse effects of no-analog faunas and technique-specific bias. We have compiled new calibration data sets for the North ($N = 862$) and South ($N = 321$) Atlantic and the Pacific Ocean ($N = 1111$). We show evidence that these data sets offer adequate coverage of the Sea-Surface Temperature (SST) and faunal variation range and that they are not affected by the presence of pre-Holocene samples and/or calcite dissolution. We have applied four transfer function techniques, including Artificial Neural Networks, Revised Analog Method and SIMMAX (with and without distance weighting) on faunal counts in a Last Glacial Maximum (LGM) data set for the Atlantic Ocean (748 samples in 167 cores; based on the GLAMAP-2000 compilation) and a new data set for the Pacific Ocean (265 samples in 82 cores) and show that three of these techniques provide adequate degree of independence for the advantage of a multi-technique approach to be realised. The application of our new approach to the glacial Pacific lends support to the contraction and perhaps even a cooling of the Western Pacific Warm Pool and a substantial ($>3\,°C$) cooling of the eastern equatorial Pacific and the eastern boundary currents. Our results do not provide conclusive evidence for LGM warming anywhere in the Pacific. The Atlantic reconstruction shows a number of robust patterns, including substantial cooling of eastern boundary currents with considerable advection of subpolar waters into the Benguela Current, a cooling of the equatorial Atlantic by ~5 °C, and steep SST gradients in the mid-latitude North Atlantic. The transfer function techniques generally agree that subtropical gyre areas in both hemispheres did not change significantly since the LGM, although the ANN technique produced

---

*Corresponding author. Tel: +44-1784-443-586; fax: +44-1784-471-780.

E-mail address: m.kucera@gl.rhul.ac.uk (M. Kucera).

0277-3791/$ - see front matter © 2004 Elsevier Ltd. All rights reserved.
doi:10.1016/j.quascirev.2004.07.014

glacial SST in the southern gyre 1–2 °C warmer than today. We have revisited the issue of sea-ice occurrence in the Nordic Seas and using the distribution of subpolar species of planktonic foraminifera in glacial samples, we conclude that the Norwegian Sea must have been ice-free during the summer.

1. Introduction

All biological processes, from the growth of individual organisms to the dynamics of large ecological systems, are affected by the physical and chemical state of their environment. If we knew how the environment modifies the basic genetic design of organisms and how it controls their spatial and temporal distribution, we could use the fossil record of such organisms to reconstruct the state and variation of past environments. Whilst most fossils record and preserve environmentally valuable and extractible information, planktonic foraminifera are by far the most important signal carriers in palaeoceanography. The physical and chemical properties of foraminiferal shells provide a multitude of palaeoproxies, including passively recorded chemical and isotopic signals, signatures of environmentally modulated metabolic processes, and ecologically controlled aspects of taxonomic abundances and shell morphology (Henderson, 2002). Physical and chemical properties of foraminiferal shells follow different taphonomic pathways and proxies based on these properties rely on different sets of assumptions. This makes it possible to derive independent estimates of palaeoceanographic parameters from the same fossil assemblage, thus providing a unique opportunity to assess the robustness of such palaeoenvironmental reconstructions.

In an ideal world, one would wish to have a full mechanistic understanding of how a proxy works. Unfortunately, in most cases we lack the detailed knowledge of the chemical, physical and biological processes that operate to record and preserve an environmental signal in a proxy record. In the absence of a direct deterministic relationship we are forced to resort to the process of empirical calibration. Here, the relationship between a relatively easily measured parameter of a fossil and an environmental variable is derived by observing and describing such relationship in the present-day situation. This process is methodologically relatively simple but it involves a number of assumptions that have significant impact on the applicability and reliability of each empirically calibrated proxy (Hutson, 1977; Birks, 1995; ter Braak, 1995).

Due to the nature of the processes involved, chemical proxies are more likely to approach the desired mechanistic level of understanding, whilst proxies based on organism size, shape and abundance remain bound to the realms of empirical calibration. Despite this limitation, the possibility to complement and cross-validate the reliability of chemical proxies, especially those derived from samples with unfavourable taphonomic histories has provided the impetus to improve existing physical palaeoproxies (ter Braak and Juggins, 1993; Pflaumann et al., 1996; Waelbroeck et al., 1998; Malmgren et al., 2001) as well as development of new ones (e.g. Bollmann et al., 2002; Schmidt et al., 2004).

The most commonly used physical palaeoproxy involves mathematical analysis of census counts of microfossil assemblages: the so-called "transfer function" approach. Although this kind of species–environment calibration has a long history in the natural sciences its application to palaeoenvironmental reconstruction was brought to prominence by the CLIMAP group in their pivotal effort to reconstruct the sea-surface temperature field of the last glacial maximum ocean (CLIMAP Project Members, 1976, 1981). The results of the CLIMAP reconstructions have been used extensively for forcing and validating ocean circulation models (e.g. Webb et al., 1997). These are the same models that are being used to evaluate the effects of man-made changes to our planet's surface and atmosphere. Many political decisions that are being taken today are based on such evaluations. It is therefore very important to understand properly how the validation data for these models are produced and, equally important, what is their reliability and precision.

Ever since the pioneering study by CLIMAP Project Members (1976), the Last Glacial Maximum (LGM), with its different, yet well defined, insolation and greenhouse gas forcing, has been the prime target for comparison of model outputs and proxy data. The international EPILOG initiative has been recently launched (Mix et al., 2001) with the view to revise the reconstruction of the glacial Earth surface. The MARGO effort is focused on providing a new LGM Sea-Surface Temperature (SST) reconstruction, using the enormous body of new, high-quality data produced in the last two decades. Transfer functions based on planktonic foraminifer counts have always played a central role in glacial palaeothermometry and the new calibration techniques developed in the last decade have contributed greatly to renewed interest in foraminifer SST reconstructions. In this study, we review the use of planktonic foraminifer counts for palaeotemperature reconstructions and present the rationale for a new approach based on the application of multiple techniques calibrated on geographically constrained calibration data sets. We then use this approach to develop a new LGM SST reconstruction for the Atlantic and

Pacific Oceans and compare our results with previous studies and other palaeothermometers.

2. Principles of ecological calibration

The process of ecological calibration as applied to the technique of transfer functions based on assemblages of planktonic foraminifera consists of three steps: firstly, a calibration data set is assembled consisting of counts of planktonic foraminifer taxa in modern ocean and of instrumentally recorded values of environmental parameters extracted at corresponding locations. Secondly, a mathematical model is developed to characterise or describe the relationship between the environmental variable and the census counts. Finally, this mathematical model is applied to fossil samples to produce an estimate of the desired environmental parameter (Hutson, 1977).

2.1. Designing a calibration data set

A correctly designed and constrained calibration data set is an essential prerequisite for any environmental calibration. Perhaps the most important property of a calibration data set is its coverage; both in terms of geographical area and the range of the environmental variable that is being calibrated. Optimally, a calibration data set should include samples representing the entire range of the environmental variable as observed today and the entire range of ecological and geographical circumstances where the taxonomic units included in the calibration data set occur. Ecological coverage takes precedence over the geographical coverage: it is more important to cover the full range of the environmental variable. If this can be achieved in a limited part of the world, than the calibration data set could be limited to such area and the resulting transfer function applied in an appropriate regional context. The rationale for this approach is that the larger the area covered by the calibration data set, the larger the amount of noise in the data. The noise can arise from several sources, including (i) increased genetic and ecological variation in the counted categories not captured by the taxonomic concept used to define these units, (ii) increase in inconsistency of recognition of categories as different taxonomic practices may be applied in different areas, (iii) an increase in the influence of secondary or "nuisance" environmental gradients such as salinity, nutrients and water column characteristics, and (iv) an increase in the influence of a range of different taphonomic processes. All of these effects can lead to a decrease in the signal to noise ratio in a calibration data set and increase in information content that the transfer function has to distil from the data. A global calibration data set will thus yield a transfer function applicable to all fossil samples, but the likely error of environmental reconstructions based on it will be larger.

The issue of geographical coverage of calibration data sets has gained further relevance with the discovery of cryptic genetic types within morphologically defined species of planktonic foraminifera (Huber et al., 1997). These genetic types have specific ecological preferences and often restricted geographical ranges (de Vargas et al., 1999, 2001; Darling et al., 2000, 2003, 2004), suggesting that they represent distinct biological species. In theory, if these cryptic species have different ecological preferences, their recognition should increase the precision of transfer functions. For example, Kucera and Darling (2002) modelled the effect of splitting *Globigerina bulloides* in the Atlantic Ocean into three genetic types and showed an error decrease by up to 30% in transfer functions incorporating the three genetic types. Whilst it will probably never be possible to avoid the lumping of cryptic species into morphologically defined taxonomic units, limiting the geographical range of the calibration data set will help to alleviate the problem. This is because individual genetic types among modern planktonic foraminifera display a much greater degree of endemicity than morphologically defined species, which often show cosmopolitan distribution (Table 1). This is not to say that geographically constrained calibration data sets will not suffer from the lumping of cryptic species at all. Some cryptic genetic types do show a high degree of geographic separation, such as the different types of *Neogloboquadrina pachyderma* (Darling et al., 2004), whereas the distribution of many other cryptic genetic types shows a large degree of overlap, such as in *Globorotalia truncatulinoides* (de Vargas et al., 2001).

Another important aspect of a calibration data set is the balance in the coverage of the ecological and geographical range by individual samples. If one region or one part of the range of the environmental parameter is over represented, the calibration equation could become less reliable or less accurate in the remaining part of the range. This problem is well-known in weighted averaging type methods (ter Braak and Juggins, 1993) and presumably affects other techniques such as PCA-regression and artificial neural networks that essentially derive a single or "global" species/environment relationship. The issue of insufficient coverage of the range of the reconstructed environmental parameter is especially pertinent of the Modern Analog Technique and its variants (see below).

The number of taxonomic (ecologic, functional) units included in the data set depends on historical praxis and the taxonomic reality of each particular group of fossils in each region of the world. Theoretically, the more taxonomic units (=variables) are included, the better the resulting transfer function will be. On the other hand, there is no added benefit of splitting a taxonomic

Table 1
Endemicity of high-latitude planktonic foraminifer faunas

Morphospecies	Genetic types			Communality		
	North Atlantic (NA)	South Atlantic (SA)	California Margin (CM)	NA/SA	SA/CM	NA/CM
G. bulloides	IIa, IIb	IIa, IIb, IIc	IIa, IId	2/3	1/4	1/3
Turborotalita quinqueloba	IIa, IIb	IIa, IIc	IIc, IId	1/3	1/3	0/3
N. pachyderma (sin)	I	II, III, IV, V	not studied	0/5		
N. pachyderma (dex)	I	I	II	1/1	0/2	0/2
Total				4/12	2/9	1/8
Proportion of shared genetic types				33 %	22 %	13 %

The four morphospecies selected are the main constituents of high-latitude assemblages in all three areas considered. Communality refers to the number of genetic types shared between two areas out of the total number of genetic types recognised in the two areas. Based on data from Kucera and Darling (2002) and Darling et al. (2000, 2003, 2004).

unit unless there are good reasons to believe that the finer units have been and will be identified consistently by most researchers. There is no theoretical reason for excluding rare species, although most of the mathematical techniques will not derive much benefit from this additional information. The main problem with abundances of rare species is the inevitably low signal to noise ratio. Pflaumann et al. (1996) attempted to address this problem in polar assemblages of North Atlantic planktonic foraminifera, where species other than *N. pachyderma* (sin) occur only sporadically.

A fossil assemblage recovered from a sediment core does not represent a single snapshot of time: it is the sum of the export production of fossil material throughout the time contained in the thickness of the sample from which the fossil assemblage has been extracted. In marine sediments, this means decades to thousands of years. It is therefore obvious that spot observations or measurements, such as plankton net counts or measurements of individual specimens kept in laboratory cultures are not appropriate for a calibration that is to be applied to a fossil data set. Data from sediment traps have been used for calibration purposes in severely undersampled regions (Ortiz and Mix, 1997); this is theoretically possible if data are available for at least one entire yearly cycle, but it should be recognised that assemblages recovered from sediment traps have a specific taphonomy that is different from sediment samples. Another issue relating to the amount of time represented in a given sample is the higher probability of the faunal count reflecting a single anomalous event not typical for the average climatic conditions at the site, such as an exceptionally productive bloom, or a short-lived warm or cold spell.

The most appropriate way of obtaining data for empirical calibration is from coretop samples. Such samples should be taken from sediment slices of similar thickness as those in the fossil samples and care must be taken to assure that they represent continuous sedimentation up to the present day, or at least that the surface sediment from which the fossils are extracted is of Holocene age. Detailed age quality assessment criteria form an integral part of the MARGO recommendations (Kucera et al., 2005). Consistent application of these criteria allows a quantitative assessment of the effects of older than Holocene samples being included in the calibration data set.

Instrumental measurements are normally used to assign values of environmental parameters to samples in the calibration data set. Since the calibration data set includes time-averaged surface sediment assemblages, the environmental parameters assigned to each samples must be averaged over the same period as the amount of time contained in the analysed sediment. Such data are rarely available, so objectively analysed data sets of instrumental measurements, such as the World Ocean Atlas (WOA, 1998), provide the best alternative. Although the coverage of certain areas is poor and the duration of measurement series varies considerably, the spatiotemporal averaging used to construct these data sets best approximates the desired time-averaged values. The accelerated effect of global warming of the last decade should be taken into account: earlier versions of the World Ocean Atlas may be better suited to assign values of environmental parameters to samples that have been collected decades ago or samples with moderate sedimentation rates that are largely dominated by foraminifera deposited prior to recent climate extremes.

An issue of particular importance is the exact definition of the temperature that is to be extracted from a fossil assemblage. In theory, a calibration can be based on any SST definition in terms of seasonal and depth-averaging, as long as it can be assumed that the SST defined in this way is an important determinant in the ecological system of the calibration data set.

Separate summer and winter temperature calibrations are normally used for planktonic foraminifer transfer functions. This approach is justified as long as one can assume the same joint distribution between seasonal SST patterns in the past as in the calibration data set. An analysis by Morey et al. (2005) shows that temperature indeed appears the strongest determinant in the ecological system of modern planktonic foraminifera, but that a sufficient residual variance remains, suggesting a higher dimensionality in the system (see also Watkins and Mix, 1998). If the reconstructed seasonal temperatures are used to extract a seasonality signal, one has to realise that the prediction errors on the estimates for each season have to be propagated accordingly. With these reservations in mind, we see no reason at present to abandon the practice of generating seasonally resolved SST reconstructions. Similarly, we adopt the MARGO standard definition of SST as temperature at 10-m level in the ocean (Kucera et al., 2005), accepting that other SST definitions may yield lower prediction errors (Pflaumann et al., 1996). Adopting a precise SST definition in calibration allows meaningful comparison among reconstructions from different proxies, which was a priority of the MARGO approach.

2.2. Predictive regression, generalisation and error rates

Most mathematical techniques derive a calibration equation (or algorithm) through an optimisation process designed to minimise the calibration error—a measure of the difference between predicted value and actual value of the desired environmental parameter(s) in all samples of the calibration data set. In predictive regression, this criterion is supplemented by a second consideration—generalisation. A predictor (in our case a transfer function) that is to be applied outside of the calibration data set, must extract the general relationship between input and output variables, rather than learning every detail of the calibration data set. It is always possible to devise a technique that will reproduce the entire calibration data set, i.e. the difference between actual and predicted values of the environmental parameter(s) in the calibration data set will be zero [or near zero in data sets, where identical configurations of input variables (census counts) correspond to different values of output variable(s) (environmental parameter)]. One must, therefore, avoid the trap of searching for a technique that simply minimises the calibration error and optimise both the calibration error and the prediction error. This procedure is referred to as validation—an attempt to simulate how the transfer function will behave outside of the calibration data set. A validation of any environmental calibration designed for predictive purposes requires a split sampling or cross-validation of the modern database (Birks,

1995). The prediction error is then expressed in terms of the Root Mean Square Error of Prediction (RMSEP)—the square root of the mean of the squared differences between the observed and predicted values for all samples from the test set. It is important to remember that RMSEP is only an estimate of how a given ecological calibration will behave outside of the modern calibration data set. RMSEP values depend on the split-sampling method (ter Braak, 1995) and no clear guidance is available as to which method is more appropriate. In this paper, we use split sampling for validation of artificial neural networks and leaving-one-out for validation of the other techniques.

2.3. Application to fossil samples

There are two issues that need to be considered when applying a transfer function to a fossil sample: how far back in time can a calibration be used and how to recognise and deal with no-analog situations. Both issues stem from the stationarity principle, the basic assumption of all predictive regression, stating that the properties of samples and the relationships among them and the environment are identical throughout the range of the application of the predictor.

The exact range in time of each calibration depends on the rate of evolution in the group of fossils on which the calibration is based. The reliability is highest in samples derived from the same time-frame as the calibration data set and decreases until the time when the signal carrier or its components first evolved. Molecular-clock ages of the most recent divergences among cryptic species in planktonic foraminifera (Darling et al., 2003, 2004; de Vargas et al., 1999, 2001) as well as morphometric observations on changes in their ecological preferences (Schmidt et al., 2003) suggest that an ecological calibration based on modern planktonic foraminifera can be used for at least the last glacial cycle, whilst its application in samples older than 1 Ma would very likely suffer significantly from the breakdown of the stationarity assumption.

The concept of no-analog situations (or conditions) is extensively discussed by Hutson (1977). A no-analog condition occurs when the combination of input and output variables in a fossil sample is not represented in the calibration data set or when the relationship between an environmental parameter and assemblage counts in the fossil sample is not represented in the calibration data set. The latter is very difficult to identify: the same combination of foraminiferal species could correspond to different states of an environmental parameter—if one of these states were not represented in the calibration data set, it will be impossible to detect that a reconstruction is in error. This source of potential error is an inevitability of every predictive regression and the only way to avoid it is by careful selection of the calibration data set.

The situation where an assemblage count in a fossil sample differs from all counts in the calibration data set occurs more frequently. It can be detected in several ways and it does not necessarily mean that the reconstructed environmental parameter in this sample is incorrect. Hutson (1977) divided no-analog conditions into benign and malignant. A benign no-analog condition may involve a taxonomic unit, which is not very important for the calibration equation, or it may be a condition, which is outside of the calibration data set, but is captured correctly by techniques that allow for extrapolation. The easiest way to detect a no-analog sample is when a percentage of one of its components exceeds the range in the calibration data set. Another way to detect a no-analog condition is by examining (dis)similarity measures between a fossil sample and the calibration data set.

A no-analog sample may yield unstable reconstructions of the environmental parameter, i.e. the reconstruction will be sensitive to small changes in the calibration data set. This situation can be easily detected by sub-sampling the calibration data set and producing environmental reconstructions using each of the subsets of the calibration data set (see Malmgren et al., 2001). Finally, a very useful and powerful way of determining malignant no-analog conditions is the multi-technique approach suggested by Hutson (1977). If transfer function techniques based on different approaches agree on the value of the reconstructed parameter, the no-

analog condition may be benign. If the techniques disagree, the condition is malignant.

If it is determined that a fossil sample is affected by a no-analog condition, an examination of the sedimentological and faunal characteristics of the sample is essential to assess the possibility of taxonomic inconsistency, taphonomic bias (carbonate dissolution, winnowing) and reworking. When these possibilities have been discounted, the assemblage in the sample can be considered a genuine no-analog. It is then possible to examine the differences between techniques and internal consistency of the estimates, in search for patterns that will allow us to understand the cause of the problem. No-analog samples can be used in palaeoenvironmental reconstructions, but the accuracy of the reconstructed value is low—the correct value is likely to be included in the range of possible estimates, but there are no means of knowing where.

A conceptual model can be devised to allow classification of no-analog samples and provide a repeatable and quantitative means to assess reliability of transfer function palaeoreconstructions (Fig. 1). A set of critical values can be designated on the basis of the properties of the calibration data set to divide the bivariate space of a measure of how well is a fossil sample represented in the calibration data set (e.g. a dissimilarity coefficient) and a measure of divergence among estimates from different techniques (e.g. the difference between the lowest and highest estimate). The highest reliability will be given to convergent SST estimates in samples that appear well represented in

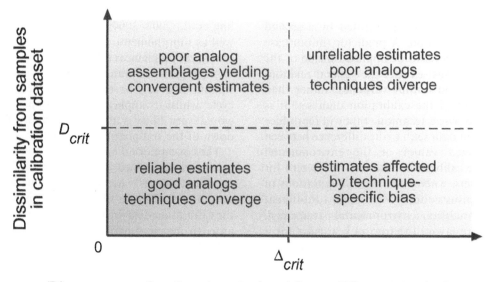

Fig. 1. A conceptual model for identification of no-analog samples and assessment of reliability of paleoreconstructions. The model is based on the division of the bivariate space of a measure of how well is a fossil sample represented in the calibration data set, such as a dissimilarity coefficient, (D) and a measure of divergence among estimates from different techniques (Δ). The division of this bivariate space is based on empirically derived critical values of both measures.

the calibration data set, whereas SST estimates in samples that are not represented in the calibration data set (no-analogs) could be rejected or given the lowest reliability rating. SST estimates in samples that appear well represented in the calibration data set, but where estimates from different techniques diverge will be given intermediate reliability rating, acknowledging the presence of an apparent technique-specific bias. This model can be expanded to include more dimensions; it can be modified to include an arbitrary number of categories and even produce a continuous reliability index.

3. MARGO framework for calibration of planktonic foraminiferal transfer functions

3.1. Geographically constrained calibration data sets

The classical CLIMAP (1976) study employed six regionally calibrated foraminifer transfer functions. This practice has since been followed in regional studies (Niebler and Gersonde, 1998; Pflaumann and Jian, 1999; Barrows et al., 2000), and to some extent in more global calibration exercises (Ortiz and Mix, 1997; Mix et al., 1999; Trend-Staid and Prell, 2002). Our study follows the CLIMAP approach in the aspect of regionally constrained calibrations, but unlike the earlier studies, we provide a conceptual basis for the regionalisation, derived from the knowledge of distribution of cryptic

genetic types among modern planktonic foraminifer species.

Since all studies of distribution of genetic types within species of modern planktonic foraminifera demonstrate a degree of ecological separation among the genetic types (de Vargas et al., 1999, 2001; Darling et al., 2000, 2004; Kucera and Darling, 2002), we may assume that a significant portion of the "noise" characteristic for SST reconstructions based on planktonic foraminifera assemblages is introduced by individual genetic types having acquired differential ecological preferences in different oceans or water masses over time. In order to minimise the noise introduced by such cryptic species we aim to separate different populations by regionalisation of transfer function calibration data sets. This approach also reduces a potential bias in accuracy estimates introduced by differential regional sample density, where less well-sampled regions would compare badly with densely sampled ones when calibrated within a common data set.

Thus, we have subdivided the world ocean into five domains as outlined in Fig. 2. For each domain an individual calibration data set is employed. Sample numbers and prediction errors for each domain are listed in Table 2. In the Atlantic with its wide latitudinal extent we use individual calibration data sets for North Atlantic and South-Atlantic, respectively, with a narrow overlap in the tropical region (Fig. 2), which represents a single faunal province.

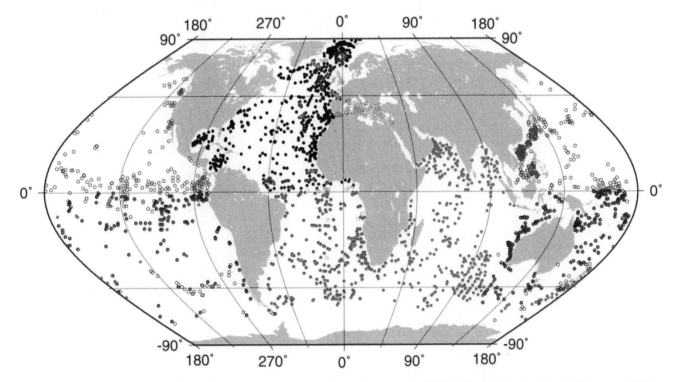

Fig. 2. Location of coretop samples with planktonic foraminifer counts used for each regional SST calibration: North Atlantic (this paper; solid blue circles), South Atlantic (this paper; open orange circles), Mediterranean (Hayes et al., 2005; open light blue circles), Indian Ocean and Australia (Barrows and Juggins, 2005; solid green circles), Pacific (this paper; open magenta circles).

Table 2
Overview of the MARGO planktonic foraminifer data sets with prediction errors for individual regions

Calibration data set				Root Mean Square Error of Prediction (°C)					LGM data set	
Region	Source	Samples	Remarks	SST	ANN	RAM	SIMMAX	MAT	Cores	Samples
North Atlantic	1	862		Summer	1.14	1.09	0.91	1.42	102	550
				Winter	0.96	0.79	0.86	1.32		
				Annual	0.96	0.85	0.81	1.26		
South Atlantic	1	321	83 tropical samples in common with North Atlantic	Summer	0.81	0.76	0.98	1.36	69	198
				Winter	0.95	0.96	0.98	1.47		
				Annual	0.83	0.62	0.93	1.36		
Pacific	1	1111	Some core tops common to IndoPacific data set 13 LGM cores common to IndoPacific data set	Summer	1.19	0.78	1.00	1.63	82	265
				Winter	1.64	1.17	1.27	1.88		
				Annual	1.35	0.93	1.10	1.67		
Mediterranean	2	274	Includes 129 samples from North Atlantic	Summer	1.14	0.74			37	273
				Winter	0.79	0.47				
				Annual	0.91	0.54				
Indian Ocean and Australia (IndoPacific)	3	1344	Some core tops common to Pacific data set 13 LGM cores common to Pacific data set	Summer	0.93	0.91		0.89	170	301
				Winter	1.01	0.98		0.98		
				Annual	0.86	0.86		0.84		
Total									460	1574

(1) This paper; (2) Hayes et al. (2004); (3) Barrows and Juggins (2004). The error rates have been determined by cross validation of the calibration data sets using ten independent partitions (ANN and MAT and RAM for Indian Ocean and Australia) and the leaving-one-out method (SIMMAX, RAM, MAT). MAT values for North and South Atlantic and Pacific data sets refer to error rates for SIMMAX without distance weighting. Note that the leaving-one-out procedure in RAM02 software, used for all regions except the Indian Ocean and Australia, underestimates the error rate as the samples are "left out" after the two-dimensional interpolation procedure.

The Mediterrananean with its narrow zonal extent between 30° and 45°N covers a limited temperature range between 18–27 °C for summer and 11–17 °C for winter. To derive SST from glacial Mediterranean faunas, Hayes et al. (2005) employ a calibration data set which extends into the eastern North Atlantic up to 70°N and 30°W marked by North Atlantic Drift and Norwegian Current water, presuming that this was a likely source area of the subpolar faunal components encountered in the glacial Mediterranean.

For the Indian Ocean and Australian domain, Barrows and Juggins (2005) employ a calibration data set covering the Indian Ocean and southern Pacific and Atlantic (Fig. 2). In this way, the calibration data set avoids all northern high-latitude faunas, whilst providing a broad range of analog samples for SST reconstructions in the large and climatically disparate Indo-Australian region. The Pacific calibration data set covers the western and eastern margins and the equatorial region; most of the samples are concentrated in the tropical and subtropical belts (Fig. 2). The subpolar Southern Ocean and, in particular, the northern high latitudes are still only poorly represented in the calibration data set. Therefore, no further subdivision by hemispheres was made for the Pacific. However, it must be noted that the accuracy of SST reconstructions in these areas are likely to suffer from the low density of calibration samples.

A different subset of taxonomic categories has been used for each domain, following historical practice in species identification and the occurrence of known endemic taxa, such as *Globigerinoides ruber* (pink). In all calibration data sets, the highest possible number of taxonomic units was used, ranging between 23 and 42 (Table 3). Modern SST values were assigned to all calibration samples using the data in the World Ocean Atlas version 2 (WOA, 1998) (1° grid version; summer = JAS, winter = JFM, annual average = 12-month average) at 10 m water depth. This is according to MARGO recommendations (Kucera et al., 2005) to ensure a common reference base for all SST proxies used by the MARGO project. The WOA98 data set has benefited from error reduction as compared to previous versions while no significant new data were included. Temperatures at the sample locations were computed as the area-weighted average of the four WOA temperature points surrounding the sample location using the WOA 98 Sample software (http://www.palmod.uni-bremen.de/~csn/woasample.html).

3.1.1. MARGO calibration data set for North and South Atlantic

The Atlantic calibration data set used in this study is largely based on the GLAMAP-2000 compilation containing faunal counts from 947 coretops from the entire Atlantic (Pflaumann et al., 2003). In addition, we have also used 129 samples from the Brown University data set (Trend-Staid and Prell, 2002) that were not included in the GLAMAP-2000 data set. The additional samples from the Brown University data set have been selected so as to exclude: (1) samples from deeper parts of the South Atlantic, which may be strongly affected by dissolution resulting in a disproportionate enrichment of dissolution resistant species like *Pulleniatina obliquiloculata*; (2) faunas with occurrence of *Globorotaloides hexagonus*, which were considered substantially older than Holocene, because this species became extinct in the Atlantic at approximately 60,000 years B.P. (Pflaumann, 1986); and (3) samples with census smaller than 300 specimens. The MARGO calibration data set includes 26 taxonomic categories counted in the > 150 μm size fraction (Table 3), following the taxonomic concepts of Pflaumann et al. (1996, 2003). Rare species were considered and counted in most samples, but due to inconsistent use and very low abundances these were not used for SST estimates. All calibration and LGM data used in this study are available on the PANGEA portal (http://www.pangea.de/Projects/MARGO/).

Different from previous studies we used separate calibration data sets for North and South Atlantic (Fig. 2). This is to avoid mixing of genetically differentiated populations of planktonic foraminifera in the two hemispheres (Darling et al., 2004) and to reduce the noise potentially introduced into SST estimates by their differential temperature adaptation. As a boundary between North and South Atlantic we chose the zone of minimum seasonality, which is located slightly north of the geographical equator (approximately following a line from 3°N to 8°N) where it separates caloric hemispheres (Fig. 2). To avoid a sharp break in the tropical faunas we allowed for a 5° latitudinal overlap between the data sets with 83 samples in the overlapping zone used in both data sets (Fig. 2).

The North Atlantic is one of the most densely sampled regions of the World Ocean. 862 core top samples in our data set cover an area ranging from the equator to the very high latitudes at 86°N into the Arctic Ocean, and covering a temperature range from 29 °C in the Caribbean to −1.8 °C in the Arctic Ocean. At the "cold end" 17 samples from the high Arctic represent conditions with perennial sea-ice cover. In contrast, the South Atlantic is more poorly and unevenly covered with 322 samples, extending down to 56°S (approximately the limit of carbonate preservation) and covering a temperature range of 3–29 °C. The majority of samples are concentrated in the tropical and subtropical South Atlantic, while only few samples from polar water-masses are available, because of carbonate dissolution. Only 10 samples in the sub-Antarctic Atlantic sector of the Southern Ocean represent annual average temperatures below 10 °C. The separate calibration in North and

Table 3
Planktonic foraminifer taxonomic categories used in each of the regional calibration databases

Species or taxonomic unit Source	North Atlantic 1	South Atlantic 1	Mediterranean 2	Pacific 1,3	IndoPacific 4
Orbulina universa	X	X	X	X	X
Globigerinoides conglobatus	X	X	X	X	X
Globigerinoides ruber (pink)	X	X			X
Globigerinoides ruber (white)	X	X			X
Globigerinoides ruber total			X	X	
Globoturborotalita tenella	X	X	X	X	X
Globigerinoides sacculifer without sac (= *trilobus*)	X	X			X
Globigerinoides sacculifer with sac	X	X			X
Globigerinoides sacculifer total			X	X	
Sphaeroidinella dehiscens	X	X	X	X	X
Globigerinella adamsi					X
Globigerinella siphonifera (= *aequilateralis*)	X	X		X	X
Globigerinella calida	X	X		X	X
Globigerinella siphonifera + *calida*			X		
Globigerina bulloides	X	X	X	X	X
Globigerina falconensis	X	X	X	X	X
Beela digitata	X	X	X	X	X
Hastigerina pelagica					X
Globoturborotalita rubescens	X	X	X	X	X
Turborotalita humilis					X
Turborotalita quinqueloba	X	X	X	X	X
Neogloboquadrina pachyderma (sin)	X	X	X	X	X
Neogloboquadrina pachyderma (dex)				X	X
P/D intergrade + *N. pachyderma* (dex)	X	X	X		
Neogloboquadrina dutertrei	X	X	X	X	X
Globoquadrina conglomerata				X	X
Globorotaloides hexagonus				X	X
Pulleniatina obliquiloculata	X	X	X	X	X
Globorotalia inflata	X	X	X	X	X
Globorotalia truncatulinoides (sin)				X	X
Globorotalia truncatulinoides (dex)				X	X
Globorotalia truncatulinoides total	X	X	X		
Globorotalia crassaformis	X	X	X	X	X
Globorotalia crassula					X
Globorotalia hirsuta	X	X	X	X	X
Globorotalia scitula	X	X	X	X	X
Globorotalia theyeri					X
Dentogloborotalia anfracta					X
Tenuitella iota					X
Berggrenia pumilio					X
Globorotalia menardii					X
Globorotalia tumida					X
Globorotalia menardii + *tumida*	X	X	X	X	
Globorotalia menardii flexuosa					X
Globorotalia ungulata					X
Candeina nitida				X	X
Globigerinita glutinata	X	X	X	X	X
Globigerinita uvula					X
Number of taxonomic categories	26	26	23	28	42

(1) This paper, (2) Hayes et al. (2004), (3) Chen et al. (2004), (4) Barrows and Juggins (2004).
P/D intergrades encompass specimens that do not satisfy the sensu stricto definitions of neither *N. pachyderma* (dex) nor *N. dutertrei*.

South Atlantic domains will also serve to reduce the bias introduced into South Atlantic estimates by the asymmetrical sample density between the high latitudes of the two regions.

Following the MARGO recommendations, all core-top samples underwent a quality ranking into levels 1–4 (Kucera et al., 2005), samples lacking any information on age control (as is the case for many CLIMAP samples) were classified into category 5. Out of 862 samples in the north Atlantic data base 116 fall into category 1 or 2 with well constrained ^{14}C dating or other age control (i.e. coiling direction of *G. hirsuta*, undisturbed sediment surfaces of box and multicorer samples) suggesting ages younger than 2000 and 4000 years, respectively. 525 samples fall into category 5. In the south Atlantic data set only 9 out of 322 data fall into category 1–2, and 175 samples were classified into category 5. Initially, all samples, including those of category 5, were included in the calibration, in order to facilitate an assessment of the potential age control bias on the transfer functions.

3.1.2. MARGO calibration data set for the Pacific Ocean

A coretop data set for the entire Pacific was compiled complementary to the MARGO data set of Barrows and Juggins (2005) for the southwestern Pacific and Indian Ocean and Chen et al. (2005) for the western Pacific. As described by Chen et al. (2005), the data were taken from two global data sets (Ortiz and Mix, 1997; Prell et al., 1999; supplemented by some unpublished data) and several regional studies (Coulbourn et al., 1980; Miao et al., 1994; Thiede et al., 1997; Chen et al., 1998; Pflaumann and Jian, 1999; Ujiié and Ujiié, 2000; Kiefer et al., 2001; H. Schulz, unpublished data).

The MARGO Pacific data set includes coretop samples from the easternmost Indian Ocean, specifically east of 115°E (Fig. 2; Ortiz and Mix, 1997; Prell et al., 1999). The region around Australia corresponds oceanographically with the Pacific via the Indonesian throughflow, the Leeuwin Current, and the Antarctic Circumpolar Current. Although the Indonesian throughflow was reduced during the LGM sea level lowstand, the exchange of surface water and hence planktonic biota persisted throughout the last glacial cycle (Müller and Opdyke, 2000). Therefore, the coretop faunal data from the eastern Indian Ocean may complement those from the western Pacific.

Samples containing the pink variety of *G. ruber* have been excluded, as this species became extinct in the Indo-Pacific during Oxygen Isotope Stage 5 (Thompson et al., 1979). Some samples with obviously erroneous faunal or geographical data were also excluded, including samples where the total count of species abundances deviated significantly from 100%. If 35 species are consistently counted and the relative abundance of each of them is reported in rounded numbers with 0.1% precision, then the maximum error due to rounding is about 2%; every sample where the total deviated from 100% by more than this number has been excluded. When duplicate faunal counts have been encountered, the counts in the Brown University data set (Prell et al., 1999) were given preference, because of their accurate data documentation (raw counts) and generally better quality of species subdivision and sample sizes.

Neither samples affected by carbonate dissolution nor samples with poor age control were categorically excluded. In fact, samples from below the lysocline were retained in the calibration database to compensate for fossil samples, which are also affected by carbonate dissolution, hence improving the applicability of the calibration data set to reconstructing SST in the Pacific. The age control quality is indicated for each sample in the database following the MARGO criteria (Kucera et al., 2005). Seventy-two per cent of the coretop samples had no (or was not known to have any) stratigraphic control (level 5). Since their exclusion from the database would significantly reduce coverage, we opted for a database with a maximum number of samples thereby accepting a reduction in average data quality.

The coverage by the 1111 samples of the final database is high in the low-latitude (35°N–35°S) western Pacific and the tropical central and eastern Pacific, adequate for the eastern boundary currents, but rather poor in the subpolar regions, particularly in the subarctic northwestern Pacific (Fig. 2). Taxonomic categories used to calibrate the transfer functions are listed in Table 3. To conform to the practice encountered in most Pacific LGM counts, the *G. sacculifer* morphotypes with and without the sack-like final chamber were grouped, as were the *G. menardii*-group species *G. menardii* and *G. tumida*. Intergrade forms between right-coiling *N. pachyderma* and *N. dutertrei* were added to the category of *N. pachyderma* (dextral), except for samples in the low-latitude western Pacific. Here, we adopted the approach of Chen et al. (2005) of adding these intergrades to *N. dutertrei* assuming that they most likely represent juvenile forms of *N. dutertrei*.

3.2. Calibration techniques

There is extensive literature dedicated to the theory and practice of ecological calibration (see Birks, 1995). This literature describes a wide range of numerical techniques, yet, of these, only a limited number have been applied to planktonic foraminifer data. Interestingly, there is little or no theoretical basis to suggest that the chosen techniques are superior or better suited to the problem—it is more for reasons of historical practice and personal preference that palaeoceanographers tend to resort to different tools than palaeolimnologists or palynologists. What is clear from the literature is that there is no single mathematical technique that is to be recommended above all other. Each technique has its strong points and its weaknesses. It is the understanding of these strengths and weaknesses and the insights into the stability of fossil estimates based on simultaneous application of different calibrations that provide most information on the reliability and accuracy of palaeoenvironmental reconstructions.

In terms of statistical theory, calibration methods can be subdivided on the basis of the following dichotomies (ter Braak 1995):

- Classical vs. inverse
- Linear vs. non-linear
- Parametric or non-parametric
- With or without an element of dimension reduction

Classical approaches involve the construction of a model where the values of variables (census counts) are expressed as a function of environmental parameters. Of the commonly used methods in palaeoceanography, only RAM (see below) falls into this category. While classical approach treats environmental parameters jointly, the inverse approach treats them separately (ter Braak, 1995). This means that techniques based on the inverse approach can calibrate a single environmental parameter independently.

Since organisms tend to respond to environmental change in a highly non-linear manner, it is desirable to use non-linear calibration techniques. Most of the modern techniques used in palaeoceanography are inherently non-linear: only the original Imbrie–Kipp method (Imbrie and Kipp, 1971) potentially suffers from the curse of linearity, although even this method employs a novel data pre-treatment that renders it more able to dealing with unimodal relationships. None of the commonly used techniques is strictly parametric.

Dimension reduction is inherent to the Imbrie–Kipp method and to some techniques commonly used in palaeolimnology, such as Weighted Averaging and its variants. The main purpose of dimension reduction in the calibration data set is to separate noise from signal: the original species variables are replaced by a small number of components that contain the environmentally relevant portion of the information, i.e. the signal. The difficulty in this approach lies in defining an objective method for extracting a sufficient number of components to maximise the signal to noise content.

Given the obvious difficulty in deciding superiority of techniques, in combination with the clear benefits of multiple estimates for each fossil sample, the MARGO approach is based on the consistent application of as many techniques as is possible or practical. We have decided not to use the Imbrie–Kipp technique, as all comparisons have consistently demonstrated that it yields substantially higher prediction errors (Pflaumann et al., 1996; Waelbroeck et al., 1998; Malmgren et al., 2001). The selection of techniques reflects the current state of the subject—it should not be interpreted as an attempt to codify the use of these techniques. On the contrary, we hope that future research will stimulate the development of new methods, which may further enhance the main aspect of the MARGO approach:

consistent application of multiple transfer function techniques as a tool to assess the reliability of palaeotemperature estimates.

3.2.1. Modern analog technique—MAT

Among statisticians, this technique is known as k-nearest neighbour. It has been introduced to palaeoceanography by Hutson (1980). MAT does not generate a single unique or "global" calibration formula between census count data and physical properties. Instead, this method searches the calibration data set for samples with assemblages that most resemble the fossil assemblage. The environment representing the fossil sample is then reconstructed from the physical properties recorded in the best modern analog samples. To identify the best analogs in the calibration data set, the square chord distance measure has been shown to be most effective (Prell, 1985). A subset of least dissimilar samples from the training set is then used to estimate the environmental parameter(s) of each sample from the test set as an average of the observed values associated with the 10 best analogs weighed by a similarity coefficient.

The number of best analogs used to calculate the estimate is usually set to a given number, reflecting the size of the calibration data set. The larger and better balanced the calibration data set, the more of the best analogs can be used. In practice, 5–12 analogs have been used and there have been attempts to use a variable number of best analogs by applying a threshold to the dissimilarity coefficient (Prell, 1985; Barrows et al., 2000; Trend-Staid and Prell, 2002) or by optimising the number of retained analogs for a given calibration data set (Barrows and Juggins, 2005). The latter approach has been further developed in the RAM technique discussed below.

Like the Imbrie and Kipp method, MAT and all its modifications allow a good understanding of the calibration process. In addition, MAT can cope with non-linear relationships between species and environment and it is easy to add new samples into the calibration data set. The weak points are in the necessity to decide arbitrarily on the exact form of the similarity and dissimilarity measures and, especially, on the number of best analogs, although cross-validation procedures can be used to guide the latter (Barrows and Juggins, 2005). By definition, MAT is a strictly interpolative technique (it cannot extrapolate) and, like all inverse methods, it will tend to perform poorly at the extremes of the range of the estimated environmental parameter(s). The technique is completely dependent on the size and coverage of the calibration data set; it is not always able to benefit from the full information in the calibration data set (Kucera and Darling, 2002) and its power to generalise may be limited.

3.2.2. SIMMAX and SIMMAXndw

Developed by Pflaumann et al. (1996), this technique follows the general strategy of MAT, but it differs from MAT in the way best analogs are defined and treated. In this technique, the scalar product of the normalised assemblage vectors is used as a measure of faunal similarity and the estimated SST is calculated as an average of the observed SSTs of the best analogs weighted by the similarity coefficient and the inverse of the geographical distance between each best analog and the unknown sample. This technique suffers from the same problems and offers the same advantages as MAT. However, it performs better in areas of low variability in the calibration data set, where the lack of changes in faunal assemblages is supplemented by the information on the geographical position of the sample. It may also perform better under some no-analog situations as it can handle geographical inconsistency among the best analogs. It is important to realise that the introduction of geographical information in the technique violates the stationarity principle: we know that the relationship between geography and environmental parameters has not been constant through time (otherwise we would not need to reconstruct it with such an effort). This violation is not large, however, as the best analogs are selected purely on the basis of the similarity coefficient. SIMMAX yields consistently very low prediction errors, but these errors are almost certainly underestimated because it is difficult to properly account for the geographical distance weighting in the cross-validation. We should therefore not be lulled into believing that it is superior to other techniques.

Non-distance-weighted SST estimates ($SIMMAX_{ndw}$) are often calculated and presented alongside the distance-weighted SIMMAX estimates (e.g. Pflaumann et al., 1996, 2003). This practice is followed in the applications presented in this paper. Although a different similarity coefficient is used to select the best analog samples, $SIMMAX_{ndw}$ is in effect a direct equivalent of the classical MAT approach. All SIMMAX and $SIMMAX_{ndw}$ reconstructions presented in this paper were derived using the same procedures as in Pflaumann et al. (1996), including the consistent use of 10 best analogs.

3.2.3. Revised analog method—RAM

The revised analog method (RAM) (Waelbroeck et al., 1998) combines the "response surface" approach successfully applied in palynology with a modified MAT procedure. It is the only classical regression technique commonly used in palaeoceanography. As a first step, RAM expands the calibration data set by addition of "virtual samples" obtained by remapping the calibration data set onto a regular grid of environmental variables with a mesh size γ. For each node on the grid, an artificial assemblage is constructed by interpolating assemblages from all calibration samples located within a given radius R from the node weighted by their Euclidean distance from the node. The parameters γ and R have to be manually optimised on the basis of the properties of the calibration data set. R is usually taken slightly larger than γ, e.g. $R = \gamma + 0.1\,°C$. These two parameters determine the number of virtual samples. It is recommended to keep this number lower than the number of samples in the original calibration data set (Waelbroeck et al., 1998).

As a second step, the expanded data set is subjected to a k-nearest neighbour search for best analogs, using the same similarity and dissimilarity coefficients as in Prell (1985). The number of most similar samples (analogs) that are to be used to reconstruct the given environmental parameter is determined using an objective criterion for the selection of the most relevant best analogs. The optimised selection relies on the assumption that a large change in dissimilarity indicates a shift from one type of fauna to another. Therefore, instead of retaining a fixed number of best analogs, as in MAT or SIMMAX, RAM keeps all best analogs encountered before the first jump in dissimilarity. A jump is defined by an increase in the dissimilarity coefficient between two of the ranked samples higher than a certain fraction α of the dissimilarity of the last retained best analog.

A modification has been recently developed and is embedded in the RAM02 software that automatically determines the value of α. In the modified algorithm, the rate of increase in dissimilarity is compared to a threshold value, set to 10% as a starting value for each new sample. If the rate of increase in dissimilarity is higher than the threshold, and if at least three best analog samples have already been selected, the average SST is based on these samples. If less than three best analogs have been selected, the value of the threshold is increased until more than three modern analogs are selected. A maximum number of β modern analogs are kept. The parameter β is usually set to 10. As shown in Waelbroeck et al. (1998), SST estimates are not very sensitive to changes in β. A cut-off value of 0.6 of the dissimilarity coefficient is used to recognise no-analog samples (see discussion below). Samples with dissimilarity values higher than 0.6 are excluded. In this paper, we use the latest RAM02 software with the following parameters for all regional calibrations: initial $\alpha = 0.1$, $\beta = 10$, $\gamma = 0.25\,°C$, $R = 0.3\,°C$.

The addition of artificial samples with assemblages interpolated in the field of two (or theoretically more) environmental parameters means that RAM does not enjoy the benefit of the inverse regression approach where individual environmental parameters can be reconstructed independently: the estimates will change if a different combination of environmental parameters is used to derive the virtual coretops. As a trade-off,

RAM is better able to generalise than MAT and is less dependent on the size and even coverage of the calibration data set.

3.2.4. Artificial neural networks—ANN

The general principles and architecture of a back propagation (BP) neural network are described by Malmgren and Nordlund (1997) and their use in palaeoceanographic reconstruction demonstrated by Malmgren et al. (2001). An artificial neural network consists of a set of interconnected processing units (neurons), which map a set of input variables (census counts) onto one or more output variables (environmental parameter(s)). Learning proceeds by adjusting the way each neuron modifies its input and is based on a procedure that minimises the prediction error for each sample in the calibration data set. A sufficiently complex network will be able to learn all details in the calibration data set. Therefore a subset of samples is removed from the calibration data set and used to calculate "prediction error" after each learning epoch. The learning of the network is stopped when no improvement in prediction error occurs (i.e. when the network begins to overfit the data and to learn details in the calibration data set rather than the general relationship between census counts and environmental parameters).

The number and types of neurons included in the network has to be determined separately for each data set. Most commercial software packages determine the network's configuration through genetic optimisation, which involves training of thousands of networks. This is absolutely essential, as different configurations yield very different error rates (Malmgren et al., 2001); it also makes the process extremely time-consuming. When properly trained, ANNs ought to be very good at generalizing. The technique is not as dependent on the size of the calibration data set as MAT and it allows extrapolation (i.e. it is possible to obtain an estimate of an environmental parameter which is outside of the range of the parameter in the calibration data set). The biggest disadvantage of ANN is that it is very difficult to interpret the resulting network, especially in terms of the contribution of individual variables to the resulting estimate.

The networks used in this paper have been trained separately on each of the three databases (North Atlantic, South Atlantic and Pacific). In each case the calibration data set was split into 10 representative pairs of training (80%) and test (20%) subsets and for each partition, the network parameters were optimised to yield the lowest error in the test set. The partitioning of the calibration data set and the optimisation was performed using a genetic algorithm implemented in the commercial software NeuroGenetic Optimiser v2.6 (Biocomp). The maximum number of neurons, which could have linear, logarithmic or hyperbolic tangent transfer functions, in each of one or two hidden layers was set to 32 for the larger North Atlantic and Pacific databases and 16 for the smaller South Atlantic database. The maximum number of passes was set to 2500 and the algorithm was instructed to stop learning when no improvement in the prediction error (based on the test set) occurred for 40 consecutive passes. The training and optimisation were done separately for each output variable (SST definition) and each time the genetic algorithm searched through a total of 3000 network configurations (30 generations of 100 networks). For each partition and output variable (SST definition) the best network (one which produced the lowest error in the test set) was retained and all ANN SST reconstructions were calculated as averages of the outputs of the 10 best networks. The configurations of the best networks vary greatly among the partitions (Table 4), demonstrating the importance of a thorough optimisation performed separately on each new training data set.

3.3. Calibration results

3.3.1. Error rates and residuals

The results of calibration of individual transfer function techniques are summarised in Table 2 and Fig. 3. The parameters of the best neural networks for each geographical domain and SST definition are listed in Table 4. It should be noted that these error rate values should not be taken literally as a measure of relative success of the individual techniques. Ter Braak (1995) showed the error rate estimates vary with the choice of cross-validation procedure, and we must reiterate that the low error values produced by RAM reflect the fact that the leaving-one-out validation procedure in the RAM02 software is implemented after the two-dimensional interpolation. Indeed, Barrows and Juggins (2005) used the same 5-fold leave-out for all techniques in their analysis of the Indo-pacific calibration data set: under this more rigorous cross-validation the apparent advantage of RAM over MAT (4 analogs) and ANN is lost (Table 2). For the trials on the other data sets conducted here the classical MAT approach, represented by the SIMMAX$_{ndw}$ technique, produced the highest error rates (Fig. 3), which may be indicative of the relatively lower power of this technique when used with the default of 10 analogs. Whilst the absolute values of the error rate should be interpreted with caution, the patterns of errors related to seasons and regional domains are more robust and informative.

All techniques produced highest error rates for the Pacific data set and the lowest error rates for the South Atlantic data set (Fig. 3). The difference between the North- and South-Atlantic error rates is clearly related to the larger number of samples in the North Atlantic

Table 4
Parameters and error rates of artificial neural networks trained on the MARGO Atlantic and Pacific databases

Summer SST

Ocean / Partition	tst	trn	1HL Lo	1HL T	1HL Li	2HL Lo	2HL T	2HL Li	Out	N
North Atlantic										
1	1.2005	1.1069	5	1	5				1Li	11
2	1.1646	1.0919	3	4	8				1Lo	15
3	1.1140	1.0883	15	2	10	1	2	4	1Li	34
4	1.0697	1.0647	12	10	7		7	24	1Li	60
5	1.1900	1.0610	9	2	13		6	3	1Li	33
6	1.2795	1.1127	9	11	9				1Li	29
7	1.1031	1.0824	16	2	3				1Li	21
8	1.1503	1.0751	9	1	17				1Li	27
9	1.0109	1.1124	9	2	6				1Li	17
10	1.1106	1.1251	9	3	14	17	3	9	1Lo	55
Average	1.1393	1.0921								30
South Atlantic										
1	0.8075	0.7124	6	4	5				1T	10
2	0.7440	0.8548	5	2	3				1T	10
3	0.8221	0.6838	8	3					1T	11
4	0.7810	0.8286	11	3	2	1	1		1T	18
5	0.6595	0.6386	5	3					1T	8
6	0.7098	0.6103	6		3			1	1Li	10
7	0.7929	0.8224	1	4	2				1Lo	7
8	0.9074	0.7967	4		2				1Li	7
9	1.1369	0.6538	3	3	6		8	4	1Li	24
10	0.7347	0.8402	1		1			1	1Li	6
Average	0.8096	0.7442								11
Pacific										
1	1.3653	0.9578	10	7	5	2	1	2	1Li	27
2	1.1156	1.0424	7	3	2				1Lo	12
3	1.1379	1.0352	5	12	14				1Lo	31
4	1.2257	1.0881	22	3	2				1Lo	27
5	1.1862	1.0303	7	2	5				1Lo	14
6	1.1733	0.9392	10	3	7	2		1	1Li	23
7	1.3116	0.9115	10	10	9			3	1Li	32
8	1.0352	1.0268	9	9	8				1Li	26
9	1.2406	0.9270	13	3	8				1Li	24
10	1.1362	1.0178	5	2	4		7	8	1Li	26
Average	1.1928	0.9976								24

Winter SST

Ocean / Partition	tst	trn	1HL Lo	1HL T	1HL Li	2HL Lo	2HL T	2HL Li	Out	N
North Atlantic										
1	0.9892	0.7675	20		12			13	1Li	45
2	0.9466	0.8348	9	4	2	2	6	17	1Li	40
3	0.9750	0.8203	13	6	12	2	2	2	1Lo	35
4	0.9117	0.8536		23	2		4	4	1Li	33
5	0.9483	0.8414	9	7	9		10	4	1Li	39
6	1.1361	0.8281	5	2	1			1	1Li	9
7	0.8402	0.8959	13	12	3		7		1Li	37
8	0.9506	0.8238	24		7	2	1	5	1Lo	39
9	0.9570	0.8819	8	2	10	8	6	6	1Li	40
10	0.9190	0.8230	6	6	6	5	1	4	1Li	28
Average	0.9574	0.8370								35
South Atlantic										
1	1.0370	0.6887	10	3					1T	13
2	0.8970	0.8588	3	10	2			1	1Li	16
3	0.8661	0.6458	14	2			5	3	1Li	24
4	1.0430	1.0311	8	1	4				1Li	9
5	1.0835	0.7725	4	1	7		5	3	1Li	17
6	0.9741	0.6687	8	1				1	1Li	17
7	0.8872	1.0492	3	4	1				1T	8
8	0.8100	0.9366	6		3	2	2	1	1Lo	20
9	0.9582	0.7750	3	2	3		3		1Li	8
10	0.9420	0.7937	7	5	3	5	6	1	1Li	27
Average	0.9498	0.8220								16
Pacific										
1	1.9575	1.4737	4	1	12				1Li	17
2	1.4271	1.3655	8	9	2	2			1Lo	21
3	1.6387	1.3172	9	5	6				1Lo	20
4	1.5794	1.4191	16	1	15	7	6	7	1Li	52
5	1.5047	1.4458	7	7	6				1Li	20
6	1.6164	1.4987	7		20				1Li	29
7	1.9392	1.5567	2	1		4	4	5	1Li	22
8	1.4816	1.5882	8	8	6				1Li	10
9	1.6882	1.6239	2	8	5	3	3	4	1Li	25
10	1.5609	1.5150	5	5	4	6	1	19	1Li	40
Average	1.6394	1.4804								26

Annual SST

Ocean / Partition	tst	trn	1HL Lo	1HL T	1HL Li	2HL Lo	2HL T	2HL Li	Out	N
North Atlantic										
1	1.0572	0.8507	7	7	7	2		4	1Li	27
2	0.9809	0.8490	9	2	7				1Li	18
3	0.9438	0.8217	18	14	1				1Lo	33
4	0.8838	0.8674	10	4	9			6	1Li	31
5	0.9644	0.9208		10	6				1Li	16
6	1.1232	0.8653	10	8	4			1	1Li	23
7	0.9055	0.8692	7	9	5				1Li	21
8	0.9768	0.8802	10	1	11			1	1Li	24
9	0.8670	0.8883	12	8	1		4	18	1Li	43
10	0.9302	0.8024	14	2	1	3	1	6	1Li	27
Average	0.9633	0.8615								26
South Atlantic										
1	0.8313	0.6415	10	3	1	8	6	1	1Li	29
2	0.7532	0.7817	3	9	2				1T	14
3	0.7978	0.6676	3	4	3				1Li	10
4	0.8825	0.8923	4	4	1				1T	9
5	0.8123	0.6763	2	2	1		2	2	1Li	9
6	0.8292	0.6522	4	2	3				1Li	9
7	0.7664	0.7884	1	1					1T	2
8	0.7710	0.8550	1	7	7		1	1	1Li	17
9	0.9990	0.6268	2	4	1				1Li	7
10	0.8292	0.7517	4		3		3	4	1Li	17
Average	0.8272	0.7334								12
Pacific										
1	1.6379	1.1443	7	7	7				1Li	14
2	1.3407	1.2547	7	4	4	1	1	1	1Li	16
3	1.2893	1.2023	5		10	2	1	1	1Li	19
4	1.3604	1.1276	8	8					1Lo	16
5	1.1627	1.2201	6	1	12				1Li	19
6	1.3023	1.1659	14	3	10			3	1Li	28
7	1.5046	1.1190	6	3	1			1	1Li	13
8	1.1999	1.3040	3	1	16				1Li	20
9	1.3835	1.1680	4	12	10		3		1Li	30
10	1.2798	1.1621	10	3	11			1	1Li	24
Average	1.3461	1.1868								20

HL = hidden layer, Out = output neuron, N = total number of neurons in hidden layers, tst = RMSE based on the test set, trn = RMSE based on the training set, Lo = neurons with logarithmic transfer function, T = neurons with hyperbolic tangent transfer function, Li = neurons with linear transfer function.

M. Kucera et al. / Quaternary Science Reviews 24 (2005) 951–998

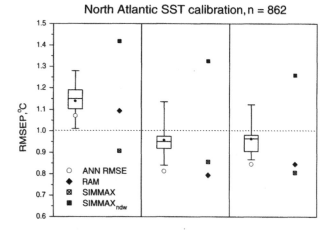

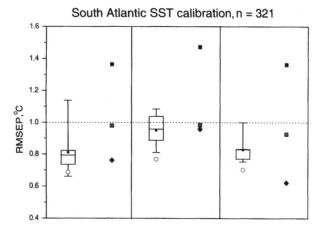

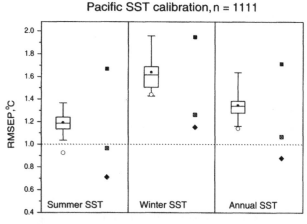

Fig. 3. Summary of error rate estimates (RMSEP = root mean square error of prediction) for North Atlantic, South Atlantic and Pacific SST calibration. The prediction errors for RAM and both SIMMAX varieties (ndw = no distance weighting) were calculated using the leaving-one-out technique. For ANN, the error rate was calculated as the RMSE of samples in the test set of each partition. The box-and-whiskers plots depict the minimum, 0.25 percentile, median, 0.75 percentile, and the maximum RMSEP values; solid circles represent the arithmetic average of the 10 partitions. Estimates of SST in all samples were also produced by applying the best network from each partition to the entire data set and averaging the 10 estimates. The error rate based on this approach is reported as ANN RMSE.

polar region (Fig. 2), where all techniques produced the highest errors (Fig. 4). This "cold end tail" in the North Atlantic is a known phenomenon (Pflaumann et al., 1996) caused by the absence of faunal variation in the polar province where a single species, *N. pachyderma* (sin.), dominates the planktonic assemblage. This effect also explains the higher summer SST error rates for the North Atlantic compared to the South Atlantic. It is interesting to note that ANN performed best on the South Atlantic data set, suggesting that it may have been most successful of the four techniques in dealing with the poor sample coverage in the southern high-latitudes. The high error rates recorded by all techniques for the Pacific data set may be explained by a combination of factors including the larger geographical coverage of the data set and hence an inherently higher amount of noise and a greater degree of lumping of cryptic species, poor sample coverage in many areas (Fig. 2), greater disparity in the origin of the census counts, and the effects of carbonate dissolution. The pattern of highest residual values in the mid-to-high latitudes of both Hemispheres (Fig. 5) would appear to support the uneven sample coverage scenario. Apart from the "cold end tail" in the North Atlantic, the techniques did not produce any significant SST-related bias, as can be seen from the distribution of the residual values (Fig. 4).

Understanding of sample characteristics that are indicative of higher prediction errors is extremely beneficial in any attempt to assess the reliability of fossil SST estimates. Similarity (RAM) and dissimilarity (SIMMAX) coefficients of the best analog samples can be used to express how well a given combination of species abundances is represented in a calibration data set. Similarly, a standard deviation of estimates from the best neural networks for each of the 10 partitions of the calibration data set indicates the sensitivity of the estimate to small changes in the calibration data set. If a fossil assemblage is well represented by a number of good analogs, the standard deviation of the 10 ANN estimates will be low. In all three calibration data sets used in this study, there is no correlation between measures of sensitivity of SST estimates and the observed errors (Fig. 6) indicating that most species abundance combinations and SST patterns are well represented in the calibration data sets and there is no bias towards larger errors being produced for more "exotic" samples. A different pattern emerges when the standard deviation of the best analog samples used to derive the SST estimate in RAM and SIMMAX is scrutinised. This measure of disparity among the selected analogs shows a clear positive correlation with prediction errors (Fig. 6), suggesting that samples with best analogs distributed across a wider SST range are more likely to be in error than samples with best analogs drawn from a narrow SST range.

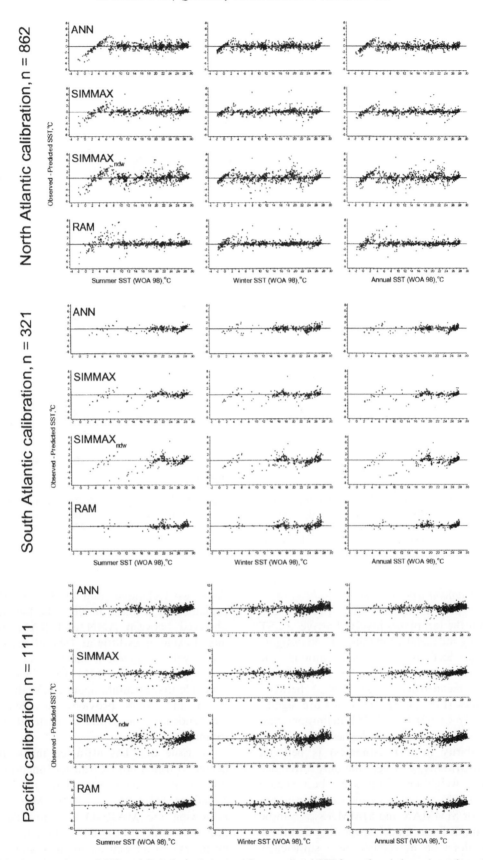

Fig. 4. Relationships between observed SST and deviations of observed from predicted SST for each technique. Apart from the well-documented "cold-end tails", more prominent in the North Atlantic data set, which includes many truly polar samples, there is no systematic bias in the residuals. We have calculated linear regressions for each plot; these are not shown because the regression lines' slopes are graphically indistinguishable from zero.

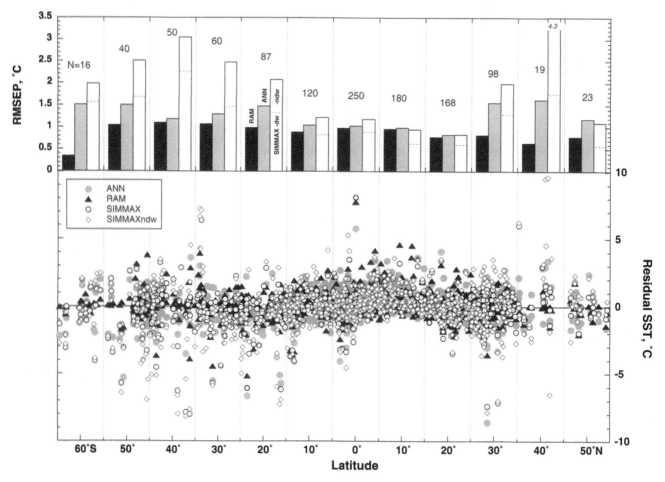

Fig. 5. Latitudinal distribution of annual SST residuals (observed minus predicted SST values) and root mean square errors of prediction (RMSEP) for all transfer function techniques in the Pacific. The prediction errors for ANN were calculated from the RMSE of samples in the test set of each partition, and for RAM and SIMMAX (ndw = no distance weighting) by the leaving-one-out technique (for RAM applied after virtual samples have been added, which explains the lower prediction errors of this technique).

3.3.2. Differences among transfer function techniques

The advantage of a multi-technique approach can only be realised if the techniques involved arrive at SST estimates in a truly independent manner. To investigate the similarity in SST reconstructions among the techniques, we have examined the similarity in residuals (observed minus reconstructed SST values) between pairs of techniques (Fig. 7). The residuals are significantly correlated ($p < 0.05$) for all SST definitions and technique pairs. As expected, the highest correlation (and hence similarity in SST estimates) is between SIMMAX and SIMMAX$_{ndw}$, where the only difference is in the distance weighting of the same best analog samples. Between 55% and 70% of the variation in the residual scatter for SIMMAX and SIMMAX$_{ndw}$ can be explained by a simple linear model, suggesting that these two techniques alone do not constitute a sufficiently independent framework.

RAM and ANN both show substantially lower correlations, with < 48% of the residual scatter between these two techniques and the two SIMMAX varieties explained by a linear model. The lowest correlation is observed between ANN and RAM, where a linear model explains as little as 18% of the residual scatter. These values suggest a much greater difference than between the two SIMMAX varieties and we conclude that a simultaneous application of SIMMAX, RAM and ANN provides an adequate degree of independence among SST estimates derived from these techniques.

An issue that has attracted much attention is the degree to which faunal transfer function techniques can provide independent seasonally resolved SST reconstructions (Mix et al., 2001; Morey et al., 2005). Following the MARGO recommendations, we have decided to follow the common practice in reconstructing mean annual as well as summer and winter SST (see Section 2.1). To investigate the degree of independence between the seasonal SST reconstructions, we have examined the correlation between summer and annual

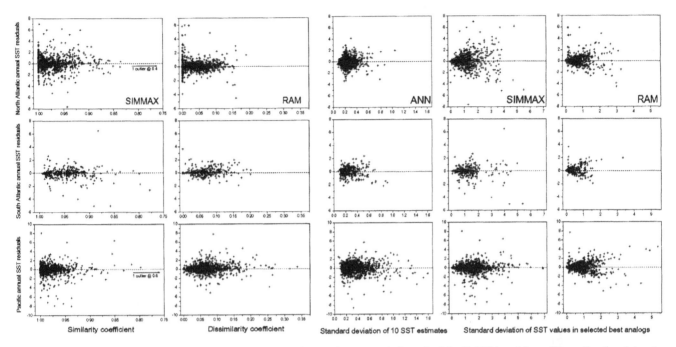

Fig. 6. The standard deviation of annual SST estimates produced by the best networks in each of the 10 ANN partitions of the calibration data sets, the standard deviations of the best selected analogs for SIMMAX and RAM and similarity (SIMMAX) and dissimilarity (RAM) coefficients for the nearest analog plotted against annual SST residuals.

SST residuals for all techniques (Fig. 8). In any *k*-nearest neighbour method, the same set of best analog samples will be identified no matter how the desired SST is defined. This means that MAT and SIMMAX will tend to reproduce the SST structure of the calibration data set, and this fact is clearly reflected in the highest correlations between summer and annual SST residuals for SIMMAX and SIMMAX$_{ndw}$ (Fig. 8). The separate training of ANN for each SST definition appears to have had a limited impact on the independence of seasonal reconstructions, but the correlation coefficients for this technique are still consistently high (Fig. 8). The two-dimensional interpolation procedure involved in RAM provides an escape from the constraints of the calibration data set, apparently leading to the least dependent seasonal SST reconstructions (Fig. 8). However, the classical calibration approach (ter Braak, 1995) represented by the interpolation procedure in RAM, does not provide unique reconstructions in a multivariate space and the degree of dependence of the seasonal estimates varies with the choice of pairs of variables used to define the interpolation space. Given the very high correlations between winter and summer SST in the calibration data sets ($r > 0.94$) it remains nearly impossible to provide any statistically sound evidence that the two variables can be reconstructed independently. With the data at hand, we tend to concede that the seasonal reconstructions may not be sufficiently independent to warrant the use of their

differential as a measure of past seasonality and we abstain from such practice in this study.

3.3.3. Bias due to coretop ages and calcite dissolution

A comparison of average residual values for samples from each of the five age quality categories (Fig. 9) did not reveal any systematic shift towards colder estimates in lower quality levels, as would have been expected from samples containing pre-Holocene faunas (see Section 3.3). This result justifies our decision to retain category 5 samples in the calibration data set, in order to facilitate a broader spatial coverage. The Pacific calibration shows the highest tendency for underestimation of SST in lower age quality categories, but the pattern is inconsistent among the techniques with ANN indicating a greater percentage of underestimates in category 4 samples than in category 5 samples (Fig. 9). Although we cannot exclude that some of the category 5 samples may represent pre-Holocene assemblages, these samples must be relatively rare.

Carbonate dissolution can significantly alter species composition of planktonic foraminiferal assemblages (Berger, 1968). To avoid this bias, all samples with signs of calcite dissolution have been removed from the Atlantic calibration data set (Pflaumann et al., 2003). A plot of residual values against water depth (Fig. 10) does not reveal any obvious dissolution bias (higher residual values for samples from deeper parts of the ocean) in the Atlantic data. However, a small dissolution bias can be

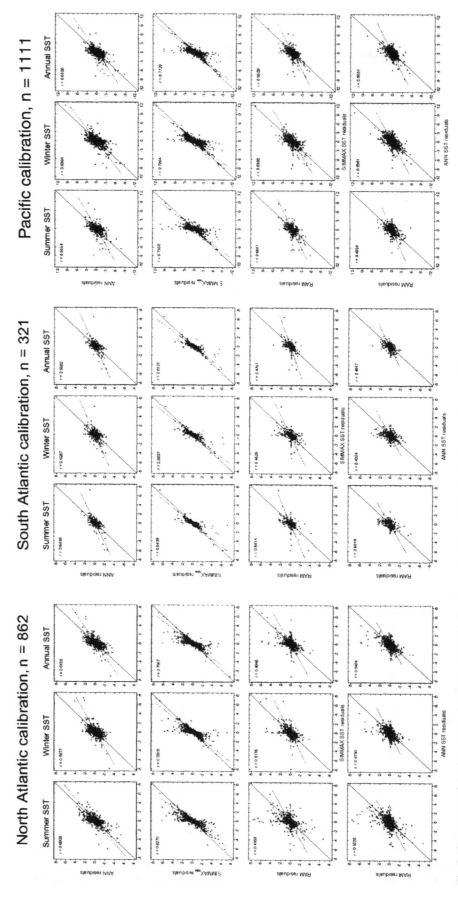

Fig. 7. Assessment of the degree of independence in SST estimates produced by individual techniques in all three calibration data sets. The residuals are significantly correlated for all SST definitions and techniques pairs, although RAM and ANN appear to produce adequately independent SST estimates.

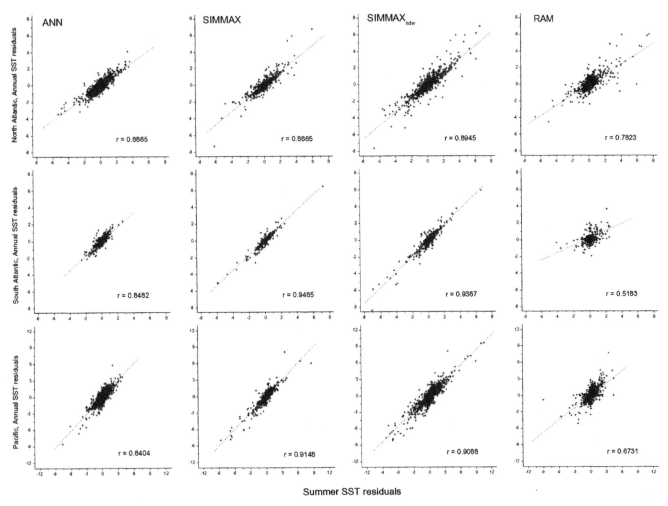

Fig. 8. Correlation between summer and annual SST residuals in all calibration data sets. The high correlations for all techniques suggest that seasonally resolved SST estimates may be largely repeating the present day SST pattern.

deduced from the distribution of residual values against depth in the Pacific data set. This is not surprising as we have specifically chosen not to exclude samples that may have been affected by calcite dissolution. Interestingly, the dissolution bias appears to be manifest in only a few samples showing excessive residual values. For the vast majority of samples, the transfer functions show an equally good performance in SST estimates from above and below the lysocline (Fig. 10), indicating that either the techniques are able to account for dissolution bias (for example, by having appropriate similarly biased samples in the calibration set), or that the bias was not sufficiently large to significantly alter the SST signal of foraminifer assemblages.

3.3.4. Effect of the use of geographically separated calibration data sets

The Atlantic calibration data set is best suited to test the merit of the use of geographically constrained calibration data sets. Both parts of the Atlantic are inhabited by the same set of morphospecies of planktonic foraminifera and the GLAMAP-2000 data on which our calibration data set is based provide a high degree of taxonomic consistency and good taphonomic control. In theory, if the planktonic foraminifer faunas in North and South Atlantic consist of the same taxonomical units with identical ecological preferences, then a transfer function calibrated in one part of the ocean should perform equally well in the other part. To test this hypothesis, we have applied the ANNs trained on the South Atlantic database to the North Atlantic coretop samples and vice versa (Fig. 11). The results show a doubling of RMSEP in the South Atlantic and a tripling of RMSEP in the North Atlantic when ANNs based on the opposite calibration data set are used. Interestingly, the increase in error rates (manifested by higher observed minus predicted SST values) is not spread evenly throughout the latitudinal range of each data set. A clear pattern emerges with a narrow tropical band between 20° North and 20° South (Fig. 11), where

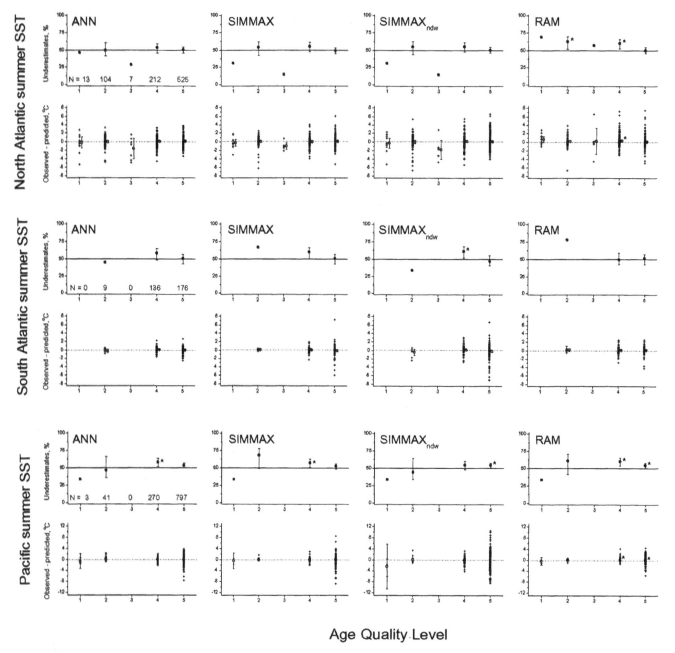

Fig. 9. The effect of age quality control in the coretop samples on SST estimates. Vertical bars represent 95% confidence intervals; asterisks highlight averages that deviate significantly from 50% (for proportion of SST underestimates) and from 0 (for mean residuals).

no increase in SST errors occurs. This cannot be an artefact of the overlap between the North and South Atlantic calibration data sets (Fig. 2), as much smaller range of samples from the equatorial region is shared between the two data sets (Fig. 11). The pattern can be explained by the presence of a tropical province in the equatorial Atlantic where identical faunas are fully intermixed allowing the transfer function to extract the relationship between abundances of taxa and tropical SST patterns from samples covering any part of this region. Beyond this tropical zone, the error rates rise remarkably, suggesting that ecological preferences of

species outside of the tropical zone differ between the North and South Atlantic. Molecular genetic data from the high latitudes of both hemispheres in the Atlantic (Table 1) indicate that in the subpolar and polar waters the differences in ecological preferences of species can be attributed to the presence of cryptic genetic types. These observations provide a strong support for the use of geographically constrained calibration data sets, and underline the possible extent of the adverse effect of lumping of crytpic genetic types into morphologically defined species (Kucera and Darling, 2002).

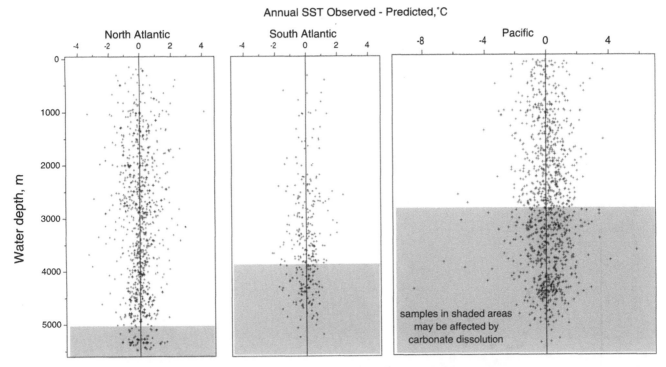

Fig. 10. Relationship between depth of coretop samples and ANN annual SST residuals. The shaded areas indicate the typical range of depths in each region where carbonate dissolution may occur (from Berger and Winterer, 1974). Most of the samples with highest SST residual values in the Pacific calibration data set derive from depths where carbonate dissolution may have modified the foraminiferal assemblage.

4. Atlantic and Pacific glacial temperatures

4.1. Atlantic LGM data base

The LGM data set used here is adopted from the GLAMAP-2000 study (Niebler et al., 2003; Pflaumann et al., 2003) which greatly benefits from a stringent quality and age control and careful selection of the LGM interval. GLAMAP-2000 presented glacial SST reconstructions using two alternative LGM definitions. The first, covering the time interval from 21,500 to 18,000 yBP, is based on the concept of the last isotopic maximum (LIM) when ice-sheets reached their maximum extent (Sarnthein et al., 2003a). The second definition, which we use herein, covers the interval from 23,000 to 19,000 yBP, as defined by the EPILOG workshop (Mix et al., 2001) taking into account the slightly earlier onset of deglaciation in the southern hemisphere. The MARGO Atlantic data set comprises LGM faunal census in 173 cores, most of the samples corresponding to quality levels 1 and 2 according to the MARGO recommendations (Kucera et al., 2005) offering detailed AMS-^{14}C-stratigraphy or other high-resolution age control. Only a few samples fall into category 3 with a less rigorous, but still solid age control. The LGM records for each core either cover the entire EPILOG LGM interval or at least fall into the 23,000 to 19,000 yBP time window. In most of the Atlantic records

the LGM interval is represented by several samples, making it possible to approximate the full range of LGM faunal variability occurring at each site. Records from the North Atlantic exhibit a better time resolution (550 samples in 104 records) than those from the South Atlantic (198 samples in 69 records). Since the coverage of this high-quality data set is deemed adequate, to avoid a quality loss, no samples with less well-constrained LGM data were added. LGM SST was estimated applying three transfer function techniques (1) ANN; (2) RAM; (3) SIMMAX (with distance weighting) described above and downcore SST estimates were averaged for each core.

Comparing the SST estimates produced by the three different computational techniques helped to identify severe no-analog faunas. As a starting point we inspected samples where 10 individually trained ANNs produced standard deviations higher than 2 °C. Indeed, these samples also yielded the highest offsets among the single methods and are generally marked by the poorest similarity/dissimilarity values. The faunas in these samples were found to show large portions of certain species or counterintuitive patterns of species abundances. The severe no-analog samples (all samples in cores CH82-24; V23-100; GIK16776-1; GeoB2116-4; 2 samples in core M35-027-1) show a random geographical distribution (Fig. 12) and are considered malignant, either as a result of miscounting due to taxonomic error

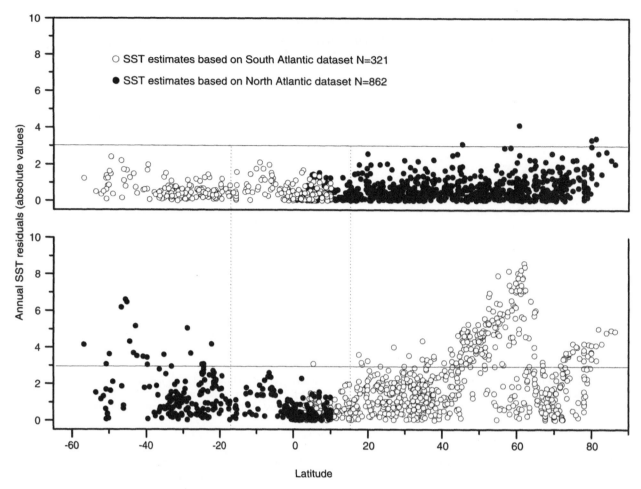

Fig. 11. The effect of applying artificial neural networks trained on the South Atlantic database to North Atlantic coretop samples and vice versa.

or as genuine no-analog faunas that accumulated near steep oceanographic gradients. The SST reconstructions based on these samples have not been used in any further analysis. In contrast, samples with the different estimates deviating less than 4 °C whilst exhibiting poor similarity/ dissimilarity and/or high standard deviation of 10 ANNs, and moreover displaying a regional distribution patterns, were considered as benign genuine non-analogs and retained in the database.

4.2. Atlantic LGM results

A state-of-the-art-reconstruction of LGM SST in the Atlantic has been recently published (GLAMAP-2000 project; Gersonde et al., 2003; Niebler et al., 2003; Pflaumann et al., 2003; Sarnthein et al., 2003a). Therefore we aim to focus on a comparison of the performance of three different computational methods (ANN; RAM; SIMMAX) on the GLAMAP-2000 LGM data set. This multi-technique approach facilitates a test of the robustness of SST estimates from foraminifer

transfer functions and provides a more powerful and rigorous means to identify potential no-analog conditions or faunas. Whilst congruent results of the three methods may be considered as robust, offsets among individual SST estimates may indicate situations where the temperature information provided by the faunas does not allow for an unambiguous SST interpretation, which is particularly interesting when occurring at a regional scale. Moreover, this multi-technique approach may help to unravel discrepancies with other, independent, SST proxies. ANN provides an innovative approach of SST estimation independent of the modern analog techniques used before. ANN LGM SST patterns are therefore reported in detail in Fig. 13. Corresponding glacial-to-modern anomalies are given in Fig. 14. To produce these maps, we chose the Eckert pseudocylindrical equiareal projection (Snyder, 1987) suitable to depict the meridional Atlantic with correct size relations, while keeping the figures intelligible for both, polar and equatorial regions. Contours were drawn by triangulation of reconstructed SST values at

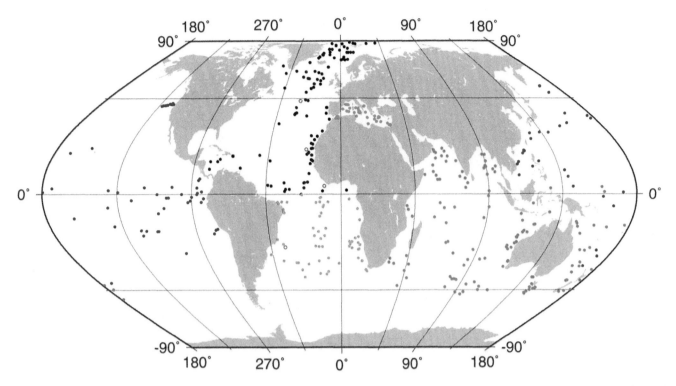

Fig. 12. Location of sediment cores with LGM samples in the North and South Atlantic and the Pacific (this paper) and the Mediterranean (Hayes et al., 2005) and Indian Ocean and Australia (Barrows and Juggins, 2005). Colours and symbols as in Fig. 2. Open circles mark severe no-analog samples which were not used for LGM SST reconstructions.

sample locations using GMT-software (Wessel and Smith, 1991, 1998). To highlight the differences in SST estimates by individual techniques, the results are also shown on a latitudinal transect across the Atlantic (Fig. 15).

The ANN SST estimates show the strongest glacial cooling in the subpolar central northern North Atlantic where temperatures dropped by 8–12 °C (Fig. 14) to a minimum of <4 °C (Fig. 13) during the glacial warm season and by the same magnitude to a minimum of 1 °C during the cold season. A steep latitudinal SST gradient from 4–6 °C at 50°N to 20 °C at approx. 38°N is seen in mid-latitude LGM North Atlantic. ANN values suggest a distinct cooling of the eastern boundary system in the north Atlantic, where the Canary Current cooled, particularly during the cold season, by more than 8 °C. In contrast, the vast area of the subtropical gyre remained as warm as today (22–25 °C). Temperatures of 20–24 °C in the eastern tropical Atlantic suggest a marked cooling of the equatorial upwelling system by 4–6 °C during glacial northern summer, and less pronounced during austral summer (2–4 °C). Extreme cooling by up to >10 °C affected the Benguela Current, both during glacial warm and cold seasons, while the rest of the South Atlantic shows little difference from present-day temperatures. ANN estimates in 6 cores from the central south Atlantic suggest temperatures reaching up to 28 °C. Compared with modern tempera-

tures these values are 2–6 °C warmer than today. This warm anomaly was particularly pronounced during the austral summer.

In general terms, the ANN patterns confirm the LGM reconstructions by Pflaumann et al. (2003) and Niebler et al. (2003). Furthermore, the Atlantic transect shows that all three techniques produced similar estimates (Fig. 15). Particularly good agreement is found at the mid-to-high latitude north Atlantic at 50–60°N, where the strong cooling by up to 11 °C is jointly reconstructed by all methods, with deviations smaller than 1 °C. Also the steep temperature gradient marking the mid-latitudes is reproduced congruently. A similarly good consensus is reached on the magnitude of tropical cooling by 4–6 °C in the tropical eastern Atlantic. Winter cooling in contrast was more pronounced according to SIMMAX and RAM.

However, there are also several distinct offsets between estimates from different techniques. In particular, north of 60° in the Nordic Seas warm season estimates from the three techniques markedly diverge. Here, RAM produced the coldest SST of −1 to 3 °C, SIMMAX produced cold temperatures of 3 °C decreasing to 1 °C north of 80° and ANN produced the warmest SST lacking any latitudinal gradient (3.4–2.7 °C). The cold season estimates from all techniques converge towards 1 to −1 °C. A distinct discrepancy also marks the eastern boundary current off Northwest Africa,

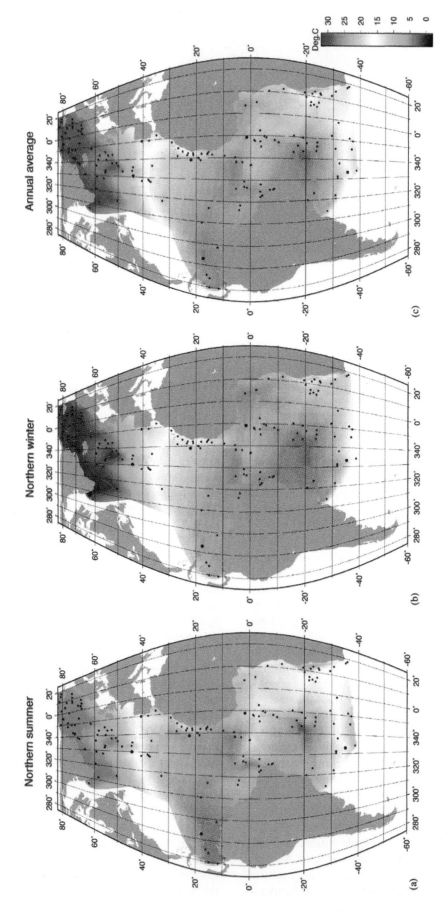

Fig. 13. Sea-Surface Temperatures (SST) in the Atlantic reconstructed by Artificial Neural Networks (ANN) trained separately on North Atlantic and South Atlantic calibration data sets. Contours are based on triangulation of SST at sample locations (dots) (a) northern summer; (b) northern winter; (c) annual average (average of 12 months). Enlarged symbols mark severe no-analog samples excluded from triangulation.

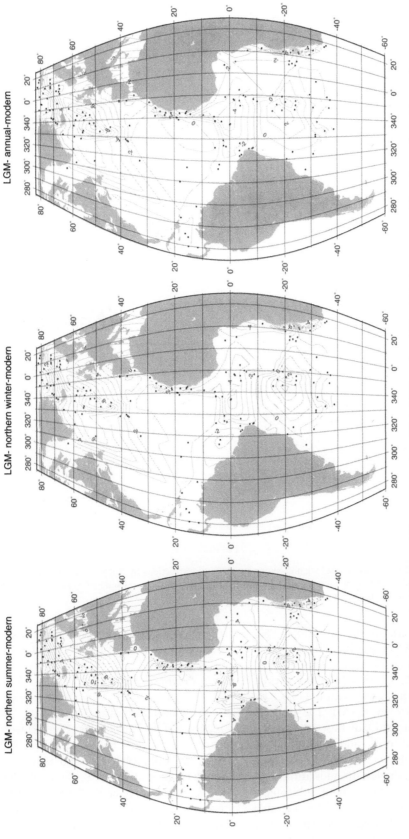

Fig. 14. Glacial-to-modern SST anomalies expressed as ANN estimates minus WOA (1998) modern SST. (a) Northern summer; (b) northern winter; (c) annual average.

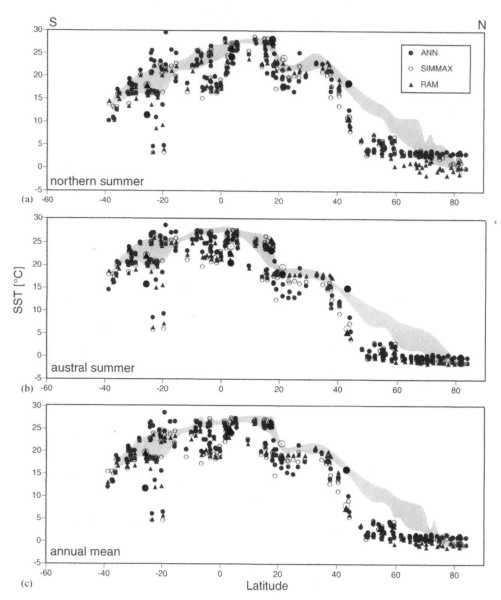

Fig. 15. Latitudinal transects of SST estimates in the Atlantic comparing SST reconstructions by ANN, SIMMAX, and RAM. Grey shaded area envelops modern SST values (WOA, 1998) at the same stations. (a) Northern summer; (b) austral summer; (c) annual average. Enlarged symbols mark severe no-analog samples excluded from triangulation.

marked by a distinct pattern in the latitudinal SST transect (Fig. 15). Here, ANN produced consistently cooler SST (16–20 °C during the warm season) than RAM and SIMMAX (20–22 °C); this difference is even more pronounced during the cold season (13 °C by ANN, 17° by SIMMAX, 15° by RAM). The transect (Fig. 15) reveals a southward shift of this structure by ~5° to 18°–23° as compared to its modern position, as also suggested by Pflaumann et al. (2003). A large scatter of SST values marks the low-to-mid-latitudes at 15–25° in the south Atlantic, where both subtropical gyre and Benguela Current were subject of considerable disagreement among the different techniques. The pattern of warmer LGM SST in the southern subtropical gyre is most pronounced in the ANN estimates;

the SIMMAX and RAM estimates show little change compared to present-day SST values in this region.

4.3. Pacific LGM data set

Foraminiferal assemblage data from the LGM sections of 81 Pacific sites (Fig. 12) were compiled from various sources (Chen et al., 1998, 2005, unpublished; Weaver et al., 1998; Lee and Slowey, 1999; Mix et al., 1999; Ortiz et al., 1999; Kiefer et al., 2001; Feldberg and Mix, 2002; Morey et al., 2005). The LGM interval in 23 cores was defined by radiocarbon dates, thus qualifying for a MARGO chronostratigraphic level 1 or 2. In addition, there are 9 cores with a biostratigraphic or lithostratigraphic age model (MARGO age control

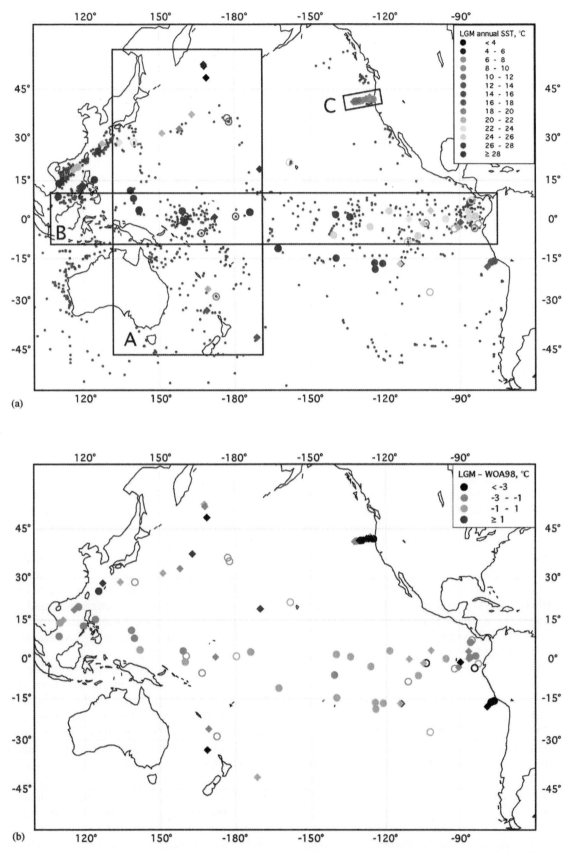

Fig. 16. (a) LGM annual SST in the Pacific estimated from foraminifer assemblages. The SST values represent averages of the estimates from ANN, SIMMAX, and RAM, all of which were calibrated using the Pacific coretop data set (small grey circles). Rectangles A–C mark the three transects shown in Fig. 19. (b) Differences between annual LGM and modern SST (WOA, 1998). Samples where the maximum difference in SST estimates among the four techniques is <2 °C are indicated by solid circles (highest reliability), between 2–4 °C by solid diamonds (medium reliability), and >4 °C by open circles (lowest reliability).

category 3; Fig. 16), with 6 of these level-3 sites located in the eastern tropical Pacific. There is no detailed age model information available for the remaining 50 sites, which include many of the original CLIMAP cores (CLIMAP Project Members, 1976) and additional data by Feldberg and Mix (2002) and Morey et al. (2005). Water depths of the cores range between 503 and 4614 m, with an average of 2930 m, suggesting that at least half of the sites may be affected by carbonate dissolution. Nevertheless, none of the deeper sites were removed from the data set, as we had addressed the dissolution problem by including coretop data from partially dissolved sediments into the calibration data set (Sections 3.1.2 and 3.3.3).

Our study area overlaps with the Indo-Australian reconstruction (Barrows and Juggins, 2005) in the western equatorial and southwestern Pacific. Sample coverage is good in the tropical Pacific and satisfactory in the western Pacific (Fig. 16). All other regions are underrepresented to an extent that no reliable picture of the LGM surface ocean can be deduced from foraminiferal assemblages alone. Therefore, our results from the Pacific are discussed in three representative transects with good sample density.

4.4. LGM results for the Pacific

Pacific glacial SST (Figs. 16 and 17) was estimated using four different techniques (ANN, SIMMAX, $SIMMAX_{ndw}$, and RAM). Annual SST were reproduced by these four techniques within a range of 0.02 to 3.88 °C (average of 1.02 °C). The nine sites with the lowest SST consistency among the methods are concentrated in the eastern equatorial Pacific and the central subtropical Pacific and the reliability of SST reconstructions in these areas may be therefore considered lower. Moreover, the offsets in SST estimates among methods are not random. Statistically, SIMMAX produced the warmest LGM estimates ($+0.7$ °C above the average; $SIMMAX_{ndw}$ $+0.4$ °C), RAM the coldest (-0.8 °C), and ANN was near the average ($+0.1$ °C). Although these offsets in SST estimates all fall within the error range for individual techniques (Fig. 3), they appear robust and may result in some bias in the assessment of glacial Pacific SST. A comparison of the LGM SST estimates with present-day SST patterns along three transects (Fig. 16(a)) reveals some major thermal features of the glacial Pacific and serves to illustrate the uncertainty of the reconstruction as revealed by deviations among the four different approaches.

In the western Pacific Ocean, thirteen cores are included in both the LGM reconstructions of Barrows and Juggins (2005) and this study. This overlap provides the opportunity to test the compatibility of SST estimates made by the two studies. Such a comparison

is not strictly an independent test since the two calibration data sets include many of the same core top data in the South Pacific, although the Indo-Australian database includes core tops from both the Indian and southern Atlantic Oceans. Although RAM, ANN and a *k*-nearest neighbour technique are used by both studies, they are based on different parameterisations. Lastly, for a few of the cores the LGM samples are not the same, because of differences in the stratigraphic models used by the two studies.

The correlation between annual SST estimated by the two studies is high (Fig. 18) varying from 0.96 (RAM) to 0.98 (ANN and consensus). Mean and maximum absolute differences between estimates are highest for RAM (1.5 and 3.2 °C, respectively) and lowest for ANN (0.7 and 2.4 °C, respectively). Importantly, there is no significant difference in the mean of the estimates for any of the techniques ($p \leqslant 0.05$), indicating that although we use (partially) different calibration data sets, different parameterisations of the numerical methods, and in some cases, different LGM samples, there is no systematic bias in the estimates from the two studies. Interestingly, the differences between studies is smallest for the ANN, suggesting again that this method may be better at generalizing whereas the analog-based methods are more dependent on the detail of the calibration data set. In summary, the results between the two reconstructions are compatible.

4.4.1. North–South transect in western Central Pacific

A North–South (N–S) transect through the western Central Pacific reveals SST differences across the major oceanographic zones of both hemispheres. The transect extends from 130°E to 170°W, thus including only open ocean sites (Fig. 16(a)). Sites from marginal basins like the South China Sea and the Eastern China Sea were avoided, since they potentially represent local hydrography affected by land-sea interaction and/or sea level controlled changes in local ocean circulation rather than large-scale patterns of climate change (Kiefer and Kienast, 2005).

The transect shows a contraction of the subtropical gyres, which is stronger in the southern than in the northern Pacific (Fig. 19(A)). In the southern Pacific, the subtropical front was shifted northward by ~8° latitude. Accordingly, the SST gradient across 26–33°S was considerably higher (see also Barrows and Juggins, 2005). This feature is consistently revealed by all methods, with the smallest SST changes reconstructed by ANN. The scatter between the methods is 3–4 °C across the shifting subtropical front, leaving some uncertainty about the actual SST gradient during the LGM. The northwestern Pacific subarctic gyre was cooler by only 1–4 °C during the LGM, as estimated within 2 °C by all methods, whereas a site at 42°S from near the Sub-Antarctic Front, where all methods agree

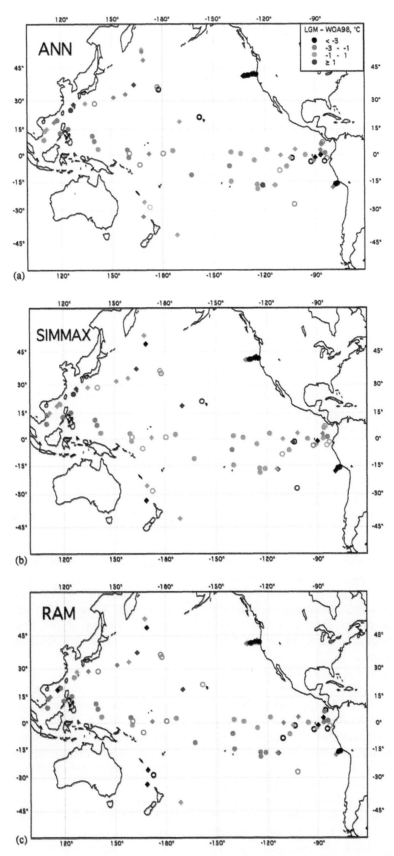

Fig. 17. Differences between annual LGM SST estimated by ANN (a), SIMMAX (b), and RAM (c) and modern SST (WOA 1998). Grey circles in (c) mark sites where RAM did not produce an LGM SST estimate because of a dissimilarity coefficient above 0.6 (see Section 3.2.3).

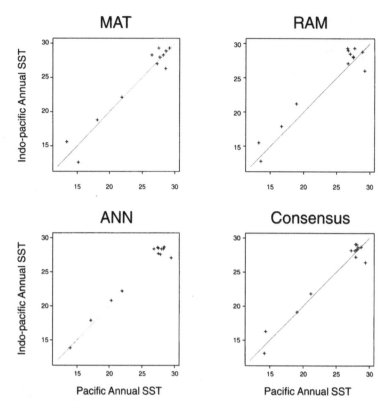

Fig. 18. Differences between Annual LGM SST for MAT (SIMMAX$_{ndw}$), RAM, ANN and a consensus estimate for 13 Pacific cores derived in this study and in Barrows and Juggins (2005).

within 1.6 °C was not much different from today (Fig. 19(A)). However, the LGM SST at this site might still be overestimated, given the paucity of calibration data from the high latitudes, and especially the northern subarctic Pacific (Fig. 16). SST in the northern subtropics were similar to today within 2 °C or even 1–3 °C warmer (site RC13-017 at 19°N). This also applies to a site east of the transect (Fig. 16(b); PC17 off Hawaii; Lee et al., 2001), which constitutes a key site for the subtropics because of its shallow water depth (503 m) and accordingly good preservation of foraminifera. Glacial SST estimates at PC17 are within 1 °C of the modern according to SIMMAX and RAM and 3 °C cooler according to ANN (Fig. 17). However, a low similarity coefficient (0.4, SIMMAX$_{ndw}$) and a relatively high standard deviation of 10 ANN estimates (1.35 °C) suggest a no-analog glacial assemblage, possibly reflecting the paucity of calibration samples from subtropical sediments.

The slight cooling of the subarctic northern Pacific and the persistently warm subtropics resulted in a steeper SST gradient across the subtropical front (Fig. 19(A)). However, LGM SST from 7 sites between 28° and 37°N across the Kuroshio extension are very similar to modern values, apart from some small-scale zonal variability. Hence, our results suggest no major changes in the path or temperature of the Kuroshio

current, in line with results by Kawahata and Ohshima (2002), although the Kuroshio was apparently weaker and cooler in its subtropical source region (Sawada and Handa, 1998; Ujiié, 2003). In the northwestern tropical Pacific at 8–11°N, annual LGM SST estimates from all techniques varied around 27 °C, i.e. consistently 2 °C cooler than today, suggesting a contracted glacial Western Pacific Warm Pool (Fig. 19(A)). Annual LGM SST between 27 and 29 °C suggest at least seasonal cooling of the equatorial western Pacific, a feature which is also resolved in the equatorial Pacific transect.

4.4.2. Equatorial Pacific E–W transect

An E–W transect along the equatorial Pacific between 10°N and 10°S is shown in Fig. 19(B). It illustrates the zonal extension of the Western Pacific Warm Pool, the intensity of the eastern equatorial upwelling, and more generally, the pattern of thermal E–W asymmetry across the equatorial Pacific associated with predominant La Niña and El Niño constellations. On a large scale, the LGM SST were cooler than today in the western and easternmost equatorial Pacific and similar to today in the central equatorial Pacific between ~180 and 105°W (Fig. 19(B)). In the western equatorial Pacific, average SST were mostly between 27 and 28 °C and hence tended to be cooler than modern by about 1 °C, suggesting a contraction of the Western Pacific Warm Pool. However, while RAM estimates an LGM cooling of up to 2.8 °C, SIMMAX suggests no change at all at most sites or even a slight warming of 0.8 °C (Figs. 17 and 19(B)). Further east, between 180° and 105°W, glacial annual SST gradually decreased eastwards from 28.5 to 25 °C, similar to today. Here, SST estimates of all four approaches approximately match modern SST within ±1 °C (Figs. 17 and 19(B)). Given a slightly cooler western equatorial Pacific, this means that the E–W SST gradient was lower during the LGM, pointing towards a more dominant El Niño constellation (stronger or more frequent El Niño events). East of 110°W, however, glacial SST were cooler than today, with small cooling by ~1 °C north of the equator, but as much as 8–11 °C south of the equator (Figs. 16(b) and 19(B)). In this area, the consistency among the methods is poor (2–5 °C) for these cold estimates, with SIMMAX tending towards extremely cold and ANN towards moderate glacial SST values (Fig. 19(B)). Nevertheless,

the thermal N–S asymmetry in the LGM reconstructions by all methods supports the idea of Feldberg and Mix (2002) that eastern equatorial Pacific SST were driven by equator-ward advection of particularly cool surface waters of the glacial Peru Current (Fig. 16), rather than by equatorial upwelling.

4.4.3. California Current transect

An E–W transect of 10 cores at 42°N across the California Current system (Ortiz et al., 1999) provides an example of small-scale glacial SST changes across an eastern boundary current (Fig. 16(a)). Furthermore, it allows comparison with earlier SST reconstructions from the identical faunal data but with different methods and calibration data sets. Ortiz et al. (1999), using an Imbrie–Kipp transfer function approach calibrated on a global database, found that LGM temperatures were uniformly ~3.3 °C cooler with the same SST gradient of 2 °C from colder near-coastal to warmer offshore waters as today. In contrast,

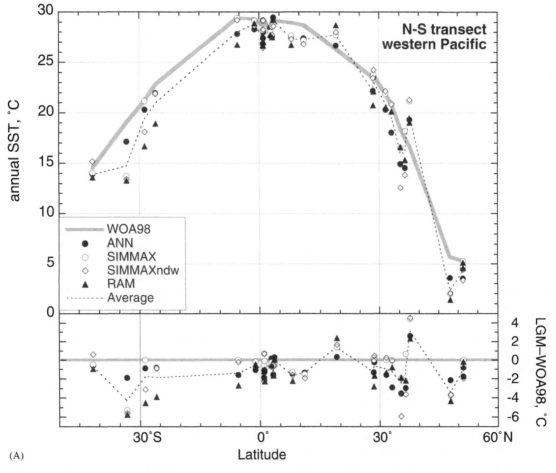

(A)

Fig. 19. Comparison between LGM and modern mean annual SST along the three transects indicated in Fig. 16. (A) North–South transect across the open-ocean western Pacific. (B) East–West transect across the equatorial Pacific. (C) East–West transect across the California Current at 42°N. LGM SST are shown as discrete estimates from ANN, SIMMAX, MAT (SIMMAX$_{ndw}$) and RAM, and an average from ANN, SIMMAX, and RAM (dashed line, in (A) and (B) only). Modern SST are 10-m water depth values from the World Ocean Atlas 1998 (WOA, 1998) from the same sites. Upper panels show absolute SST values, lower panels the differences between LGM and WOA (1998).

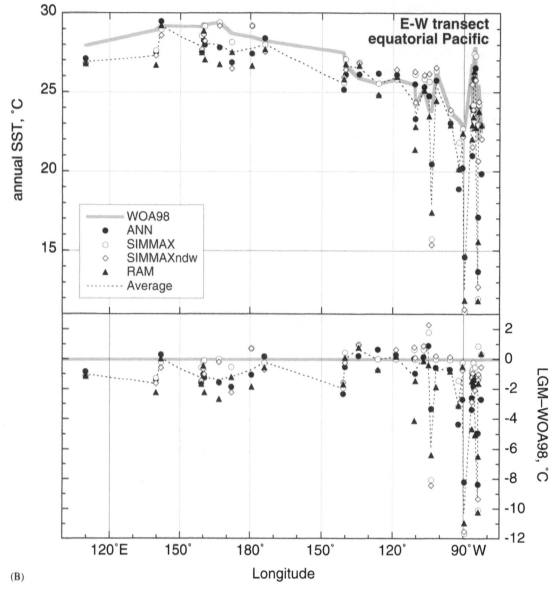

(B)

Fig. 19. *(Continued)*

Trend-Staid and Prell (2002) in their revision of the same faunal data using MAT and a different global calibration data set found greater cooling of 6 °C further offshore, resulting in a reduced E–W temperature gradient during the LGM. In contrast to both studies, our SST estimates are based only on Pacific calibration data and employ different techniques in parallel (Fig. 19(C)). Glacial SST calculated with ANN are cooler than today by about 3.4–4 °C without changing the offshore gradient significantly, thus generally supporting the results of Ortiz et al. (1999). SST estimates from SIMMAX and RAM result in a similar average cooling across the transect. However, in contrast to Ortiz et al. (1999) and Trend-Staid and Prell (2002), SIMMAX and RAM calculate a cooling of 4–5 °C near the coast but only less than 3 °C further offshore,

resulting in an increased E–W SST gradient of ~3.5 °C. Our results clearly demonstrate the influence of the choice of transfer function technique and calibration data set on detailed reconstructions of small-scale SST variability, and the value of the multi-technique approach in identifying the nature and magnitude of technique-specific bias in SST reconstructions.

5. Discussion

5.1. Reliability of LGM SST reconstructions

Fig. 20 shows how the conceptual model for assessment of reliability of SST estimates outlined in Section 2.3 (Fig. 1) applies to glacial Atlantic and Pacific SST

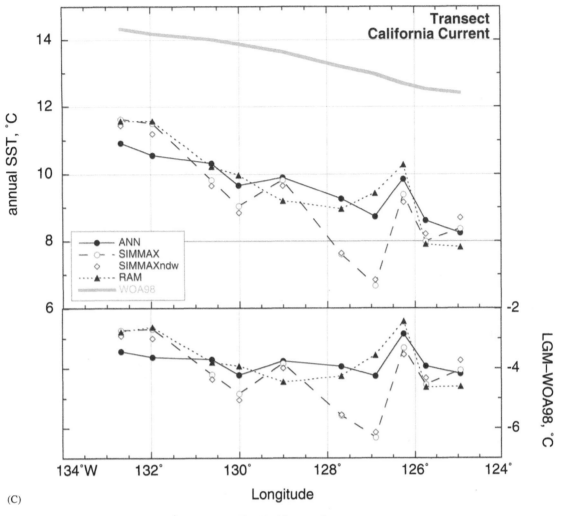

(C)

Fig. 19. (*Continued*)

reconstructions. The standard deviation of the 10 ANN SST estimates was <1 °C in >98% of the calibration data set samples. Since the distribution of prediction errors suggests that all data sets offer sufficient coverage of SST range and faunal variation (see Section 3.3.1, Fig. 6), we have adopted the 1 °C value as a threshold separating good-analog and poor-analog samples and the double of this value as a threshold for severe no-analog samples (Fig. 20). Based on these criteria, less than 5% of North Atlantic and Pacific samples were considered as poor analogs, whereas 13% of South Atlantic LGM samples fell into this category. Six Atlantic LGM samples were classified as severe no-analogs and excluded from further analysis (Section 4.2). The higher abundance of poor-analog samples in the South Atlantic most likely reflects the relatively poorer spatial coverage and size of this calibration data set (Fig. 2). High correlation between the standard deviation of 10 ANN estimates and SIMMAX similarity coefficients in the LGM samples (Fig. 20) suggests that both measures of sample representation in the calibra-

tion data set would yield similar reliability ratings (Fig. 21).

In the calibration data sets, SST estimates from ANN, RAM and SIMMAX diverged by <2 °C (measured as the standard deviation of the three estimates) in >99% of samples. We have adopted this value as an indicator of low or high degree of technique-specific bias in the SST estimates. Only <7% of North Atlantic and Pacific LGM samples were affected by technique specific bias, whereas >13% of South Atlantic LGM samples showed a significant divergence among the techniques (Fig. 20). The measures of sample representation and techniques divergence show a weak positive correlation ($r = 0.3$–0.4, $p < 0.001$) in all data sets (Fig. 20) suggesting a tendency for divergent estimates to occur in poorly represented samples. In general, it can be said that most of the LGM samples exhibit similar properties as samples in the calibration data sets and these LGM samples can be expected to yield robust SST estimates. The spatial distribution of these measures among LGM samples can be used to identify regionally consistent patterns that

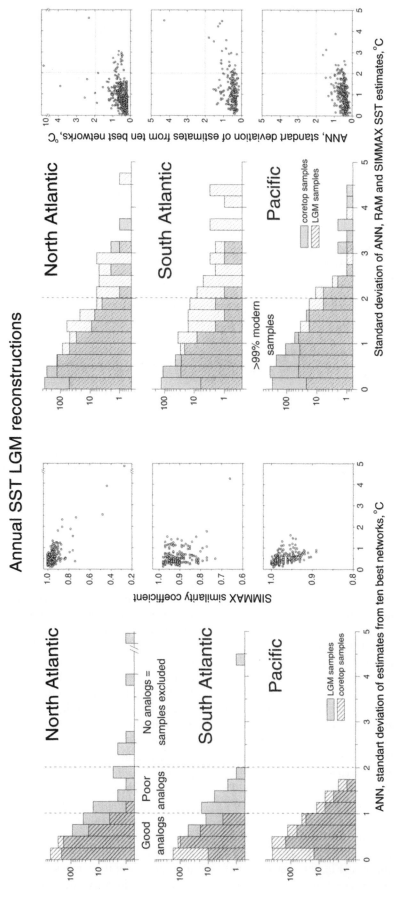

Fig. 20. The conceptual model for assessment of reliability of SST estimates (Section 2.3, Fig. 1) applied to glacial Atlantic and Pacific SST reconstructions. Only annual SST data are shown.

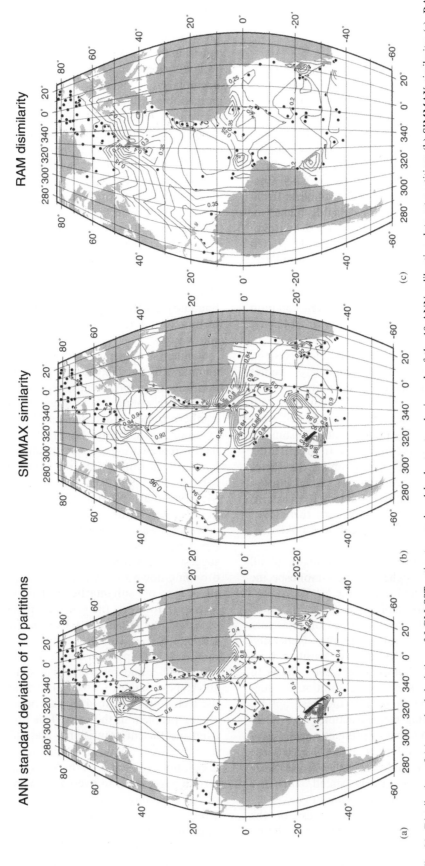

Fig. 21. Distribution of (a) standard deviation of LGM SST estimates produced by best networks in each of the 10 ANN calibration data set partition; (b) SIMMAX similarity; (c). RAM dissimilarity in glacial samples from the Atlantic Ocean.

may lead to systematic offsets or higher prediction errors.

5.2. Atlantic LGM SST patterns

As shown in Fig. 15, all methods produced fairly robust glacial Atlantic SST, both in absolute values and in distribution patterns, but there also distinct regional offsets. The estimates from the two variants of the modern analog technique (SIMMAX and RAM) are more closely related, while the ANN estimates appear more different. On average, the coldest temperatures were estimated by RAM and the warmest by ANN (Fig. 22). The overall deviations between estimates from different techniques were on average slightly higher in the south Atlantic (standard deviation of 1.02 °C) than in the north Atlantic (standard deviation of 0.76 °C) (Fig. 22). Comparing the distribution of dissimilarity and similarity coefficients and standard deviations of estimates produced by 10 individually trained ANNs reveals that the three techniques encountered common regional problems, providing a means to robustly identify no-analog situations (Fig. 22). This seems to apply in particular for the eastern boundary system, where poor dissimilarity and similarity values (<0.8; >0.2) and high standard deviations of 10 individually trained ANNs (0.8–1.6 °C) coincide with considerable offsets between estimates of different techniques. Values suggesting poor quality of estimates are also found at 20–30°N off Argentine. In the Atlantic, the regional distribution of offsets among the three methods (Fig. 22) reveals pronounced regional discrepancies in the very high latitudes and in the eastern boundary system in general. Moreover, there is a disagreement whether the subtropical south Atlantic gyre area was warmer than today.

While overestimation is a problem common to all transfer functions when applied on near-monospecific polar faunas at the "cold end" (i.e. >95% of N. pachyderma sin.), the offsets of as much as 2 °C must result from differential interpretation of the rare subpolar faunal elements by the techniques. Generally, ANN appears least- and RAM most sensitive (and sometimes oversensitive, Weinelt et al., 2001, 2003) to the latter. Perhaps the intermediate solution as suggested by SIMMAX values here is the best option. Considering the conflicting reconstructions from independent proxies, which all tend to produce considerably warmer than expected SST for the glacial Nordic Seas (de Vernal et al., 2005; Meland et al., 2005; Rosell-Melé et al., 2004), it must be clearly stated, that the foraminifer glacial SST estimates cannot be easily reconciled with any temperatures in excess of the reconstructed values. Moreover, the foraminifer-based SST reconstructions are consistent with glacial oxygen isotope values in the Nordic Seas (Weinelt et al., 1996;

Meland et al., 2005), suggesting a plausible pattern of the salinity/temperature field in the glacial Nordic seas.

The discrepancy marking the eastern boundary current off Northwest Africa where ANN produced consistently cooler SST than RAM and SIMMAX probably results from a general problem with upwelling assemblages. Glacial assemblages in this region are marked by large portions of G. bulloides, a species showing a bimodal (or multimodal) temperature preference. Enlisting modern subpolar analogs for the interpretation of glacial upwelling faunas at lower latitudes may therefore result in SST underestimation. Pflaumann et al. (2003) have pointed out that in an upwelling area off northwest Africa SIMMAX estimates clearly benefit from the distance weighting employed by this method, alleviating underestimation introduced by subpolar analogs. RAM partly solves the problem by reducing the number of selected analogs to 6–7, arriving at similar values as SIMMAX. The consensus estimate would thus appear to indicate that the northern eastern boundary cooled only moderately by 2–4 °C during the warm season and more strongly by 2 to 6 °C during the cold season.

A much stronger cooling is found for the southern boundary Benguela Current. Pflaumann et al. (2003) ascribed the strong hemispheric asymmetry in eastern boundary cooling to a larger strengthening of the southern trade wind system causing a stronger increase in upwelling intensity. Based on the distribution of a factor representing eastern boundary assemblage extending in a tongue into the south equatorial current, Mix et al. (1999) lend support to the suggestion by McIntyre et al. (1989) that advection of subpolar waters in the eastern boundary currents rather than intensified upwelling was responsible for tropical cooling. Our reconstructions based on the South Atlantic calibration suggests even stronger cooling by 5 to >10 °C with SST dropping down to 5 °C, also pointing towards a strong advection of subantartic water into the Benguela Current. However, the generally patchy pattern of SST estimates in the Benguela Current (neighbouring cores yielding SST values differing by as much as 5 °C) combined with poor dissimilarity and similarity values (<0.8; >0.2) and relatively high standard deviations of 10 ANN estimates (>0.8 °C) and large offsets among the techniques (up to 4 °C), all point to a marked no-analog problem in this area. This problem likely results from insufficient transitional and subantarctic analogs in the South Atlantic database due to carbonate dissolution. Whereas both ANN and RAM are to a certain degree able to overcome the problem, SIMMAX is strictly bound to the given range of calibration samples. In particular, with a poor coverage of calibration data the benefits of distance weighting are not realised and this method may even be disadvantageous (Pflaumann et al., 1996). Thus, SIMMAX is likely

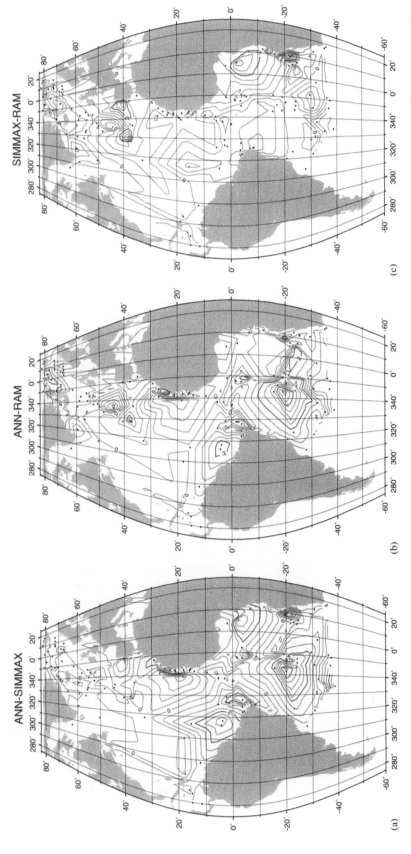

Fig. 22. Distribution of differences between LGM SST reconstructions by individual techniques in the Atlantic. Enhanced contours mark areas with maximum offsets. (a) ANN–SIMMAX; (b) ANN–RAM; (c) SIMMAX–RAM.

to produce the most biased estimates in this region, as also suggested from its performance in the calibration database. The severe no-analog problems in the Benguela Current region may be further aggravated by the high temporal SST variability during the LGM (Fig. 23). An alternative hypothesis for this no-analog situation might be that Benguela Current faunas are partly advected from the Indian Ocean, a hypothesis to be tested by extending the south Atlantic calibration into the Indian Ocean. We conclude that in this particular case, our reconstruction may suffer from the lack of analogs in the South Atlantic data set, and a consensus Benguela Current reconstruction should be nearer the colder end of the range of our glacial SST estimates.

Foraminifera-based reconstructions generally agree that the subtropical gyre areas in either hemisphere were subject to little glacial change (CLIMAP Project Members, 1976; Mix et al., 1999; GLAMAP-2000). The pronounced warm anomaly in the central South Atlantic as produced by ANN suggesting glacial SST of

27–29 °C is not supported by RAM and SIMMAX estimates, both suggesting cold season SST of 20° to 23° and warm season SST of 22° to 24°, matching modern values. The standard deviation of 10 partitions of ANN in these samples is low (< 0.6 °C), SIMMAX similarity is > 0.9 and RAM dissimilarity < 0.2, suggesting that the fauna in these samples has good representation in the calibration data set. Interestingly, a (less pronounced + 1 to 2 °C) warm anomaly in the South Atlantic has previously been reconstructed by CLIMAP Project Members (1976, 1981) based on the Imbrie–Kipp technique. This issue clearly calls for independent evidence by geochemical SST proxies, although at present no Mg/Ca or U^{k}_{37} data are available for cores from the deep basins of the South Atlantic to support either scenario. If real, such a warm anomaly may be a further expression of a weakened glacial thermohaline circulation retaining heat in the South Atlantic and reducing cross-equatorial heat flow as predicted by climate models (Stocker, 1998; Seidov and Maslin, 2001) and observed for short term D–O variability, where data display strong interhemispheric asymmetry (Charles et al., 1996; Blunier et al., 1998; Rühlemann et al., 1999).

Robust results include the excellent agreement of SST estimates in the mid-to-high latitude north Atlantic (50–60°N) where ANN, SIMMAX, and RAM jointly report a strong cooling by up to 11 °C (Fig. 21) and the steep temperature gradient at mid-latitudes. Good similarity/dissimilarity and ANN standard deviation values indicate a good representation of these faunas in the calibration data set. This consensus underlines the value of foraminifer based SST reconstructions in the mid-latitude North Atlantic. The same holds true for the consensus on the magnitude of tropical cooling by 4–6 °C in the tropical eastern Atlantic supported by all three techniques, which confirms the results by GLA-MAP-2000 (Sarnthein et al., 2003a), and provides further evidence for an agreement with the continental record, suggesting distinct cooling and high lapse rates in the glacial tropics (e.g. Farrera et al., 1999; Pinot et al., 1999).

In summary, our analyses reinforce the robustness of state-of-the-art foraminifer SST reconstruction techniques. Glacial SST patterns observed in the South Atlantic, suggest a more pronounced interhemispheric asymmetry with a very cool Benguela Current on the one hand and perhaps a warm anomaly in the central South Atlantic. No significant change was found for the North Atlantic as compared to the GLAMAP-2000 reconstruction (Sarnthein et al., 2003a).

5.3. Pacific LGM SST patterns

Using the degree of convergence between SST estimates from the three transfer function techniques as an indicator of robustness of SST reconstructions,

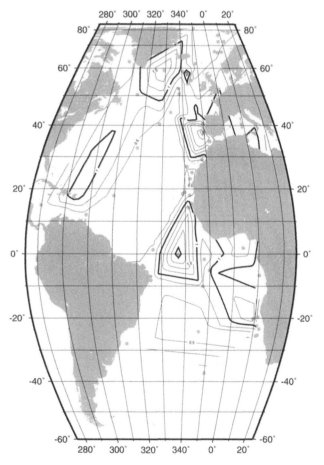

Fig. 23. Temporal LGM SST variability, expressed as averages of ANN, SIMMAX and RAM standard deviations of SST estimates in the 23–19 ky EPILOG LGM interval (Mix et al., 2001) Only records with 4 and more samples are considered. Contours mark areas with standard deviations > 1 °C.

one can attempt to identify consistent glacial SST features in some areas of the Pacific Ocean and assess the degree of uncertainty in others. Our results indicate that the consistency of absolute SST reconstructions between techniques and hence the robustness of glacial SST patterns is high in the central and western tropical Pacific and the eastern boundary currents but rather low in the subtropical gyres of both hemispheres and in the eastern equatorial Pacific (EEP) (Figs. 16 and 19(A–C)). However, when the LGM climate is depicted as anomalies relative to the present-day, the usefulness of reconstructions in the EEP is increased by the consistently large size of the negative anomalies in this area, while the significance of the SST reconstructions appears somewhat lower in the western equatorial Pacific due to the small LGM-modern difference.

A broadly consistent LGM SST trend is recorded in the tropical Pacific. The three techniques generally agree on a cooler LGM in the western and the easternmost Pacific and on a similar to modern SST in the central equatorial Pacific (Fig. 19(B)). Whilst a high degree of convergence among SST estimates from the three methods in the central equatorial Pacific provides sound evidence for LGM SST being similar to modern in this area, some uncertainty remains with respect to the extension of this zone of constant SST towards both the western and eastern Pacific (Fig. 17). RAM, and to a less extent also ANN, suggest major large-scale cooling at both ends of the equatorial Pacific, while the systematically higher SIMMAX SST estimates indicate only a minor LGM cooling in restricted regions. For example, although general agreement exists on some cooling in the western Pacific, the methods disagree if LGM cooling caused only a minor contraction of the Western Pacific Warm Pool towards the open Pacific (SIMMAX; also see Barrows and Juggins, 2005) or an additional general cooling of the Warm Pool by ~1–2 °C (ANN, RAM) as far east as 180°E. In the eastern equatorial Pacific the methods also disagree in the magnitude and extent of the cooling. SIMMAX estimates indicate a <1 °C cooling in the easternmost Pacific east of 95°W (apart from some extremely cold samples). On the contrary, RAM reconstructs a heterogeneous pattern of cooling by several degrees as far west as 115°W, approximately over the area of the cold tongue of eastern equatorial upwelling (Fig. 17). Mg/Ca records from the open-ocean western Pacific indicate LGM SST generally cooler than modern by 1–3 °C (Lea et al., 2000; Stott et al., 2002; Palmer and Pearson, 2003; Rosenthal et al., 2003) supporting the cooling scenario. Independent SST evidence from only one Mg/Ca record from the EEP (Koutavas et al., 2002) indicates a slightly (~1 °C) cooler LGM. This evidence is certainly insufficient to favour one specific scenario. However, in analogy to the western Pacific it appears most likely, that the actual LGM SST was cooler than the conservative SIMMAX estimate.

A major issue in the equatorial Pacific transect concerns the asymmetry associated with changing predominance of La Niña or El Niño conditions, which should result in an increased or decreased average thermal E–W gradient (Clement and Cane, 1999; Rosenthal and Broccoli, 2004). Our results indicating cooler LGM SST at both ends (eastern and western) of the equatorial Pacific do not tilt the SST gradient towards either side and therefore do match neither an intensified (stronger and/or more frequent) El Niño nor an intensified La Niña scenario. It might be, however, that the foraminifer assemblages over-represent times (seasons or anomalous years) of upwelling and/or shallow thermocline and associated increased productivity and export flux of foraminifer shells. If this were the case, one could argue that the ANN and RAM glacial SST estimates are biased towards cold temperatures at sites of high inter-annual or seasonal SST variability. If, on the other hand, the glacial SST pattern inferred from foraminifer assemblages holds true, the reconstructed LGM cooling of the eastern and western Pacific would support a general intensification of El Niño/La Niña oscillations, where the equatorial Pacific was cooled in the west during El Niño and in the east during La Niña. However, an intensified glacial El Niño/La Niña oscillation would oppose evidence from corals, which indicate weaker El Niño during glacials (Tudhope et al., 2001).

The foraminifer data may be regarded more conservatively as reflecting a general cooling of the eastern and western equatorial Pacific without invoking any El Niño analogy (Rosenthal and Broccoli, 2004). Cooling of the EEP could partly be the consequence of equatorward advection of particularly cool surface waters from the glacial Peru Current as suggested by Feldberg and Mix (2002). Evidence for that is the N–S asymmetry of SST in the eastern equatorial Pacific, with a small cooling of ~1 °C north of the equator, increasing to as much as 8–11 °C further south (Fig. 16(b)). Furthermore, stronger LGM trade winds in response to a steepened meridional SST gradient might have increased equatorial upwelling and cooling of the eastern equatorial Pacific (Fig. 16). Such steepened SST gradients would require an equatorward shift of the subtropical frontal systems. No large change is recorded for the northern Pacific subtropical front (Fig. 19(A)). In the southwestern Pacific, on the other hand, a possible northward shift of the subtropical front by some 8° latitude is inferred from average annual SST (Fig. 19). The contraction of the southern subtropical gyre was small according to ANN reconstructions, but pronounced according to RAM with a presumably major effect on trade wind strength.

Another robust observation supported by all three transfer function techniques is a major cooling of the two eastern boundary currents (California Current and Peru Current) by 2.5–6 °C (Figs. 16(b) and 17). Cooling of the Peru Current at the three sites off Peru where the cooling is documented may be associated with an increased upwelling of cold intermediate water in response to increased trade wind strength. This explanation is supported by ANN results showing least cooling further offshore and consistent with the previously inferred contraction of the southern subtropical gyre. In contrast, the uniform cooling across the California Current off Oregon as shown by ANN data (Fig. 19(C)) was ascribed to open ocean upwelling driven by higher wind stress curl (Ortiz and Mix, 1997). However, if the additional near coastal cooling suggested by RAM and especially SIMMAX is correct (Fig. 19(C)), then some increased coastal upwelling may be inferred to explain the increased LGM SST gradient across the California Current.

Poor regional sample coverage and divergence of SST reconstructions leave some open questions. High subarctic and subantarctic latitudes are clearly under-represented, especially in the eastern Pacific (Fig. 16(a)). Furthermore, no consensus can be reached about the LGM SST of the northern subtropical gyre, in spite of the analysis of two samples from the central gyre (Fig. 16(b)). Taken together, the two samples cover an LGM SST range between 22.2 and 28.7 °C, equivalent to a 3 °C colder or 2.7 °C warmer LGM. However, a multi-proxy reconstruction done on one of these cores suggests that the subtropical gyre was 2.5 °C cooler during winter and 1 °C cooler during summer (Lee et al., 2001). If the northern subtropics were in fact cooler (and if the warm LGM results from the Okinawa Trough were ascribed to a local sea-level related effect), then one general conclusion of this study is that the surface waters were not significantly warmer than at present anywhere in the Pacific Ocean during the LGM (Figs. 16(b) and 17).

5.4. Sea-ice distribution in the Nordic Seas

One of the incompletely resolved questions in glacial high-latitude reconstructions concerns sea-ice distribution. Its large impact on the climate system via albedo effects, oceanic atmospheric heat exchange and productivity calls for a fresh look at this issue. Even in the modern ocean, sea-ice is a highly variable parameter, its distribution controlled not only by the SST regime, but strongly affected by wind and current regimes. According to satellite observations between 1978 and 1987 (Gloersen et al., 1992) sea-ice concentrations of >50% coverage occurred from freezing temperatures (−1.7 °C) to SST values 2 °C above freezing. In addition, transfer functions derived from planktonic foraminifer assem-

blages generally tend to overestimate SST at the "cold end", making it impossible to deduce sea-ice extent in the past directly from SST estimates. Neither is it possible to derive sea-ice distribution directly from the faunal patterns, because virtually monospecific assemblages of the polar species N. pachyderma (sin.) may occur in seasonally ice-covered areas (e.g. East Greenland Current) as well as in the almost perennially ice-covered central Arctic Ocean. Sarnthein et al. (2003b) adopted an indirect approach identifying a threshold value of 2.5 °C in SIMMAX SST estimates in the modern Arctic ocean below which sea-ice may occur and above which sea-ice may be widely excluded. Estimates >3 °C were considered as perennially ice-free. Winter sea-ice occurred at winter SST estimates below 0.4 °C when comparing with the data of Gloersen et al. (1992), and below 0.7 °C when considering more severe winter conditions of the Little ice age (Kellogg, 1980). A maximum summer ice edge for the LGM reconstructed following these criteria was found to be delimiting an ice-free area south of 75°N at Fram Strait, whereas the Arctic ocean was perennially ice covered. Only during coldest glacial summers did sea-ice also expand along the Greenland margin down to Scorseby Sund (72°N) and along the northern Norwegian margin (67°N) (Weinelt et al., 2003). During glacial winter, sea-ice extended south to approximately 60°N, following a line from Southern Norway to south of Iceland, to the southern tip of Greenland, and towards New Found-land delimiting Labrador Sea. This finding coincides well with results based on dinocyst transfer functions (de Vernal and Hillaire-Marcel, 2000). Ice-free conditions in the central and eastern Nordic seas combined with the occurrence of high $\delta^{18}O$ values of 4.7‰, suggest high-density surface water in the central Nordic Seas and thus ongoing influence of saline Atlantic surface water maintaining convection and deep-water export (Weinelt et al., 1996).

However, it may be argued that the distinct north-to-south SST gradient produced by SIMMAX partly results from the distance weighting procedure where SIMMAX emphasises the weight of nearby analog samples rather than extracting the subtle information provided by the faunas themselves. Thus, temperatures <2.5 °C (indeed only sparsely represented in the calibration data set by 17 core tops), would be reconstructed more likely in areas with similarly cold modern SST associated with sea-ice, but outside the modern range the presence of sea-ice is less likely to be identified.

In order to find a more robust solution, we have explored the potential of ANN and RAM, both independent of geographical weighting, to similarly identify such sea-ice SST thresholds. The comparison of estimated and measured SST in Fig. 24 shows that RAM indeed estimated SST of most modern faunas related to sea-ice colder than 2 °C (Fig. 24). Still, for quite a few warmer samples cold SST estimates were produced. A less conclusive pattern was found for ANN, which generally tends to overestimate SST below 4 °C and does not generate reconstructions colder than 2.5–3 °C. This appears to be a result of ANN relying solely on the abundance of *N. pachyderma* (sin.), regardless of the composition of the small subpolar fraction.

As shown in Fig. 15 (compare also Fig. 22) the Nordic Seas are one of the regions where LGM estimates by the three techniques notably diverge, ANN yielding hardly any SST gradients with estimates ranging from 2.8 to 3.2 °C (5 °C in one exception), RAM yielding a large SST range from −1.8 to 3 °C, and SIMMAX estimates ranging from −0.3 to 4.0 °C, with the coldest samples indeed concentrating geographically north of 75°N. In detail, RAM SST estimates show a pattern (Fig. 22) where coldest estimates (below 2 °C) concentrate in the Arctic ocean north of 74° (here fitting the SIMMAX results) and along the Greenland margin (not reflected in the SIMMAX estimates), warmer samples consis-

tently occur in the eastern Norwegian Sea, again roughly coinciding with the SIMMAX pattern. Interestingly, the only ANN estimates falling slightly below 3 °C are consistently estimated for monospecific *N. pachyderma* (sin.) faunas from the Greenland margin. In contrast, ANN estimates provide no clue for sea-ice at Fram Strait.

The discrepancies produced by the three different techniques may be partly due to differential interpretation of small "subpolar" faunal portions, obviously lacking a clear relationship with SST. A possible cause for a biased (non-linear) faunal–temperature relationships may be the circumstance that rare morphological *N. pachyderma* (dex.) from polar environments in Fram Strait and Greenland Sea have been recently genetically identified as *N. pachyderma* (sin.) (Bauch et al., 2003). To test this hypothesis we mapped the portions of "true" subpolar species, which include *T. quinqueloba*, *G. bulloides* and *G. glutinata*. These species are strictly bound to seasonally open conditions guaranteeing food and light supply, in particular the symbiont-bearing *T. quinqueloba*. Thus, presence of even small portions of

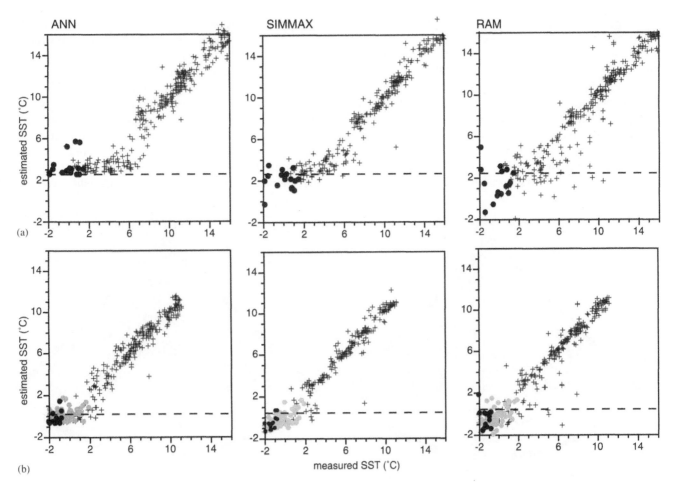

Fig. 24. Measured versus estimated SST with samples occurring beneath sea-ice highlighted: (a) in summer, (b) in winter. Black dots are samples below sea-ice according to Gloersen et al. (1992), grey dots according to severe Little Ice Age conditions (Kellogg, 1980). Stippled lines mark threshold SST values defined by Sarnthein et al. (2003b).

these "true" subpolar species should indicate ice-free conditions. This is the case for instance for modern surface water in the Arctic Domain, ice free during summer, where subpolar species occur at negligible abundances (0.3–1%), whereas in the perennially ice-covered regions "true" subpolar species are totally absent. In the LGM reconstruction a pattern emerges, where subpolar species occurring at abundances > 1% are limited to the Norwegian Sea, where they reach maximum abundances of 8.7% (Fig. 25(a)) in single samples and 3.4% in core averages (Fig. 25(b)), thus resembling contemporaneous faunas living south of Iceland and west of Ireland. Absence of subpolar species marks the East Greenland margin. An area with very rare subpolar species (<1%) covers the central Nordic Seas and extends into the eastern Fram Strait. From this more meridional pattern as compared to the GLAMAP-2000 reconstruction (Sarnthein et al., 2003b), summer sea-ice in the Norwegian Sea may be safely ruled out. Based on the lack of "true" subpolar species, and further supported by the RAM results, we regard it as likely that LGM sea-ice extended also along the Greenland margin. Very low portions of "true" subpolar species in the central Nordic Seas up to Fram Strait perhaps indicate Arctic Domain-like watermass, which may be occasionally ice-covered.

The more meridional pattern resembles the modern situation, with cold and ice-covered polar water to the west and is in harmony with the LGM reconstruction by Meland et al. (2005) based on $\delta^{18}O$, which implies close-to-freezing temperatures off Greenland by assuming constant salinity. Winter estimates of high-latitude environments are generally questionable, because growth of planktonic foraminifera here is strictly limited to summer months (Carstens and Wefer, 1992; Simstich

et al., 2003). Winter estimates are mainly reflecting the seasonality pattern occurring today at the modern analog sites. Nevertheless, for glacial high-latitude winter all three techniques produced a very consistent pattern with estimates ranging from −1.5 to <1 °C suggesting sea-ice covering the Nordic Seas down to 60°N (based on either winter sea-ice scenario), a narrow lobe even extending into the central northern Atlantic (Figs. 13 and 14).

5.5. Temporal variability of LGM SST

The high per-core sample density in the Atlantic (5 samples per core in the North Atlantic; 3 samples per core in the South Atlantic) allows us to explore the millenial-scale variability during the LGM time interval. Though generally considered a stable climatic end-member, high-resolution SST records document considerable internal variability during the LGM. Based on a high-resolution SST transect from the northern North Atlantic, Weinelt et al. (2003) have shown that Dansgaard-Oeschger SST variability continued south of the Iceland–Faroe Ridge throughout the LGM, whereas the Nordic Seas remained remarkably stable. Following a suggestion by the EPILOG workshop (Mix et al., 1999) we use the standard deviation of downcore SST estimates as a measure of LGM SST variability, using only sites where the LGM interval is covered by 4 or more samples (i.e. corresponding to a time resolution equal or better than 1000 years). Interestingly, the standard deviations were in general of the same magnitude in the South and North Atlantic (0.75 °C) and no systematic deviations were found between individual transfer function techniques. Fig. 23 depicts the regional patterns of temporal variability averaged

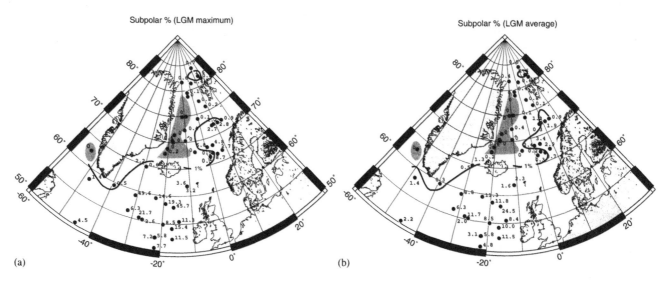

Fig. 25. Distribution of "true" subpolar species in the LGM faunas of the Nordic Seas (including *T. quinqueloba*, *G. bulloides*, *G. glutinata* but excluding *N. pachyderma* dex.). Contour marks 1% abundance limit suggesting seasonally ice-free conditions. Shaded area marks zero occurrences, suggesting perennial ice cover. (a) Maximum values of LGM samples. (b) Averaged values of LGM samples per core.

from ANN, RAM, and SIMMAX. Accordingly, the highest LGM SST variability (standard deviations > 1 °C) marks indeed the zone south of the Iceland–Faroe Ridge, the eastern boundary system in general, and the Iberian margin in particular (1–2.8 °C). Some variation is observed in the northwest and southwest African boundary currents (0.5–1.5 °C), and the south equatorial current is marked by high SST variability. This pattern suggests a scenario with highly dynamic eastern boundary currents marked by alternating transport of cool water towards low latitudes.

6. Conclusions

- In an attempt to minimise bias associated with the presence of cryptic species of planktonic foraminifera and to provide an objective tool for assessing reliability of environmental estimates in fossil samples, we have developed the conceptual framework for a multi-technique approach of environmental calibration of planktonic foraminifer census counts in geographically constrained calibration data sets. Using calibration data sets from the Atlantic and Pacific Oceans, we demonstrate the merit of this approach and conclude that it should be considered for any future environmental reconstructions based on planktonic foraminifera.

- A new SST reconstruction of the LGM Atlantic Ocean based on updated calibration and coretop databases and the application of the regionalised, multi-technique approach reveals a number of robust patterns, including the magnitude of cooling in the subpolar North Atlantic by more than 10 °C, steep SST gradients in the mid-latitude North Atlantic, and pronounced cooling of the equatorial region by 4–6 °C. Substantially colder glacial eastern boundary currents suggest considerable advection of subpolar waters into the Benguela Current during the LGM whereas subtropical gyre areas in both hemispheres did not change significantly since the LGM.

- A multi-technique approach allowed the identification of areas where SST reconstructions may be subject to larger uncertainties, as reflected in larger offsets among SST reconstructions produced by different techniques exceeding the prediction errors defined on the calibration data set. This is in particular the case for samples from the eastern boundary systems and from the South Atlantic subtropical gyre.

- Some disagreement also remains for the Nordic Seas, where the different techniques show differential sensitivity to low-diversity polar faunas. After revisiting the issue of sea-ice occurrence in the Nordic Seas, we find that the distribution of subpolar species of planktonic foraminifera in glacial samples yields robust support that the Norwegian Sea must have remained ice-free during glacial summers.

- In the Pacific LGM reconstruction, SST estimates by all three techniques applied clearly indicate a substantial cooling of the eastern boundary currents (Chile Current and California Current) and suggest a general contraction of the subtropical gyres which was more distinct in the southern than in the northern Pacific. In the tropical Pacific, a contraction of the Western Pacific Warm Pool and a cooling of the eastern equatorial Pacific are evident, whilst the central equatorial Pacific was not much different from today. The apparent cooling pattern of the tropical Pacific during the LGM may be related to an increased amplitude or frequency of the El Niño Southern Oscillation. Analyses of foraminiferal assemblages do not provide compelling evidence for a warmer-than-today glacial sea surface anywhere in the Pacific.

Acknowledgements

This research has been supported by grants from The Leverhulme Trust (MK, AH), Nuffield Foundation (MK), the UK Natural Environmental Research Council (MK, TK) and by the International Marine Past Global Changes Study (IMAGES) program. We thank James Casford and two anonymous referees for constructive criticism of an earlier draft of this paper.

References

Barrows, T.T, Juggins, S., 2005. Sea-surface temperatures around the Australian margin and Indian Ocean during the last glacial maximum. Quaternary Science Reviews, this issue (doi:10.1016/j.quascirev.2004.07.020).

Barrows, T.T., Juggins, S., De Deckker, P., Thiede, J., Martinez, J.I., 2000. Sea-surface temperatures of the southwest Pacific Ocean during the Last Glacial Maximum. Paleoceanography 15, 95–109.

Bauch, D., Darling, K.F., Simstich, J., Bauch, H.A., Erlenkeuser, H., Kroon, D., 2003. Palaeoceanographic implications of genetic variation in living North Atlantic Neogloboquadrina pachyderma. Nature 424, 299–302.

Berger, W.H., 1968. Planktonic foraminifera: Selective solution and paleoclimatic interpretation. Deep-Sea Research 15, 31–43.

Berger, W.H., Winterer, E.L., 1974. Plate stratigraphy and the fluctuating carbonate line. International Association of Sedimentologists (Special Publication) 1, 11–48.

Birks, H.J.B., 1995. Quantitative palaeoenvironmental reconstructions. In: Maddy, D., Brew, J.S. (Eds.), Statistical Modelling of Quaternary Science Data. Technical Guide 5. Quaternary Research Association, Cambridge, pp. 161–254.

Blunier, T., Chappellaz, J., Schwander, J., Dällenbach, A., Stauffer, B., Stocker, T.F., Raynaud, D., Jouzel, J., Clausen, H.B., Hammer, C.U., Johnsen, S.J., 1998. Asynchrony of Antarctic and Greenland climate change during the last glacial period. Nature 394, 739–743.

Bollmann, J., Henderiks, J., Brabec, B., 2002. Global calibration of *Gephyrocapsa* coccolith abundance in Holocene sediments for paleotemperature assessment. Paleoceanography 17, doi:10.1029/2001PA000742.

Carstens, J., Wefer, G., 1992. Recent distribution of planktonic foraminifera in the Nansen Basin, Arctic Ocean. Deep-Sea Research 39, 507–524.

Charles, C.D., Lynch-Stieglitz, J., Ninnemann, U.S., Fairbanks, R.G., 1996. Climate connections between the hemisphere revealed by deep sea sediment core/ice core correlations. Earth and Planetary Science Letters 142, 19–27.

Chen, M.-T., Ho, H.-W., Lai, T.-D., Zheng, L., Miao, Q., Shea, K.-S., Chen, M.-P., Wang, P., Wei, K.-Y., Huang, C.-Y., 1998. Recent planktonic foraminifers and their relationships to surface ocean hydrography of the South China Sea. Marine Geology 146, 173–190.

Chen, M.-T., Huang, C.-C., Pflaumann, U., Waelbroeck, C., Kucera. M., 2005. Estimating glacial western Pacific sea-surface temperature: methodological overview and data compilation of surface sediment planktic foraminifer faunas. Quaternary Science Review, this issue (doi:10.1016/j.quascirev.2004.07.013).

Clement, A.C., Cane, M., 1999. A role for the tropical Pacific coupled ocean–atmosphere system on milankovitch and millennial time-scales. Part I: a modeling study of tropical Pacific variability. In: Clark, P.U., Webb, R.S., Keigwin, L.D. (Eds.), Mechanisms of Global Climate Change at Millennial Time Scales. Geophys. Monog. American Geophysical Union, Washington, DC, pp. 363–371.

CLIMAP Project Members, 1976. The surface of the ice-age Earth. Science 191, 1131–1137.

CLIMAP Project Members, 1981. Seasonal Reconstructions of the Earth's Surface at the Last Glacial Maximum. Geol. Soc. Am. Map and Chart Ser., MC-36, Geol. Soc. of Am., Boulder, Colorado.

Coulbourn, W.T., Parker, F.L., Berger, W.H., 1980. Faunal and solution patterns of planktonic foraminifera in surface sediments of the North Pacific. Marine Micropaleontology 5, 329–399.

Darling, K.F., Wade, C.M., Stewart, I.A., Kroon, D., Dingle, R., Leigh Brown, A.J., 2000. Molecular evidence for genetic mixing of Arctic and Antarctic subpolar populations of planktonic foraminifers. Nature 405, 43–47.

Darling, K.F., Kucera, M., Wade, C., von Langen, P., Pak, D., 2003. Seasonal distribution of genetic types of planktonic foraminifer morphospecies in the Santa Barbara Channel and its paleoceanographic implications. Paleoceanography 18, doi:10.1029/2001PA000723.

Darling, K.F., Kucera, M., Pudsey, C., Wade, C., 2004. Molecular evidence links cryptic diversification in polar plankton to Quaternary climate dynamics. Proceedings of the National Academy of Sciences USA 101, 7657–7662.

de Vargas, C., Norris, R., Zaninetti, L., Gibb, S.W., Pawlowski, J., 1999. Molecular evidence of cryptic speciation in planktonic foraminifers and their relation to oceanic provinces. Proceedings of the National Academy of Sciences USA 96, 2864–2868.

de Vargas, C., Renaud, S., Hilbrecht, H., Pawlovski, J., 2001. Pleistocene adaptive radiation in *Globorotalia truncatulinoides*: genetic, morphologic, and environmental evidence. Paleobiology 27, 104–25.

de Vernal, A., Hillaire-Marcel, C., 2000. Sea-ice cover, sea-surface and halo-thermocline structure of the northwest North Atlantic: modern versus full glacial conditions. Quaternary Science Review 19, 65–85.

de Vernal, A., Eynaud, F., Henry, M., Hillaire-Marcel, C., Londeix, L., Mangin, S., Matthiessen, J., Marret, F., Radi, T., Rochon, A., Solignac, S., Turon, J.-L., 2005. Reconstruction of sea-surface conditions at middle to high latitudes of the Northern Hemisphere during the Last Glacial Maximum (LGM) based on dinoflagellate cyst assemblages. Quaternary Science Review, this issue (doi:10.1016/j.quascirev.2004.06.014).

Farrera, I., Harrison, S.P., Prentice, I.C., Ramstein, G., Guiot, J., Bartlein, P.J., Bonnefille, R., Bush, M., Cramer, W.v., Grafenstein, U., Holmgren, K., Hooghiemstra, H., Hope, G., Jolly, D., Lauritzen, S.-E., Ono, Y., Pinot, S., Stute, M., Yu, G., 1999. Tropical climates at the Last Glacial Maximum: a new synthesis of terrestrial paleoclimate data. I. Vegetation, lake levels, and geochemistry. Climate Dynamics 15, 823–856.

Feldberg, M.J., Mix, A.C., 2002. Sea-surface temperature estimates in the Southeast Pacific based on planktonic foraminiferal species: modern calibration and last glacial maximum. Marine Micropaleontology 44, 1–29.

Gersonde, R., Abelmann, A., Brathauer, U., Becquey, S., Bianchi, C., Cortese, G., Grobe, H., Kuhn, G., Niebler, H.-S., Segl, M., Sieger, R., Zielinski, U., Fütterer, D. K., 2003. Last glacial sea surface temperatures and sea-ice extent in the Southern Ocean (Atlantic-Indian sector): a multiproxy approach. Paleoceanography 18, doi:10.1029/2002PA000809.

Gloersen, P., Campbell, W. J., Cavalieri, D.J., Comiso, J.C., Parkinson, C.L., Zwally, H.J., 1992. Arctic and Antarctic sea ice, 1978–1987: satellite passive-microwave observations and analysis, NASA SP-511, Sci. and Technol. Inf. Progr., NASA, 290pp.

Hayes, A., Kucera, M., Kallel, N., Sbaffi, L., Rohling, E.J., 2005. Glacial Mediterranean sea surface temperatures based on planktonic foraminiferal assemblages. Quaternary Science Reviews, this issue (doi:10.1016/j.quascirev.2004.02.018).

Henderson, G.M., 2002. New oceanic proxies for paleoclimate. Earth and Planetary Science Letters 203, 1–13.

Huber, B.T., Bijma, J., Darling, K.F., 1997. Cryptic speciation in the living planktonic foraminifer *Globigerinella siphonifera* (d'Orbigny). Paleobiology 23, 33–62.

Hutson, W.H., 1977. Transfer functions under no-analog conditions: experiments with Indian Ocean planktonic foraminifera. Quaternary Research 8, 355–367.

Hutson, W.H., 1980. The Agulhas current during the late Pleistocene: analysis of modern faunal analogs. Science 207, 64–66.

Imbrie, J., Kipp, N.G., 1971. A new micropaleontological method for quantitative paleoclimatology: application to a late Pleistocene Caribbean core. In: Turekian, K.K. (Ed.), The Late Cenozoic Glacial Ages. Yale University Press, New Haven, pp. 71–181.

Kawahata, H., Ohshima, H., 2002. Small latitudinal shift in the Kuroshio Extension (Central Pacific) during glacial times: evidence from pollen transport. Quaternary Science Reviews 21, 1705–1717.

Kellogg, T.B., 1980. Paleoclimatology and paleooceanography of the Norwegian and Greenland seas: Glacial–interglacial contrasts. Boreas 9, 115–137.

Kiefer, T., Kienast, M., 2005. Patterns of deglacial warming in the Pacific Ocean: a review with emphasis on the time interval of Heinrich event 1. Quaternary Science Reviews, this issue (doi:10.1016/j.quascirev.2004.02.021).

Kiefer, T., Sarnthein, M., Erlenkeuser, H., Grootes, P., Roberts, A., 2001. North Pacific response to millennial-scale changes in ocean circulation over the last 60 ky. Paleoceanography 16 (2), 179–189.

Koutavas, A., Lynch-Stieglitz, J., Marchitto Jr., T.M., Sachs, J.P., 2002. El Niño-like pattern in ice age tropical Pacific sea surface temperature. Science 297, 226–230.

Kucera, M., Darling, K.F., 2002. Genetic diversity among modern planktonic foraminifer species: its effect on paleoceanographic reconstructions. Philosophical Transactions of the Royal Society London A 360, 695–718.

Kucera, M., Rosell-Melé, A., Schneider, R., Waelbroeck, C., Weinelt, M., 2005. Multiproxy approach for the reconstruction of the glacial ocean surface (MARGO). Quaternary Science Reviews, this issue (doi:10.1016/j.quascirev.2004.07.017).

Lea, D.W., Pak, D.K., Spero, H.J., 2000. Climate impact of late Quaternary equatorial Pacific sea surface temperature variations. Science 289, 1719–1724.

Lee, K.E., Slowey, N., 1999. Cool surface waters of the subtropical North Pacific Ocean during the Last Glacial. Nature 397, 512–514.

Lee, K.E., Slowey, N.C., Herbert, T.D., 2001. Glacial SSTs in the subtropical North Pacific: a comparison of U^{K}_{37} $\delta^{18}O$ and foraminiferal assemblage temperature estimates. Paleoceanography 16, 268–279.

Malmgren, B.A., Nordlund, U., 1997. Application of artificial neural networks to paleoceanographic data. Palaeogeography, Palaeoclimatology, Palaeoecology 136, 359–373.

Malmgren, B.A., Kucera, M., Nyberg, J., Waelbroeck, C., 2001. Comparison of statistical and artificial neural network techniques for estimating past sea-surface temperatures from planktonic foraminifer census data. Paleoceanography 16, 520–530.

McIntyre, A., Ruddiman, W.F., Karlin, K., Mix, A.C., 1989. Surface water response of the equatorial Atlantic Ocean to orbital forcing. Paleoceanography 4, 19–55.

Meland, M.Y.E., Jansen, E., Elderfield, H., 2005. Constraints on SST estimates for the northern north Atlantic/Nordic Seas during the LGM. Quaternary Science Reviews, this issue (doi:10.1016/j.quascirev.2004.05.011).

Miao, Q., Thunell, R.C., Anderson, D.M., 1994. Glacial–Holocene carbonate dissolution and sea surface temperatures in the South China and Sulu seas. Paleoceanography 9, 269–290.

Mix, A.C., Morey, A., Pisias, N.G., 1999. Foraminiferal faunal estimates of paleotemperature: circumventing the no-analog problem yields cool ice age tropics. Paleoceanography 14, 350–359.

Mix, A.C., Bard, E., Schneider, R., 2001. Environmental processes of the ice age: land, oceans, glaciers (EPILOG). Quaternary Science Review 20, 627–657.

Morey, A.E., Mix, A.C., Pisias, N.G., 2005. Planktonic foraminiferal assemblages preserved in surface sediments correspond to multiple environmental variables. Quaternary Science Reviews, this issue (doi:10.1016/j.quascirev.2003.09.011).

Müller, A., Opdyke, B., 2000. Glacial–interglacial changes in nutrient utilization and paleoproductivity in the Indonesian throughflow-sensitive Timor Trough, easternmost Indian Ocean. Paleoceanography 15, 85–94.

Niebler, H.-S., Gersonde, R., 1998. A planktic foraminiferal transfer function for the southern South Atlantic Ocean. Marine Micropaleontology 34, 213–234.

Niebler, H.S., Mulitza, S., Donner, B., Arz,, H., Pätzold, J., Wefer, G., 2003. Sea-surface temperatures in the equatorial and South Atlantic Ocean during the Last Glacial Maximum (23–19 ka). Paleoceanography 18, doi:10.1029/2003PA000902.

Ortiz, J.D., Mix, A.C., 1997. Comparison of Imbrie–Kipp transfer function and modern analog temperature estimates using sediment trap and core top foraminiferal faunas. Paleoceanography 12, 175–190.

Ortiz, J.D., Mix, A.C., Hostetler, S., Kashgarian, M., 1999. The California Current of the last glacial maximum: reconstruction at 42°N based on multiple proxies. Paleoceanography 12, 191–206.

Palmer, M.R., Pearson, P.N., 2003. A 23,000-year record of surface water pH and pCO$_2$ in the Western Equatorial Pacific Ocean. Science 300, 480–482.

Pflaumann, U., 1986. Sea-surface temperatures during the last 750,000 years in the eastern equatorial Atlantic: planktonic foraminiferal record of "Meteor"-cores 13519, 13521, and 16415. "Meteor" Forschungsergebnisse 40, 137–161.

Pflaumann, U., Jian, Z., 1999. Modern distribution patterns of planktonic foraminifera in the South China Sea and western Pacific: a new transfer technique to estimate regional sea surface temperatures. Marine Geology 156, 41–83.

Pflaumann, U., Duprat, J., Pujol, C., Labeyrie, L., 1996. SIMMAX: A modern analog technique to deduce Atlantic sea surface temperatures from planktonic foraminifera in deep-sea sediments. Paleoceanography 11, 15–35.

Pflaumann, U., Sarnthein, M., Chapman, M., Duprat, J., Huels, M., Kiefer, T., Maslin, M., Schulz, H., van Kreveld, S., Vogelsang, E., Weinelt, M., 2003. North Atlantic: sea-surface conditions reconstructed by GLAMAP-2000. Paleoceanography 18, doi:10/1029/2002PA000774.

Pinot, S., Ramstein, G., Harrison, S.P., Prentice, I.C., Guiot, J., Stute, M., Joussaume, S., 1999. Tropical paleoclimates at the Last Glacial Maximum: comparison of Paleoclimate Modeling Intercomparison Project (PMIP) simulations and paleodata. Climate Dynamics 15, 857–874.

Prell, W.L., 1985. The stability of low-latitude sea-surface temperatures, an evaluation of the CLIMAP reconstruction with emphasis on the positive SST anomalies. Rep. TR025, US Dept. of Energy, Washington, DC.

Prell, W., Martin, A., Cullen, J., Trend, M., 1999. The Brown University Foraminiferal Data Base. IGBP PAGES/World Data Center—A for Paleoclimatology Data Contribution Series # 1999-027.

Rosell-Melé, A., Bard, E., Emeis, K.-C., Grieger, B., Hewitt, C., Müller, P.J., Schneider, R.R., 2004. Sea surface temperature anomalies in the oceans at the LGM estimated from the alkenone-U^{K}_{37} index: comparison with GCMs. Geophysical Research Letters 31, L03208.

Rosenthal, Y., Broccoli, A.J., 2004. In Search of Paleo-ENSO. Science 304, 219–221.

Rosenthal, Y., Oppo, D.W., Linsley, B.K., 2003. The amplitude and phasing of climate change during the last deglaciation in the Sulu Sea, western equatorial Pacific. Geophysical Research Letters 30, 1428.

Rühlemann, C., Mulitza, S., Müller, P.J., Wefer, G., Zahn, R., 1999. Warming of the tropical Atlantic Ocean and slowdown of thermohaline circulation during the last deglaciation. Nature 402, 511–514.

Sarnthein, M., Gersonde, R., Niebler, S., Pflaumann, U., Spielhagen, R., Thiede, Wefer, G., Weinelt, M., 2003a. Preface: Glacial Atlantic Ocean Mapping (GLAMAP-2000). Paleoceanography 18, doi:10/1029/2002PA000770.

Sarnthein, M., Pflaumann, U., Vogelsang, E., Weinelt, M., 2003b. Past extent of sea-ice in the northern North Atlantic inferred from foraminiferal paleotemperature estimates. Paleoceanography 18, doi:10/1029/2002PA000771.

Sawada, K., Handa, N., 1998. Variability of the path of the Kuroshio ocean current over the past 25,000 years. Nature 392, 592–595.

Seidov, D., Maslin, M., 2001. Atlantic Ocean heat piracy and the bipolar climate see-saw during Heinrich and Dansgaard-Oeschger events. Journal of Quaternary Science 16, 321–328.

Schmidt, D.N., Renaud, S., Bollmann, J., 2003. Response of planktic foraminiferal size to late Quaternary climate change. Paleoceanography 18, doi 10.1029/2002PA000831.

Schmidt, D.N., Renaud, S., Bollmann, J., Schiebel, R., Thierstein, H.R., 2004. Size distribution of Holocene planktic foraminifer assemblages: biogeography, ecology and adaptation. Marine Micropaleontology 50, 319–338.

Simstich, J., Sarnthein, M., Erlenkeuser, H., 2003. Paired $\delta^{18}O$ signals of *Neogloboquadrina pachyderma* (s) and *Turborotalita quinqueloba* show thermal stratification structure in Nordic Seas. Marine Micropaleontology 48, 107–125.

Snyder, J.P., 1987. Map Projections—A Working Manual. USGS Professsional Paper 1395, 383pp.

Stocker, T.F., 1998. The seesaw effect. Science 282, 61–62.

Stott, L., Poulsen, C., Lund, S., Thunell, R., 2002. Super ENSO and global climate oscillations at millennial time scales. Science 297, 222–226.

ter Braak, C.J.F., 1995. Non-linear methods for multivariate statistical calibration and their use in palaeoecology: a comparison of inverse (*k*-nearest neighbours, partial least squares and weighted averaging partial least squares) and classical approaches. Chemometrics and Intelligent Laboratory Systems 28, 165–180.

ter Braak, C.J.F., Juggins, S., 1993. Weighted averaging partial least squares regression (WA-PLS): an improved method for reconstructing environmental variables from species assemblages. Hydrobiologia 269/270, 485–502.

Thiede, J., Nees, S., Schulz, H., De Deckker, P., 1997. Oceanic surface conditions recorded on the sea floor of the Southwest Pacific Ocean through the distribution of foraminifers and biogenic silica. Palaeogeography, Palaeoclimatology, Palaeoecology 131, 207–239.

Thompson, P.R., Bé, A.W.H., Duplessy, J.-C., Shackleton, N.J., 1979. Disappearance of pink-pigmented *Globigerinoides ruber* at 120,000 yr BP in the Indian and Pacific Oceans. Nature 280, 554–558.

Trend-Staid, M., Prell, W.L., 2002. Sea surface temperature at the Last Glacial Maximum: a reconstruction using the modern analog technique. Paleoceanography 17, doi:10.1029/2000PA000506.

Tudhope, A.W., Chilcott, C.P., McCulloch, M.T., Cook, E.R., Chappell, J., Ellam, R.M., Lea, D.W., Lough, J.M., Shimmield, G.B., 2001. Variability in the El Nino-Southern Oscillation through a glacial–interglacial cycle. Science 291, 1511–1517.

Ujiié, H., 2003. 370-ka paleoceanographic record from the Hess Rise, central North Pacific Ocean, and an indistinct 'Kuroshio Extension'. Marine Micropaleontology 49, 21–47.

Ujiié, Y., Ujiié, H., 2000. Distribution and oceanographic relations of modern planktonic foraminifera in the Ryukyu Arc region, northwest Pacific Ocean. Journal of Foraminiferal Research 30, 336–360.

Waelbroeck, C., Labeyrie, L., Duplessy, J.-C., Guiot, J., Labracherie, M., Leclaire, H., Duprat, J., 1998. Improving past sea surface temperature estimates based on planktonic fossil faunas. Paleoceanography 13, 272–283.

Watkins, J.M., Mix, A.C., 1998. Testing the effects of tropical temperature, productivity, and mixed-layer depth on foraminiferal transfer function. Paleoceanography 13, 96–105.

Weaver, A.J., Eby, M., Fanning, A.F., Wiebe, E.C., 1998. Simulated influence of carbon dioxide, orbital forcing and ice sheets on the climate of the Last Glacial Maximum. Nature 394, 847–853.

Webb, R.S., Rind, D.H., Lehman, S.J., Healy, R.J., Sigman, D., 1997. Influence of ocean heat transport on the climate of the Last Glacial Maximum. Nature 385, 695–699.

Weinelt, M., Sarnthein, M., Pflaumann, U., Schulz, H., Jung, S., 1996. Ice-free Nordic Seas during the Last Glacial Maximum?—Potential sites of deepwater formation. Paleoclimates, Data and Modelling 1, 283–309.

Weinelt, M., Kuhnt, W., Sarnthein, M., Altenbach, A., Costello, O., Erlenkeuser, H., Pflaumann, U., Simstich, J., Struck, U., Thies, A., Trauth, M., Vogelsang, E., 2001. Paleoceanographic Proxies in the Northern North Atlantic. In: Schlüter, M., Schäfer, P., Ritzrau, W., Thiede, J. (Eds.), The Northern North Atlantic: A Changing Environment. Springer, Berlin, pp. 319–352.

Weinelt, M., Vogelsang, E., Kucera, M., Pflaumann, U., Sarnthein, M., Völker, A., Erlenkeuser, H., Malmgren, B.A., 2003. Variability of North Atlantic heat transfer during MIS 2. Paleoceanography 18, doi:10/1029/2002PA000772.

Wessel, P., Smith, W.H.F., 1991. Free software helps map and display data,. EOS Transactions of American Geophysical Union 72, 445–446.

Wessel, P., Smith, W.H.F., 1998. New, improved version of the Generic Map Tools Released. EOS Transactions of American Geophysical Union 79, 579.

WOA, 1998. World Ocean Atlas 1998, Version 2, http://www.nodc.noaa.gov/oc5/woa98.html. Tech. rep., National Oceanographic Data Center, Silver Spring, Maryland.

Quaternary Science Reviews 24 (2005) 999–1016

Glacial Mediterranean sea surface temperatures based on planktonic foraminiferal assemblages

Angela Hayes[a,b,*], Michal Kucera[b], Nejib Kallel[c], Laura Sbaffi[d], Eelco J. Rohling[e]

[a]*Department of Geography, Mary Immaculate College, University of Limerick, Limerick, Ireland*
[b]*Department of Geology, Royal Holloway University of London, Egham, Surrey TW20 0EX, UK*
[c]*Laboratoire E08/C10, Faculte des sciences de Sfax, Route de Soukra, B.P. 763, 3038 Sfax, Tunisia*
[d]*Geoscience Australia, Petroleum & Marine division, GPO Box 378, Canberra ACT 2601, Australia*
[e]*Southampton Oceanography Centre, School of Ocean & Earth Science, Southampton SO14 3ZH, UK*

Received 8 September 2003; accepted 10 February 2004

Abstract

We present a new reconstruction of Mediterranean sea surface temperatures (SST) during the last glacial maximum (LGM). A calibration data set based on census counts of 23 species of planktonic foraminifera in 129 North Atlantic and 145 Mediterranean core top samples was used to develop summer, winter and annual average SST reconstructions using artificial neural networks (ANNs) and the revised analogue method (RAM). Prediction errors determined by cross-validation of the calibration data set ranged between 0.5 and 1.1 °C, with both techniques being most successful in predicting winter SSTs. Glacial reconstructions are based on a new, expanded data set of 273 samples in 37 cores with consistent minimum level of age control.

The new LGM reconstructions suggest that the east–west temperature gradient during the glacial summer was 9 °C, whereas during the glacial winter, the gradient was 6 °C, both some 4 °C higher than that existing today. In contrast to earlier studies, our results tend to suggest much cooler SST estimates throughout the glacial Mediterranean, particularly in the eastern basin where previous SST reconstructions indicated a decrease of only 1 °C. Our new SST reconstructions will provide the modelling community with a detailed and updated portrayal of the Mediterranean Sea during the LGM, setting new targets on which glacial simulations can be tested.

© 2004 Elsevier Ltd. All rights reserved.

1. Introduction

The last glacial maximum (LGM) is a period in geological time that represents a climate dramatically different from that of today. Sea surface temperature (SST) reconstructions during this interval have become the main target of many palaeoceanographic studies since they provide valuable information required for modelling climate and ocean circulation. The use of planktonic foraminifera as a SST proxy was recognised much earlier. Murray (1897) identified that specific faunal assemblages provided clues to the temperature of the water they lived in. Schott (1935) introduced quantitative counting of species within the assemblages. He recognised that species compositions vary downcore, concluding that surface water temperatures changed as the climate fluctuated between glacial and interglacial periods. Since then, planktonic foraminifera have become a widely used proxy for determining SST reconstructions on a global scale (CLIMAP, 1976; Sarthein et al., 2003).

Situated in a subtropical semi-arid climate, the Mediterranean Sea is a semi-enclosed basin with the Strait of Gibraltar providing the only connection to the open ocean. Due to its subtropical location, the Mediterranean Sea receives much solar radiation, which along with the prevailing eastward surface circulation

---

*Corresponding author. Department of Geography, Mary Immaculate College, University of Limerick, Limerick, Ireland. Tel.: +353-61-204577; fax: +353-61-313632.

E-mail address: angela.hayes@mic.ul.ie (A. Hayes).

0277-3791/$ - see front matter © 2004 Elsevier Ltd. All rights reserved.
doi:10.1016/j.quascirev.2004.02.018

causes a strong west–east temperature gradient (Fig. 1). An excess of evaporation over precipitation furthermore results in a strong salinity increase from west to east (Wüst, 1961). Net buoyancy loss maintains a two-layer flow regime through the Strait of Gibraltar, consisting of Atlantic surface water inflow and Mediterranean subsurface water outflow (Lacombe et al., 1981; Lacombe and Richez, 1982; Pistek et al., 1985). The well-constrained hydrological balance in the Mediterranean Sea makes the basin an ideal location for testing mesoscale climate and ocean circulation models under both modern and past climate forcing conditions (Bigg, 1994; Myers et al., 1998; Myers and Rohling, 2000).

Mediterranean planktonic foraminifera have been extensively studied since the recovery of the first piston cores during the Swedish Deep Sea Expedition of 1946–47. Many authors have since produced qualitative palaeoclimatic reconstructions based on the distribution of planktonic faunal assemblages throughout the Quaternary (e.g. Parker, 1958; Todd, 1958; Olausson, 1960, 1961; Herman, 1972; Cita et al., 1977; Vergnaud-Grazzini et al., 1977; Thunell, 1978; Rohling et al., 1993a, 1995; Hayes et al., 1999; Buccheri et al., 2002). Several methods have been developed to quantify the relationship between planktonic faunal assemblages and the physical characteristics of overlying water masses (Malmgren et al., 2001). Two of these methods have been used for SST reconstructions in the Mediterranean Sea. The first of these, known as the Imbrie–Kipp transfer function (IKTF), uses the Q-mode principal component analysis to reduce species abundances into statistically independent end-member assemblages that are then regressed onto the environmental parameters (Imbrie and Kipp, 1971). The second technique, known as the modern analogue technique (MAT) (Hutson, 1980) uses an alternative approach: rather than generating a unique calibration formula, this method searches a database of modern faunal assemblages, extracting those that best resemble the fossil assemblage. The environmental parameters associated with the modern assemblages are then used to reconstruct those for the fossil sample (Hutson, 1980). Both methods have been fundamental in efforts to estimate Quaternary sea surface temperatures in the Mediterranean Sea (e.g. Thiede, 1978; Thunell, 1979; Kallel et al., 1997b; Gonzalez-Donoso et al., 2000; Sbaffi et al., 2001; Perez-Folgado et al., 2003). However, most of these recent studies have been limited to single-core palaeoclimatic reconstructions rather than multi-core analysis on a basin-wide scale.

CLIMAP (1976) presented the first attempt to quantify glacial sea surface temperatures across the entire Mediterranean Sea. The LGM interval was defined at 18,000 yr BP based on the maximum extent of continental glaciers and a combination of biostratigraphic, geochemical and isotopic evidence. The calibration data set was composed of both Mediterranean (Parker, 1958; Todd, 1958) and Atlantic (CLIMAP, 1976) core top samples and included counts of 22 planktonic foraminiferal species. Using the IKTF, CLIMAP reconstructed glacial SSTs producing a standard error of estimate of ±1.38 °C. However, due to taxonomic discrepancies and uncertain LGM stratigraphy, the reliability and precision of the reconstructions has been disputed (Thunell, 1979). Thiede (1978) used a similar approach to produce glacial winter and summer sea surface temperature estimates based on 33 cores (Figs. 2A and B). However, his reconstructions were derived from transfer functions based entirely on a North Atlantic calibration data set (Molina-Cruz and Thiede, 1978), which only provides modern analogues for glacial faunal assemblages in the western Mediterranean Sea. Conversely, a calibration data set composed entirely of Mediterranean samples will only provide analogues for the eastern Mediterranean glacial assemblages and exclude the western basin. Thiede's (1978)

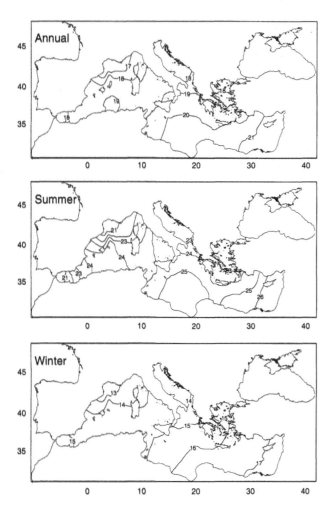

Fig. 1. Present day distribution of annual, summer and winter sea-surface temperatures at 10 m depth in the Mediterranean (data from the World Ocean Atlas 98, vol. 2).

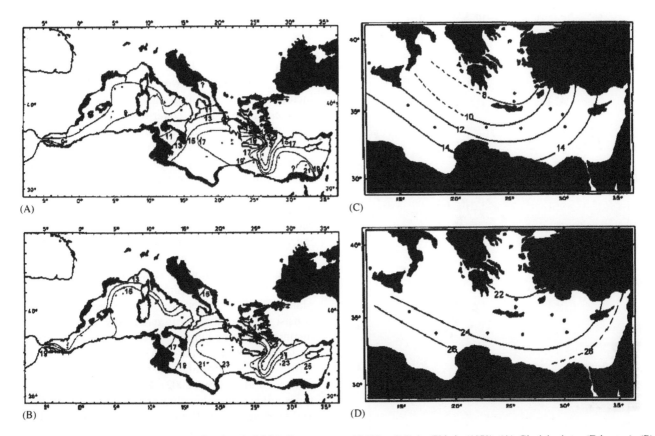

Fig. 2. Sea surface temperatures estimates for the glacial Mediterranean at 18,000 yr B.P. by Thiede (1978). (A) Glacial winter (February). (B) Glacial summer (August) and Thunell (1979) (C) Glacial winter. (D) Glacial summer. Black dots indicate core localities used for glacial reconstruction.

results displayed temperature values ranging from 7 °C in the north–west of the basin to >20 °C in the south–east during the glacial winter (Fig. 2A). During the summer, the reconstructed temperatures were significantly higher recording a range between 13 °C in the north–west to approximately 25 °C in the south–east (Fig. 2B). In a subsequent study, Thunell (1979) established a calibration data set composed of 66 Mediterranean and 8 North Atlantic trigger-weight cores. Applying the IKTF, faunal counts (21 species) from 10 piston cores enabled the reconstruction of eastern Mediterranean SSTs during the LGM (Figs. 2C and D). The standard error of estimates for summer and winter temperatures were 1.0 and 1.2 °C, respectively, and the LGM interval was defined on the basis of faunal stratigraphy, oxygen isotope stratigraphy and radiometric dates, providing a better age control than that used by Thiede (1978). Thunell's (1979) glacial summer temperatures were similar to those suggested by Thiede (1978), ranging between 22 and 26 °C (Fig. 3D). However, discrepancies between the two studies are more noticeable during the glacial winter with Thunell (1979) estimating temperatures between 8 and 14 °C (Fig. 3C), some 5 °C lower than those of Thiede (1978).

In recent years, the need for improvement in the precision and reliability of SST estimates has led to the development of new transfer function techniques. Specifically, improvements in the MAT include the modern analogue with a similarity index (SIMMAX) method (Pflaumann et al., 1996) and the revised analogue method (RAM) (Waelbroeck et al., 1998). A further approach relies on the concept of artificial neural networks (ANN), a computer-intensive method based on unsupervised learning of a relationship between two sets of variables (Malmgren and Nordlund, 1997; Malmgren et al., 2001).

The purpose of this paper is to provide an updated, detailed and comprehensive reconstruction of Mediterranean SSTs during the LGM. For the first time in a quarter of a century, we aim to provide SST reconstructions of consistent quality and reliability across the entire Mediterranean Sea. We use the ANN technique as a basis for our reconstructions, and the RAM for a comparison. The ANN has proved effective in providing reliable palaeoestimates compared to conventional computational techniques (Malmgren et al., 2001). Using a larger, more constrained calibration data set, a new, expanded fossil data set and a combination of two computational techniques (ANN and RAM), we

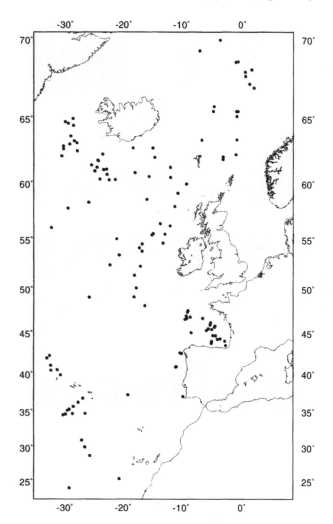

Fig. 3. Map illustrating the core localities in the North Atlantic portion of the calibration data set used in this study (taken from North Atlantic database presented in Kucera et al., 2004b).

believe that our reconstruction will provide an accurate portrayal of the glacial Mediterranean, providing more reliable surface forcing fields for numerical modelling studies.

2. Materials

2.1. Calibration data set

The calibration data set used in this study is based on census counts of planktonic foraminiferal species in 274 core tops from the North Atlantic Ocean (Fig. 3) and the Mediterranean Sea (Fig. 4). The Mediterranean data set contains 145 core tops, 32 of which are additional to those presented by Thunell (1978) and Kallel et al. (1997b) (Tables 1 and 2). The remaining 129 cores are taken from the North Atlantic database presented in Kucera et al. (2004b), which is an updated version of the

GLAMAP data set presented in Pflaumann et al. (2003). The North Atlantic core tops are included to provide analogues for glacial Mediterranean assemblages, which contain species that do not occur in the Mediterranean today. Intuitively, it would seem advantageous to include in the training database as many Atlantic samples as possible, covering the widest range of environments encountered in this part of the world today. There are, however, two main reasons to avoid such approach. Firstly, all the commonly used transfer function techniques derive an environmental calibration on the basis of finding the minimum total error of prediction. If one part of the ocean or one part of the range of an environmental variable is over represented, the techniques may produce calibrations preferentially minimising the error in the over represented set of samples, whilst allowing larger errors to occur elsewhere. Secondly, morphologically defined species of planktonic foraminifera (morphospecies) are known to consist of complexes of genetically and ecologically distinct types; the lumping of these types into morphospecies is a likely source of a significant portion of the error rate of foraminifer-based transfer functions (Kucera and Darling, 2002). Many of the cryptic genetic types have geographically distinct distributions (Kucera and Darling, 2002). Therefore, limiting the spatial extent of the calibration data set should decrease the degree of lumping of these ecologically distinct units and improve the error rate of a transfer function.

With these two assumptions in mind, all of the Atlantic core tops that were included into the calibration database were selected within the latitudinal range of approximately 25–70°N and longitudinal range of 5°E–30°W. This encompasses an area small enough within the North Atlantic to reduce the effect of lumping of cryptic genetic types but large enough to include assemblages from subpolar–polar water masses which may have existed in the glacial Mediterranean (Rohling et al., 1998). Samples from upwelling areas were excluded since they produce specific opportunistic faunal assemblages determined by uncharacteristic physical parameters (e.g. Ufkes et al., 1998). Similarly, core tops recording high percentages (>70%) of *Neogloboquadrina pachyderma* (left-coiling) were also omitted since such assemblages have never been observed in the Mediterranean Sea.

All samples within the calibration data set were picked from the >150 μm size fraction. Total planktonic foraminiferal assemblages were quantified in relative abundances from splits containing approximately 300 specimens. A total of 23 species were counted in each sample (Table 5). Species selection was determined by the history of taxonomic practice of individual authors. For example, *Globigerinoides ruber* is commonly categorised into pink and white varieties. However, Thunell (1978) recorded both varieties collectively, therefore to

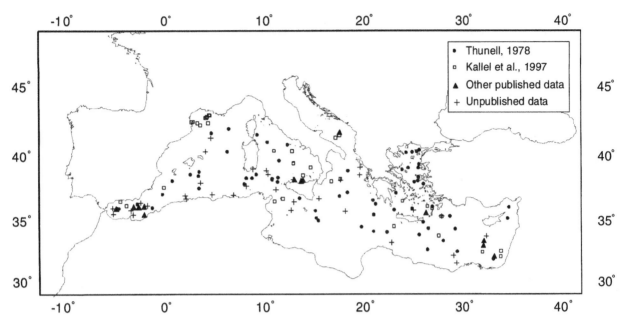

Fig. 4. Map illustrating the core localities in the Mediterranean portion of the calibration data set used in this study. Refer to Tables 1 and 2 for source details.

Table 1
Summary of core top samples used in the Mediterranean calibration data set taken from published sources in addition to samples published by Thunell (1978) and Kallel (1997b)

Core	Latitude	Longitude	Water depth (m)	Age quality	Data source
MD84-627	32.14	33.45	1185	3	Paterne et al. (1986)
MD84-639	33.40	32.42	870	3	Paterne et al. (1986)
MD84-641	33.02	32.38	1375	3	Paterne et al. (1986)
KS8232	36.11	−2.12	1920	3	Pujol and Vergnaud-Grazzini (1989)
KC8241	35.98	−4.40	1282	2	Pujol and Vergnaud-Grazzini (1989)
IN68-9	41.79	17.91	1234	2	Jorissen et al. (1993)
KS310	35.55	−1.57	1900	4	Rohling et al. (1995)
LC21	35.66	26.58	1522	2	Hayes et al. (1999)
BS79-22	38.23	14.23	1449	4	Sbaffi et al. (2001)
BS79-33	38.16	14.02	1282	4	Sbaffi et al. (2001)
BS79-38	38.25	13.35	1489	4	Sbaffi et al. (2001)
MD95-2043	36.14	−2.62	1841	1	Perez-Folgado et al. (2003)
ODP 977	36.19	−1.57	1984	1	Perez-Folgado et al. (2003)

achieve consistency within the data set we have grouped the pink and white varieties of *G. ruber* into one taxonomic category. For the same reasons, we have also grouped *Globigerinella siphonifera* with *Globigerinella calida*, *Globorotalia menardii* with *Globorotalia tumida* and "P/D intergrades" with *Neogloboquadrina pachyderma* (dextral).

For the purpose of calibration, winter (January–March), summer (July–September) and annual SST data were extracted at a depth of 10 m from the World Ocean Atlas (WOA, vol. 2) (Kucera et al., 2004a), using the WOA 98 sample software (http://www.palmod. uni-bremen.de/~csn/woasample.html). The temperatures are calculated as the area-weighted averages of the four

WOA temperature points surrounding the sample location. An age quality level was assigned to each core top within the calibration data set following the criteria defined by the MARGO group (Kucera et al., 2004a). This is to investigate and eliminate the possible effects of residual glacial assemblages in surface sediments. The age quality levels ranged between 1 and 5 with level 1 indicative of core tops with the best chronological control, and level 5 the poorest. In instances where the quality of the core was borderline between two levels, the lower Chronozone quality level was selected. All of the data used in the calibration and LGM data sets are stored in the PANGAEA database (www.pangaea.de/ Projects/MARGO).

Table 2
Details of unpublished core top samples added to the Mediterranean calibration data set

Core	Latitude	Longitude	Water depth (m)	Age quality	Faunal counts provided by
BS79-37	38.22	13.27	1458	4	Sbaffi
T87-11B	39.20	19.94	1322	2	Hayes
T87-14B	38.60	19.92	1999	2	Hayes
T87-43B	35.90	13.03	1514	2	Hayes
T87-45B	36.50	13.32	1716	2	Hayes
T87-65B	38.60	10.78	904	2	Hayes
T87-71B	38.90	10.59	2654	2	Hayes
T87-95B	39.00	9.28	499	2	Hayes
T87-108B	36.70	2.64	212	2	Hayes
T87-118B	37.00	2.55	2000	2	Hayes
T87-128B	35.50	−2.67	296	2	Hayes
T87-133B	35.80	−2.89	1100	2	Hayes
T83/20	31.60	29.61	790	2	Hayes
T83/25	32.20	29.29	2034	2	Hayes
T83/67	33.20	23.09	2075	2	Hayes
Ki06	37.46	11.55	885	5	Kallel
Ki201	41.35	5.13	2450	5	Kallel
Ki202	40.35	4.60	1870	5	Kallel
Ki203	37.08	5.26	2245	5	Kallel
Ki206	38.20	18.01	950	5	Kallel
Ki211	35.85	25.33	1835	5	Kallel
Ki307	36.43	−1.90	1420	5	Kallel
Ki320	36.01	−4.71	950	5	Kallel
1Ki02	37.98	4.10	1870	5	Kallel
1Ki04	37.05	7.37	2500	5	Kallel
1Ki08	35.75	18.45	3733	5	Kallel
1Ki11	33.76	32.71	946	5	Kallel
Ki5 & Ks5	37.73	8.60	1520	5	Kallel
K7 & Ks7	36.75	15.83	2500	5	Kallel
Ks 09	35.35	28.13	1330	5	Kallel
KS8233	36.20	−1.23	2500	5	Kallel
KC8240	35.55	−4.66	533	5	Kallel

2.2. Last glacial maximum data set

For the purpose of this paper, the LGM Chronozone is defined as the interval between 19,000 and 23,000 cal-yr BP (Mix et al., 2001) that includes the centre of the LGM event previously defined by CLIMAP (1976). Fig. 5 illustrates the 37 Mediterranean cores that were selected for this time slice. This data set includes 13 cores taken from Kallel et al. (1997b), 10 other previously published cores (Table 3) and new data from an additional 14 cores (Table 4). The cores used in Thiede's (1978) and Thunell's (1979) studies were not incorporated into our LGM data set since we were unable to trace them. Whilst the addition of these samples would have allowed a higher spatial resolution of the LGM reconstruction, by using only the 37 new cores we were able to ensure taxonomic consistency and a consistent minimum level of age control across this new LGM data set.

All samples falling within the LGM interval were selected in each core. A total of 273 samples were used ranging between 1 and 23 samples per core. The census counts of planktonic foraminifera from all the samples included in the LGM data set are based on the >150 μm size fraction using the same taxonomic concepts as applied in the calibration data set (Table 5). Counts were based, where possible, on suitable aliquots of approximately 300 specimens obtained using a random split. Each core was assigned an age quality level (1 indicating the highest quality level) according to the EPILOG criteria (Mix et al., 2001) expanded by the MARGO group (Kucera et al., 2005a). The LGM SST estimates for each core were obtained by calculating the average SST from all samples within the defined LGM interval. The standard deviation was calculated for those cores that contained two or more samples within the glacial interval, this included 35 cores within the LGM data set.

3. Methods

3.1. Artificial neural network

ANN, a branch of artificial intelligence, are computer systems that have the ability of unsupervised learning of

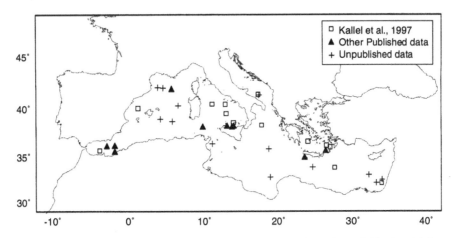

Fig. 5. Map illustrating the core localities in the LGM data set used in this study. Refer to Tables 3 and 4 for source details.

Table 3
Summary of cores used in the LGM data set taken from published sources in addition to those in Kallel (1997b)

Core	Latitude	Longitude	Water depth (m)	No. of samples	Age quality	Age model	Data source
T87/2/20G	34.97	23.75	707	3	2	a	Rohling and Gieskes (1989)
KS310	35.55	−1.57	1900	7	2	a	Rohling et al. (1995)
BC15	41.95	5.93	2500	3	2	a, b	Rohling et al. (1998)
LC07	38.14	10.07	488	3	3	b	Hayes et al. (1999)
LC21	35.66	26.58	1522	8	3	a, b	Hayes et al. (1999)
MD95-2043	36.14	−2.62	18.41	7	2	a, b	Cacho et al. (1999)
BS79-22	38.23	14.23	1449	21	3	b	Sbaffi et al. (2001)
BS79-33	38.16	14.02	1282	13	3	a, b	Sbaffi et al. (2001)
BS79-38	38.25	13.35	1489	24	3	b	Sbaffi et al. (2001)
ODP 977	36.19	−1.57	1984	5	3	a, b, c	Perez-Folgado et al. (2003)

a, AMS ^{14}C data; b, Oxygen isotope stratigraphy; c, biostratigraphy.

Table 4
Details of unpublished cores added to the LGM data set

Core	Latitude	Longitude	Water depth (m)	No. of samples in LGM interval	Age quality	Age model	Faunal counts provided by
ODP 969	33.84	24.88	2200	5	4	c	Hayes
ODP 973	35.78	18.94	3695	2	4	c	Hayes
ODP 975	38.89	4.50	2426	5	4	c	Hayes
LC01	40.26	6.89	2845	4	3	b	Hayes
LC04	38.65	6.11	2855	3	3	b	Hayes
BS79-37	38.22	13.27	1458	3	3	a	Sbaffi
MD84-627	32.14	33.45	1185	3	3	b	Kallel
MD84-632	32.47	34.23	14.5	19	3	b	Kallel
MD84-641	33.02	32.38	1375	6	1	a, b	Kallel
MD99-2346	42.04	4.15	2089	23	3	a, b	Kallel
MD90-917	41.30	17.61	1010	17	2	a, b	Kallel
MD99-2344	42.00	4.84	2326	4	3	a, b	Kallel
KET80-39	36.32	11.41	290	15	4	c	Kallel
KL96	32.77	19.19	1390	2	3	b	Hayes

a, AMS ^{14}C data; b, Oxygen isotope stratigraphy; c, biostratigraphy.

a relationship between two sets of variables (Wasserman, 1989). The general principles and architecture of backpropagation (BP) neural networks and its application to palaeoceanographic data are described in Malmgren and Nordlund (1997) and Malmgren et al. (2001). A trained ANN can be best compared to a complicated, recurrent mathematical formula transforming input variables (species abundances) into

Table 5
List of planktonic foraminiferal species and taxonomic categories used in the calibration and LGM data sets

Species
Orbulina universa
Globigerinoides conglobatus
Globigerinoides ruber (pink and white)
Globoturborotalita tenella
Globigerinoides sacculifer (with and without sac)
Sphaeroidinella dehiscens
Globigerinella siphonifera + *Globigerinella calida*
Globigerina bulloides
Globigerina falconensis
Beella digitata
Globoturborotalita rubescens
Turborotalita quinqueloba
Neogloboquadrina pachyderma sin.
Neogloboquadrina dutertrei
Neogloboquadrina pachyderma dex. + "P/D intergrades"
Pulleniatina obliquiloculata
Globorotalia inflata
Globorotalia truncatulinoides
Globorotalia crassaformis
Globorotalia hirsuta
Globorotalia scitula
Globorotalia menardii + *Globorotalia tumida*
Globigerinita glutinata

desired output variables (SST). A BP ANN consists of a series of processing units (neurons) arranged in layers where each neuron in a preceding layer is connected to all neurons of the following layer. The training of a BP ANN is accomplished by iterative adjustment of arrays of coefficients defining the output of each of the neurons. The adjustment of the coefficients is based on the change of total error rate between two individual iterations. ANNs can successfully learn complex, non-linear relationships and the technique is not as dependent on the size, coverage and balance of the calibration data set as the k-nearest-neighbour techniques (MAT, SIMMAX, RAM).

In this application, the calibration data set (274 samples) was randomly split into two representative sub-sets using the NeuroGenetic Optimizer (v2.6) program. The larger sub-set, comprising 80% of the samples (219 samples), was used for training, whilst the remaining 20% (55 samples) constituted the test sub-set. This splitting procedure was repeated 10 times to allow for a reliable estimate of prediction error and provide a means of assessing the dependency of the estimated values on small changes in the calibration data set. Training of the ANN was performed separately for annual, winter (January–March) and summer (July September) SSTs. The best network in each of the 10 partitions has been retained and all SST reconstructions were calculated as averages of reconstructions produced by these 10 networks. In each partition, the prediction error of the

ANN technique, the root mean square error of prediction (RMSEP), was calculated as the square root of the sum of the squared differences between the observed and predicted values for all samples within the test set, divided by the number of samples (55 samples).

The NGO software uses a genetic algorithm to search for the parameters of the network that produce the lowest error of prediction (Malmgren et al., 2001). In this application, the maximum number of neurons per hidden layer was set at 16 in each of one or two hidden layers. Given the small number of samples in the calibration data set, training of larger networks would be severely underdetermined. For each partition, the algorithm searched through 30 generations with each population containing 100 neural network configurations, recording a total of 3000 network configurations per partition. The maximum number of learning epochs was set to 2500 for each configuration. If no improvement in the prediction error in the test set occurred after 30 consecutive epochs the network is instructed to stop. The parameters of the best networks for each temperature definition and partition are listed in Table 6.

3.2. Revised analogue method

Of the alternative methods available, we opted to use the RAM as a comparison for the ANN technique. IKTF is known to produce the largest errors amongst all the techniques (Malmgren et al., 2001), whilst both MAT and SIMMAX depend on large numbers of samples in the calibration data set. Additionally, SIMMAX includes a weighting procedure of the best analogues' SSTs according to the inverse geographical distances of the most similar samples (Pflaumann et al., 1996). This is problematic in the glacial Mediterranean where some of the best analogues are expected to derive from the North Atlantic. The RAM technique appears the most suitable alternative to ANN in the Mediterranean, as it may overcome the lack of samples in the calibration data set by the addition of virtual coretops and a restrictive selection of the number of best analogue samples.

RAM is a variant of MAT with several significant modifications (Waelbroeck et al., 1998). Like MAT, this method searches the calibration data set for the best analogues on the basis of a dissimilarity coefficient. In addition, for each sample RAM attempts to constrain the number of best analogues by searching for 'jumps' in the dissimilarity coefficient. The best analogues encountered prior to the first jump are retained; however, if no jumps are detected then a given number ($\beta = 10$ in this application) of analogues are kept. The second modification allows the calibration data set to be artificially expanded by mapping the original data onto a grid of environmental variables with a given mesh size (γ). The technique creates virtual faunal assemblages at each

Table 6
Configurations of the best artificial neural networks for each partition of the calibration dataset

Partition	RMSEP		1HL			2HL			Out	
	tst	trn	Lo	T	Li	Lo	T	Li		N
Summer										
1	1.1539	1.1027	7	5	5				1Li	17
2	1.1984	1.0302	6	6		2		2	1Li	16
3	1.0139	1.3570	1	1	1	4	2	3	1Li	12
4	1.2653	0.9637		6	5				1Li	11
5	0.9447	1.0918	1	2	8				1Li	11
6	1.2747	1.1227	4	5		3	2		1T	14
7	0.9818	1.1159	4	1	1				1Li	6
8	1.2495	1.0266	3	5	4	6	4	1	1Li	23
9	1.2325	1.1031	1	3					1Li	4
10	1.0784	1.0355		3	2				1Li	5
Average	1.1393	1.0949								12
Winter										
1	0.8486	0.6804	7	5					1Li	12
2	0.8030	0.6847	5	4	3				1Li	12
3	0.6524	0.6964	3	4	2				1Li	9
4	0.7444	0.617	11	2	2		2	14	1Li	31
5	0.6987	0.8582	6	3	2	7	5		1T	23
6	1.0584	0.805	2		2	1	1		1T	6
7	0.7415	0.6724	6	6	2		1	1	1Li	16
8	0.7686	0.6725	12	2		2	1	1	1Li	18
9	0.8868	0.6993	1	2	2				1Li	5
10	0.7286	0.8232	1	1					1Li	2
Average	0.7931	0.7209								13
Annual										
1	0.9947	0.844	9	2	4				1Li	15
2	0.9666	0.8374	6	6	2	1		1	1Lo	16
3	0.7891	0.9882	1	1	1	2	2	2	1Li	9
4	0.9598	0.8451	4	4					1Lo	8
5	0.742	0.8803		6	2				1Li	8
6	1.0708	0.7936	9	1	6		3	2	1Li	21
7	0.7677	0.804	4	4	7	3	2	2	1Li	22
8	0.9397	0.7816	2	3	2	1	1	2	1Li	11
9	0.9998	0.7931	11		5	2	3	1	1Li	22
10	0.8916	0.8434	2	2	3				1Li	7
Average	0.9122	0.8411								14

tst, test data (20% of data set); trn, training data (80% of data set); Lo, neurons with logarithmic transfer function; T, neurons with hyperbolic tangent transfer function; Li, neurons with linear transfer function; HL, hidden layer, Out, output layer; N, number of neurons in hidden layers.

node on the grid by interpolating abundances from the calibration data set within a given radius (R). A more detailed explanation of the RAM is described in Waelbroeck et al. (1998). For the purpose of this study we have opted to use the modified RAM02 software, which automatically selects the parameter related to the selection of the best analogues (α) (Kucera et al., 2005b). The interpolation procedure in RAM02 is constrained to two dimensions. In this study, we have used $\gamma = 0.25$ and $R = 0.3$, for summer vs. winter SST interpolation (following Waelbroeck et al., 1998) and an annual vs.

seasonality SST interpolation to produce independent estimates of annual SST. The RMSEP for RAM has been calculated by using the 'leaving-one-out' method; however, the RAM02 software leaves out the best analogue (i.e. itself) after the interpolation. Since the virtual core tops are based on the assemblages of the 'left out' sample the error estimates are inevitably underestimated.

4. Results and discussion

4.1. Calibration

The results of the calibration of both ANN and RAM are displayed in Figs. 6–8. For ANN, the average values of the 10 networks are shown for each SST reconstruction. The errors produced in this way (show as RMSE in Fig. 6) are slightly underestimated when compared to the average RMSEP values, which provide a better estimate of how the technique will perform on samples from outside of the calibration data set. Because of the way the RAM02 software performed the leaving-one-out validation (see above) the error rates reported for RAM (Fig. 6) are not fully comparable to ANN RMSEP values. Nevertheless, we observe a similar pattern where both techniques appear most successful in reconstructing winter SSTs and least successful in reconstructing summer SSTs. It is interesting to note that for all SST definitions, the error rate calculated by RAM for the entire data set is lower than that for the Mediterranean data set. The reverse pattern is observed for the ANN results, which record a lower error rate for the Mediterranean portion of the calibration data set (Fig. 6).

An examination of the residual values (i.e. observed minus predicted SST values) shows very little bias towards a systematic under or overestimating at any part of the SST range (Fig. 7A). As noted previously, the winter season temperatures are easier to reconstruct by both ANN and RAM, compared to summer or annual average values. Using the ANN technique, approximately 12% of the data set produced errors of $\pm 1\,^\circ C$, compared to 7% produced by RAM. The relatively weak correlation between residuals of each technique (Fig. 7B) suggests that ANN and RAM are predicting SSTs in a different manner. To assess the effect of age quality on the SST estimates, Fig. 8 shows the relationship between the residuals from each technique and the age quality of each core top. It is immediately obvious that age quality has no bias towards reducing the incidence of underestimated SST reconstructions. The only exception is observed in summer SST estimates by ANN, where quality level 3 cores show evidence of underestimation; the lack of core numbers within this category could account for this discrepancy. We

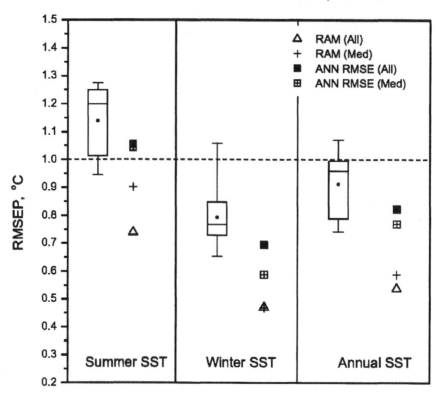

Fig. 6. Summary of the prediction errors produced by ANN and RAM. The box and whiskers plots show the minimum and maximum RMSEP values for ANN, along with the 0.25 percentile, median, mean (black dot) and 0.75 percentile. The ANN RMSE values were calculated from SST estimates based on average values of the ten best networks. The RAM error rate is based on leaving-one-out procedure applied after the two-dimensional interpolation; such error rate is therefore expected to be an underestimate of how the technique will perform outside of the calibration data set.

conclude that the calibration data set does not appear to be affected by samples with residual glacial assemblages.

4.2. Application to glacial samples

The SST estimates discussed in the following sections are determined by calculating the average SSTs from all samples contained within the LGM interval in each core. It should be noted that core KL96 (Table 4) does not contain SST estimates from RAM since no suitable analogues were found for the assemblages encountered in this core.

4.2.1. Glacial summer

The results of both RAM and ANN suggest that the SST gradient between the western and eastern basins was significantly greater during the LGM than today (Fig. 9). During the summer, with the exception of the Adriatic and Aegean Seas, the eastern Mediterranean was the warmer of the two basins with temperatures in the far east reaching ~23 °C. Surface waters at the Strait of Sicily and along the North African coast as far as the Balearic basin are estimated at ~17 °C. Cooler water conditions (~13 °C) are observed extending southwards

from the Gulf of Lions (Fig. 9A). Westward into the Alboran Sea (western Mediterranean) SSTs were up to ~14 °C. The ANN estimates suggest a warm surface water anomaly (~16 °C) off the east coast of Spain (Fig. 9A). The coldest part of the basin is observed in the Gulf of Lions where temperatures drop to ~10 °C, coinciding with the area of western Mediterranean deep water formation (WMDW) (MEDOC group, 1970). The southerly displacement of the isotherms in the eastern basin reflects an extension of cool conditions over the open basin from the Aegean Sea (~16 °C) (Fig. 9A). The differences between the glacial and present day (WOA 98, vol. 2) summer SSTs expressed as a temperature anomaly are illustrated in Fig. 9B. From the distribution of summer anomalies it is evident that major temperature changes occurred in the western basin with temperatures dropping to 11 °C below modern temperatures. Other significant changes are observed in the Adriatic and Aegean Seas recording temperature decreases of 6 and 7 °C, respectively. Both of these areas contribute to the formation of eastern Mediterranean deep water (EMDW) today (Wüst, 1961; Miller, 1963; Roether et al., 1996), and given these coolings, likely did so in glacial times.

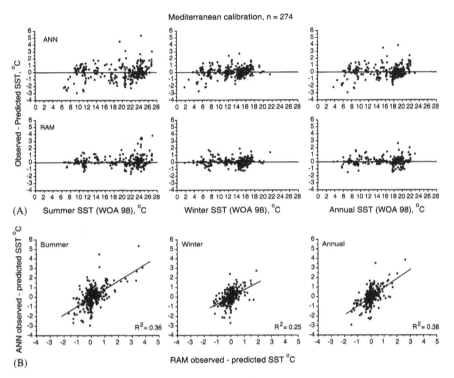

Fig. 7. (A) Relationships between observed and residual (i.e. observed–predicted) SSTs for summer, winter and annual temperatures in the calibration data set. (B) Relationship between SST residuals for ANN and RAM summer, winter and annual temperatures.

4.2.2. Glacial winter

Although the west–east temperature gradient was significantly stronger during the glacial winter than today, it was somewhat reduced compared to reconstructions for the glacial summer (Fig. 9A). The warmer eastern basin has a maximum temperature of ~16 °C, with cooler (13 °C) surface water conditions observed in the Aegean Sea extending southwards into the Levantine basin. The Gulf of Lions remained the coldest part of the Mediterranean with temperatures dropping as low as ~7 °C. Winter temperature anomalies are less dramatic than those observed during the glacial summer (Fig. 9B). The major changes occurred in the western basin, particularly in the Alboran Sea and Gulf of Lions where SSTs are recorded at 6 °C lower than modern winter temperatures. Once again the Aegean Sea was much cooler than today displaying a 5 °C decrease during the glacial winter.

4.3. Comparison between ANN and RAM

In general, both ANN and RAM reconstructed similar isotherm patterns in the glacial Mediterranean. However, the ANN technique tends to reconstruct higher SSTs for both the summer and winter, by 1–2 °C (Fig. 9). The mean standard deviations of SST estimates between the 10 partitions (ANN) and the best analogues (RAM) are illustrated in Fig. 10. RAM produces higher standard deviations, particularly during

the glacial summer, with 91.6% of the cores recording a standard deviation above 1. This is only slightly reduced to 86.1% for the glacial winter. By comparison, the standard deviations above 1 for ANN are 62.2% and 40.5% for summer and winter, respectively. Generally, the ANN appears more consistent (i.e. lower standard deviations) in predicting warmer SSTs (Fig. 10—see samples indicated by a cross). Conversely, RAM produces its lowest standard deviations for cores located in the cooler Gulf of Lions (Fig. 10—see samples indicated by an open triangle). The high standard deviations produced by RAM reflect the variability of SSTs among the selected best analogues. Although RAM adds a number of virtual samples to expand the calibration data set, there still appears to be an insufficient number of samples with consistent SSTs between the selected best analogues.

By calculating the standard deviation of the mean SST predictions within the LGM interval, we can observe the stability of the LGM at each core location (Fig. 11A). It should be noted that there is no relationship between the standard deviation and the number of samples within the LGM interval. Both techniques suggest a similar degree of variability within the LGM, only during the glacial summer RAM produced higher standard deviations (paired t-test, $p < 0.05$) indicating greater SST variability, especially at two locations (Fig. 11A). Despite the limitations of this analysis given by the uneven numbers of samples

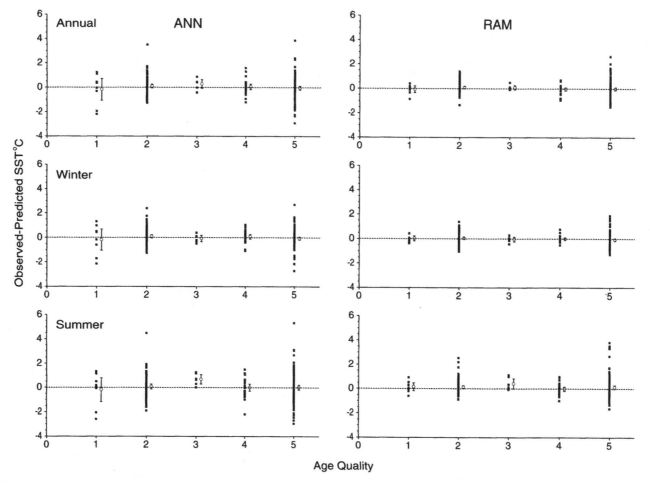

Fig. 8. Relationship between SST residuals and the age quality level of each core top sample in the calibration data set. Plots illustrate annual, summer and winter temperatures for both ANN and RAM methods. Open circles indicate the mean residual SST for each quality level. The error bars indicate the 95% confidence intervals. Quality levels range from 1 to 5 (1 being the best quality—see Kucera et al. (2004a) for full criteria).

per core in the LGM interval, the fact that some of the LGM variability clearly exceeds the expected prediction errors (Fig. 6) suggests a significant LGM instability in the Mediterranean, centred on areas with SST values in the middle of the glacial Mediterranean range.

Although the lack of consistency among the selected best analogues seems to be affecting RAM SST reconstructions, both the ANN and RAM techniques have produced similar SST patterns during the LGM (Fig. 11B). With this in mind, further discussions will relate only to SST estimates derived using ANN. The effect of the quality of dating of glacial samples on the coherence of SST reconstructions is shown in Fig. 12. In the southeastern Mediterranean, a core with a well-constrained chronological timeframe (level 1) produced a similar SST prediction to three other cores from the same region with age quality level 3. This highlights a general observation that samples from less well-dated cores do not deviate from regional patterns suggested by samples from cores with the best available chronology.

4.4. Comparison with previous LGM SST reconstructions

Although there are obvious similarities, our glacial SST reconstructions display some significant differences compared to previous studies based on the IKTF technique in the Mediterranean Sea. During the glacial summer, using the ANN technique, SST reconstructions range from 14 °C in the Alboran Sea to 23 °C in the south east of the basin, giving an SST gradient of 9 °C. This differs from initial reconstructions proposed by CLIMAP (1976) who estimated the SST gradient at 12 °C, while Thiede (1978) suggested that it was between 6 and 12 °C. The ANN SST estimates in the eastern basin tend to be 2–3 °C lower than those suggested by both CLIMAP (1976) and Thiede (1978). A similar deficit is noted in comparison with glacial reconstructions conducted by Thunell (1979). Although the ANN estimates in the western basin are in general agreement with CLIMAP (1976), there are some significant disagreements with Thiede's (1978) data. In the Gulf

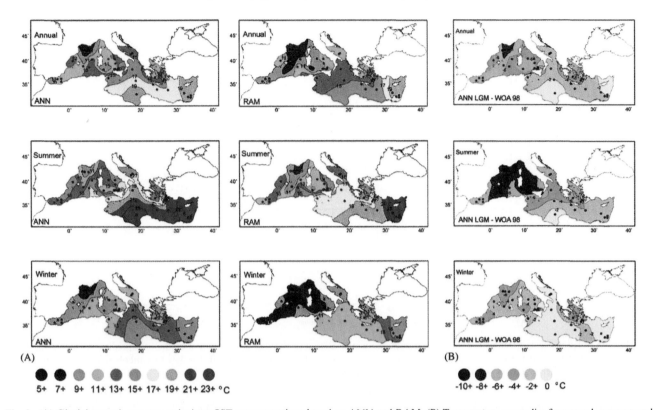

Fig. 9. (A) Glacial annual, summer and winter SST reconstructions based on ANN and RAM. (B) Temperature anomalies for annual, summer and winter SSTs during the last glacial maximum. Values are derived by subtracting modern day SSTs (World Ocean Atlas 98 v2) from glacial values, based on ANN reconstructions. The contour lines were created using the nearest neighbour option in the MapInfo Professional software (version 6.5). Black dots indicate location of cores used for reconstructions. Since we only aimed to produce a schematic representation of LGM SSTs, it should be noted that the base map does not take into account the lowered sea level during the LGM.

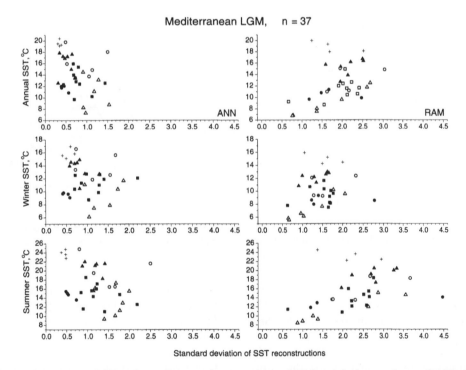

Fig. 10. The mean standard deviations of SST estimates between the 10 partitions (ANN) and the best analogues (RAM) plotted against glacial annual, summer and winter SSTs. Closed circle = Alboran Sea; open triangle = Gulf of Lions/Balearic basin; closed square = Tyrrhenian Sea/Strait of Sicily; open circle = Central Mediterranean; closed triangle = Southern Aegean Sea/Levantine basin; cross = Far eastern Mediterranean.

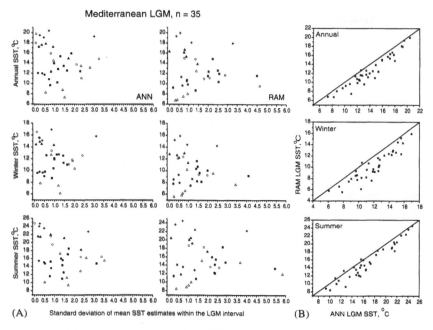

Fig. 11. (A) Relationship between glacial annual, summer and winter SSTs and the standard deviation of the SST predictions within the LGM interval for both ANN and RAM. Cores containing only one sample within the LGM are omitted. For key to symbols refer to Fig. 10 caption. (B) Relationship between LGM SSTs for ANN and RAM summer, winter and annual temperatures.

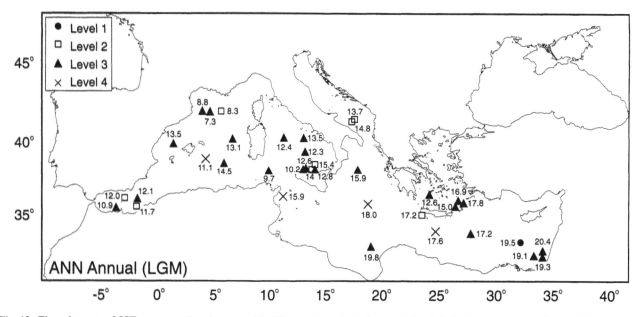

Fig. 12. The coherency of SST reconstructions in cores with different chronological control. Symbols indicate the age quality level (1–4) (Kucera et al., 2004a) of each core within the LGM data set. Values indicate annual SSTs based on ANN estimates.

of Lions, Thiede (1978) estimates an SST between 13–15 °C in contrast to ANN estimates of ~10 °C. The most notable differences occur in the Alboran Sea. According to Thiede (1979) a north–south gradient of 6 °C (13–19 °C) existed in the glacial summer in the Alboran Sea, equalling that estimated for the entire basin (Fig. 2B). In contrast, the ANN summer reconstructions suggest an average SST of 14 °C in the

entire Alboran Sea. This is in agreement with Perez-Folgado et al. (2003) who estimated a glacial summer SST of ~14 °C based on two cores in the Alboran Sea, using the modern analogue technique. Similarly, estimates produced by alkenone studies suggest SSTs between 11.5 and 13 °C during the LGM in the Alboran Sea (Cacho et al., 2000; Perez-Folgado et al., 2003). These compare well to our annual estimates of ~12 °C.

All reconstructions observe a SST decrease in the Aegean Sea. Both CLIMAP (1976) and Thunell (1979) suggest a reduction of approximately 1 °C (22 °C) relative to the present day, whereas Thiede (1978) and our ANN estimates indicate a more significant decrease of 7 °C (Fig. 9B).

The west–east SST gradient in the Mediterranean was significantly reduced during the glacial winter, with ANN estimating a value of just 6 °C compared to between 10 and 12 °C suggested by Thiede (1978). The ANN estimates in the eastern basin tend to be 2–3 °C higher than temperatures proposed by Thunell (1979), yet some 2–4 °C lower than Thiede's (1978) reconstructions. The Aegean Sea provides the exception with ANN SST estimates of 11 °C, comparable to that of Thiede (1978) but approximately 3 °C higher than those suggested by Thunell (1979). Smaller discrepancies occur within the western basin with both the ANN and Thiede's (1979) reconstructions differing within 1–2 °C, which is close to the estimated error of prediction for the ANN (Fig. 6).

All previous glacial SST reconstructions, including data from this study are summarised in Fig. 13. During the glacial summer it is interesting to note that CLIMAP (1976) and Thiede (1978) estimate that SSTs in the eastern basin were the same or within 1 °C of modern day values. In the western basin, with the exception of Thiede (1978), both CLIMAP (1976) and this study indicate a cooling of approximately 6 °C compared to

the present day. Conversely, during the glacial winter there is more variability within the SST reconstructions across the entire basin (Fig. 13). From all previous reconstructions, our results generally compare well with estimates proposed by CLIMAP (1976). However, a more precise comparison is difficult since the geographical resolution of cores used by CLIMAP (1976) is poor, therefore providing only a broad overview of the glacial Mediterranean. Compared to Thiede (1978), our ANN reconstructions only agree with winter reconstructions in the western basin and in the Aegean Sea during both seasons. This is expected since the calibration data set on which Thiede (1978) based his reconstructions only contained North Atlantic samples. This poses a problem for SST reconstructions in the eastern Mediterranean since the calibration data set contains no modern analogues for this region. As North Atlantic faunal assemblages are more similar to those in the cooler western basin, it is not surprising that Thiede's (1978) SST reconstructions from this region are more comparable with our results. Finally when compared with Thunell's (1979) reconstructions in the eastern Mediterranean, the ANN estimates are slightly lower in the summer and higher in the winter. These differences can probably be attributed to the core coverage of the reconstructions by Thunell (1979), which did not include any cores from the far eastern Mediterranean (see Figs. 2 and 5).

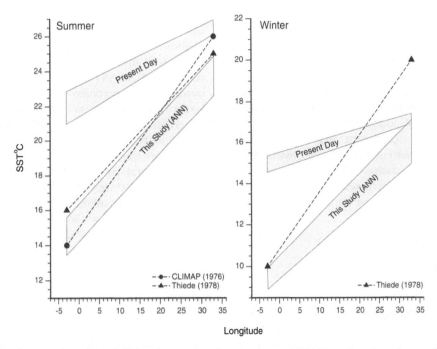

Fig. 13. A summary of Mediterranean east–west LGM temperature gradients reconstructed in this study and previous studies compared to present day gradients. Gradients for both summer and winter seasons are based on SSTs in the Alboran Sea and the south east Mediterranean. The grey shaded areas illustrate the range of SST gradients for both this study and the present day, whereas the gradients reconstructed for CLIMAP (1976) and Thiede (1978) are based on the average SST values.

4.5. Implications for models

Glacial SST reconstructions provide a means of testing ocean general circulation models. Previous attempts to model the circulation of the Mediterranean during the LGM have been conducted by Bigg (1994) and Myers et al. (1998). Bigg (1994) based his model on composite LGM SST reconstructions taken from Thiede (1978) and Thunell (1979). By comparison, Myers et al. (1998) opted to use the most extreme LGM SST estimates, using SST anomalies of −8 °C in the north western basin (Rohling et al., 1998), −7 °C in the central basin (Paterne et al., 1998) and −1 °C in the southern extremity of the eastern basin (Paterne et al., 1998). Our LGM SST reconstructions, as explained previously, differ from Thiede (1978) and Thunell (1979) but are more comparable to those by Rohling et al. (1998) in the northwestern basin and Paterne et al. (1998) in the central Mediterranean. However, in the southeastern basin our results observe LGM SSTs 2–3 °C colder compared to 1 °C colder as estimated by Paterne et al. (1998).

The LGM simulation produced by Bigg (1994) suggests an annual cooling in the eastern basin of 2 °C. This compares well with our estimates; however, the simulation fails to observe the dramatic cooling that occurs in the Aegean Sea. On the other hand, the western basin is significantly underestimated observing an annual cooling of only 4 °C compared to 7 °C based on the ANN. By comparison, Myers et al. (1998) concluded that simulations for the LGM only produced significant SST changes in the western Mediterranean. It is difficult to compare our SST estimates with those of Myers et al. (1998) since the LGM simulation is specific to January, whereas the ANN calculates average winter SSTs. However, a strong similarity exists between both SST estimates in the western basin, although in the far south east of the basin the simulation suggests a maximum SST of 13 °C compared to 16 °C based on ANN. The absence of a consistent and reliable LGM data set is evident when using SST reconstructions within modelling applications. Previous attempts at modelling the glacial Mediterranean are either based on outdated SST reconstructions or a combination of individual studies.

5. Conclusions

To produce more precise and reliable reconstruction of glacial SSTs in the Mediterranean we compiled a new, more comprehensive calibration data set consisting of 129 core tops from the North Atlantic (Fig. 3) and 145 from the Mediterranean Sea (Fig. 4). Annual, summer and winter SSTs have been estimated using both ANN

and RAM methodologies, with average error rates ranging between 1.1 and 0.5 °C (Fig. 6).

Glacial reconstructions are based on the largest ever LGM data set (Fig. 5), consisting of 273 samples in 37 cores, including new data for 12 cores. During the glacial summer, our reconstructions indicate a basin-wide east–west gradient of ~9 °C, whilst a gradient of 6 °C is suggested for the glacial winter. Unlike previous studies, our ANN SST estimates suggest a significantly cooler (3 °C) eastern basin during the glacial summer compared to the present day. In the western basin, ANN results are comparable to CLIMAP (1976) but disagree significantly with Thiede (1978). A significant decrease in SSTs has been confirmed in the Aegean Sea during the LGM. ANN SST estimates are much lower in the eastern basin compared to Thiede (1978) yet significantly higher (2 °C) than values suggested by Thunell (1979). SST anomalies in the western basin are much lower during the winter and compare well to estimates proposed by Thiede (1978). SST estimates following the RAM technique produce comparable LGM reconstructions to ANN estimates in terms of magnitude and gradients; however, estimates tend to be ~1 °C lower than those using the ANN technique. This discrepancy is comparable to our estimate of prediction errors of both techniques and we attribute it to an insufficient number of core tops within the calibration data set leading to the lack of consistency in SST values among the best-selected analogues. The reliability of our results is strengthened by the concordance of both techniques in extracting similar glacial SST patterns within the Mediterranean Sea. However, transfer function techniques such as RAM and MAT would benefit from larger calibration data sets.

Previous model simulations of the glacial Mediterranean relied on SST reconstructions that have become outdated. To overcome this problem, we constructed a larger calibration data set with a better constrained geographical range and a defined age quality control. Additionally, we used two modern transfer function techniques that produce lower RMSEP estimates and better generalisations than traditional techniques. Finally, the introduction of the largest-ever Mediterranean LGM data set combined with an enhanced age control on the definition of the LGM allowed us to provide modellers with a new set of targets on which to test simulations.

Acknowledgements

We would like to thank Simon Troelstra, Babette Hoogakker and Ralf Schiebel for providing samples and Mara Weinelt for running RAM on all the data. This project was funded by the Leverhulme Trust, the Nuffield Foundation and NERC. Thanks are due to three anonymous referees for their constructive and helpful criticism of the first version of this paper.

References

Bigg, G.R., 1994. An ocean general circulation model view of the glacial Mediterranean thermohaline circulation. Paleoceanography 9, 705–722.

Buccheri, G., Capretto, G., Donato, V., Esposito, P., Ferruzza, G., Pescatore, T., Ermolli, E., Senatore, M., Sprovieri, M., Bertoldo, M., Carella, D., Madonia, G., 2002. A high resolution record of the last deglaciation in the southern Tyrrhenian Sea: environmental and climatic evolution. Marine Geology 186, 447–470.

Cacho, I., Grimalt, J.O., Pelejero, C., Canals, M., Sierro, F.J., Flores, J.A., Shackleton, N., 1999. Heinrich event imprints in Alboran Sea paleotemperatures. Paleoceanography 14, 698–705.

Cacho, I., Grimalt, J.O., Sierro, F.J., Shackleton, N., Canals, M., 2000. Evidence for enhanced Mediterranean thermohaline circulation during rapid climatic coolings. Earth and Planetary Science Letters 183, 417–429.

Cita, M.B., Vergnaud-Grazzini, C., Robert, C., Chamley, H., Ciaranfi, N., d'Onofrio, S., 1977. Palaeoclimatic record of a long deep sea core from the eastern Mediterranean. Quaternary Research 8, 205–235.

CLIMAP, 1976. The surface of the Ice-Age Earth. Science 191, 1131–1137.

Gonzalez-Donoso, J., Serrano, F., Linares, D., 2000. Sea surface temperature during the Quaternary at ODP sites 976 and 975 (western Mediterranean). Palaeogeography, Palaeoclimatology, Palaeoecology 162, 17–44.

Hayes, A., Rohling, E.J., De Rijk, S., Kroon, D., Zachariasse, W.J., 1999. Mediterranean planktonic foraminiferal faunas during the last glacial cycle. Marine Geology 153, 239–252.

Herman, Y., 1972. Quaternary eastern Mediterranean sediments: micropalaeontological climatic record. In: Stanley, D.J. (Ed.), The Mediterranean Sea. Dowden, Hutchinson & Ross, Stroudsburg, PA, pp. 129–147.

Hutson, W.H., 1980. The Agulhas current during the Late Pleistocene: analysis of modern faunal analogs. Science 207, 64–66.

Imbrie, J., Kipp, J.Z., 1971. A new micropalaeontological method for quantitative paleoclimatology: application to a Late Pleistocene Caribbean core. In: Turekian, K.K. (Ed.), The Late Cenozoic Glacial Ages. Yale University Press, New Haven, pp. 71–181.

Jorissen, F.J., Asioli, A., Borsetti, A.M., Capotondi, L., De Visscher, J.P., Hilgen, F.J., Rohling, E.J., Van der Borg, K., Vergnaud-Grazzini, C., Zachariasse, W.J., 1993. Late Quaternary central Mediterranean biochronology. Marine Micropaleontology 21, 169–189.

Kallel, N., Paterne, M., Duplessy, J., Vergnaud-Grazzini, C., Pujol, C., Labeyrie, L., Arnold, M., Fontugne, M., Pierre, C., 1997b. Enhanced rainfall in the Mediterranean region during the last sapropel event. Oceanologica Acta 20, 697–712.

Kucera, M., Darling, K.F., 2002. Cryptic species of planktonic foraminifera: their effect on palaeoceanographic reconstructions. Philosophical Transactions of the Royal Society of London 360, 695–718.

Kucera, M., Rosell-Melé, A., Schneider, R., Waelbroeck, C., Weinelt, M., 2005a. Multiproxy approach for the reconstruction of the glacial ocean surface (MARGO). Quaternary Science Reviews, this issue (doi:10.1016/j.quascirev.2004.07.017).

Kucera, M., Weinelt, Mara, Kiefer, T., Pflaumann, U., Hayes, A., Weinelt, Martin, Chen, M.-T., Mix, A.C., Barrows, T.T., Cortijo, E., Duprat, J., Juggins, S., Waelbroeck, C., 2005b. Reconstruction of the glacial Atlantic and Pacific sea-surface temperatures from assemblages of planktonic foraminifera: multi-technique approach based on geographically constrained calibration datasets. Quaternary Science Reviews, this issue (doi:10.1016/j.quascirev.2004.07.014).

Lacombe, H., Gascard, J.C., Gonella, J., Bethoux, J.P., 1981. Response of the Mediterranean to the water and energy fluxes across its surface, on seasonal and inter-annual scales. Oceanologica Acta 4, 247–255.

Lacombe, H., Richez, C., 1982. The regimes of the Straits of Gibraltar. In: Nihoul, J.C.J. (Ed.), Hydrodynamics of Semi-enclosed Seas. Elsevier, Amsterdam, pp. 13–74.

Malmgren, B.A., Nordlund, U., 1997. Application of artificial neural networks to paleoceanographic data. Palaeogeography, Palaeoclimatology, Palaeoecology 136, 359–373.

Malmgren, B.A., Kucera, M., Nyberg, J., Waelbroeck, C., 2001. Comparison of statistical and artificial neural networks for estimating past sea surface temperatures from planktonic foraminifer census data. Paleoceanography 16 (5), 520–530.

MEDOC group, 1970. Observation of deep water formation in the Mediterranean Sea. Nature 277, 1037–1040.

Miller, A.R., 1963. Physical oceanography of the Mediterranean Sea: a discourse. Rapp. PV CIESM 17, 857–871.

Mix, A.C., Bard, E., Schneider, R., 2001. Environmental processes of the ice age: land, oceans, glaciers (EPILOG). Quaternary Science Reviews 20, 627–657.

Molina-Cruz, A., Thiede, J., 1978. The glacial eastern boundary current along the Atlantic Eurafrican continental margin. Deep-Sea Research 25, 337–356.

Murray, J., 1897. On the distribution of the pelagic foraminifera at the surface and on the floor of the ocean. Natural Science (Ecology) 11, 17–27.

Myers, P.G., Haines, K., Rohling, E.J., 1998. Modeling the paleocirculation of the Mediterranean: The last glacial maximum and the Holocene with emphasis on the formation of sapropel S_1. Paleoceanography 13 (6), 586–606.

Myers, P.G., Rohling, E.J., 2000. Quaternary Research 53, 98–104.

Olausson, E., 1960. Descriptions of sediment from the Mediterranean and Red Sea. Reports of the Swedish Deep Sea Expedition, 1947–48 8 (5), 287–334.

Olausson, E., 1961. Studies of deep sea cores. Reports of the Swedish Deep Sea Expedition, 1947–48 8 (4), 353–391.

Parker, F.L., 1958. Eastern Mediterranean foraminifera. Reports of the Swedish Deep Sea Expedition, 1947–48 8, 217–283.

Paterne, M., Guichard, F., Labeyrie, J., Gillot, P.Y., Duplessy, J.C., 1986. Tyrrhenian Sea tephrochronology of the oxygen isotope record for the past 60,000 years. Marine Geology 72, 259–285.

Paterne, M., Duplessy, J.C., Kallel, N., Labeyrie, J., 1998. CLIVAMP last glacial maximum sea salinities and temperatures, Technical Reports CLIVAMP-MAS3-CT95-0043, European Union Marine Science and Technology Program, Gif-sur-Yvette, France.

Perez-Folgado, M., Sierro, F.J., Flores, J.A., Cacho, I., Grimalt, J.O., Zahn, R., Shackleton, N., 2003. Western Mediterranean planktonic foraminifera events and millennial climatic variability during the last 70 kyr. Marine Micropaleontology 48, 49–70.

Pistek, P., De Strobel, F., Montanari, C., 1985. Deep Sea circulation in the Alboran Sea. Journal of Geophysical Research 90, 4969–4976.

Pflaumann, U., Duprat, J., Pujol, C., Labeyrie, L., 1996. SIMMAX: a modern analog technique to deduce Atlantic sea surface temperatures from planktonic foraminifera in deep-sea sediments. Paleoceanography 11 (1), 15–35.

Pflaumann, U., Sarnthein, M., Hapman, M., d'Abreu, L., Funnell, B., Huels, M., Kiefer, T., Maslin, M., Schulz, H., Swallow, J., van Kreveld, S., Vautravers, M., Vogelsang, E., Weinelt, M., 2003. Glacial North Atlantic: Sea-surface conditions reconstructed by GLAMAP 2000. Paleoceanography 18 (3), 1065.

Pujol, C., Vergnaud-Grazzini, C., 1989. Paleoceanography of the last deglaciation in the Alboran Sea (Western Mediterranean). Stable isotopes and planktonic foraminiferal records. Marine Micropaleontology 15, 253–267.

Roether, W., Manca, B.B., Klein, B., Bregant, D., Georgopoulus, D., Beitzel, V., Kovacevic, V., Luchetta, A., 1996. Recent changes in eastern Mediterranean deep waters. Science 271, 333–336.

Rohling, E.J., Gieskes, W.W.C., 1989. Late Quaternary changes in Mediterranean intermediate water density and formation rate. Paleoceanography 4, 531–545.

Rohling, E.J., Jorissen, F.J., Vergnaud-Grazzini, C., Zachariasse, W.J., 1993a. Northern Levantine and Adriatic Quaternary planktic foraminifera; reconstruction of paleoenvironmental gradients. Marine Micropaleontology 21, 191–218.

Rohling, E.J., den Dulk, M., Pujol, C., Vergnaud-Grazzini, C., 1995. Abrupt hydrographic changes in the Alboran Sea (Western Mediterranean) around 8000 yrs BP. Deep-Sea Research 42, 1609–1619.

Rohling, E.J., Hayes, A., Kroon, D., De Rijk, S., Zachariasse, W.J., 1998. Abrupt cold spells in the Late Quaternary NW Mediterranean. Paleoceanography 13 (4), 316–322.

Sarthein, M., Gersonde, R., Niebler, S., Pflaumann, U., Spielhagen, R., Thiede, J., Wefer, G., Weinelt, M., 2003. Overview of Glacial Atlantic Ocean Mapping (GLAMAP 2000). Paleoceanography 18 (2), 1030.

Sbaffi, L., Wezel, F.C., Kallel, N., Paterne, M., Cacho, I., Ziveri, P., Shackleton, N., 2001. Response of the pelagic environment to Palaeoclimatic changes in the central Mediterranean Sea during the Late Quaternary. Marine Geology 178, 39–62.

Schott, W., 1935. Die Foraminiferen aus dem aequatorialen Teil des Atlantischen Ozeans. Deutsch. Atl. Exped. Meteor. 1925–1927 3, 34–134.

Thiede, J., 1978. A glacial Mediterranean. Nature 276, 680–683.

Thunell, R.C., 1978. Distribution of recent planktonic foraminifera in surface sediments of the Mediterranean Sea. Marine Micropaleontology 3, 147–173.

Thunell, R.C., 1979. Eastern Mediterranean Sea during the last glacial maximum; an 18,000-years B.P. reconstruction. Quaternary Research 11, 353–372.

Todd, R., 1958. Foraminifera from western deep-sea cores. Report of the Swedish Deep Sea Expedition, 1946–47 8, 167–215.

Ufkes, E., Jansen, J.H.F., Brummer, G.J., 1998. Living planktonic foraminifera in the eastern South Atlantic during spring: indicators of water masses, upwelling and Congo (Zaire) River plume. Marine Micropaleontology 33, 27–53.

Vergnaud-Grazzini, C., Ryan, W.B.F., Cita, M.B., 1977. Stable isotope fractionation, climatic change and episodic stagnation in the eastern Mediterranean during the Late Quaternary. Marine Micropaleontology 2, 353–370.

Waelbroeck, C., Labeyrie, L., Duplessy, J., Guiot, J., Labracherie, M., Leclaire, H., Duprat, J., 1998. Improving past sea surface temperature estimates based on planktonic fossil faunas. Paleoceanography 13 (3), 272–283.

Wasserman, P.D., 1989. Neural Computing—Theory and Practice. Van Nostrand Reinhold, New York 230pp.

Wüst, G., 1961. On the vertical circulation of the Mediterranean Sea. Journal of Geophysical Research 66, 3261–3271.

ELSEVIER

Quaternary Science Reviews 24 (2005) 1017–1047

Sea-surface temperatures around the Australian margin and Indian Ocean during the Last Glacial Maximum

Timothy T. Barrows[a],*, Steve Juggins[b]

[a]Department of Nuclear Physics, Research School of Physical Sciences and Engineering, Australian National University, Canberra, ACT 0200, Australia
[b]School of Geography, Politics and Sociology, University of Newcastle, Newcastle upon Tyne, England, NE1 7RU, UK

Received 5 November 2003; accepted 15 July 2004

Abstract

We present new last glacial maximum (LGM) sea-surface temperature (SST) maps for the oceans around Australia based on planktonic foraminifera assemblages. To provide the most reliable SST estimates we use the modern analog technique, the revised analog method, and artificial neural networks in conjunction with an expanded modern core top database. All three methods produce similar quality predictions and the root mean squared error of the consensus prediction (the average of the three) under cross-validation is only ±0.77 °C. We determine LGM SST using data from 165 cores, most of which have good age control from oxygen isotope stratigraphy and radiocarbon dates. The coldest SST occurred at 20,500 ± 1400 cal yr BP, predating the maximum in oxygen isotope records at 18,200 ± 1500 cal yr BP. During the LGM interval we observe cooling within the tropics of up to 4 °C in the eastern Indian Ocean, and mostly between 0 and 3 °C elsewhere along the equator. The high latitudes cooled by the greatest degree, a maximum of 7–9 °C in the southwest Pacific Ocean. Our maps improve substantially on previous attempts by making higher quality temperature estimates, using more cores, and improving age control.
© 2004 Elsevier Ltd. All rights reserved.

1. Introduction

In the two decades since the CLIMAP project (CLIMAP project members, 1981; hereafter 'CLIMAP'), considerable interest has been maintained in the state of the Earth during the height of the last glaciation. The original CLIMAP concept of a geographical map of climate variables from a narrow window of time continues to have relevance today. As a climate end-member, the last glacial maximum (LGM) strongly influenced the biogeography of modern flora, fauna, and humans, and shaped much of the modern landscape. The LGM demonstrates how the Earth responds during maximum global cooling, and conversely, comparison

with the present describes how the Earth heats up during global warming. Global warming since the LGM is the greatest seen in recent geological history.

The major conclusions of CLIMAP (1981) (which updated CLIMAP, 1976) were based around the subtropical gyres; the surface circulation cells in the mid latitudes of each ocean. CLIMAP's reconstruction found these gyres to be thermally and geographically 'stable'. Climatic belts were displaced equatorward and compressed towards the centres of these gyres, with a concomitant steepening of thermal gradients. There was modest cooling along the equator and in the eastern boundary currents, with increased upwelling and advection of cool waters. There was limited cooling, or a warming, within the western boundary currents and stronger flow. The high latitudes experienced the greatest cooling. Overall, it was perceived that there was more energetic and rapid turnover of surface waters (CLIMAP, 1981).

*Corresponding author. Tel.: +61-2-6125-2077; fax: +61-2-6125-0748.
E-mail address: tim.barrows@anu.edu.au (T.T. Barrows).

0277-3791/$ - see front matter © 2004 Elsevier Ltd. All rights reserved.
doi:10.1016/j.quascirev.2004.07.020

Since the early 1980s our knowledge of the paleoceanography of the Indo-Pacific region has grown dramatically. The CLIMAP view of LGM sea-surface temperature (SST) has been refined by using improved methodologies and by taking cores in areas poorly covered by CLIMAP, which included much of the Australian region. In many cases it has been found that SST estimated by CLIMAP was too warm. In one classic example, a large warm anomaly off the eastern coast of Australia originally interpreted as a 'stronger' western boundary current was removed when better age control was applied to one core (Anderson et al., 1989). However, a large-scale revision by Prell (1985) found cooling in the tropics and the subtropical eastern Indian Ocean to be even less than CLIMAP. This conclusion was supported by the most recent review (Trend-Staid and Prell, 2002). One of the broadest conclusions of CLIMAP that remains robust today is that the tropical sea surface cooled to a lesser degree than the polar latitudes. For planktonic foraminifera, this observation appears independent of the technique or the training set used for estimating SST and is therefore an intrinsic feature of the data itself (Prell, 1985; Barrows et al., 2000).

A large number of new studies have been published since CLIMAP and provide an excellent opportunity to revise the original CLIMAP maps, especially in the sparsely covered Indo-Pacific region. Better data exists in the form of more core sites with better age control, more accurate modern SST maps, and a greatly expanded modern core top database. This paper derives new SST estimates based on planktonic foraminifera assemblages for the LGM time slice in the Australian region with the coverage extended to include the Indian Ocean. Our intention is to provide a snapshot of the temperature of the sea surface, both annual mean and seasonal extremes, during the coldest phase of the last glacial cycle. The reconstruction improves on previous work by including new results, reevaluating stratigraphy, and recalculating temperatures using the best available techniques to provide a comprehensive review of the original CLIMAP conclusions. This paper constitutes part of the MARGO project and contributes to the global reconstruction by that working group.

The paper is presented in three parts. First, we describe a new core top database of planktonic foraminifera assemblages. We use and rigorously evaluate state-of-the-art statistical approaches to ensure we make the highest quality SST estimates. Second, we investigate the stratigraphy of the LGM and establish the timing of minimum SST versus the $\delta^{18}O$ proxy for maximum ice volume. Last, we apply the techniques from the first section to the LGM data. We construct maps of SST across the Indian Ocean and around the Australian margin and interpret these patterns in terms of likely changes in oceanography.

2. Training set and evaluation of techniques

2.1. Training set

The training set used here (AUSMAT-F4) builds on earlier versions of the AUSMAT database (AUSMAT-F1 (Martinez et al., 1999); AUSMAT-F2 (Barrows et al., 2000), AUSMAT-F3 (Martinez et al., 2003)). In constructing the database, we excluded core tops from the North Pacific and the Atlantic Oceans because of faunal endemism (Darling et al., 2000) and differences in oceanography. The database consists of 1344 Southern Hemisphere core tops from a global database of 2619 core tops. The core of this is the Brown planktonic foraminifera database (Prell et al., 1999).

For AUSMAT-F4, we added 46 new core tops counted by J. Duprat and N. Kallel (unpublished data), together with the published data of Cayre et al. (1999), Feldberg and Mix (2002) and Niebler et al. (2003) (the latter replaces the data set of Niebler and Gersonde, 1998). We excluded 46 core tops because they contained the pink variety of *Globigerinoides ruber*, a species extinct in the Indo-Pacific Ocean since Oxygen Isotope Chronozone 5e (Thompson, 1981). The presence of this species indicates the core top is not late Holocene in age and is probably an old surface, contaminated or is reworked. We applied a simple quality control procedure before using the training set by eliminating outliers with an absolute standardized residual of greater than three standard deviations using the modern analog technique with the five closest analogs. Of the 31 samples eliminated (3% of the database), some had low counts, were from below the lysocline, or had anomalous faunas suggestive of contamination, a non-contemporary origin, or typographical errors. The remaining 1303 samples formed the final training set (Fig. 1).

We harmonized the taxonomy of AUSMAT-F4 to be consistent with the nomenclature adopted by the MARGO working group. We included the rare species *Hastigerina pelagica*, *H. digitata*, *Globorotalia bermudezi*, *Gr. cavernula*, and *Gr. ungulata*, but their impact on estimating SST is small. Three temperature variables were chosen (at 0 m water depth) at each core site to align with the foraminifera assemblages; mean annual average (T_{mean}); the warmest month (T_{max}), and the coldest month (T_{min}). It is important to remember that the latter two do not represent a single calendar month, but the warmest or coldest month whenever they occur during summer or winter. Foraminifera respond to the temperature of these months regardless of the time of the year at which they occur. We also included the five MARGO temperature variables (at 10 m water depth); annual average (T_{ann}), and the austral calendar seasons of summer (January–February–March), autumn (April–May–June), winter (July–August–September),

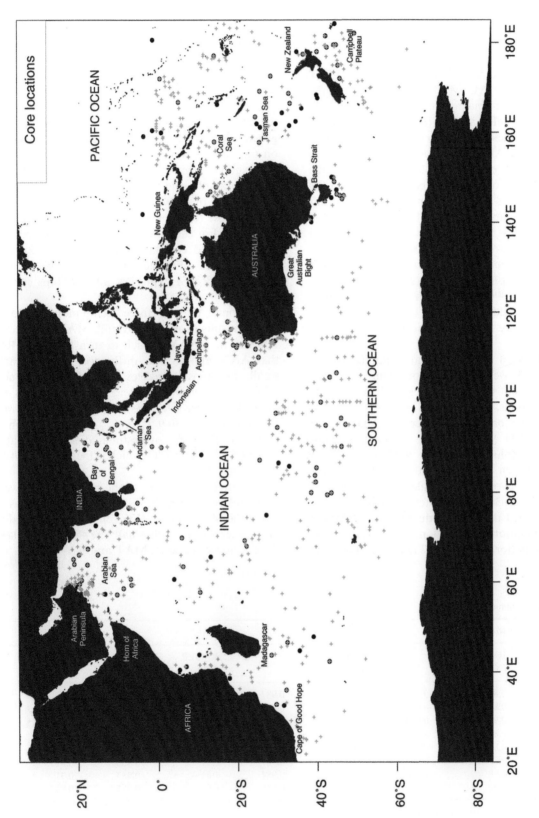

Fig. 1. Regional map of the Australian region and Indian Ocean. Crosses are core top positions and circles are core locations used in the LGM reconstruction.

and Spring (October–November–December). All temperatures were calculated from the World Ocean Atlas 1998 (Conkright et al., 1998).

2.2. Performance of SST estimation methodologies

We chose three statistical methods for evaluating SST estimation using the AUSMAT-F4 training set: the conventional modern analog technique (MAT; Prell, 1985), the revised analog method (RAM; Waelbroeck et al., 1998) and artificial neural networks (ANN; Malmgren et al., 2001). Although the first two methods are similar in their use of a dissimilarity coefficient, the inclusion of ANN provides an independent method for the calculation of SST. Previous studies have demonstrated the Imbrie and Kipp (1971) transfer function method produces less reliable results and is not pursued here (Prell, 1985; Barrows et al., 2000; Malmgren et al., 2001). We have also chosen not to use a technique that weights with geographical distance (such as SIMMAX; Pflaumann et al., 1996), as such an approach unfairly weights towards modern SST at the core site, producing unrealistically precise errors (see also Pflaumann et al., 2003).

Specific details of the numerical methods are as follows. For MAT, we followed the procedure described in Prell (1985) except we used the weighted mean of the four closest analogs, because this number gave the smallest reconstruction error under cross-validation (see below). For RAM we used the parameters gamma $= R = 0.2\,°C$, alpha $= 0.1$ and a maximum of five analogs, also chosen by cross-validation. Finally, for ANN we used a network with a single hidden layer of 15 neurons trained with 300 learning iterations, because our experiments showed that this produced lowest RMSEP under cross-validation. Calculations for MAT, RAM and ANN were performed using C2 (Juggins, 2003), the

WinRAM version of RAM98 (Waelbroeck et al., 1998) and NNET neural network library for SPLUS (Venables and Ripley, 2002) respectively.

The ability of each method to reconstruct modern SST was assessed by a five-fold leave-out cross-validation or split-sampling of the modern database. This involved splitting the data into five random subsets of which each contained approximately 20% of the observations. Each subset was removed from the database in turn and the remaining core tops used to generate the reconstruction models that were then tested on the omitted samples. This process was repeated for each subset to give a leave-out SST estimate for each modern sample, and the whole data-splitting procedure repeated 10 times with different random partitions of the database to provide a more reliable estimate of the error rate. For each method we report the performance as the squared correlation between the observed and predicted SST and the root mean squared error of prediction (RMSEP) across all core-tops for the 10 partitions.

We found that all three techniques produced very similar results (Table 1). Overall, MAT produces lower RMSEP than RAM, which in turn has lower RMSEP than ANN, although the differences in performance are small. For most variables and most techniques, the RMSEP is under 1°C, and as low as 0.84°C. All methods are able to reconstruct SST across the range of the training set (30.9°C in T_{max} to -1.6°C in T_{min}). However, SSTs are overestimated by 1–2°C at mean annual SSTs below 2°C, which is the lowest limit we can confidently predict to.

A relative comparison of the predictions between the three different methods show no systematic bias, either along the SST gradient (Fig. 2), or geographically, and differences are generally small: the standard deviation of the differences between MAT and RAM is 0.53°C, and

Table 1
Cross-validation performance statistics

Variable	MAT[a]		RAM[b]		ANN[c]		Consensus	
	RMSEP (°C)	r^2	RMSEP (°C)	r^2	RMSEP (°C)	r^2	RMSEP (°C)	r^2
T_{mean}	0.84	0.988	0.86	0.987	0.87	0.987	0.77	0.990
T_{max}	0.87	0.986	0.88	0.986	0.91	0.985	0.80	0.989
T_{min}	0.98	0.984	0.98	0.984	1.01	0.983	0.89	0.986
T_{10mean}	0.84	0.988	0.86	0.987	0.86	0.987	0.77	0.990
Summer (JFM)	0.89	0.985	0.91	0.984	0.93	0.983	0.82	0.988
Autumn (AMJ)	0.89	0.987	0.89	0.987	0.91	0.986	0.81	0.989
Winter (JAS)	0.98	0.984	0.98	0.984	1.01	0.983	0.91	0.986
Spring (OND)	0.90	0.987	0.92	0.986	0.95	0.985	0.82	0.987

RMSEP = root mean squared error of prediction (°C).
[a]Weighted mean of 4 analogs.
[b]Gamma = 0.2, R = 0.2, alpha = 0.2, maximum of five analogs.
[c]Three hundred epochs, weight decay = 1.0e−5, mean and SD of 10 networks, 15 neurons in hidden layer.

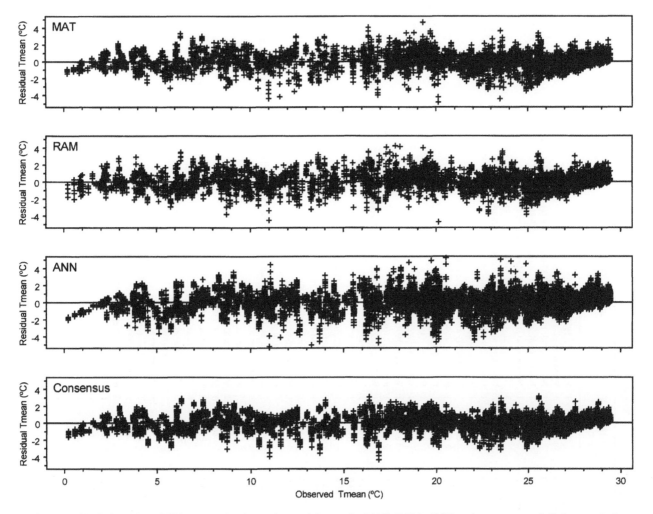

Fig. 2. Residuals from the 5-fold leave out for the modern training set for MAT, RAM, ANN, and a consensus of all three methods.

between MAT and ANN is 0.92 °C. Given that all three methods have similar performance and that there is no systematic bias between methods we developed a consensus reconstruction, calculated as the average of the three SST predictions. The consensus predictions have significantly lower cross-validation RMSEP for the modern training set than the individual techniques alone (i.e. 0.77 °C for T_{mean}, compared to 0.84 °C (MAT), 0.86 °C (RAM) and 0.87 °C (ANN): Table 1, Fig. 2). This prediction of SST is impressive (3% of the T_{mean} temperature range) considering the number of other variables, such as water column structure, that could potentially affect assemblage composition. The prediction error also incorporates errors associated with sample preparation, counting, taxonomic differences between workers and core top age.

Mean annual temperature is the variable predicted most precisely. T_{max} and summer SST also have low RMSEP whereas T_{min} and winter SST are the least well predicted. These results are perhaps unsurprising considering that the accumulation of foraminifera occurs throughout the year in most places, thereby integrating

an annual temperature signal. In some areas, particularly the high latitudes, foraminiferal productivity is often biased towards the warmest months, thereby relating more to summer temperatures. Conversely, an increase in foraminiferal productivity as a result of advection of cold water into the equatorial Pacific or upwelling often occurs during the winter months. Therefore, we also include these seasonal temperature variables in our reconstruction.

3. Data for LGM reconstruction

The compilation of Barrows et al. (2000) forms the basis for the reconstruction in the western Pacific Ocean. Overall we used the original data of CLIMAP (1981), published in Moore et al. (1980), and Prell et al. (1980) together with published data from subsequent studies listed in Table 2. Only the previously chosen LGM level data were available from the compilation of Prell (1985). Data from Thunell et al. (1994) and Trend-Staid and Prell (2002) were unavailable. Data from this paper is

Table 2
Site data and stratigraphy for cores used in LGM reconstruction

Core	Longitude (°E)	Latitude (°N)	Depth (m)	Isotopes sp.	Event 2.2 (cm)	SST$_{min}$ (cm)	^{14}C	Sed. rate (cm ka^{-1})	Sampl. int. (yr)	N	References for faunal, δ^{18}O and ^{14}C data
A015558	51.73	8.98	3985	G. sacculifer	28	58		2.8	2900	1	Prell et al. (1980)
CHAT-1K	188.50	−41.58	3556	Uvigerina spp.	54	70	4	3.4	2600	1	Weaver et al. (1998)
DSDP site 588	161.23	−26.11	1533	Uvigerina spp.	31	42		2.0	5100	1	Nelson et al. (1993b) and Martinez (1994)
DSDP site 591	164.45	−31.58	2131	Uvigerina spp.	31	40	1	2.0	5100	1	Nelson et al. (1993b) and Martinez (1994)
DSDP site 592	165.44	−36.47	1088	Uvigerina spp.	30	31		1.5	6800	1	Nelson et al. (1993b) and Martinez (1994)
DSDP site 593	167.67	−40.51	1068	Uvigerina spp.	38	38		1.9	5100	1	Nelson et al. (1993b) and Martinez (1994)
DSDP site 594	174.95	−45.52	1204	Uvigerina spp.	191	252	4	12.3	1600	2	Nelson et al. (1993a) and Wells and Okada (1997)
E26-001	168.34	−40.28	914	Uvigerina spp.	30	61		3.0	4100	1	Hesse (1994) and Barrows et al. (2000)
E27-030	147.23	−45.07	3552	G. bulloides	60	60	2	2.9	3400	1	Passlow et al. (1997)
E36-023	150.05	−43.92	2533	G. bulloides	19	19	1	0.9	6800	1	Nees (1994) and Martinez (1994)
E45-027	105.58	−43.23	3779	G. bulloides	20	30		1.5	6800	1	Prell et al. (1980)
E45-029	106.52	−44.88	3863	G. bulloides	49	39		1.9	5100	1	Howard and Prell (1992)
E45-078	114.35	−45.04	4031	—		31		1.5	6800	1	Prell et al. (1980)
E45-102	113.58	−33.62	1980	—	—	24	1	1.2	4100	1	Wells and Wells (1994)
E48-003	100.01	−41.02	3868	—	—	22		1.1	10300	1	Williams (1976)
E48-011	97.58	−29.70	3462	G. ruber	22	22		1.1	6800	1	Prell et al. (1980)
E48-022	85.41	−39.90	3324	G. truncatulinoides	82	82	6	4.0	2600	1	Williams (1976) and De Deckker, P. (unpub)
E48-023	83.72	−39.52	3459	G. truncatulinoides	32	55		2.7	5100	1	Williams (1976)
E48-027	79.90	−38.54	3231	G. bulloides	42	62		3.0	3400	1	Williams (1976)
E48-028	79.92	−38.55	3173	—		73	1	3.6	3400	1	Williams (1976) and De Deckker, P. (unpub)
E48-035	86.49	−30.35	1800	G. bulloides	19	35.5	4	1.7	2900	1	De Deckker, P. (unpub.)
E49-018	90.17	−46.05	3291	G. bulloides	40	50		2.4	4100	1	Howard and Prell (1992)
E49-021	94.88	−42.18	3328	G. bulloides	38	49		2.4	3400	1	Howard and Prell (1992)
E49-023	95.08	−47.12	3206	G. bulloides	28	59		2.9	2900	1	Howard and Prell (1992)
E55-006	141.06	−38.85	2346	Uvigerina spp.	40	40	1	2.0	2600	1	Passlow et al. (1997)
FR1/94-GC03	150.00	−44.26	2667	G. bulloides	35	30	5	1.5	3400	1	Barrows et al. (2000), Hiramatsu and De Deckker (1997) and De Deckker, P. (unpub)
FR10/95-GC05	121.03	−14.01	2472	G. sacculifer	57	86		4.2	500	8	Martinez et al. (1999)
FR10/95-GC11	115.00	−17.64	2458	G. sacculifer	29	30		1.5	1400	2	Martinez et al. (1999)
FR10/95-GC14	112.66	−20.05	997	G. sacculifer	31	41		2.0	1400	2	Martinez et al. (1999)
FR10/95-GC17	113.50	−22.13	1093	G. sacculifer	129.5	150		7.3	400	10	Martinez et al. (1999)
FR10/95-GC20	111.83	−24.74	841	G. sacculifer	54	62		3.0	900	4	Martinez et al. (1999)
FR10/95-GC29	114.59	−30.99	1220	G. sacculifer	97	109		5.3	900	4	Martinez et al. (1999)
FR2/96-GC10	108.51	−24.46	2852	G. sacculifer	22	25		1.2	1900	2	Martinez et al. (1999)
FR2/96-GC17	112.74	−12.25	2571	G. sacculifer	47	47		2.3	700	5	Martinez et al. (1999)
FR2/96-GC27	116.27	−18.56	1024	G. sacculifer	121	109		5.3	300	13	Martinez et al. (1999)
G6-4	118.07	−10.78	3510	G. ruber	121	126	4	6.1	1600	1	Wang et al. (1999) and Martinez, Barrows and Wang (unpub)
H214	177.44	−36.93	2045	G. bulloides	210.5	210.5	19	10.3	800	5	Samson (1999)
K12	127.74	2.69	3510	G. ruber	316.75	296.75		14.5	800	5	Barmawidjaja et al. (1993)
KH90-3-P2	160.01	−1.17	2494	P. obliquiloculata	58.5	63.1		3.1	2600	1	Kimoto et al. (2003)
KH92-1-3a	139.64	8.02	2830	G. sacculifer	30.1	32.4		1.6	2600	1	Kimoto et al. (2003)
KH92-1-5a	141.86	3.53	2283	G. sacculifer	19.8	17.5		0.9	6800	1	Kimoto et al. (2003)
MD76-131(C)	72.57	15.53	1230	Uvigerina spp.	200	206		10.0	1000	3	Cayre et al. (1999)
MD76-131(P)	72.57	15.53	1230	?	?	190		9.3			Prell (1985)
MD76-135	50.52	14.44	1895	C. wuellerstorfi	220	204		10.0	1000	3	Cayre et al. (1999)
MD77-169	95.03	10.13	2360	?	?	151		7.4			Prell (1985)
MD77-171	94.09	11.46	1760	?	?	261		12.7			Prell (1985)
MD77-179	91.01	18.22	1986	?	?	52		2.5			Prell (1985)

Core				Species							Reference
MD77-180	89.51	18.28	1986	?	?	181		8.8		1	Prell (1985)
MD77-191	76.43	7.30	1254	?	?	672	11	32.8		1	Prell (1985) and Ruddiman (1992)
MD77-194	75.14	10.28	1222	?	?	311		15.2		1	Prell (1985)
MD77-203	59.34	20.42	2442	?	?	341		16.6		1	Prell (1985)
MD79-254	38.67	−17.88	1934	Cibicides spp.	178	240		11.7	900	8	van Campo et al. (1990)
MD88-769	90.11	−46.07	3420	Cibicides spp.	50	107	17	5.2	900	7	Salvignac (1998)
MD88-770	96.47	−46.02	3290	C. wuellerstorfi	100	110	18	5.4	400	10	Labeyrie et al. (1996)
MD90-963	73.88	5.07	2446	C. wuellerstorfi	40	49		2.4	2600	1	Cayre et al. (1999) and Bassinot et al. (1994)
MD94-101	79.42	−42.50	2920	Cibicides spp.	30	55		2.7	1900	3	Salvignac (1998)
MD94-102	79.84	−43.51	3205	Cibicides spp.	65	82.5	22	4.0	1200	2	Salvignac (1998)
NGC098	162.51	−35.00	1338	G. sacculifer	28	28		1.4	5100	1	Kawagata (2001)
NGC099	162.00	−30.00	1158	G. sacculifer	20	20		1.0	4100	1	Kawagata (2001)
NGC100	162.00	−25.27	1299	G. sacculifer	28	24		1.2	5100	1	Kawagata (2001)
ODP site 723	57.61	18.05	800	U. excellens	329.68	580		28.3	1700	2	Anderson and Prell (1993) and Niitsuma et al. (1991)
ODP site 828	166.28	−15.29	3087	G. sacculifer	1202	1230		60.0	400	10	Martinez (1994)
ODP122-760A	115.54	−16.92	1970	C. wuellerstorfi	40	39		1.9	10300	1	Wells and Wells (1994)
ODP122-762B	112.25	−19.89	1360	Uvigerina spp.	55	80		3.9	6800	1	Wells and Wells (1994)
P69	178.00	−40.40	2195	Uvigerina spp.	315	465		22.7	1400	2	Weaver et al. (1998)
Q200	172.03	−46.00	1370	Uvigerina spp.	34	59		2.9	2100	1	Weaver et al. (1998)
Q585	182.08	−49.70	4354	Uvigerina spp.	63	73		3.6	2900	1	Weaver et al. (1998)
R657(S)	181.51	−42.53	1408	Cibicides spp.	50	50	5	2.4	4100	1	Sikes et al. (2002)
R657(W)	181.51	−42.53	1408	Uvigerina spp.	60	50		2.4	4100	1	Weaver et al. (1998)
RC08-039	42.35	−42.88	4330	G. bulloides	85	70		3.4	2600	1	Prell et al. (1980)
RC08-078	184.23	−44.78	1756	—	—	50	1	2.4	4100	1	Barrows et al. (2000)
RC09-110	187.98	−42.87	1917	—	—	72		3.5	2900	1	Thiede (unpub.)
RC09-124	172.6	−28.75	2540	Uvigerina spp.	30	40		2.0		1	Moore et al. (1980) and Prell (1985)
RC09-126	168.73	−33.23	2060	—	—	30		1.5		1	Moore et al. (1980) and Prell (1985)
RC09-148	113.70	−29.68	2056	—	—	19		0.9	6800	1	Wells and Wells (1994)
RC09-150(P)	114.55	−31.28	2703	G. inflata	50	80	5	3.9	2600	1	Prell et al. (1980), CLIMAP (1981), Bé and Duplessy (1976) and Conolly (1967)
RC09-150(W)	114.55	−31.28	2703	C. wuellerstorfi	60	90		4.4	2300	1	Wells and Wells (1994)
RC09-161	59.60	19.57	3332	—	160	160		7.8	1300	3	Prell et al. (1980)
RC09-162	60.42	19.08	3092	G. sacculifer	50	50		2.4	4100	1	Prell et al. (1980)
RC10-131	157.97	−14.53	2933	G. sacculifer	40	30	1	1.5	6800	1	Anderson et al. (1989)
RC11-086	18.45	−35.78	2829	Uvigerina spp.	45	40		2.0	5100	1	Prell et al. (1980)
RC11-120	79.87	−43.52	3193	benthic sp.	70	74		3.6	2900	1	Prell et al. (1980) and Martinson et al. (1987)
RC11-121	82.25	−39.72	3426	—	—	40	1	2.0	4100	1	Prell et al. (1980)
RC11-126	94.42	−30.07	2336	G. sacculifer	20	20		1.0	10300	1	Prell et al. (1980)
RC11-134	110.55	−33.07	2345	G. sacculifer	20	10	1	0.5	10300	1	CLIMAP (1981)
RC11-145	110.02	−25.48	3911	G. sacculifer	50	29	1	1.4	6800	1	Wells and Wells (1994) and Prell et al. (1980)
RC11-147	112.75	−19.06	1953	Uvigerina sp.	45	60	2	2.9	2900	1	Prell et al. (1980)
RC12-107	169.20	−26.00	3115	—	—	40		2.0		1	Moore et al. (1980) and Prell (1985)
RC12-109	157.87	−25.88	2930	G. sacculifer	40	40		2.0	5100	1	Anderson et al. (1989)
RC12-113(A)	163.52	−24.88	2454	G. sacculifer	40	50		2.4	4100	1	Anderson et al. (1989)
RC12-113(M)	163.52	−24.88	2454	G. sacculifer	40	40	1	2.0	2600	1	Martinez (1994) and Anderson et al. (1989)
RC12-328	60.60	−3.95	3087	?	?	30		1.5		1	Prell (1985)
RC12-339	90.03	9.13	3010	G. sacculifer	80	70		3.4	2900	1	Prell et al. (1980)
RC12-340	90.02	12.70	3012	?	?	70		3.4		1	Prell (1985)
RC12-341	89.58	13.05	2988	?	?	60		2.9		1	Prell (1985)
RC12-343	90.57	15.17	2666	?	?	60		2.9		1	Prell (1985)
RC12-344	96.07	12.77	2140	?	?	290		14.1		1	Prell (1985)
RC13-038	177.10	−14.52	2867	—	—	40		2.0	5100	1	Barrows et al. (2000)
RC14-007	44.75	−35.52	3288	G. inflata	20	25		1.2	5100	1	Prell et al. (1980)
RC14-009	47.88	−39.02	2692	G. inflata	70	70		3.4	2600	1	Prell et al. (1980)

Table 2 (continued)

Core	Longitude (°E)	Latitude (°N)	Depth (m)	Isotopes sp.	Event 2.2 (cm)	SST$_{min}$ (cm)	^{14}C	Sed. rate (cm ka^{-1})	Sampl. int. (yr)	N	References for faunal, δ^{18}O and ^{14}C data
RC14-029	88.32	-10.92	2869	G. sacculifer	20	20	1	1.0	6800	1	Prell et al. (1980)
RC14-035	89.95	-0.83	3021	G. sacculifer	31	40	4	2.0	3400	1	Prell et al. (1980)
RC14-037	90.17	1.47	2226	G. sacculifer	40	40		2.0	4100	1	Prell et al. (1980)
RC14-039	90.52	-5.86	2952	?	?	70		3.4		1	Prell (1985)
RC17-069	32.60	-31.50	3380	G. inflata	40	60	2	2.9	2900	1	Prell et al. (1980) and Bé and Duplessy (1976)
RC17-073	36.02	-32.12	2021	G. sacculifer	20	30		1.5	6800	1	Prell et al. (1980)
RC17-098	65.62	-13.22	3409	G. sacculifer	42	53		2.6	4100	1	Prell et al. (1980)
RC17-108	57.72	-10.51	4051	—	—	40		2.0	5100	1	CLIMAP (1981)
RC17-113	66.10	15.03	3874	—	—	40		1.5	6800	1	Prell et al. (1980)
RS067-GC10	140.10	-38.86	3332	—	—	50.5		2.5	3400	1	Passlow et al. (1997)
RS096-GC21	108.50	-23.77	2100	C. wuellerstorfi	19.5	29	1	1.4	5100	1	Wells and Wells (1994)
RS147-GC07	146.28	-45.15	3307	planktonic sp.	89.5	108		5.3	700	5	Samson (1999)
RS147-GC14	145.24	-46.45	3360	—	—	25		1.2	4100	1	Howard et al. (2002)
RS147-GC31	149.05	-44.53	3402	G. bulloides	58	69		3.4	2100	1	Howard et al. (2002)
S794	178.00	-35.31	2195	G. bulloides	100	144		7.0	1700	2	Weaver et al. (1998) and Wright et al. (1995)
SH9006	117.60	-4.33	1999	G. ruber	650	574		28.0	400	10	Ding et al. (2002)
SH9016	128.24	-8.46	1805	G. ruber	58	55	9	2.7	1900	2	Spooner et al. (in press)
SH9022	122.04	-11.35	2313	G. ruber	569	559	8	27.3	700	5	Ding et al. (2002)
SH9034	111.01	-9.10	3330	G. ruber	459	780	8	38.0	300	13	Ding et al. (2002)
SO36-07	144.75	-42.25	1085	Uvigerina spp.	65.5	67	1	3.3	1200	3	Lynch-Stieglitz et al. (1994) and Wells and Connell (1997)
SO36-21	145.50	-43.80	2829	G. bulloides	30	20	1	1.0	10300	1	Wells and Connell (1997)
SO36-39	145.99	-45.87	2497	G. bulloides	30	30	1	1.5	6800	1	Wells and Connell (1997)
SO36-47	145.98	-46.87	2497	G. bulloides	30	50		2.4	5100	1	Wells and Connell (1997)
U938	179.50	-45.08	2700	Uvigerina spp.	81	130	5	6.3	1900	1	Weaver et al. (1998) and Sikes et al. (2002)
U939	179.50	-44.50	1300	Uvigerina spp.	56	57.5	6	2.8	1700	2	Sikes et al. (2002)
V14-077	32.86	-29.64	1818	—	—	29		1.4	5100	1	Prell et al. (1980)
V14-081	43.78	-28.43	3634	—	—	35		1.7	5100	1	Prell et al. (1980)
V14-101	58.57	8.65	2849	G. sacculifer	70	70		3.4	2300	1	Prell et al. (1980)
V14-102	57.18	10.25	3915	G. sacculifer	40	40		2.0	5100	1	Prell et al. (1980)
V16-089	85.78	-33.02	3416	G. ruber	30	40		2.0	5100	1	CLIMAP (1981)
V18-191	46.57	-32.43	2946	—	—	30		1.5	5100	1	CLIMAP (1981)
V18-207	87.12	-25.63	2434	G. sacculifer	80	40		2.0	4100	1	Prell et al. (1980)
V19-096	172.00	-0.87	4252	—	—	15		0.7		1	Moore et al. (1980) and Prell (1985)
V19-178	73.25	8.12	2188	—	—	80		3.9	2300	1	Prell et al. (1980)
V19-185	59.33	6.70	2867	G. ruber	30	30		1.5	5100	1	Prell et al. (1980)
V19-188	60.67	6.87	3356	G. sacculifer	55	60	14	2.9	2900	1	Prell et al. (1980)
V19-201	40.43	-5.33	1846	G. ruber	30	70		3.4	2900	1	Prell et al. (1980)
V19-202	41.18	-6.98	2589	G. ruber	30	40		2.0	6800	1	Prell et al. (1980)
V19-204	43.82	-10.23	3524	P. obliquiloculata	10	30		1.5	6800	1	Prell et al. (1980)
V20-170	69.23	-21.80	2479	G. sacculifer	15	30	2	1.5	5100	1	Prell et al. (1980)
V20-175	68.00	-22.30	3526	G. sacculifer	20	30		1.5	6800	1	Prell et al. (1980)
V24-157	147.92	-14.95	1212	G. sacculifer	30	30		1.5	6800	1	Anderson et al. (1989)
V24-161	151.45	-18.20	1670	G. sacculifer	40	50		2.4	4100	1	Anderson et al. (1989)
V24-170	146.88	-13.52	2243	G. sacculifer	30	40		2.0	5100	1	Anderson et al. (1989)
V24-184	146.20	-12.87	2708	G. sacculifer	20	30		1.5	6800	1	Anderson et al. (1989)
V28-201	186.10	2.77	3217	—	—	20		1.0		1	Moore et al. (1980) and Prell (1985)
V28-203	180.60	0.95	3243	G. sacculifer	40	40		2.0		1	Moore et al. (1980) and Prell (1985)
V28-230	166.75	-5.50	2992			20		1.0		1	Moore et al. (1980) and Prell (1985)

V28-238	160.48	1.02	3120	Uvigerina spp.	35	39	58	1.9	5100	1	Moore et al. (1980) and Thompson (1981)
V28-239	159.18	3.25	3490	G. sacculifer	29	52		2.5	3400	1	Moore et al. (1980) and Thompson (1981)
V28-342(P)	120.50	−14.10	2730	?		90		4.4		1	Prell (1985)
V28-342(W)	120.50	−14.10	2730	—		89		4.3	2100	1	Wells and Wells (1994)
V28-345	117.95	−17.67	1904	?	—	90		4.4		1	Prell (1985)
V29-015	88.73	11.95	3173	?		70		3.4		1	Prell (1985)
V29-029	77.58	5.12	2673	G. sacculifer	80	80		3.9	2600	1	Prell et al. (1980)
V29-030	76.25	3.08	3651	—		80		3.9	2600	1	Prell et al. (1980)
V29-045	69.82	−6.00	2860	—		40		2.0	5100	1	CLIMAP (1981)
V29-048	63.43	−6.27	3882	G. sacculifer	30	20		1.0	6800	1	Prell et al. (1980)
V29-064	74.87	−27.28	3404	—		30		1.5	6800	1	CLIMAP (1981)
V33-065	112.58	−22.73	1692	?		40		2.0	3400	1	Wells and Wells (1994)
V34-087	57.35	13.27	2144	?	?	100		4.9		1	Prell (1985)
V34-088	59.76	16.54	2171	?	?	100		4.9		1	Prell (1985)
V34-091	64.03	20.94	3393	?	?	70		3.4		1	Prell (1985)
V34-092	65.12	21.15	3166	?	?	85		4.1		1	Prell (1985)
V34-101	67.42	17.49	3038	?	?	40		2.0		1	Prell (1985)
V34-109	66.08	19.73	2742	?	?	80		3.9		1	Prell (1985)
V34-111	63.88	17.62	3623	?	?	60		2.9		1	Prell (1985)
W268	178.97	−42.85	980	Uvigerina spp.	24	25	7	1.2	2300	1	Sikes et al. (2002)
Z2108	161.61	−33.38	1448	Uvigerina spp.	46	75	3	3.7	2600	1	Nelson et al. (1993b) and Barrows et al. (2000)
Z2112	166.53	−33.53	2858	planktonic sp.	90.5	88.5		4.3	700	5	Samson (1999)

Abbreviations are as follows:

Sed. rate = Sedimentation rate; Sampl. int. = sampling interval; N = number of samples in LGM interval.
G. sacculifer=Globigerinoides sacculifer; G. bulloides=Globigerina bulloides; G. ruber=Globigerinoides ruber; G. truncatulinoides=Globorotalia truncatulinoides; P. obliquiloculata=Pulleniatina obliquiloculata; G. inflata=Globorotalia inflata.

available as an electronic appendix from the PAN-GAEA data repository, the WDC-A for palaeoclimatology, and the Australian Quaternary Data Archive.

When selecting data for the LGM SST reconstruction, we preferred cores with down core foraminiferal counts, $\delta^{18}O$ stratigraphy and radiocarbon dates. Additional cores were included where core coverage was sparse. A total of 165 cores were used providing good geographical distribution, except for the abyssal plains where carbonate preservation is poor (Fig. 1).

4. Age model

4.1. Timing of the LGM

We follow the definition of the LGM as the most recent event when terrestrial ice volume was at maximum as an integrated average across the globe (Barrows, 2000). The time slice chosen to represent the LGM in this study is $21,000 \pm 2000$ cal yr BP (Mix et al., 2001), as adopted by the MARGO group for its global reconstruction. The choice of a chronostratigraphic definition for the LGM means that good absolute dating is essential for the reconstruction. However, many of the cores used here do not have radiocarbon dates and only a minority has dates within this time range. This leads to the need for a suitable synchronous stratigraphy.

Faced with a similar problem, the CLIMAP project focused on the maximum value of oxygen isotope analyses on foraminiferal carbonate as a datum for the LGM level within deep-sea cores. This is event 2.2 within Oxygen Isotope Chronozone 2 (OIC 2) in the stratigraphy of Prell et al. (1986), which is ascribed an absolute age of 17,850 yr in Martinson et al. (1987). Event 2.2 correlates with the maximum storage of water on land as glacial ice, which is preferentially richer in the light isotope (^{16}O) compared to the oceans. It was assumed by CLIMAP that this signal is synchronous within the mixing time of the world's oceans of a few thousand years. Prell et al. (1980) tested this assumption by dating cores around the Indian Ocean. The resulting date for the horizon ($18,600 \pm 1500$ ^{14}C yr BP) is within the error of a previous date determined by Pisias (1976) for cores in the Atlantic and Indian Oceans ($17,900 \pm 1300$ ^{14}C yr BP). Both of these dates are within the error of the time of maximum glaciation on land ($18,000$ ^{14}C yr BP), as chosen by CLIMAP (1976).

Whereas the majority of cores in this study have oxygen isotope records, most are measured on planktonic and not benthic foraminifera. This means that many of the $\delta^{18}O$ records are subject to the effects of surface temperature changes (e.g., a $4\,°C$ degree cooling is similar to the entire ice volume effect) and changes in the hydrological budget at the site (e.g., an increase in

evaporation increases $\delta^{18}O$), both of which can mask or produce a false oxygen isotope event. All cores here have planktonic foraminiferal data, but there is no single taxon or assemblage that can be used as a biostratigraphic marker for the LGM. However, the level, or levels, corresponding to minimum SST (SST_{min}) during OIC 2 can be calculated from the faunal data and can potentially be used to locate the LGM.

Since CLIMAP (1981), the number of cores with high quality age control and high-resolution sampling has increased considerably. This creates the opportunity to accurately determine the ages of event 2.2 and the SST_{min} using a subset of the highest quality cores to assess their stratigraphic significance. We selected seven widely distributed cores with a detailed oxygen isotope record (Table 3). These cores come from diverse oceanographic settings from within the tropics, the temperate latitudes and the subantarctic region providing two meridional transects in the eastern Indian Ocean and the western Pacific Ocean (Table 2). As pointed out by Alley et al. (2002), extrema of curves are sensitive to measurement errors (here, the sampling interval and the errors in the SST estimate and the $\delta^{18}O$ measurement), so we have tried to include as many independent cores as possible, without sacrificing quality. Where available, the level with the highest $\delta^{18}O$ was chosen from a benthic foraminiferal species to minimise the temperature effect on $\delta^{18}O$. Only AMS ^{14}C dates on the shallowest dwelling planktonic species were used to minimise the reservoir effect. Ages were calculated directly from the AMS ^{14}C date where this was determined at the level of interest, or an age was interpolated from closely spaced dates in the date range of $13,000–20,000$ ^{14}C yr BP, to minimise any error from fluctuating sedimentation rates. If the date was interpolated, then the error on the final age includes propagated errors from both dates. Radiocarbon dates were calibrated using CALIB 4.4 and the marine INTCAL98 curve (Stuiver et al., 1998). Conventional radiocarbon dates are expressed in years (^{14}C yr BP) and calibrated dates are reported in calibrated years (cal yr BP). Reservoir corrections were estimated using the CALIB Marine Reservoir Correction Database. For the subantarctic sites, these corrections are likely to be a minimum during the LGM (Bard, 1988). The level(/s) with the coldest mean SST in each core was calculated using the techniques described earlier in the paper.

The timing of the two events in each of the seven cores is listed in Table 3. In all seven cores, the SST_{min} occurs before event 2.2. The age of the SST_{min} is $20,500 \pm 1400$ cal yr BP ($17,800 \pm 1300$ ^{14}C yr BP) and the age of event 2.2 is $18,200 \pm 1500$ cal yr BP ($15,800 \pm 1300$ ^{14}C yr BP). The external error on the final ages is about an order of magnitude higher than the error on the original radiocarbon dates. This scatter is to be expected from stratigraphic uncertainty (the

Table 3
LGM age control

Core	ΔR^a	±	$\delta^{18}O_{max}$ level date (yr BP)	±	$\delta^{18}O_{max}$ level age (cal yr)	$\pm^b$	SST_{min} level date (yr BP)	±	SST_{min} level age (ca yr)	$\pm^b$	Reference
Pacific Ocean											
V28-238	50	31	13 840	450	15 890	570/780	16 090	450	18 570	600/580	Broecker et al. (1988a)
Z2112	10	7	17 890	220	20 680	410/400	18 320	220	21 180	420/410	Samson (1999)
RS147-GC07	−1	70	16 450	130	19 040	350/330	18 150	180	21 000	400/390	Samson (1999)
DSDP site 594	120	90	15 180	190	17 440	360/340	17 350	190	19 940	400/390	Wells and Okada (1997)
Indian Ocean											
SH9022	64	24	15 280	190	17 620	340/330	17 630	190	20 320	390/380	Ding et al. (2002)[c]
FR10/95-GC17	64	24	16 160	300	18 630	460/440	20 150	700	23 010	470/870	van der Kaas et al. (2002)
MD88-770	200	100	16 110	210	18 420	390/380	17 100	210	19 560	410/400	Labeyrie et al. (1996)
Mean			15 800	1300	18 200	1500	17 800	1300	20 500	1400	

[a]Value for MD88-770 estimated from an apparent surface age of 700 yr (Bard, 1988). Value for DSDP site 594 estimated from Sikes et al. (2002).
[b]68.3% confidence interval.
[c]Benthic $\delta^{18}O$.

combined effects of sampling interval, sedimentation rate and bioturbation). There is no significant geographical dependence on the lead; the difference in the subantarctic is statistically the same as for the tropics using modern reservoir corrections. These ages are very similar to the ages for the same events as determined previously in the western Pacific Ocean (Barrows et al., 2000). The date on the oxygen isotope maximum is younger than the date calculated by Prell et al. (1980) previously in the Indian Ocean. However in three of their five cores, Prell et al. (1980) chose a level deeper than event 2.2 as the LGM level.

The finding that minimum SST leads the maximum $\delta^{18}O$ signal by 2300 years is a general feature of the cores in the Indo-Pacific region. Where oxygen isotope records are available (117 SST estimates), SST leads in 70 cases (60%), 25 estimates are on the same level (21%) and SST lags in a minority of cases (19%). The percentage of SST leads is higher where benthic isotope data is available (81%), indicating a confounding of the planktonic foraminiferal isotope data by SST.

The lead of SST over $\delta^{18}O$ in deep-sea cores has been previously observed in the region studied here by employing a spectrum of SST estimation techniques. The lead is observed in the Southern Ocean using foraminifera (Williams, 1976), and diatoms (Labeyrie et al., 1996); in the Indian Ocean using radiolaria (Hays et al., 1976); in the southeast Atlantic and Indian Oceans using alkenones (Bard et al., 1997; Sonzogni et al., 1998; Sachs et al., 2001); and in the tropical Pacific and Indian Oceans and subantarctic Pacific and Southern Oceans using Mg/Ca in foraminiferal calcite (Mashiotta et al., 1999; Lea et al., 2000; Pahnke et al., 2003; Visser et al., 2003). Such a widely observed phenomenon across this sector of the Southern Hemisphere is consistent with our finding.

The age of event 2.2 ($18,200 \pm 1500$ cal yr BP) is slightly younger than the timing of minimum sea level, for which it is an indirect proxy. Hanebuth et al. (2000) places the sea level minimum before 21,000 cal yr BP and Yokoyama et al. (2000) determine the minimum lasted from 22,000 to 19,000 cal yr BP. The time delay may represent a lag propagating the geochemical signal (Barrows et al., 2000) and deep-sea temperature change (Mix et al., 2001) into the Southern Hemisphere during the LGM. The lag observed here (of the order of 2000 years) is close to expected given the ventilation time of the deep ocean (Broecker et al., 2004). A lag has been detected in the arrival of the meltwater signal in the Pacific and Indian Oceans of about 1000 years (Broecker et al., 1988b; Duplessy et al., 1991) and this figure is likely to have been larger during the LGM because of slower thermohaline circulation (Shackleton et al., 1988). Sea level rise apparently lagged the high northern insolation minimum by about 3000 yr and possibly by as much as 5000 yr (Alley et al., 2002). Mixing time would have further increased the lag, and the lag would be greater still if the isotopic composition of the ice sheets lagged ice volume (Mix and Ruddiman, 1984).

Within the error of the age determination (± 1400 years), SST is at a minimum at the same time across a vast region around Australia. Whereas there may be a phase lead by high latitude southern hemisphere SST over high latitude northern hemisphere SST (Imbrie et al., 1992), no low latitude lead is detectable here within the error of the dating. The timing of minimum SST is very similar to the temperature minimum on the Antarctic continent as recorded by the $\delta^{18}O$ and δD of ice. Lea et al. (2000) observed no discernable phase lag between equatorial eastern Pacific Ocean SST and Antarctica surface temperatures. Alley et al. (2002) rated the Byrd ice core as the best palaeotemperature record in Antarctica and determined that minimum temperature occurred at 22,000 yr, which is within the error of the age of the SST_{min} calculated here. The regional extent of the simultaneous cooling indicates cooling transmitted rapidly by the atmosphere (e.g., greenhouse gases) rather than by deep-sea circulation. However, we cannot rule out that some areas did not cool simultaneously, perhaps due to advection of heat associated with local currents, fronts or upwelling. Our conclusion does not extend west of our westernmost site (80°E) but the undated lead of SST over $\delta^{18}O$ is strong circumstantial evidence that the conclusion extends over this region as well. Despite these caveats, the variability of the age of the SST_{min} (± 1400 years) is at least as low as for event 2.2 (± 1500 years).

The age of the SST_{min} falls entirely within our target chronozone whereas that of event 2.2 lies at the younger margin. Consequently, we have chosen to use the SST_{min} as the datum on which to base our reconstruction. Apart from being available for all cores, the SST_{min} datum has the advantage by being an in situ generated signal. However, it has the disadvantage of not being measured as precisely as $\delta^{18}O$. In effect the SST_{min} is a quantitative biostratigraphic horizon: SST_{min} numerically represents the fauna responding to maximum cold.

4.2. Stratigraphy

Most of the 165 cores we selected for this study have oxygen isotope records (79%), where the data was available (141 cores). A total of 270 radiocarbon dates constrain the stratigraphy of the upper 50,000 yr in 43 cores (30% of the total). This represents a considerable improvement since CLIMAP (1981) when a total of 24% of cores had $\delta^{18}O$ records and 15% were ^{14}C dated. In most cases the selection of the SST_{min} level was unambiguous: we chose the depth level within Oxygen Isotope Chronozone 2, guided by any radiocarbon dates, with the coldest T_{mean}. Where dating information was not present, the uppermost SST_{min} in the core was chosen, guided by regional sedimentation rates and faunal changes down core.

Where the sample resolution was sufficiently high (sampling interval <2000 yr) we averaged multiple levels to calculate SST within the chronozone. Unavoidably, lower resolution cores with only one sample represent only a small amount of time within the LGM, whereas the cores with multiple levels represent more of an integration of SST during the interval and the SST estimate is statistically more 'stable'.

Based on the age determined above, sedimentation rates since the SST_{min} (the middle depth in a multiple sample average) range from 0.5 to 60 cm 1000 yr with a median of 2.5 cm 1000 yr^{-1}. The resultant sampling interval (the average time span between each sample above the SST_{min} level, or just through the LGM if

significantly different) ranges from 300 to 10,300 yr with a median value of 3400 yr. It is likely that some of the cores with sample resolutions greater than 4000 yr (64 estimates) do not have a sample that falls within the $21,000 \pm 2000$ yr interval. Hence these SST estimates are potentially minima only. Cores with high sedimentation rates combined with regular sampling, resulting in a sampling interval <1000 yr, provide the best medium for SST estimates.

After the LGM interval, SST increases rapidly after 18,000 cal yr BP in most cases. There is no evidence of a widespread return to glacial conditions during deglaciation (such as during the Younger Dryas Chronozone). Where a temperature reversal does occur, it is usually either before the Younger Dryas or is within the error of the SST estimate.

The primary limitation on this reconstruction is the sampling resolution of the cores. Historically cores have been sampled every 10 cm (10% of the core being processed), which often results in only 2 or 3 samples down to the LGM level in cores from the subtropics where sedimentation rates are low (1–2 cm 1000 yr^{-1}). In cores where the SST change is high (>3 times the prediction error), sampling will be the biggest potential source of error.

5. LGM sea surface temperature estimates

5.1. Methods

Based on the evaluation results from the first part of this paper, we calculated SST using all three techniques (MAT, RAM, ANN). For each level we used the consensus reconstruction, i.e. the average of the three methodological estimates without weighting, and calculated the standard deviation between these values as a measure of reliability. Where different researchers had worked on the same core levels, the SST estimates were averaged. We focused on reconstructing mean annual temperature (T_{mean}) because planktonic foraminifera best predict this variable. We also included the warmest and coldest months (T_{max} and T_{min}) in the reconstructions because, as noted before, these temperatures may be more applicable than T_{mean} in some areas. This inclusion also allows us to make a comparison with previous studies where February and August monthly temperature were used (e.g., CLIMAP, 1981). We also estimated the five MARGO temperature variables from the data to contribute to a global LGM reconstruction using other methodologies.

Temperature anomalies were calculated relative to World Ocean Atlas 1998 values and not to core top values, because (1) no systematic variation was found in the distribution of residuals for core tops during the calibration, which would call for a relative SST estimate (see for example Sachs et al., 2001), (2) the LGM SST estimate is not dependent on the quality of a second surface sample (its age, faunal composition, etc.), or the accuracy and precision of a second SST estimate, and (3) modern samples were not available for all cores.

5.2. Quality control

A range of criteria was used to assess the quality of the LGM SST estimates. The mathematical squared chord 'distance' to the nearest analog (MAT) is a good measure of how well the modern training set represents the LGM fauna. No estimates have a distance greater than 0.25 (Trend-Staid and Prell, 2002), indicating that good analogs are available for all LGM faunas. Many of the larger distances come from cores in subtropical and temperate settings where planktonic foraminifera assemblages are most diverse, and hence average distances tend to be greater (Fig. 3a). The standard deviation for each MAT estimate is not reported here because it is only a measure of variability between the analogs, and not an error on the SST measurement.

The standard deviation between the estimates from the three techniques provides a measure of methodological consistency. For T_{mean}, the average standard deviation is 0.5 °C, only 21 estimates are greater than 1.0 °C, and the maximum standard deviation is 1.7 °C, about twice the prediction error. There is no geographic pattern corresponding to the magnitude of the standard deviation (Fig. 3b), and the difference does not vary significantly for multiple level estimates. For the latter, there is the potential to inadvertently capture variability within the LGM and for the inclusion of samples outside the chronozone because of fluctuations in the sedimentation rate.

Five of the cores had counts from two independent workers. The standard deviation between these replicates averaged 0.7 °C, the same as the overall reconstruction. Finally, none of the LGM faunas contained more than 95% *Neogloboquadrina pachyderma* (sinistral), the practical limit for temperature estimation where the assemblage becomes monospecific.

5.3. Mapping

We contoured the LGM SST estimates into maps to interpret the geographic pattern of cooling. We objectively gridded the SST estimates using the kriging interpolation method (searching all the data using a linear variogram, including data outside the grid), on a grid resolution of 5° × 5°. This grid spacing was chosen because it provided an optimum between smoothing local variations, and preserving oceanographic patterns. To minimise the danger of extrapolation error (Broccoli and Marciniak, 1996) we restricted contouring to within the geographical limits of our core coverage. The LGM

coastline was drawn at –120 m (Yokoyama et al., 2000; Hanebuth et al., 2000) using the global ETOPO5 digital terrain model. We chose the equidistant cylindrical projection because our data is concentrated at low latitudes where this projection does not distort the map significantly.

To assess any mapping biases we gridded modern SST using only the core sites (Fig. 5a). As expected, the reproduction of modern isotherms is worst in areas of poor core coverage. However, the major oceanographic features are preserved, only finer scale patterns (of the order of 2° × 2°) are not seen (such as around New Zealand). The least reliable parts of the map are the margins, particularly along the southern boundary. The same will apply to the LGM maps. To minimise edge effects, the map was edited using five control points along the southern margin to preserve SST patterns within the map where unrealistic isotherms had been extrapolated. The SST ascribed to these points was such as to preserve the trends between nearby points. These control point locations were subsequently also used for the LGM maps. The inclusion of control points was solely for mapping purposes and does not affect our conclusions. In the western Pacific Ocean we included 12 nearby estimates from Kucera et al. (2004) to guide isotherms along the northern and eastern map edges. Kucera et al. (2004) found no systematic bias between estimates from the two reconstructions, justifying their inclusion in our grid.

After the initial gridding process, the maps were screened for estimates that were regionally discordant and produced unrealistic oceanography and isotherm patterns. Our aim is to produce the most consistent picture from the majority of the data rather than to focus on regional outliers. High-quality cores with good sample resolution were used as reference points. Estimates that were different to those in neighbouring cores by more than two standard deviations of the consensus prediction, out of trend with higher-quality cores, or produced isotherms creating unreasonable oceanography, were flagged with a quality level of 2, and were removed from the grid. Only cores with a regional concordance quality level of 1 were used to construct the map (Table 4). Four estimates were found to be regionally discordant (2% of the data). All but one of these cores had a sampling interval greater than 4000 yr, indicating that the level may not be in the LGM chronozone. One estimate (FR10/96-GC27) is much warmer than neighbouring estimates and may record very local conditions. The exclusion of these sites will need to be reassessed with the inclusion of new data in the future.

5.4. Interpretation of the maps

In the following section, we highlight the major features of the SST maps. Previously it has been popular

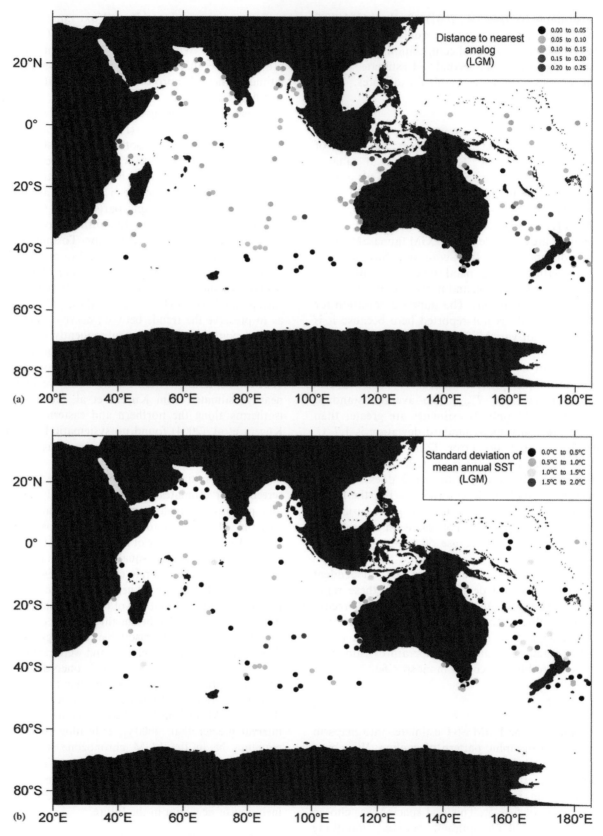

Fig. 3. (a,b) Quality measures for the LGM SST estimates, mapped by core location.

to try and determine past oceanic frontal movements by tracking isotherm movements. Similarly, there have been many attempts to determine the 'strength' of a surface current or of upwelling/divergence on the basis of temperature alone. However, it is difficult to determine movements in frontal systems, and changes in the volume of surface currents or upwelling/divergence on the basis of SST alone. First, isotherms correlating with fronts, upwelling zones and current positions vary seasonally and zonally (e.g., by 4 °C or more for the Subtropical Front) and so are only coincidental, and therefore the same correlation might not exist in the past. Wind patterns responsible for frontal systems and upwelling/divergence are unlikely to have varied coevally with surface temperature in the past because they are subject to different forcing. Second, because of interannual fluctuations, SST gradients that are signatures of modern fronts (such as the Subtropical Front) are not preserved at multi-year timescales (such as on long-term SST maps), and will therefore will not be preserved in sediment records. Third, bathymetry and landmasses contribute to front, current and upwelling positions, independently of climate change (Belkin and Gordon, 1996). Last, SST does not give 'strength' of a current or upwelling for which an estimate of mass transport is required.

The modern and LGM T_{mean} maps are presented in Fig. 4, and the anomaly map between Figs. 4a and b is Fig. 5b. Modern and LGM T_{max}, and T_{min} maps are given in Figs. 6 and 7 respectively. Seasonality maps (the difference between the warmest and coolest months) are given in Fig. 8. The thermal equator is drawn on Figs. 4a, b and 5a along the axis of highest mean temperatures. It is important to note that the thermal equator cannot be drawn on the T_{min} and T_{max} maps, because these maps are month transgressive. Rather than being limited to two calendar months (such as February and August), the T_{min}/T_{max} maps offers the advantage of identifying the thermal maximum/minimum whenever it occurs through the year. On the T_{max} map, the warmest month will be a month in austral summer. Conversely, the coolest month will be during the austral winter. Points north of the thermal equator will be the warmest (T_{max}) or coolest (T_{min}) month in the opposite season.

6. Results

6.1. Tropical SST (20°N to 20°S)

The most striking change in the tropics is the contraction of the Indo-Pacific 'warm pool', the region straddling the two oceans where mean temperatures are above 28 °C (Fig. 4a). In the Indian Ocean, the 28 °C isotherm retreated into the Bay of Bengal, and the Arabian Sea (Fig. 4b). This resulted in a slight warming

in the northern part of the map. The warmest part of the Indian Ocean becomes the Andaman Sea (28.3–28.8 °C). Two cores from the central Bay of Bengal (RC12-339,V29-015) with anomalies of –1.9 and –2 °C might be associated with upwelling off the adjacent islands to the east.

Tropical cooling resulted in significant shifts in the position of the thermal equator, the axis of highest mean temperature. East of 120°E, the thermal equator is close to the modern position. West of 100°E, the thermal equator lies further north in a boreal summer position, and is close to the modern position in the western Indian Ocean.

West of 100°E, tropical SST change ranges from a cooling of 2.5 °C to a warming of 0.5 °C. Of the 51 cores in this region, 20 cool by more than the methodological error. Most of the colder anomalies lie close to the equator. Isotherms bend towards the east in the path of the Southwest Monsoon Current, which flows from the west along the equator (Fig. 4b). The Indian South Equatorial Current is visible as westward warming isotherms at 15°S, ending in a western Indian Ocean 'warm pool' (Fig. 4b). The surface water then cooled as it was transported south into the Mozambique Current and north across the equator into the Somali Current. We do not observe the central Indian Ocean warm anomaly reconstructed by Trend-Staid and Prell (2002).

Upwelling is clear off the Horn of Africa and the Arabian Peninsula (Figs. 4b and 7b) as a pool of cooler water (Fig. 4b). Southwesterly winds blew across this region as they do in the African Monsoon today sufficiently to produce negative Ekman pumping. Unchanged to slightly warmer SST off the Arabian Peninsula supports the conclusion of Prell (1985), Anderson and Prell (1993) and Trend-Staid and Prell (2002) that the Southwest component of the African Monsoon was 'weaker' or was at least similar to present. However, SST anomalies are greater off the Horn of Africa suggesting upwelling here was more significant. Higher primary productivity from phytoplankton is consistent with stronger upwelling than present (Sonzogni et al., 1998), but may just reflect higher nutrient loads of the upwelled water.

Sonzogni et al. (1998) reviewed SST estimates made in the northern Indian Ocean using the alkenone unsaturation index ($U_{37}^{K'}$). A mean LGM cooling of 1.5–2.5 °C was calculated from 20 cores in the equatorial zone of the tropical Indian Ocean. These results are in close agreement with ours considering the errors. However, SST is about 1–2 °C systematically colder than the faunal estimates in the Arabian Sea and the Bay of Bengal.

South of the Indonesian Archipelago, the Java upwelling system was operating throughout the year and extended further to the east. The strongest tropical cooling is observed along the western Australia coast

Table 4
Modern and estimated LGM sea-surface temperatures

Core	Modern data				Quality			LGM estimates							Residuals			
	T_{mean}	T_{max}	T_{min}	Range	SCD_{min}	Sampl. int.	Reg. conc.	T_{mean}	±	T_{max}	±	T_{min}	±	Range	ΔT_{mean}	ΔT_{max}	ΔT_{min}	Δrange
A015558	26.3	29.0	23.4	5.5	0.204	2	1	25.8	0.4	28.1	0.3	23.6	0.3	4.5	−0.5	−0.9	0.2	−1.1
CHAT-1K	14.6	17.8	11.9	5.9	0.067	2	1	13.0	1.0	15.6	0.8	10.8	0.6	4.8	−1.6	−2.2	−1.1	−1.1
DSDP site 588	23.1	25.4	20.7	4.7	0.180	3	1	19.8	0.9	22.8	0.8	17.4	0.7	5.4	−3.2	−2.6	−3.3	0.7
DSDP site 591	20.7	23.3	18.4	4.8	0.156	3	1	17.5	1.1	20.2	1.4	14.8	1.6	5.4	−3.2	−3.0	−3.6	0.6
DSDP site 592	17.8	20.5	15.3	5.2	0.091	4	1	13.5	0.6	16.1	0.1	11.3	0.5	4.7	−4.3	−4.5	−4.0	−0.5
DSDP site 593	15.7	18.5	13.2	5.3	0.038	3	1	11.2	0.8	13.7	0.6	8.8	0.5	4.8	−4.5	−4.8	−4.4	−0.5
DSDP site 594	11.1	14.0	8.7	5.3	0.028	1	1	2.8	0.2	4.4	0.1	1.2	0.0	3.2	−8.2	−9.6	−7.5	−2.1
E26-001	15.8	18.7	13.3	5.4	0.087	3	1	11.3	1.2	13.8	0.9	8.9	0.7	4.9	−4.5	−4.9	−4.4	−0.5
E27-030	12.5	14.5	10.7	3.8	0.028	2	1	9.3	0.1	10.8	0.3	7.5	0.3	3.3	−3.2	−3.7	−3.2	−0.5
E36-023	13.5	15.9	11.4	4.6	0.059	4	1	10.3	1.2	11.7	1.2	8.7	1.7	3.0	−3.2	−4.2	−2.6	−1.6
E45-027	10.7	12.1	9.9	2.2	0.016	4	1	8.7	0.0	10.3	0.0	7.0	0.0	3.3	−1.9	−1.8	−2.9	1.1
E45-029	9.7	10.9	8.8	2.1	0.028	3	1	6.4	1.1	8.4	0.8	4.4	0.3	4.0	−3.3	−2.5	−4.4	1.9
E45-078	9.8	11.1	8.8	2.3	0.139	4	1	8.5	0.4	10.0	0.1	7.0	0.3	3.0	−1.3	−1.1	−1.8	0.7
E45-102	19.7	21.6	17.8	3.8		3	1	18.0	0.1	20.1	0.3	16.1	0.3	4.0	−1.8	−1.5	−1.7	0.2
E48-003	12.1	14.0	10.6	3.4	0.022	4	1	8.8	0.4	10.1	0.5	6.9	1.6	3.2	−3.3	−3.9	−3.7	−0.2
E48-011	19.6	22.6	17.1	5.5	0.232	4	1	19.5	1.7	22.7	1.5	16.8	1.3	5.9	−0.1	0.1	−0.3	0.4
E48-022	13.3	16.8	11.2	5.6	0.065	2	1	10.3	0.4	12.2	0.8	8.7	1.2	3.6	−3.0	−4.6	−2.6	−2.0
E48-023	13.7	17.1	11.5	5.6	0.067	3	1	11.0	0.3	12.8	0.5	9.4	0.7	3.4	−2.6	−4.3	−2.1	−2.2
E48-027	14.6	17.6	12.2	5.4	0.053	2	1	11.1	0.5	13.8	0.3	8.8	0.6	5.1	−3.5	−3.7	−3.4	−0.3
E48-028	14.6	17.6	12.2	5.4	0.059	2	1	12.0	0.2	14.6	0.1	9.9	0.2	4.7	−2.6	−3.0	−2.3	−0.7
E48-035	19.8	24.1	16.6	7.5	0.145	2	1	18.8	0.7	21.7	0.5	16.5	0.5	5.2	−1.0	−2.4	−0.1	−2.3
E49-018	8.9	9.8	7.5	2.4	0.014	3	2	4.7	0.4	6.8	0.4	2.9	0.3	4.0	−4.2	−3.0	−4.6	1.6
E49-021	11.3	12.9	9.3	3.6	0.036	2	1	8.6	0.4	9.7	0.4	6.9	0.3	2.8	−2.6	−3.2	−2.3	−0.9
E49-023	7.1	9.1	5.8	3.2	0.016	2	1	3.2	0.2	4.8	0.2	1.6	0.3	3.3	−3.9	−4.3	−4.3	0.0
E55-006	15.4	17.5	13.7	3.8	0.207	2	1	11.5	1.5	14.1	2.1	9.3	1.2	4.8	−3.9	−3.4	−4.4	1.0
FR1/94-GC03	13.2	15.6	11.2	4.4	0.046	2	1	10.8	0.7	13.1	0.3	8.7	0.6	4.4	−2.4	−2.4	−2.5	0.0
FR10/95-GC05	28.3	29.6	26.0	3.6	0.122	1	1	26.5	0.9	28.5	0.6	24.5	1.2	4.0	−1.8	−1.1	−1.5	0.4
FR10/95-GC11	27.0	29.2	24.9	4.3	0.093	1	1	24.3	0.2	26.7	0.3	22.2	0.1	4.5	−2.7	−2.5	−2.6	0.1
FR10/95-GC14	25.7	28.2	23.6	4.6	0.159	1	1	21.7	0.8	23.9	0.6	19.6	0.9	4.3	−4.1	−4.3	−4.0	−0.3
FR10/95-GC17	25.0	27.7	22.7	5.0	0.086	1	1	21.4	1.0	23.7	0.8	19.6	1.0	4.1	−3.6	−4.0	−3.1	−0.9
FR10/95-GC20	23.5	26.1	21.4	4.6	0.080	1	1	21.7	0.8	23.9	0.6	19.8	0.9	4.1	−1.8	−2.2	−1.6	−0.5
FR10/95-GC29	21.2	23.0	19.4	3.6	0.104	1	1	21.1	1.2	23.0	1.2	19.5	1.5	3.5	−0.1	0.1	0.1	0.0
FR2/96-GC10	22.8	24.6	21.1	3.5	0.086	1	1	21.9	0.6	24.0	0.6	19.9	0.7	4.1	−1.0	−0.6	−1.2	0.6
FR2/96-GC17	28.1	29.2	26.7	2.5	0.160	1	1	26.0	0.2	27.9	0.2	24.1	0.2	3.7	−2.1	−1.3	−2.6	1.3
FR2/96-GC27	26.8	29.1	24.4	4.7	0.094	1	2	27.7	0.5	29.3	0.5	25.9	0.3	3.4	0.8	0.2	1.5	−1.3
G6-4	28.1	29.5	25.9	3.6	0.160	1	1	25.0	1.0	26.8	0.4	23.1	0.5	3.7	−3.1	−2.7	−2.8	0.1
H214	18.1	20.9	15.6	5.3	0.133	1	1	15.3	1.4	17.8	1.4	13.4	1.0	4.4	−2.8	−3.1	−2.2	−0.9
K12	28.5	29.1	27.6	1.5	0.181	1	1	26.7	0.3	28.3	0.3	25.0	0.4	3.4	−1.8	−0.8	−2.6	1.8
KH90-3-P2	29.4	29.5	29.1	0.5	0.051	1	1	28.7	0.3	29.7	0.1	28.0	0.1	1.7	−0.6	0.2	−1.1	1.3
KH92-1-3a	29.0	29.4	28.2	1.3	0.053	2	1	27.3	0.4	29.1	0.3	25.6	0.3	3.6	−1.6	−0.3	−2.6	2.3
KH92-1-5a	29.2	29.7	28.8	0.9	0.079	2	1	26.6	0.6	28.7	0.4	24.5	0.4	4.2	−2.7	−1.0	−4.2	3.3
MD76-131(C)	28.3	29.8	27.1	2.6	0.149	1	1	28.6	0.4	29.9	0.1	27.6	0.1	2.3	0.3	0.1	0.5	−0.4
MD76-131(P)	28.3	29.8	27.1	2.6	0.087	5	1	27.7	0.1	29.5	0.1	26.1	0.2	3.3	−0.6	−0.3	−1.0	0.7

MD76-135	27.2	29.8	25.1	4.7	0.162	1	1	27.0	0.1	29.2	0.1	24.8	0.0	4.4	-0.2	-0.5	-0.3	-0.2
MD77-169	28.4	29.9	27.4	2.5	0.090	5	1	28.4	0.3	29.8	0.1	27.2	0.4	2.6	0.0	-0.1	-0.2	0.1
MD77-171	28.5	29.8	27.2	2.6	0.079	5	1	28.3	0.4	29.6	0.4	27.1	0.4	2.5	-0.2	-0.2	-0.1	-0.1
MD77-179	28.0	29.5	25.7	3.8	0.126	5	1	28.2	0.3	29.7	0.4	26.6	0.3	3.1	0.2	0.2	0.9	-0.7
MD77-180	28.0	29.6	25.6	3.9	0.094	5	1	28.1	0.3	29.7	0.3	26.7	0.3	3.0	0.1	0.1	1.0	-0.9
MD77-191	28.3	29.9	27.2	2.7	0.152	5	1	27.5	0.3	29.3	0.2	26.1	0.5	3.3	-0.8	-0.6	-1.2	0.6
MD77-194	28.6	30.5	27.3	3.3	0.110	5	1	28.4	0.2	29.8	0.1	27.3	0.4	2.4	-0.2	-0.8	0.1	-0.8
MD77-203	26.0	28.5	23.8	4.7	0.226	5	1	25.6	0.7	28.2	0.6	22.8	0.9	5.5	-0.5	-0.3	-1.1	0.8
MD79-254	26.8	28.7	24.9	3.8	0.155	1	1	24.4	0.7	26.5	0.4	22.2	0.7	4.4	-2.4	-2.2	-2.8	0.6
MD88-769	8.9	9.8	7.4	2.4	0.018	1	1	3.1	0.1	4.8	0.2	1.5	0.0	3.3	-5.8	-5.0	-5.9	0.9
MD88-770	7.8	9.6	7.0	2.7	0.017	2	1	2.9	0.1	4.4	0.1	1.1	1.1	3.3	-4.9	-5.2	-5.8	0.6
MD90-963	28.7	29.9	28.0	1.9	0.156	1	1	26.5	1.2	28.2	0.8	25.0	0.5	3.2	-2.2	-1.8	-3.0	1.3
MD94-101	11.8	13.4	9.8	3.6	0.024	1	1	9.2	0.6	10.8	0.8	7.8	0.9	2.9	-2.6	-2.7	-2.0	-0.7
MD94-102	10.8	12.1	8.8	3.3	0.019	1	1	7.6	0.7	9.7	1.4	6.0	0.9	3.7	-3.1	-2.3	-2.7	0.4
NGC098	19.2	22.0	16.7	5.2	0.150	3	2	17.4	1.1	20.0	0.4	14.8	0.4	5.2	-1.8	-2.0	-2.0	0.0
NGC099	21.6	24.3	19.4	4.9	0.166	3	2	20.3	0.3	22.8	0.4	17.6	0.3	5.2	-1.3	-1.5	-1.7	0.3
NGC100	23.3	25.6	21.1	4.5	0.133	3	1	21.6	0.3	24.1	0.2	19.2	0.1	4.9	-1.8	-1.5	-1.9	0.4
ODP site 723	26.0	28.7	23.1	5.6	0.089	1	1	26.0	0.0	28.5	0.2	23.2	0.5	5.3	0.0	-0.1	0.2	-0.3
ODP site 828	27.8	29.4	26.3	3.2	0.113	1	1	27.1	0.4	28.8	0.3	25.5	0.9	3.3	-0.7	-0.6	-0.8	0.1
ODP122-760A	27.3	29.3	25.2	4.1	0.132	4	1	23.9	0.4	26.1	0.4	21.8	0.8	4.4	-3.4	-3.1	-3.4	0.3
ODP122-762B	25.7	28.2	23.6	4.5	0.100	4	1	22.5	0.7	25.1	0.5	20.8	0.7	4.2	-3.2	-3.1	-2.8	-0.3
P69	16.7	19.6	13.7	5.9	0.072	1	1	12.0	0.5	14.6	0.1	9.7	0.1	4.9	-4.7	-5.1	-4.1	-1.0
Q200	10.6	13.1	8.2	4.9	0.020	2	1	2.7	0.1	4.3	0.3	0.8	0.1	3.6	-7.9	-8.8	-7.4	-1.4
Q585	9.7	12.0	6.6	5.4	0.012	2	1	2.5	0.3	4.0	0.9	0.9	0.4	3.1	-7.2	-8.0	-5.8	-2.2
R657 (S)	14.5	17.6	11.9	5.7	0.076	3	1	12.0	0.3	14.9	1.4	9.8	0.1	5.1	-2.4	-2.7	-2.1	-0.6
R657 (W)	14.5	17.6	11.9	5.7	0.096	3	1	12.3	0.6	14.7	0.4	9.9	0.3	4.8	-2.2	-2.9	-2.0	-0.9
RC08-039	9.9	12.7	6.6	6.1	0.048	2	1	3.6	0.2	5.5	0.2	2.2	0.2	3.3	-6.4	-7.3	-4.5	-2.8
RC08-078	12.8	15.6	10.1	5.5	0.080	3	1	11.4	0.4	14.1	0.6	8.6	0.2	5.4	-1.4	-1.5	-1.5	0.0
RC09-110	14.0	17.0	11.7	5.3	0.053	2	1	11.3	0.2	14.6	1.2	8.8	0.1	5.8	-2.7	-2.4	-2.8	0.5
RC09-124	21.4	24.2	19.0	5.2	0.155	5	1	19.4	1.1	22.5	1.5	17.0	1.1	5.5	-2.0	-1.7	-2.0	0.3
RC09-126	19.2	21.7	16.9	4.8	0.109	5	1	16.2	1.7	19.2	0.3	14.0	1.1	5.2	-3.0	-2.5	-2.9	0.4
RC09-148	21.5	23.5	19.7	3.8	0.130	4	1	21.4	0.4	23.4	0.9	19.6	0.5	3.8	-0.1	-0.1	-0.1	0.0
RC09-150 (P)	21.0	22.7	19.2	3.5	0.132	2	1	19.6	0.5	21.7	0.9	17.2	0.6	4.5	-1.4	-1.0	-2.0	1.0
RC09-150 (W)	21.0	22.7	19.2	3.5	0.133	2	1	17.4	0.7	19.8	0.3	15.4	0.7	4.4	-3.6	-2.9	-3.8	0.9
RC09-161	26.1	28.6	24.1	4.5	0.123	1	1	26.6	0.3	29.0	0.0	24.3	0.4	4.7	0.4	0.5	0.2	0.3
RC09-162	26.5	28.9	24.5	4.3	0.097	3	1	26.4	0.0	28.8	0.1	24.2	0.6	4.6	-0.1	0.0	-0.3	0.3
RC10-131	27.7	29.0	26.2	2.8	0.038	4	1	28.1	0.3	29.3	0.6	26.8	0.1	2.6	0.4	0.3	0.5	-0.3
RC11-086	18.8	21.1	16.4	4.8	0.138	3	1	17.3	0.5	19.6	0.3	14.5	0.6	5.1	-1.5	-1.5	-1.8	0.3
RC11-120	10.8	12.1	8.8	3.3	0.034	2	1	6.4	0.1	8.2	0.9	4.8	0.3	3.4	-4.4	-3.9	-4.0	0.1
RC11-121	13.7	16.9	11.6	5.3	0.084	3	1	10.7	0.8	12.5	0.5	8.8	0.9	3.7	-3.0	-4.4	-2.7	-1.7
RC11-126	19.5	22.5	16.7	5.8	0.075	4	1	18.7	0.4	21.8	0.2	16.1	0.5	5.7	-0.8	-0.7	-0.6	-0.1
RC11-134	18.8	20.8	17.1	3.7	0.094	4	1	17.9	0.4	20.5	0.4	15.8	0.3	4.7	-0.9	-0.3	-1.3	1.0
RC11-145	22.7	24.7	20.9	3.8	0.123	4	1	21.5	0.3	23.6	0.6	19.8	0.2	3.7	-1.2	-1.1	-1.1	0.0
RC11-147	26.2	28.6	24.1	4.5	0.148	2	1	22.6	0.9	24.9	0.4	20.6	0.8	4.3	-3.6	-3.7	-3.5	-0.2
RC12-107	23.0	25.5	20.7	4.8	0.102	5	1	21.9	0.4	24.6	0.6	19.8	0.5	4.8	-1.1	-0.9	-0.9	0.0
RC12-109	23.5	26.1	21.1	5.0	0.062	3	1	20.9	0.5	23.3	0.4	18.6	0.0	4.7	-2.6	-2.8	-2.5	-0.3
RC12-113 (A)	23.5	25.7	21.2	4.5	0.071	3	1	22.1	0.7	24.6	0.5	19.9	0.6	4.7	-1.4	-1.1	-1.3	0.2
RC12-113 (M)	23.5	25.7	21.2	4.5	0.121	2	1	22.4	0.3	24.8	0.0	20.2	0.6	4.6	-1.1	-0.9	-1.0	0.1
RC12-328	28.5	29.7	27.3	2.4	0.146	5	1	27.4	0.8	29.2	0.3	25.6	1.2	3.6	-1.1	-0.5	-1.7	1.2
RC12-339	28.5	29.8	27.7	2.2	0.119	2	1	26.6	0.8	28.7	0.6	24.8	0.9	3.9	-2.0	-1.1	-2.9	1.8

Table 4 (continued)

Core	Modern data				Quality			LGM estimates							Residuals			
	T_{mean}	T_{max}	T_{min}	Range	SCD_{min}	Sampl. int.	Reg. conc.	T_{mean}	±	T_{max}	±	T_{min}	±	Range	ΔT_{mean}	ΔT_{max}	ΔT_{min}	Δrange
RC12-340	28.5	29.9	27.4	2.5	0.147	5	1	28.2	0.5	29.5	0.6	26.9	0.7	2.5	-0.4	-0.4	-0.5	0.0
RC12-341	28.5	30.0	27.3	2.7	0.151	5	1	28.2	0.6	29.5	0.8	26.9	0.9	2.5	-0.3	-0.5	-0.4	-0.1
RC12-343	28.3	29.6	26.9	2.7	0.127	5	1	28.5	0.9	29.2	1.0	27.5	1.3	1.7	0.1	-0.4	0.7	-1.1
RC12-344	28.5	30.2	27.0	3.3	0.132	5	1	28.8	0.2	29.7	0.1	27.9	0.5	1.8	0.3	-0.5	0.9	-1.5
RC13-038	28.2	29.3	26.9	2.4	0.066	3	1	27.3	0.2	28.9	0.2	25.7	0.3	3.1	-0.9	-0.5	-1.2	0.7
RC14-007	18.9	21.6	16.6	5.1	0.105	3	1	15.6	0.3	18.1	0.5	13.3	0.0	4.8	-3.3	-3.5	-3.3	-0.2
RC14-009	16.6	18.9	13.4	5.5	0.063	2	1	13.0	1.3	15.6	1.2	10.7	0.8	4.9	-3.5	-3.3	-2.6	-0.7
RC14-029	27.7	28.7	26.7	2.1	0.129	4	1	25.4	0.5	27.4	0.5	23.6	0.6	3.7	-2.3	-1.4	-3.0	1.7
RC14-035	28.9	29.7	28.3	1.4	0.085	2	1	27.6	0.6	29.0	0.4	26.5	0.6	2.5	-1.3	-0.6	-1.7	1.1
RC14-037	28.9	29.8	28.3	1.5	0.079	3	1	28.2	0.3	29.6	0.3	27.3	0.4	2.3	-0.7	-0.2	-1.0	0.8
RC14-039	28.6	29.2	27.9	1.4	0.105	5	1	27.8	0.1	29.1	0.2	26.7	0.3	2.4	-0.7	-0.1	-1.2	1.0
RC17-069	22.9	25.3	20.7	4.6	0.124	2	1	20.3	0.8	23.0	1.1	17.9	0.9	5.1	-2.6	-2.3	-2.8	0.5
RC17-073	21.8	24.7	19.4	5.3	0.105	4	1	20.2	0.3	22.9	0.4	17.8	0.3	5.1	-1.7	-1.7	-1.5	-0.2
RC17-098	26.9	28.6	24.8	3.8	0.132	3	1	26.5	0.3	28.2	0.3	24.7	0.6	3.5	-0.4	-0.5	-0.2	-0.3
RC17-108	27.3	28.8	25.1	3.7	0.136	3	1	27.1	0.6	29.2	0.4	25.1	1.3	4.2	-0.2	0.4	-0.1	0.5
RC17-113	27.7	29.9	26.0	3.9	0.127	4	1	27.4	0.4	29.1	0.3	25.5	0.5	3.5	-0.3	-0.8	-0.4	-0.4
RS067-GC10	15.3	17.4	13.6	3.8	0.029	2	1	10.8	0.8	12.8	1.0	8.7	0.5	4.1	-4.5	-4.6	-4.9	0.3
RS096-GC21	23.2	25.0	21.4	3.5	0.109	3	1	21.7	0.1	24.0	0.2	19.8	0.1	4.2	-1.5	-1.0	-1.7	0.6
RS147-GC07	12.2	14.1	10.5	3.6	0.048	1	1	9.6	1.1	11.3	1.1	8.0	1.1	3.3	-2.5	-2.8	-2.5	-0.3
RS147-GC14	11.0	12.7	9.4	3.3	0.040	3	1	8.2	0.6	9.3	0.7	6.1	0.4	3.1	-2.8	-3.4	-3.2	-0.2
RS147-GC31	13.0	15.3	11.0	4.3	0.046	2	1	10.0	0.3	12.0	0.3	8.0	0.6	4.0	-3.0	-3.4	-3.0	-0.3
S794	18.4	21.3	15.8	5.4	0.107	1	1	14.1	0.5	17.0	1.0	11.8	0.4	5.2	-4.3	-4.3	-4.0	-0.2
SH19006	28.9	30.6	27.7	2.9	0.155	1	1	28.1	0.4	29.2	0.8	26.6	0.6	2.6	-0.8	-1.4	-1.1	-0.3
SH19016	28.3	29.7	26.6	3.1	0.108	1	1	27.8	0.4	29.1	0.4	26.5	0.5	2.6	-0.6	-0.6	-0.1	-0.5
SH19022	28.4	29.8	26.4	3.4	0.177	1	1	27.7	0.1	29.2	0.1	26.1	0.3	3.1	-0.7	-0.7	-0.3	-0.4
SH19034	27.9	28.9	26.4	2.5	0.177	1	1	25.8	0.6	27.9	0.4	24.2	0.4	3.7	-2.1	-1.1	-2.2	1.2
SO36-07	13.8	15.9	12.0	3.9	0.081	1	1	11.1	0.1	13.2	0.3	9.4	0.3	3.7	-2.8	-2.8	-2.6	-0.2
SO36-21	13.0	14.8	11.3	3.5	0.032	4	1	9.5	0.4	10.8	0.4	7.8	0.5	3.0	-3.4	-4.0	-3.5	-0.5
SO36-39	11.6	13.3	9.9	3.4	0.052	4	1	8.2	0.2	10.0	0.2	6.4	0.4	3.6	-3.3	-3.3	-3.5	0.2
SO36-47	10.8	12.4	9.1	3.3	0.023	3	1	7.2	0.7	9.7	0.7	5.3	1.1	4.4	-3.6	-2.7	-3.8	1.1
U938	11.8	15.0	9.4	5.7	0.063	1	1	4.0	0.5	5.7	0.5	2.2	0.6	3.4	-7.8	-9.4	-7.2	-2.2
U939	12.4	15.6	10.0	5.6	0.032	1	1	8.8	0.6	10.4	1.0	7.2	0.5	3.2	-3.6	-5.2	-2.8	-2.4
V14-077	23.8	26.4	21.4	4.9	0.171	3	1	21.1	0.9	23.9	1.0	19.0	0.8	4.9	-2.7	-2.4	-2.5	0.0
V14-081	23.6	26.1	21.3	4.7	0.147	3	1	21.5	0.7	24.0	0.5	19.2	0.6	4.7	-2.1	-2.1	-2.1	0.0
V14-101	27.7	29.9	26.4	3.5	0.147	2	1	25.9	0.7	28.2	0.6	23.6	0.5	4.6	-1.8	-1.7	-2.8	1.1
V14-102	27.2	29.7	25.5	4.2	0.136	3	1	26.1	0.6	28.5	0.8	23.2	0.2	5.3	-1.1	-1.2	-2.3	1.1
V16-089	18.2	23.4	15.0	8.4	0.093	3	1	15.0	1.7	17.0	2.6	12.3	1.6	4.7	-3.2	-6.5	-2.7	-3.7
V18-191	20.5	23.5	18.2	5.3	0.110	3	1	19.2	0.1	22.3	0.6	16.8	0.2	5.5	-1.3	-1.2	-1.4	0.2
V18-207	22.0	24.7	19.5	5.1	0.105	3	1	22.5	0.2	25.4	0.5	20.1	0.2	5.2	0.5	0.7	0.6	0.1
V19-096	28.9	29.4	28.0	1.3	0.164	5	1	28.3	0.2	29.0	0.2	27.4	0.3	1.6	-0.6	-0.3	-0.7	0.3
V19-178	28.7	30.1	28.0	2.2	0.093	2	1	27.4	0.4	29.0	0.4	26.4	0.4	2.6	-1.3	-1.2	-1.6	0.4
V19-185	28.2	30.0	27.2	2.8	0.170	3	1	25.7	0.9	28.0	1.1	23.4	0.8	4.6	-2.5	-2.0	-3.8	1.8
V19-188	28.3	30.1	27.4	2.6	0.090	2	1	26.6	0.5	28.7	0.4	24.5	1.0	4.1	-1.7	-1.4	-2.9	1.5
V19-201	27.3	29.3	25.2	4.1	0.153	2	1	26.3	1.0	28.4	0.9	23.9	1.4	4.4	-1.0	-0.9	-1.3	0.4

Core	T_{mean}	T_{max}	T_{min}	range	SCD_{min}	Sampl. int.	Reg. conc.	T_{mean}		T_{max}		T_{min}		range	ΔT_{mean}	ΔT_{max}	ΔT_{min}	Δrange
V19-202	27.4	29.5	25.1	4.3	0.133	4	1	26.1	0.2	28.3	0.2	23.8	0.3	4.4	-1.2	-1.2	-1.3	0.1
V19-204	27.1	28.8	24.9	3.9	0.146	4	1	26.9	0.2	28.7	0.2	24.7	0.4	4.0	-0.3	-0.1	-0.2	0.1
V20-170	24.4	26.8	21.6	5.2	0.096	3	1	23.9	0.1	26.4	0.2	21.5	0.3	4.9	-0.5	-0.5	-0.1	-0.4
V20-175	24.3	26.6	21.6	5.0	0.112	4	1	24.8	0.6	27.3	0.3	22.4	0.7	4.9	0.5	0.7	0.9	-0.1
V24-157	27.0	29.0	25.0	4.0	0.042	4	1	26.6	0.2	28.7	0.1	24.4	0.3	4.3	-0.4	-0.4	-0.7	0.3
V24-161	26.5	28.6	24.7	3.9	0.050	3	1	26.6	0.4	28.5	0.3	24.5	0.5	4.1	0.1	-0.1	-0.2	0.2
V24-170	27.1	29.1	25.0	4.1	0.049	3	1	26.2	0.8	28.2	0.5	23.9	0.9	4.3	-0.9	-0.9	-1.1	0.2
V24-184	27.1	29.2	24.9	4.3	0.105	4	1	25.8	0.4	27.7	0.1	23.6	0.5	4.1	-1.3	-1.5	-1.3	-0.2
V28-201	28.3	29.1	27.2	1.8	0.110	5	1	29.2	0.3	29.7	0.3	28.8	0.5	0.9	0.9	0.7	1.6	-0.9
V28-203	28.5	29.1	27.8	1.4	0.109	5	1	29.1	0.5	29.5	0.1	28.7	0.6	0.8	0.5	0.4	1.0	-0.6
V28-230	29.4	29.8	29.1	0.8	0.073	5	1	28.7	1.0	29.5	0.4	27.9	1.3	1.6	-0.8	-0.3	-1.2	0.8
V28-238	29.3	29.6	28.8	0.8	0.102	3	1	28.5	0.1	29.5	0.2	27.5	0.1	2.0	-0.9	-0.1	-1.3	1.3
V28-239	29.2	29.5	28.8	0.6	0.112	2	1	28.3	0.3	29.5	0.2	26.9	0.5	2.5	-0.9	0.0	-1.9	1.9
V28-342 (P)	28.2	29.5	26.0	3.5	0.122	5	1	28.0	0.7	29.6	0.3	26.5	1.1	3.0	-0.2	0.0	0.5	-0.5
V28-342 (W)	28.2	29.5	26.0	3.5	0.171	2	1	26.9	0.8	28.9	0.5	24.7	1.0	4.2	-1.3	-0.6	-1.3	0.7
V28-345	27.3	29.2	24.8	4.4	0.145	5	1	24.7	1.2	26.9	0.8	22.5	1.6	4.5	-2.6	-2.3	-2.3	0.0
V29-015	28.6	30.1	27.3	2.8	0.081	5	1	26.8	0.9	28.4	0.9	25.4	0.8	3.0	-1.9	-1.7	-1.9	0.3
V29-029	28.4	29.7	27.7	2.0	0.108	2	1	27.1	0.5	29.2	0.5	25.4	0.4	3.7	-1.3	-0.6	-2.3	1.7
V29-030	28.7	29.8	28.1	1.8	0.123	3	1	27.4	0.4	29.2	0.2	25.8	0.3	3.3	-1.2	-0.6	-2.2	1.6
V29-045	28.2	29.4	27.2	2.2	0.138	3	1	26.5	1.0	28.0	1.1	24.6	1.1	3.4	-1.7	-1.4	-2.6	1.2
V29-048	28.1	29.4	26.7	2.7	0.106	4	1	27.4	0.5	29.1	0.4	26.1	0.7	3.0	-0.7	-0.3	-0.6	0.3
V29-064	22.2	25.0	19.0	6.0	0.117	4	1	22.4	0.1	25.4	0.1	19.8	0.1	5.6	0.2	0.4	0.7	-0.4
V33-065	24.5	27.2	22.4	4.8	0.120	2	1	22.1	0.2	24.3	0.2	20.3	0.2	4.0	-2.4	-2.9	-2.1	-0.8
V34-087	26.8	29.7	25.0	4.7	0.093	5	1	25.9	0.2	28.5	0.1	23.3	0.1	5.2	-1.0	-1.2	-1.7	0.5
V34-088	26.7	29.4	24.8	4.6	0.101	5	1	26.2	1.1	28.6	0.6	23.9	1.2	4.7	-0.6	-0.8	-0.9	0.1
V34-091	26.9	29.2	24.4	4.9	0.123	5	1	27.5	0.6	29.2	0.5	25.6	1.0	3.7	0.5	0.0	1.2	-1.2
V34-092	27.0	29.3	24.3	5.0	0.228	5	1	24.7	0.5	26.9	0.6	22.8	0.6	4.1	-2.3	-2.4	-1.6	-0.9
V34-101	27.6	29.4	26.0	3.4	0.103	5	1	26.9	0.2	28.9	0.1	25.0	0.2	3.8	-0.7	-0.5	-1.0	0.4
V34-109	27.3	29.4	25.0	4.4	0.145	5	1	26.8	0.7	28.8	0.4	24.9	0.7	3.8	-0.5	-0.6	-0.1	-0.5
V34-111	27.2	29.4	25.1	4.3	0.101	5	1	26.6	0.4	28.8	0.2	24.5	0.6	4.2	-0.6	-0.6	-0.6	0.0
W268	14.0	17.1	11.8	5.4	0.037	2	1	9.9	0.2	11.6	0.3	7.6	0.1	4.0	-4.1	-5.5	-4.2	-1.3
Z2108	20.3	23.1	18.0	5.2	0.092	2	1	17.0	0.1	19.9	0.1	14.5	0.0	5.4	-3.3	-3.2	-3.4	0.2
Z2112	19.3	21.8	17.1	4.7	0.095	1	1	16.9	1.4	19.6	1.3	14.6	1.1	5.0	-2.4	-2.2	-2.5	0.3

Abbreviations are as follows:

T_{max}, T_{min} = warmest and coolest monthly SST (°C); T_{mean} = annual mean SST (°C); range = seasonality of SST (°C);

ΔT_{max}, ΔT_{min} and ΔT_{mean} = LGM minus modern residuals;

SCD_{min} = squared chord distance to nearest analog;

Sampl. int. = sampling interval; 1 = <2000 yr, 2 = 2000–4000 yr, 3 = 4000–6000 yr, 4 = >6000 yr, 5 = unknown;

Reg. conc. = Regional concordance; 1 = included in map, 2 = excluded from map.

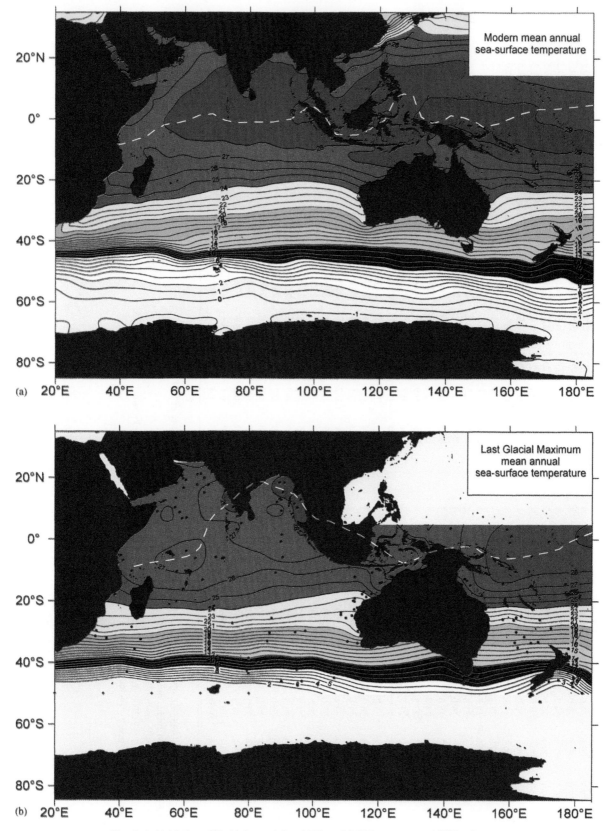

Fig. 4. (a,b) Modern (World Ocean Atlas, 1998) and LGM mean annual SST estimates.

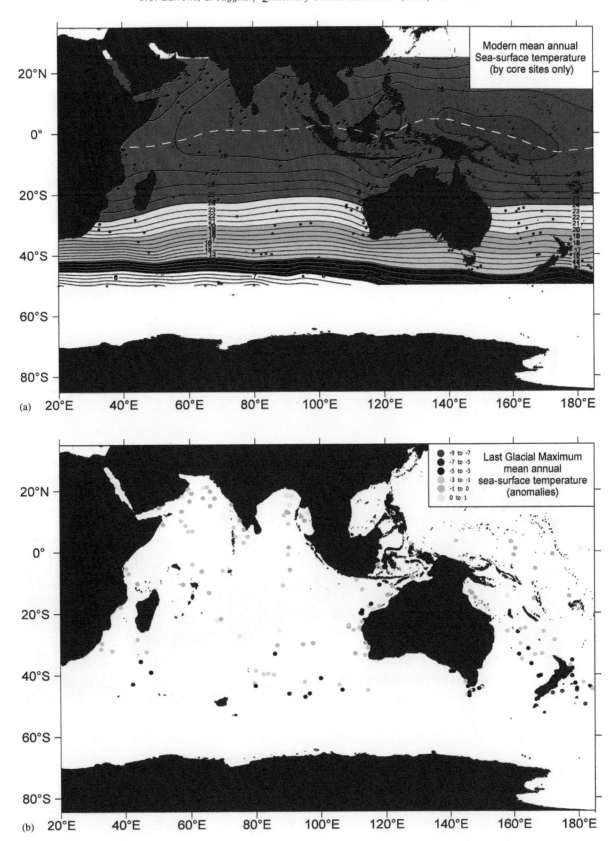

Fig. 5. (a) Modern (World Ocean Atlas, 1998) mean annual SST mapped using only the core locations. (b) Mean annual SST anomalies (LGM-modern).

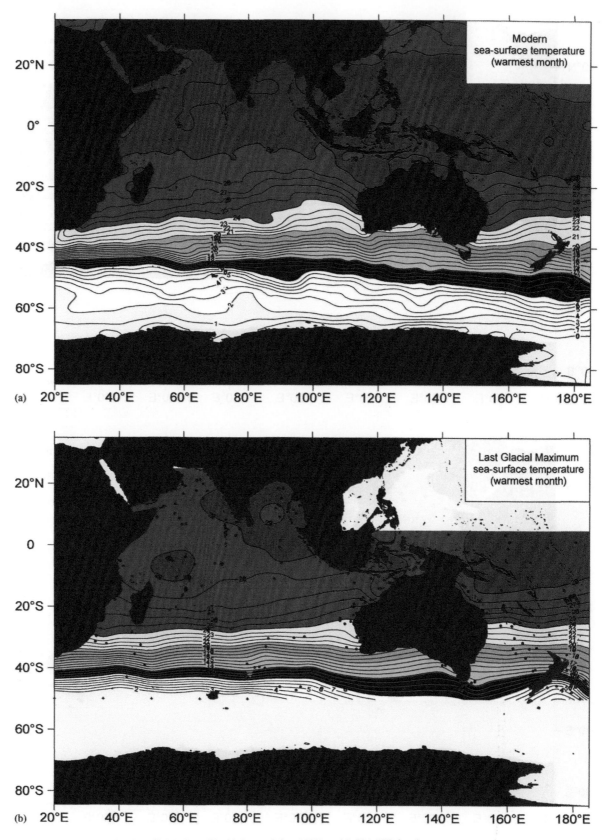

Fig. 6. (a,b) Modern (World Ocean Atlas, 1998) and LGM SST for the warmest month.

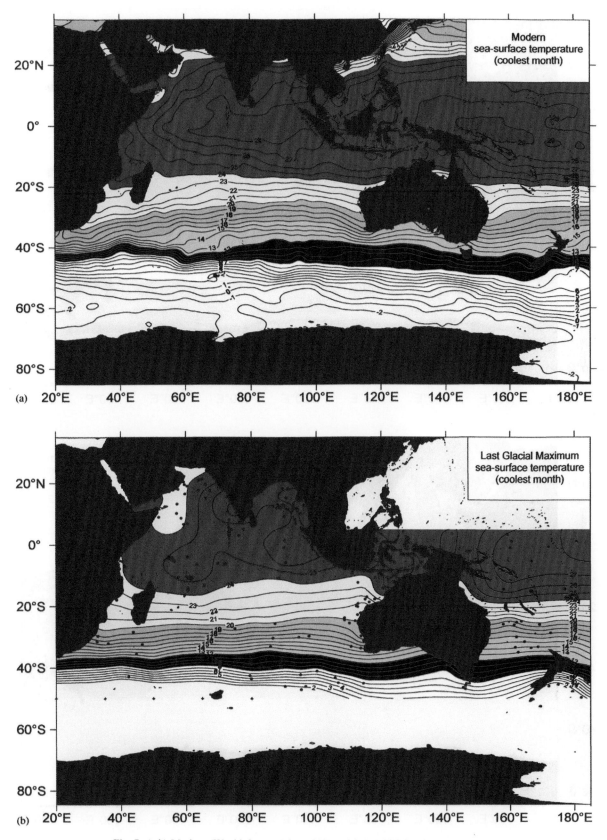

Fig. 7. (a,b) Modern (World Ocean Atlas, 1998) and LGM SST for the coolest month.

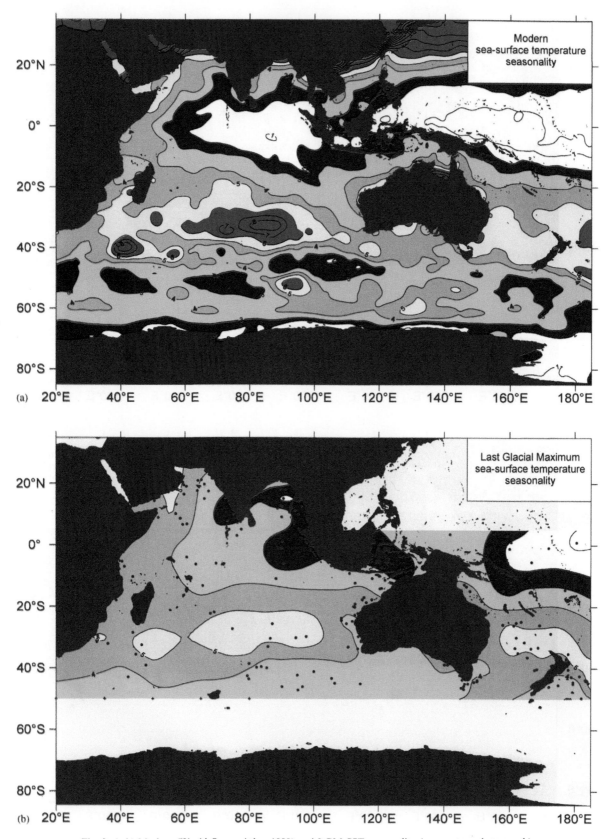

Fig. 8. (a,b) Modern (World Ocean Atlas, 1998) and LGM SST seasonality (warmest–coolest month).

(up to 4.1 °C). In the northern Molucca Sea, Barmawidjaja et al. (1993) explained the large planktonic $\delta^{18}O$ amplitude in core K12 as a change in LGM salinity. Our SST estimate for this core (−1.8 °C) demonstrates that temperature explains this anomaly (Martinez et al., 2002).

The 28 °C isotherm contracted to the east in the western equatorial Pacific Ocean. Of the 15 cores in this region, the anomalies of 7 are cooler by more than the methodological error. Our estimates are systematically warmer by 1–2 °C than the tropical SST estimates of Lea et al. (2000) and Visser et al. (2003) using Mg/Ca in *Globigerinoides ruber*. The coldest anomalies (−1.4 to −2.7 °C) are adjacent to the island of New Guinea. The latter estimate is very close to the box core of Ohkouchi et al. (1994), where these authors estimated cooling of 1.5 °C during the LGM using $U^{K'}_{37}$. In the Coral Sea, we find no evidence to support the conclusion of Anderson et al. (1989) and Trend-Staid and Prell (2002) that cold water flowed equatorward along the eastern coast of Australia.

The central core of the warm pool, which is above 29 °C, migrated from about 150°E, 5°S to the northeast at 175°W, 5°N, which resulted in a slight warming (V28-201, V28-203). This area lies in the path of the Pacific South Equatorial Current, which was transporting water from the eastern Pacific initially 5 °C colder than present (TR163-38; Martinez et al., 2003). The resulting zonal gradient of 10 °C is in strong contrast to the modern gradient of 4 °C between the same points. This might be explained by a slight southward shift in the axis of the Pacific South Equatorial Current. The two cores showing warming would then lie in the path of the eastward flowing North Equatorial Counter Current, which recycles tropical water from the west. The South Pacific Convergence zone appears to have been absent, probably because of the influx of cold water from the Pacific South Equatorial Current.

6.2. Temperate SST (20°S to 40°S)

Widely spaced low-resolution cores provide only sparse coverage of the western and central Indian Ocean. There is a band between 20°S and 35°S where 7 cores in the central Indian Ocean change by less than the methodological error. Within this band, there is a zone of high seasonality between 27–33°S (Fig. 8). This is slightly north of the modern zone of highest seasonality (31–35°S), where the mid-latitude high-pressure belt of the Hadley cell is situated.

The isotherm pattern along the eastern coast of Africa suggests that the Agulhas Current was reaching at least 30°S, before turning east into the South Indian Current. The 20 °C isotherm hits the African coast at 30°S, whereas at present it reaches the Cape of Good Hope at 38°S. This would restrict the potential for surface water exchange of heat and fauna with the Atlantic Ocean during the LGM. The cooling of 2–3 °C recorded by cores beneath the Argulhas Current, even at temperate latitudes, is consistent with only limited cooling of the tropical Indian Ocean. The isotherm pattern suggests that the Mozambique Current was supplying most of the water to the Agulhas Current, whereas the East Madagascar Current may have turned east and south into the South Indian Current, as it occasionally does today. The reduction of the influence of the Agulhas Current along the east coast of Africa created more zonal isotherms than present across the southern Indian Ocean.

On the eastern side of the Indian Ocean, deflected isotherms indicate colder water was being transported further up the western coast of Australia, penetrating into the low latitude tropics. This suggests the West Australian Current was a more developed eastern boundary current than at present (CLIMAP, 1981). Our SST estimates are significantly warmer than those produced by the FI transfer function (CLIMAP, 1981; Wells and Wells, 1994), which tends to underestimate SST (Barrows et al., 2000). The greatest cooling of 3–4 °C occurs between 17–22°S off North West Cape. Southward bowing isotherms close to the western Australian coast indicate that the Leeuwin Current was still flowing, but was transporting colder water than today. Increased upwelling south of Indonesia with increased southward Ekman transport would enhance the steric height difference that drives the Leeuwin Current (Martinez et al., 1999). The presence of the Leeuwin Current apparently prevented a typical eastern boundary current upwelling regime developing. Veeh et al. (2000) found little evidence for increased productivity off the western Australian coast during the LGM, which supports the continued presence of the Leeuwin Current.

The temperate southwestern Pacific Ocean cooled by 3–5 °C with no evidence of warm anomalies (c.f. Anderson et al., 1989). From the isotherm pattern, the East Australian Current was transporting water as far south as the east coast of Tasmania. This surface water was only 2–3 °C colder, reflecting minimal cooling in the tropics. A strong zonal temperature contrast developed across the southern Tasman Sea, as observed previously (Barrows et al., 2000). The 13 °C isotherm, currently at the southern tip of Tasmania and South Island of New Zealand, was 3° further north in the west but 8° further north in the east. A temperature contrast developed across Tasmania with the closing of the Bass Strait. The Flinders Current, a minor current today transporting cold water north along the southern shelves of Australia from western Tasmania (Middleton and Cirano, 2002), was probably more important for transporting water into the Great Australian Bight. On the other side of the Tasman Sea, the Tasman Current was transporting much cooler waters of the Circumpolar Current up the west coast of New Zealand.

A band of high seasonality runs along an axis of 29 to 32°S off the eastern coast of Australia, trending south to 40°S on the eastern side of New Zealand. The axis of highest modern seasonality begins at 34–37°S, and relates to the position of the Tasman Front where half the volume of East Australian Current is jetted east. Although not conclusive, this suggests that the Tasman Front was about 5° further north during the LGM.

6.3. Subantarctic SST (>40°S)

During the LGM, isotherms presently within the Circumpolar Current trend zonally east from a position south of Africa. The southward deflection below the Tasman Sea is probably due to the presence of the East Australian Current. Part of the West Wind Drift is deflected up the coasts of western Australia, Tasmania and New Zealand. In general, SST gradients become much steeper in the Southern Ocean, especially in the Indian Ocean sector and to the east of New Zealand, reflecting the northward expansion of sea ice.

The largest SST anomalies occur south of 40°S. Temperature estimates south of ~50°S, were not possible because the fauna essentially becomes mono-specific Neogloboquadrina pachyderma (sinistral). The greatest cooling occurred over the Campbell Plateau to the east of New Zealand (4 cores from −7.2 to −8.2 °C). This represents considerable equatorward displacement of isotherms at 175°E of ~15°, whereas at 40°E the displacement was only ~5°.

Bathymetry is very important around New Zealand for determining the position of surface fronts. Isotherms typical of the modern Subtropical Front on the Chatham Rise were displaced north but the Front itself may have been bathymetrically locked to the Rise (Weaver et al., 1998). Isotherms typical of the modern Polar Front are present on the Campbell Plateau, but the Subantarctic Front and Polar Front would have been compressed against the Plateau edge to the south in a situation analogous to the modern Drake Passage (Weaver et al., 1998). All these Fronts would be characterized by colder isotherms than present.

Independent SST estimates in the Southern Ocean are in close agreement with ours. Ikehara et al. (1997) observed 4 °C of cooling south of Tasmania using $U_{37}^{K'}$ which is very close to the faunal estimates nearby (−3 to −4 °C). In the Indian Ocean, Mashiotta et al. (1999) estimated 4 °C cooling for RC11-120 using Mg/Ca of Globigerina bulloides, which is very similar to our estimate on the same core (−4.4 °C). Using similar methodology, Pahnke et al. (2003) estimated cooling of 6 °C for core MD97-2120 east of New Zealand, where our faunal estimates range from −4 to −8 °C.

7. Discussion

Our LGM SST reconstruction makes several substantial improvements on the CLIMAP maps. We make more precise SST estimates than previous studies using planktonic foraminifera by employing a superior database of modern core tops, better quality modern SST maps, and state-of-the-art techniques. Our LGM SST estimates are more reliable because we use higher quality cores, with better age control and shorter sampling intervals. We have more comprehensive core coverage, about three times the number of cores used by CLIMAP and almost twice as many as Trend-Staid and Prell (2002) in this region. Our reconstruction bears first order similarities with earlier studies, but there are some significant differences. Our reconstruction varies most from CLIMAP in the high latitudes, where our estimates are as much as 8 °C cooler.

7.1. The meridional pattern of cooling and associated feedbacks

Cooling of the tropics has long been a contentious issue with the CLIMAP maps (e.g., Prell, 1985). The cooling we reconstruct is more extensive, but not much greater in magnitude than estimated previously using planktonic foraminifera. The area covered by the 28 °C isotherm is much less than found by CLIMAP and Trend-Staid and Prell (2002). A large region exceeds 28 °C during the warmest month, sometime during the year (Fig. 6), but only a small area is warmer than 28 °C during the coolest month (Fig. 7). The SST reconstruction is internally consistent. Cooling in the western boundary currents reflects minimal cooling in the tropics, despite the faunal assemblages being different in both areas.

Because of high incoming solar radiation at the equator, which varies little over a glacial cycle, it is physically difficult to dramatically cool the tropical sea surface. A slight decrease in SST would be accompanied by a decrease in cloud cover, allowing more insolation to reach the sea surface and warm it. This acts like a thermostat. Two ways to effectively cool the Indo-Pacific tropics are for either the surface water transported into the tropics to be cooler, or the heat to be more effectively exported to higher latitudes. The first scenario seems to have been the most important; the incoming Pacific South Equatorial and West Australia Currents were cooler, and upwelling was more important in the Indonesian Archipelago. The same surface currents appear to have been in place to export tropical heat. The exposure of continental shelves in the Indonesian Archipelago is important because land has higher albedo, thereby reflecting more insolation back into space and contributing to cooling.

The subtropical zone of the Pacific Ocean is much cooler than reconstructed by CLIMAP, without any warm anomalies suggestive of more vigorous export of heat from the tropics in the East Australian Current. Heat was being exported to the southern Tasman Sea like today, but a component of the East Australian Current parted the coast at lower latitude as the Tasman Front, probably forced by a change in the latitude of the subtropical high-pressure ridge. The Agulhas Current was not exporting heat as far south and into the Atlantic Ocean, and was instead being incorporated at lower latitudes into the South Indian Current. This potentially explains the minimal cooling to slight warming downstream in the subtropical central Indian Ocean. The cooling and contraction of the subtropical zone towards the equator is significant because it is the largest source of latent heat to the atmosphere because high insolation and low cloud cover at these latitudes promotes high evaporation. A reduced subtropical zone contributes to global cooling because it lowers the atmospheric content of water, an important greenhouse gas. The increase in land area through the exposure of continental shelves in Australasia also reduces this flux.

The strong cooling we reconstruct at the high latitudes is important for several reasons. First, the expansion of sea ice dramatically increases the albedo of the sea surface, further promoting cooling. Second, it reduces an important zone of sea–air energy exchange, especially during the winter months when the net flux is positive (into the atmosphere from the ocean). Third, strong cooling of the high latitudes will increase solution of the greenhouse gas carbon dioxide into the ocean. As the largest sink of carbon dioxide, this acts as a positive feedback mechanism to amplify global cooling. Last, the zone of formation of Antarctic Intermediate Water is moved equatorward making it more important at lower latitudes.

Comparison of the LGM state to the present demonstrates how the Earth responds to considerable global warming. Temperature anomalies indicate that the high latitudes are the most sensitive to warming and that warming in these latitudes will amplify further warming when sea-ice retreats, and acts as a reduced sink of carbon dioxide. The tropics are the zone least sensitive to global warming and as a persistent source of heat to both the atmosphere and the higher latitude oceans.

7.2. LGM atmospheric circulation

Because of frictional wind stress, surface ocean currents closely resemble surface wind patterns. Although current flow is complicated by deflection associated with continental landmasses, isotherm patterns reveal much about wind patterns. The LGM zonal westerly winds migrated north, most significantly in the

east, and these were responsible for driving more surface water up the western coasts of the landmasses. The axis of the subtropical high-pressure ridge of the Hadley Cell, corresponding to the band of highest SST seasonality, moved north in both the Indian and Pacific Oceans.

The intertropical convergence zone (ITCZ), the tropical low-pressure trough, follows the thermal equator and its interannual movements generate the modern Asian and African monsoons. The SST data indicate that there was no monsoon operating in northern Australia and the ITCZ stayed close to a position near the equator. This is supported by the seasonal extreme data that indicate upwelling occurred perennially in the Java upwelling system, probably under persistent southeast Trade Winds. Continental cooling probably reduced the heat lows that contribute to the incursion of the ITCZ across northern Australia (Sturman and Tapper, 1996). There is current consensus that the Northwest monsoon was not operating in northern Australia during the LGM and it only resumed after ~14,000 cal yr BP (Wyrwoll and Miller, 2001). West of 120°E and to the west coast of the Indian subcontinent, the ITCZ again stayed close to its mean position throughout the year, indicating no Asian monsoon operating. However, evidence for seasonal upwelling along the African coast indicates the ITCZ traversed a similar meridional range as today in the western Indian Ocean. Therefore the operation of the African monsoon was similar to present.

The continued presence of the western Pacific warm pool (albeit reduced) provided a heat source for the zonal Walker Circulation and the cooler eastern Pacific provided an area of relatively higher surface pressure upwind. The higher zonal temperature gradient along the equatorial Pacific created a situation where the Trade winds could be stronger than at present. The meridional temperature gradient from the equator to Antarctica was also higher. On the basis of conservation of energy, a contraction of circulation towards the equator and an increase in pole–equator temperature gradient would also produce stronger zonal westerly winds in the mid-latitudes.

7.3. Comparison with terrestrial records

The thermal configuration of the LGM equatorial Pacific Ocean was not analogous to any situation today. In the eastern equatorial Pacific Ocean, LGM cooling created conditions similar to an extreme modern La Niña mode of the El Niño-Southern Oscillation (ENSO) phenomenon. This mode is associated with upwelling, increased divergence along the equator, and dry coastal conditions in western South America. In the western equatorial Pacific, conditions resembled a severe El

Niño where there is mild cooling of the sea surface and the core of the warm pool moves east. El Niño conditions are associated with drier conditions in Indonesia, and northern and eastern Australia, with droughts and fires more likely. The freezing level lowers in the New Guinea Highlands bringing frosts to low elevations. These trends are consistent with reconstructions based on pollen analysis and lower snowlines (e.g., Hope and Peterson, 1976; Wang et al., 1999; van der Kaas et al., 2000). Our data are not of sufficient resolution to determine whether the inter-annual ENSO phenomenon operated.

The observed SST contrast across the Tasman Sea is also seen in the terrestrial records. The climatically important summer 10 °C isotherm, which determines the position of the alpine treeline, is south of Tasmania but intersects the middle of New Zealand. Trees remained along the Tasmanian coast during the LGM (Macphail, 1979), but the South Island of New Zealand was largely treeless (McGlone et al., 1993).

During OIC 2 in southeastern Australia, cold climate landforms were active very close to the age of the SST_{min}. Maximum periglacial activity (those conditions associated with frost, snow and ice) occurred at $21,900 \pm 500$ yr (Barrows et al., 2004) and maximum glacial extent occurred after 20,000 yr (Barrows et al., 2002). Existing dates for maximum glacial extent from New Zealand and New Guinea (Barrows et al., 2001), also point to a similar timing to the SST_{min}. During the LGM, minimum temperatures occurred at a similar time at both the sea and land surface in this sector of the southern hemisphere.

Our results reinforce our previous findings that sea-surface cooling was less than the cooling observed on adjacent landmasses (Barrows et al., 2000). This situation extends from the tropics down to temperate southern Australia. Our SST reconstruction covers a period of 4000 yr and the temporal resolution of many cores prevents a detailed analysis within this timeframe. This period is longer than the duration of the coldest conditions in southeast Australia (Barrows et al., 2004). This means that a significant brief cold period occurred within this time frame, enough for the advance and retreat of mountain glaciers. Therefore, a direct comparison with the controversial conditions responsible for mountain glaciation in the New Guinea is problematic. Accurate dating of glaciation in the western Pacific Ocean is needed to reassess the situation.

Acknowledgements

We thank J. Duprat, N. Kallel, C. Waelbroeck, E. Cortijo, C. Samson, P. De Deckker, and W. Howard for providing access to data.

References

Alley, R.B., Brook, E.J., Anandakrishnan, S., 2002. A northern lead in the orbital band: north–south phasing of Ice-Age events. Quaternary Science Reviews 21, 431–441.

Anderson, D.M., Prell, W.L., 1993. A 300 kyr record of upwelling off Oman during the Late Quaternary: evidence of the Asian Southwest Monsoon. Paleoceanography 8 (2), 193–208.

Anderson, D.M., Prell, W.L., Barratt, N.J., 1989. Estimates of sea surface temperature in the Coral Sea at the last glacial maximum. Paleoceanography 4 (6), 615–627.

Bard, E., 1988. Correction of accelerator mass spectrometry ^{14}C ages measured in planktonic foraminifera: paleoceanographic implications. Paleoceanography 3 (6), 635–645.

Bard, E., Rostek, F., Sonzogni, C., 1997. Interhemispheric synchrony of the last deglaciation inferred from alkenone paleothermometry. Nature 385, 707–710.

Barmawidjaja, B.M., Rohling, E.J., van der Kaars, W.A., Vergnaud Grazzini, C., Zachariasse, W.J., 1993. Glacial conditions of the northern Molucca Sea region (Indonesia). Palaeogeography, Palaeoclimatology, Palaeoecology 101, 147–167.

Barrows, T.T., 2000. The timing and impact of the last glacial maximum in Australia. Unpublished PhD thesis. Research School of Earth Sciences, The Australian National University, Canberra, Australia. 344pp.

Barrows, T.T., Juggins, S., De Deckker, P., Thiede, J., Martinez, J.I., 2000. Sea-surface temperatures of the southwest Pacific Ocean during the last glacial maximum. Paleoceanography 15, 95–109.

Barrows, T.T., Stone, J.O., Fifield, L.K., Cresswell, R.G., 2001. Late Pleistocene glaciation of the Kosciuszko Massif, Snowy Mountains, Australia. Quaternary Research 55, 179–189.

Barrows, T.T., Stone, J.O., Fifield, L.K., Cresswell, R.G., 2002. The timing of the last glacial maximum in Australia. Quaternary Science Reviews 21, 159–173.

Barrows, T.T., Stone, J.O., Fifield, L.K., 2004. Exposure ages for Pleistocene periglacial deposits in Australia. Quaternary Science Reviews 23, 697–708.

Bassinot, F.C., Labeyrie, L.D., Vincent, E., Quidelleur, X., Shackleton, N.J., Lancelot, Y., 1994. The astronomical theory of climate and the age of the Brunhes-Matuyama magnetic reversal. Earth and Planetary Science Letters 126, 91–108.

Bé, A.W.H., Duplessy, J.C., 1976. Subtropical convergence fluctuations and Quaternary climates in the middle latitudes of the Indian Ocean. Science 194, 419–422.

Belkin, I.M., Gordon, A.L., 1996. Southern Ocean fronts from the Greenwich meridian to Tasmania. Journal of Geophysical Research 101 (C2), 3675–3696.

Broccoli, A.J., Marciniak, E.P., 1996. Comparing simulated glacial climate and paleodata: a reexamination. Paleoceanography 11 (1), 3–14.

Broecker, W.S., Klas, M., Raganobeavan, N., Mathieu, G., Mix, A., Andree, M., Oeschger, H., Wolfli, W., Suter, M., Bonani, G., Hofmann, H., Nessi, M., Morenzoni, E., 1988a. Accelerator mass spectrometry radiocarbon measurements on marine carbonate samples from deep-sea cores and sediment traps. Radiocarbon 30 (3), 261–295.

Broecker, W.S., Oppo, D., Peng, T.-H., Curry, W., Andree, M., Wolfli, W., Bonani, G., 1988b. Radiocarbon-based chronology for the (^{18}O/^{16}O) record for the last deglaciation. Paleoceanography 3 (4), 509–515.

Broecker, W.S., Clark, E., Hajdas, I., Bonani, G., 2004. Glacial ventilation rates for the deep Pacific Ocean. Paleoceanography 19, PA2002.

Cayre, O., Beaufort, L., Vincent, E., 1999. Paleoproductivity in the equatorial Indian Ocean for the last 260,000 yr: a transfer function based on planktonic foraminifera. Quaternary Science Reviews 18, 839–857.

Climate: long-Range Investigation, Mapping and Prediction (CLIMAP) Project Members, 1976. The surface of the Ice-Age Earth. Science 191, 1131–1137.

Climate: long-Range Investigation, Mapping and Prediction (CLIMAP) Project Members, 1981. Seasonal reconstructions of the Earth's surface at the last glacial maximum. Geological Society of America Map Chart Series MC-36.

Conkright, M., Levitus, S., O'Brien, T., Boyer, T., Antonov, J., Stephens, C., 1998. World ocean atlas 1998 CD-ROM data set documentation. Technical Report 15, NODC Internal Report, Silver Spring, MD, 16pp.

Conolly, J.R., 1967. Postglacial–glacial change in climate in the Indian Ocean. Nature 214, 873–875.

Darling, K.F., Wade, C.M., Stewart, I.A., Kroon, D., Dingle, R., Leigh Brown, A.J., 2000. Molecular evidence for genetic mixing of Arctic and Antarctic subpolar populations of planktonic foraminifers. Nature 405, 43–47.

Ding, X., Guichard, F., Bassinot, F., Labeyrie, L., Fang, N.Q., 2002. Evolution of heat transport pathways in the Indonesian Archipelago during the last deglaciation. Chinese Science Bulletin 47 (22), 1912–1917.

Duplessy, J.C., Bard, E., Arnold, M., Shackleton, N.J., Duprat, J., Labeyrie, L., 1991. How fast did the ocean–atmosphere system run during the last deglaciation. Earth and Planetary Science Letters 103, 27–40.

Feldberg, M.J., Mix, A.C., 2002. Sea-surface temperature estimates in the Southeast Pacific based on planktonic foraminiferal species, modern calibration and last glacial maximum. Marine Micropaleontology 44, 1–29.

Hanebuth, T., Stattegger, K., Grootes, P.M., 2000. Rapid flooding of the Sunda Shelf: a late-glacial sea-level record. Science 288, 1033–1035.

Hays, J.D., Lozano, J.A., Shackleton, N.J., Irving, G., 1976. Reconstruction of the Atlantic and western Indian Ocean sectors of the 18,000 B.P. Antarctic Ocean. In: Cline, R.M., Hays, J.D. (Eds.), Investigation of Late Quaternary Paleoceanography and Paleoclimatology. Geological Society of America Memoir, vol. 145. pp. 337–372.

Hesse, P.P., 1994. The record of continental dust from Australia in Tasman Sea sediments. Quaternary Science Reviews 13, 257–272.

Hiramatsu, C., De Deckker, T., 1997. The late Quaternary nanofossil assemblages from three cores from the Tasman Sea. Palaeogeography, Palaeoclimatology, Palaeoecology 131 (4), 391–412.

Hope, G.S., Peterson, J.A., 1976. Palaeoenvironments. In: Hope, G.S., Peterson, J.A., Radok, U., Allison, I. (Eds.), The Equatorial Glaciers of New Guinea. A.A. Balkema, Rotterdam, pp. 173–205.

Howard, W.R., Prell, W.L., 1992. Late Quaternary surface circulation of the Southern Indian Ocean and its relationship to orbital variations. Paleoceanography 7, 79–118.

Howard, W.R., Samson, C.R., Sikes, E.L., 2002. Glacial and Holocene sea-surface temperature change and circulation at South Tasman Rise. Transactions—American Geophysical Union (West. Pac. Geophys. Meet. Suppl.) 83 (22) Abstract A51A-04.

Ikehara, M., Kawamura, K., Ohkouchi, N., Kimoto, K., Murayama, M., Nakamura, T., Oba, T., Taira, A., 1997. Alkenone sea surface temperature in the Southern Ocean for the last two deglaciations. Geophysical Research Letters 24, 679–682.

Imbrie, J., Kipp, N.G., 1971. A new micropaleontological method for quantitative paleoclimatology: application to a late Pleistocene Caribbean core. In: Turekian, K.K. (Ed.), The Late Cenozoic Glacial Ages. Yale University Press, New Haven, pp. 71–181.

Imbrie, J., et al., 1992. On the structure and origin of major glaciation cycles. 1. Linear responses to Milankovich forcing. Paleoceanography 7, 701–738.

Juggins, S., 2003. C2 User guide. Software for ecological and palaeoecological data analysis and visualisation. University of Newcastle, Newcastle upon Tyne, UK 69pp.

Kawagata, S., 2001. Tasman Front shifts and associated paleoceanographic changes during the last 250,000 years: foraminiferal evidence from the Lord Howe Rise. Marine Micropaleontology 41, 167–191.

Kimoto, K., Takaoka, H., Oda, M., Ikehara, M., Matsuoka, H., Okada, M., Oba, T., Taira, A., 2003. Carbonate dissolution and planktonic foraminiferal assemblages observed in three piston cores collected above the lysocline in the western equatorial Pacific. Marine Micropaleontology 47, 227–251.

Kucera, M., Weinelt, M., Kiefer, T., Pflaumann, U., Hayes, A., Weinelt, M., Chen, M.-T., Mix, A.C., Barrows, T.T., Cortijo, E., Duprat, J., Waelbroeck, C., Juggins, S., 2004. Reconstruction of the glacial Atlantic and Pacific sea-surface temperatures from assemblages of planktonic foraminifera: multi-technique approach based on geographically constrained calibration datasets. Quaternary Science Reviews, this volume.

Labeyrie, L., Labracherie, M., Gorfti, N., Pichon, J.J., Vautravers, M., Arnold, M., Duplessy, J.-C., Paterne, M., Michel, E., Duprat, J., Caralp, M., Turon, J.L., 1996. Hydrographic changes of the Southern Ocean (southeast Indian sector) over the last 230 kyr. Paleoceanography 11, 57–76.

Lea, D.W., Pak, D.K., Spero, H.J., 2000. Climate impact of late quaternary equatorial Pacific Sea surface temperature variations. Science 289, 1719–1724.

Lynch-Stieglitz, J., Fairbanks, R.G., Charles, C.D., 1994. Glacial–interglacial history of Antarctic intermediate water: relative strengths of Antarctic versus Indian Ocean sources. Paleoceanography 9 (1), 7–29.

Macphail, M.K., 1979. Vegetation and climates in Southern Tasmania since the last glaciation. Quaternary Research 11, 306–341.

Malmgren, B.A., Kucera, M., Nyberg, J., Waelbroeck, C., 2001. Comparison of statistical and artificial neural network techniques for estimating past sea-surface temperatures from planktonic foraminifer census data. Paleoceanography 16, 520–530.

Martinez, J.I., 1994. Late Pleistocene palaeoceanography of the Tasman Sea: implications for the dynamics of the warm pool in the western Pacific. Palaeogeography, Palaeoclimatology, Palaeoecology 112, 19–62.

Martinez, J.I., De Deckker, P., Barrows, T.T., 1999. Palaeoceanography of the last glacial maximum in the eastern Indian Ocean: planktonic foraminiferal evidence. Palaeogeography, Palaeoclimatology, Palaeoecology 147, 73–99.

Martinez, J.I., De Deckker, P., Barrows, T.T., 2002. Palaeoceanography of the western Pacific warm pool during the last glacial maximum—of significance to long term monitoring of the maritime continent. In: Bishop, P., Kershaw, A.P., Tapper, N. (Eds.), Environmental and Human History and Dynamics of the Australian Southeast Asian Region. Catena Verlag, pp. 147–172.

Martinez, J.I., Keigwin, L., Barrows, T.T., Yokoyama, Y., Southon, J., 2003. La Niña-like conditions in the eastern Equatorial Pacific and a stronger Choco jet in the northern Andes during the last glaciation. Paleoceanography 18 (2), 1033.

Martinson, D.G., Pisias, N.G., Hays, J.D., Imbrie, J., Moore, Jr., T.C., Shackleton, N.J., 1987. Age dating and the orbital theory of the ice ages: development of a high-resolution 0 to 300,000-year chronostratigraphy. Quaternary Research 27, 1–29.

Mashiotta, T.A., Lea, D.W., Spero, H.J., 1999. Glacial–interglacial changes in Subantarctic sea surface temperature and δ^{18}O-water using foraminiferal Mg. Earth and Planetary Science Letters 170, 417–432.

McGlone, M.S., Salinger, M.J., Moar, N.T., 1993. Palaeovegetation studies of New Zealand's climate since the last glacial maximum. In: Wright, H.E., et al. (Eds.), Global Climates Since the Last

Glacial Maximum. University of Minnesota Press, USA, pp. 294–317 (Chapter 12).

Middleton, J.F., Cirano, M., 2002. A northern boundary current along Australia's southern shelves: the Flinders Current. Journal of Geophysical Research 107 (C9), 3129.

Mix, A.C., Ruddiman, W.F., 1984. Oxygen-isotope analyses and Pleistocene ice volumes. Quaternary Research 21, 1–20.

Mix, A.C., Bard, E., Schneider, R., 2001. Environmental processes of the ice age: land, ocean, glaciers (EPILOG). Quaternary Science Reviews 20, 627–657.

Moore, Jr., T.C., et al., 1980. The reconstruction of sea surface temperatures in the Pacific Ocean of 18,000 B.P. Marine Micropaleontology 5, 215–247.

Nees, S., 1994. A stable-isotope record for the late Quaternary from the East Tasman Plateau. In: van der Lingen, G.J., Swanson, K.M., Muir, R.J. (Eds.), Evolution of the Tasman Sea Basin. A.A. Balkema, Brookfield, VT, pp. 197–201.

Nelson, C.S., Cooke, P.J., Hendy, C.G., Cuthbertson, A.M., 1993a. Oceanographic and climatic changes over the past 160,000 years at Deep Sea Drilling Project Site 594 off southwestern New Zealand, southwest Pacific Ocean. Paleoceanography 8 (4), 435–458.

Nelson, C.S., Hendy, C.H., Cuthbertson, A.M., 1993b. Compendium of stable oxygen and carbon isotope data for the late Quaternary interval of deep-sea cores from the New Zealand sector of the Tasman Sea and southwest Pacific Ocean, Occident. Report 16, Department of Earth Sciences, University of Waikato, Hamilton, New Zealand, pp. 1–87.

Niebler, H.-S., Gersonde, R., 1998. A planktonic foraminiferal transfer function for the southern South Atlantic Ocean. Marine Micropaleontology 34, 187–211.

Niebler, H.-S., Arz, H.W., Donner, B., Mulitza, S., Pätzold, J., Wefer, G., 2003. Sea surface temperatures in the equatorial and South Atlantic Ocean during the last glacial maximum (23–19 ka). Paleoceanography 18 (3), 1069.

Niitsuma, N., Oba, T., Okada, M., 1991. Oxygen and carbon isotope stratigraphy at Site 723, Oman Margin. In: Prell, W.L., Niitsuma, N., et al. (Eds.), Proceedings of Ocean Drill. Program Scientific Results, vol. 117. pp. 321–342.

Ohkouchi, N., Kawamura, K., Nakamura, T., Taira, A., 1994. Small changes in the sea surface temperature during the last 20,000 years: molecular evidence from the western tropical Pacific. Geophysical Research Letters 21 (20), 2207–2210.

Pahnke, K., Zahn, R., Elderfield, H., Schulz, M., 2003. 340,000-year centennial-scale marine record of Southern Hemisphere climatic oscillation. Science 301, 948–952.

Passlow, V., Pinxian, W., Chivas, A.R., 1997. Late Quaternary palaeoceanography near Tasmania, southern Australia. Palaeogeography, Palaeoclimatology, Palaeoecology 131 (4), 433–463.

Pisias, N.G., 1976. Late Quaternary sediment of the Panama Basin: sedimentation rates, periodicities, and controls of carbonate and opal accumulation. In: Cline, R.M., Hays, J.D. (Eds.), Investigation of Late Quaternary Paleoceanography and Paleoclimatology, Geological Society of America Memoir, vol. 145. pp. 375–391.

Pflaumann, U., Duprat, J., Pujol, C., Labeyrie, L., 1996. SIMMAX: a modern analog technique to deduce Atlantic sea surface temperatures from planktonic foraminifera in deep-sea sediments. Paleoceanography 11, 15–35.

Pflaumann, U., Sarnthein, M., Chapman, M., Duprat, J., Huels, M., Kiefer, T., Maslin, M., Schulz, H., van Kreveld, S., Vogelsang, E., Weinelt, M., 2003. North Atlantic: sea-surface conditions reconstructed by GLAMAP-2000. Paleoceanography 18 (3), 1065.

Prell, W.L., 1985. The stability of low-latitude sea surface temperatures: an evaluation of the CLIMAP reconstruction with emphasis on the positive SST anomalies. Technical Report TR025, US Department of Energy, Washington, DC.

Prell, W.L., Hutson, W.H., Williams, D.F., Bé, A.W.H., Geitzenauer, K., Molfino, B., 1980. Surface circulation of the Indian Ocean during the last glacial maximum, approximately 18,000 yr B.P. Quaternary Research 14, 309–336.

Prell, W.L., Imbrie, J., Martinson, D.G., Morley, J.J., Pisias, N.G., Shackleton, N.J., Streeter, H.F., 1986. Graphic correlation of oxygen isotope stratigraphy: application to the late Quaternary. Paleoceanography 1 (2), 137–162.

Prell, W., Martin, A., Cullen, J., Trend, M., 1999. The Brown University foraminiferal data base. IGBP PAGES/World Data Center-A for Paleoclimatology Data Contribution Series # 1999-027.

Ruddiman, W., 1992. Calcium carbonate database. IGBP PAGES/World Data Center-A for Paleoclimatology Data Contribution Series # 92-001. NOAA/NGDC Paleoclimatology Program, Boulder, CO, USA.

Sachs, J.P., Anderson, R.F., Lehman, S., 2001. Glacial surface temperatures of the Southeast Atlantic Ocean. Science 293, 2077–2079.

Salvignac, M.E., 1998. Variabilité hydrologique et climatique dans l'Océan Austral au cours du Quaternaire terminal. Océanographie, Université de Bordeaux, vol. I, 354pp.

Samson, C.R., 1999. Structure and timing of the last deglaciation in the subtropical and subpolar southwest Pacific: implications for driving forces of climate. Antarctic Cooperative Research Centre and Institute of Antarctic and Southern Ocean Studies, University of Tasmania, Hobart, Australia, unpublished data.

Shackleton, N.J., Duplessy, J.-C., Arnold, M., Maurice, P., Hall, M.A., Cartlidge, J., 1988. Radiocarbon age of last glacial Pacific deep water. Nature 335, 708–711.

Sikes, E.L., Howard, W.R., Neil, H.L., Volkman, J.K., 2002. Glacial–interglacial sea surface temperature changes across the subtropical front east of New Zealand based on alkenone unsaturation ratios and foraminiferal assemblages. Paleoceanography 17 (2), 1–13.

Sonzogni, C., Bard, E., Rostek, F., 1998. Tropical sea-surface temperatures during the last glacial period: a view based on Alkenones in Indian Ocean sediments. Quaternary Science Reviews 17, 1185–1201.

Spooner, M.I., Barrows, T.T., Deckker, P.De., Paterne, M. (in press). Late Quaternary Palaeoceanography of the Banda Sea, with implications for past monsoonal climates. Global and Planetary Change.

Stuiver, M., Reimer, P.J., Bard, E., Beck, J.W., Burr, G.S., Hughen, K.A., Kromer, B., McCormac, F.G., van der Plicht, J., Spurk, M., 1998. INTCAL98 radiocarbon age calibration, 24,000-0 cal BP. Radiocarbon 40 (3), 1041–1083.

Sturman, A.P., Tapper, N.J., 1996. The Weather and Climate of Australia and New Zealand. Oxford University Press, Oxford 475pp.

Thompson, P.R., 1981. Planktonic foraminifera in the western north Pacific during the past 150 000 years: comparison of modern and fossil assemblages. Palaeogeography, Palaeoclimatology, Paleoecology 35, 241–279.

Thunell, R., Anderson, D., Gellar, D., Qingmin, M., 1994. Sea-surface temperature estimates for the tropical western Pacific during the last glaciation and their implications for the Pacific warm pool. Quaternary Research 41, 255–264.

Trend-Staid, M., Prell, W.L., 2002. Sea surface temperature at the last glacial maximum: a reconstruction using the modern analog technique. Paleoceanography 17 (4), 1065.

van Campo, E., Dupplessy, J.C., Prell, W.L., Barratt, N., Sabatier, R., 1990. Comparison of terrestrial and marine temperature estimates for the past 135 kyr off southeast Africa: a test for GCM simulations of palaeoclimate. Nature 348, 209–212.

van der Kaas, S., Wang, X., Kershaw, P., Guichard, F., Setiabudi, D.A., 2000. A late Quaternary palaeoecological record from the Banda Sea, Indonesia: patterns of vegetation, climate and biomass burning in Indonesia and northern Australia. Palaeogeography, Palaeoclimatology, Palaeoecology 155, 135–153.

Veeh, H.H., McCorkle, D.C., Heggie, D.T., 2000. Glacial/interglacial variations of sedimentation on the West Australian continental margin: constraints from excess [230]Th. Marine Geology 166, 11–30.

Venables, W.N., Ripley, B.D., 2002. Modern Applied Statistics with, S, fourth ed. Springer, New York 495pp.

Visser, K., Thunell, R., Stott, L., 2003. Magnitude and timing of temperature changes in the Indo-Pacific warm pool during deglaciation. Nature 421, 152–155.

Waelbroeck, C., Labeyrie, L., Duplessy, J.-C., Guiot, J., Labracherie, M., Leclaire, H., Duprat, J., 1998. Improving past sea surface temperature estimates based on planktonic fossil faunas. Paleoceanography 13, 272–283.

Wang, X., van der Kaars, S., Kershaw, P., Bird, M., Jansen, F., 1999. A record of fire, vegetation and climate through the last three glacial cycles from Lombok Ridge core G6-4, eastern Indian Ocean, Indonesia. Palaeogeography, Palaeoclimatology, Palaeoecology 147, 241–256.

Weaver, P.P., Carter, L., Neil, H., 1998. Response of surface water masses and circulation to late Quaternary climate change east of New Zealand. Paleoceanography 13 (1), 70–83.

Wells, P.E., Connell, R., 1997. Movement of hydrological fronts and widespread erosional events in the southwestern Tasman Sea during the late Quaternary. Australian Journal of Earth Science 44, 105–112.

Wells, P., Okada, H., 1997. Response of nanoplankton to major changes in sea-surface temperature and movements of hydrological fronts over Site DSDP 594 (south Chatham Rise, southeastern New Zealand), during the last 130 kyr. Marine Micropaleontology 32, 341–363.

Wells, P.E., Wells, G.M., 1994. Large-scale reorganization of ocean currents offshore Western Australia during the Late Quaternary. Marine Micropaleontology 24, 157–186.

Williams, D.F., 1976. Late Quaternary fluctuations of the polar front and subtropical convergence in the southeast Indian Ocean. Marine Micropaleontology 1, 363–375.

Wright, I.C., McGlone, M.S., Nelson, C.S., Pillans, B.J., 1995. An integrated latest Quaternary (stage 3 to present) paleoclimatic and paleoceanographic record from offshore nothern New Zealand. Quaternary Research 44, 283–293.

Wyrwoll, K.-H., Miller, G.H., 2001. Initiation of the Australian summer monsoon 14,000 years ago. Quaternary International 83–85, 119–128.

Yokoyama, Y., Lambeck, K., De Deckker, P., Johnston, P., Fifield, L.K., 2000. Timing of the last glacial maximum from observed sea-level minima. Nature 406, 713–716.

Quaternary Science Reviews 24 (2005) 1049–1062

Estimating glacial western Pacific sea-surface temperature: methodological overview and data compilation of surface sediment planktic foraminifer faunas

M.-T. Chen[a],*, C.-C. Huang[a], U. Pflaumann[b], C. Waelbroeck[c], M. Kucera[d]

[a]*Institute of Applied Geosciences, National Taiwan Ocean University, Keelung 20224, Taiwan*
[b]*Institut für Geowissenschaften, Universität Kiel, Olshausenstr. 40-60, 24118 Kiel, Germany*
[c]*Laboratoire des Sciences du Climat et de l'Environnement, Laboratoire mixte CNRS-CEA, Domaine du CNRS, Gif-sur-Yvette, France*
[d]*Department of Geology, Royal Holloway, University of London, Egham, Surrey TW20 0EX, UK*

Received 24 October 2003; accepted 15 July 2004

Abstract

We present a detailed comparison of five "transfer function" techniques calibrated to reconstruct sea-surface temperature (SST) from planktic foraminifer counts in western Pacific surface sediments. The techniques include the Imbrie–Kipp method (IKM), modern analog technique (MAT), modern analog technique with similarity index (SIMMAX), revised analog method (RAM), and the artificial neural network technique (ANN). The calibration is based on a new database of 694 census counts of planktic foraminifers in coretop samples from the western Pacific, compiled under a cooperative effort within the MARGO (multiproxy approach for the reconstruction of the glacial ocean surface) project. All five techniques were used to reconstruct SST variation in a well-dated Holocene to last glacial maximum interval in core MD972151 from the southern South China Sea (SCS) to evaluate the magnitude of cooling in the western tropical Pacific during the LGM. Our results suggest that MAT, SIMMAX, RAM and ANN show a similar level of performance in SST estimation and produce $\leqslant 1\,°C$ uncertainties in coretop SST calibrations of the western Pacific. When applying these techniques to the downcore faunal record, the IKM, which performed significantly worst in the calibration exercise, produced glacial SST estimates similar to present-day values, whereas the other four techniques all indicated ~1 °C cooler glacial SST. Because of their better performance in the calibration dataset, and because of the convergence among the techniques in the estimated magnitude of glacial cooling in the studied core, we conclude that MAT, SIMMAX, RAM and ANN provide more robust planktic foraminifer paleo-SST estimates than traditional IKM techniques in western Pacific paleoceanographic studies.
© 2004 Elsevier Ltd. All rights reserved.

1. Introduction

Foraminifers growing in the oceans provide some of the richest paleoclimate archives in the world. For the purpose of reconstructing past sea surface temperatures (SSTs), planktic foraminifer faunas are particularly useful paleoclimate proxies because they are widely distributed and are abundant enough for quantitative

and statistical analysis. The first extensive application of planktonic foraminifers in paleoclimatology was the pioneering study by CLIMAP (1976) in which the distribution data of planktic foraminifer faunal assemblages and their relationships with SSTs were used for predicting past surface ocean-climate patterns during the last glacial maximum (LGM). Despite an intensive effort to develop new climatic and geochemical proxies for reconstructions of glacial surface ocean conditions, planktic foraminifer faunas continue to play an important role in providing essential information on past SSTs and for inferring water mass and circulation changes.

*Corresponding author. Tel.: +886-2-2462-2192x6503; fax: +886-2-2462-5038.
E-mail address: mtchen@mail.ntou.edu.tw (M.-T. Chen).

0277-3791/$ - see front matter © 2004 Elsevier Ltd. All rights reserved.
doi:10.1016/j.quascirev.2004.07.013

Tropical oceans, especially those in the western Pacific, serve as a heat engine for Earth's climate and as a vapor source for its hydrological cycle. On annual to inter-annual time scales, the western Pacific has a great impact on global climate through propagation of El Niño—Southern Oscillation (ENSO) events. There is a need to understand more precisely the role that the western Pacific plays in glacial–interglacial cycles and in the more rapid millennial- to centennial-scale oscillations of global climate change (Cane and Clement, 1999; Clement and Cane, 1999; Clement et al., 1999). Until recently, there have been debates on the magnitude of SST cooling during the LGM in different parts of the western Pacific. CLIMAP (1976, 1981) estimated ~2–3 °C LGM cooling in most western equatorial Pacific and western Australian current regions and a relatively large cooling (~4–8 °C) was estimated for the main Kuroshio pathway region near the Pacific margin of Japan. In the South China Sea (SCS) and East China Sea (ECS), where many well-dated, high-resolution paleoceanographic records have been collected and analyzed recently, CLIMAP (1981) had obtained no data which might have been used as a control.

One of the most remarkable results of CLIMAP (1976, 1981) were the reconstructed positive SST anomalies (the LGM warmer than the present) occurring in the mid-latitude gyre regions. These positive anomalies are difficult to reconcile with some terrestrial proxy-based or climate modeling results (Webster and Streeten, 1978; Rind and Peteet, 1985), in which the glacial tropics SSTs were estimated to be as much as 5–6 °C colder than the present. Moreover, geochemically based estimates using tropical coral skeletons Sr/Ca (Beck et al., 1992; Guilderson et al., 1994) have suggested a large glacial cooling, which is consistent with the terrestrial evidence. These disagreements in different types of SST reconstructions still remain to be reconciled.

The Imbrie–Kipp transfer function method (IKM) (Imbrie and Kipp, 1971), the first rigorous environmental calibration technique developed for use with marine microfossils, formed the backbone of CLIMAP LGM SST reconstructions. For reconstructing LGM SSTs in the western Pacific, the CLIMAP project developed a regional IKM transfer function FP-12E (Thompson, 1976, 1981) based on census counts of planktic foraminifers in 165 coretop samples, whereas the SSTs of the western Australian current region near the eastern Indian Ocean were estimated by another IKM transfer function, FI-2 (Hutson and Prell, 1980). The validity of CLIMAP SST reconstructions was later assessed by another, methodologically different, method: the modern analog technique (MAT) (Hutson, 1980; Prell, 1985). In the southwestern tropical Pacific near eastern Australia, a small LGM cooling of ~0–2 °C was reported from a new coretop compilation study using MAT (Barrows et al., 2000). Although the MAT results reduced the difference between the observed and estimated SSTs, this method still suggested the existence of positive anomalies in the western Pacific. It is difficult to assess how reliable the warm glacial SST estimates in the western Pacific are, since the limited coverage and uneven distribution of coretop samples as well as the different levels of clacite preservation in these samples make SST reconstructions from planktic foraminifers in this area rather difficult. To complete a better global map of glacial SSTs and to provide modeling groups with a better basis for simulating glacial and present climate conditions, the western Pacific was identified as a region where there was a great need for improvement in the collection of more high-quality coretop samples and for development of better regional estimation techniques. While a testing of SST estimation techniques against an entire Pacific planktic foraminifer coretop database is presented in Kucera et al. (2004), a separate calibration for the western part of the Pacific is needed, as this testing excludes faunas from equatorial divergence/upwelling regions which may bias the estimation.

Many new SST estimation techniques have been proposed recently to improve traditional techniques and reduce their biases. These new methods include: SIMMAX (modern analog technique using a similarity index) (Pflaumann et al., 1996); revised analog method (RAM) (Waelbroeck et al., 1998); and Artificial Neural Networks (ANN) (Malmgren and Nordlund, 1997; Malmgren et al., 2001). Each of these methods adopts a different approach for improved performance in SST reconstructions: SIMMAX uses a new similarity index and incorporates geographic information, RAM increases the number and range of calibration samples and introduces more rigorous criteria for selection of best analog samples, and ANN establishes highly nonlinear equations which optimize fauna-SST relationships using artificial intelligence techniques. All of these new techniques have been or are being applied to an Atlantic data set and have been demonstrated to reduce the biases of previous SST estimates.

Working within the framework of multiproxy approach for the reconstruction of the glacial ocean surface (MARGO) objectives, the current study was designed to compile a well-organized, high-quality surface sediment planktic foraminifer faunal database for the western Pacific. We used this database to develop better estimates of the SST, based on available, traditional or newly developed techniques. Our aim was to compare and evaluate the performance of these techniques, by analyzing the relationships between the biases and possible factors in the calibration data set. Here we present SST reconstructions using five techniques in a high-quality and well-dated International Marine Past Global Change Study (IMAGES) core from the SCS and provide the SST estimates and the uncertainty ranges of the estimates for the Holocene (0–4 ka) and LGM (19–23 ka) intervals (EPILOG chronozone) (Mix et al., 2001).

2. Data and methods

2.1. Calibration data sets

To obtain a data set of high-quality, well-organized coretop planktic foraminifer fauna data in surface sediments from the western Pacific, we cooperated as one of the MARGO working groups and compiled a western Pacific modern planktic foraminifera data base composed of 694 samples (data available on the internet [http://www.pangaea.de/Projects/MARGO/data.html]). The set of 694 coretop data used in this study has benefited from the contributions of various publication sources: global compilation efforts from Prell et al. (1999) and Ortiz and Mix (1997); East China Sea and Okinawa Trough studies by Ujiie and Ujiie (2000); SCS data collections from Miao et al. (1994), Chen et al. (1998), and Pflaumann and Jian (1999); Southwestern Pacific investigations by Thiede et al. (1997); several individual surveys of the North Pacific (Coulbourn et al., 1980; Ye and Yang, unpublished data; Kiefer et al., 2001); and unpublished data from M. Schulz. The geographical range of the 694 coretop data is from a latitude of 65°N to 65°S and a longitude of 105°E to 180°E (Fig. 1); the samples provide good coverage of the SST range between ~30 and 20 °C, but are relatively sparse in the SST range of 20–5 °C (Fig. 1). The data points are of especially high resolution in the western equatorial Pacific and in the marginal basins of the western Pacific such as the SCS, the East China Sea, and Western Australia. We think that this calibration data set is most appropriate for developing SST estimation techniques for the low- to mid-latitude western Pacific and should be suitable for reconstructing high-resolution SST records from high sedimentation rate sites.

The relative abundance of each planktic foraminifer species is expressed as a percentage of the total faunal count. Descriptive statistics of the relative abundances of 28 species of planktic foraminifers were calculated for the data set (Table 1). The taxonomy of planktic foraminifers used in this study was determined by the MARGO group and generally followed the working schemes of Parker (1962), Be (1967) and Kipp (1976), with some modifications. We combined the two different morphotypes Globigerinoides sacculifer (no sac) and G. sacculifer (with sac) together as one unit. We also combined Globorotalia menardii and G. tumida into one category to overcome an identification difficulty that vexed previous studies. In this western Pacific data set, the most important taxonomic problem is the definition and identification of morphologically similar groups

consisting of Neogloboquadrina dutertrei, N. pachyderma-dutertrei intergrade (P-D intergrade)" (Kipp, 1976), and N. pachyderma (right coiling). Most western Pacific coretop studies did not recognize the "P-D intergrade" and different studies use different criteria for recognizing N. dutertrei and N. pachyderma (right coiling); thus, we wanted to minimize this "taxonomic noise" during the processing of the coretop data for our SST estimation. We have assumed that most "P-D intergrade" specimens identified in previous western Pacific studies represent juvenile forms of N. dutertrei and could be placed in the category of N. dutertrei. In this study we therefore included specimens of "P-D intergrade" with N. dutertrei in coretop as well as downcore data. According to the sample processing procedures described in the published studies, all samples used to generate faunal counting data were weighed and sieved through a 150 μm sieve. Census counts of planktonic foraminifers were made on fractions ⩾150 μm for each sample and at least 300 whole specimens were counted for most samples. These coretop samples range in water depth between 42 and 5351 m, a wide interval that overlaps the depth of the regional carbonate lysocline and represents various calcite preservation levels.

Modern SST values (summer, winter and annual) were assigned to each of the 694 coretop samples using the data in the World Ocean Atlas version 2 (WOA, 1998), in the same way as in Kucera et al. (2004), to ensure a common reference base for all SST proxies used by the MARGO project. Temperatures at the sample locations were computed using the WOA 98 sample software (http://www.palmod.uni-bremen.de/~csn/woa-sample.html).

2.2. SST estimation techniques

Five SST estimation techniques (IKM, MAT, SIMMAX, RAM and ANN) were applied to the new calibration data set in the present study, to determine the accuracy of SST estimates produced by different techniques through a validation in a coretop data set with known SST values. The accuracy of SST estimates based on a validation of a coretop data set is the best available measure of the likely performance of a technique in fossil assemblages. For the purpose of the downcore application we have chosen a SCS IMAGES core MD972151 from which many good AMS C^{14} dates and age control data, and planktic foraminifer faunas and alkenone SST are available (Huang et al., 1999; Lee, in preparation; Lee et al., 1999; Huang et al., 2002). Since our calibration data set has good coverage in the SCS, the LGM reconstructions of this core should be highly accurate.

In developing the IKM transfer function we used the standard statistical technique of VARIMAX Q-mode factor analysis (Imbrie and Kipp, 1971) to decompose

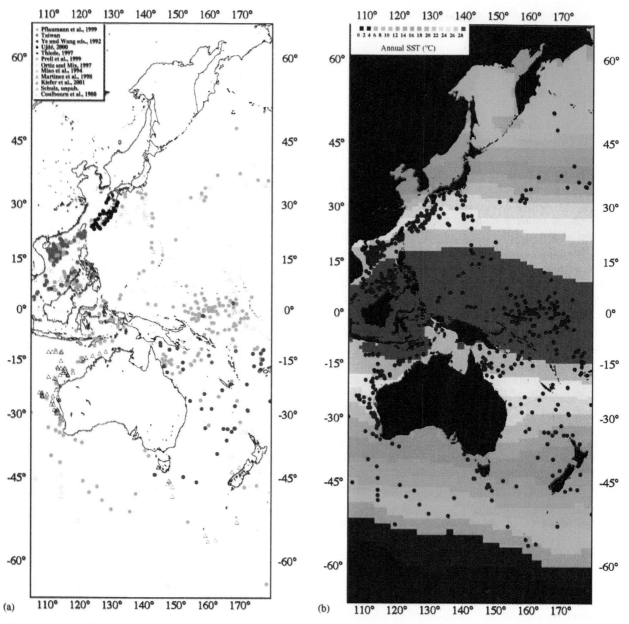

Fig. 1. A western Pacific map indicating the site locations of a newly compiled coretop data set of planktic foraminifer faunal abundances ($N = 694$) and sea-surface temperature (SST, annual average, WOA, 1998) distribution in the western Pacific.

the 694 faunal variation data into a reduced number of factor variables that were independent from each other. Our analyses of the faunal data were based on a log transformation (ln[species percentage $+1$]) that was first applied in an Atlantic and eastern equatorial Pacific study (Mix et al., 1999). In the factor analysis, the log transform amplifies the signals of less dominant species with a compression of dominant species abundance variations. The log transformation also has the effect of making the species distribution more Gaussian, which is an essential assumption of the next step in the IKM that involves multiple regression. The log transform is applied to all faunal percentage data (coretop and

downcore samples) in this study before calculating factor scores or loadings. Four factors (Table 1) that explain ~89% of the original but log-transformed coretop fauna data are derived from the factor analysis. Planktic foraminifer species *G. ruber*, *G. sacculifer*, *Globigerina bulloides*, *N. pachyderma* (right coiling), *Pulleniatina obliquiloculata*, *G. inflata*, *G. menardii+ tumida* and *Globigerinita glutinata* exhibit high factor scores that are highly correlated with these factors. We calibrated the coretop factors loadings of these four factors with their cross-product and squared terms, to annual average, winter and summer SSTs by following a standard procedure of multiple regression techniques

Table 1
Factor score assemblage matrix of 694 coretop fauna percentage data from the western Pacific area

Foraminifer species	Factor 1	Factor 2	Factor 3	Factor 4
O.universa	0.063	0.078	0.003	0.046
G.conglobatus	0.144	−0.013	0.037	0.163
G.ruber	**0.597**	0.030	−0.141	0.291
G.tenellus	0.184	−0.002	−0.107	0.015
G.sacculifer	**0.409**	−0.086	0.076	0.140
S.dehiscenes	−0.053	0.001	0.195	0.056
G.aequilateralis	0.222	−0.040	0.149	0.011
G.calida	0.182	0.006	−0.033	0.035
G.bulloides	0.196	**0.409**	0.027	**−0.540**
G.falconensis	0.088	0.195	−0.094	0.056
B.digitata	0.009	0.021	0.061	0.037
G.rubescens	0.152	−0.024	−0.073	−0.066
T.quinqueloba	−0.001	0.141	−0.056	−0.138
N.pachyderma (L)	−0.069	0.250	−0.010	**−0.443**
N.pachyderma (R)	−0.094	0.377	0.039	0.149
N.dutertrei	0.136	0.196	0.388	0.085
G.conglormerata	0.022	−0.013	0.035	−0.024
G.hexagona	0.026	−0.002	−0.020	0.001
P.obliquiloculata	0.057	−0.066	**0.661**	−0.187
G.inflata	−0.148	**0.661**	0.008	0.392
G.truncatulinoides (L)	−0.025	0.163	−0.029	−0.004
G.truncatulinoides (L)	−0.011	0.144	0.008	0.182
G.crassaformis	0.016	0.065	−0.019	0.044
G.hirsuta	0.005	0.054	−0.014	0.061
G.scitula	0.036	0.046	−0.011	−0.006
G.menardii+tumida	0.016	−0.011	**0.530**	0.151
C.nitida	0.015	−0.004	−0.009	−0.001
G.glutinata	**0.448**	0.132	−0.068	−0.259
Variance (%)	51.52	16.38	18.43	2.40
Cumulative variance	51.52	67.90	86.33	88.74

(VARIMAX solution by CABFAC factor analysis).

Table 2
Multivariate regression coefficients

	Variable	Annual SST	Winter SST	Summer SST
1	F1	36.559	48.199	26.465
2	F2	15.575	22.285	8.441
3	F3	29.943	37.463	22.657
4	F4	12.159	*	18.028
5	F1*F1	−17.477	−23.613	−12.658
6	F2*F2	−15.159	−17.752	−11.399
7	F3*F3	−13.108	−15.275	−11.065
8	F4*F4	−8.674	−19.590	−5.422
9	F1*F2	−23.893	−34.321	−14.089
10	F1*F3	−26.581	−34.093	−21.016
11	F1*F4	−17.218	−9.198	−23.261
12	F2*F3	−10.573	−21.528	*
13	F2*F4	3.345	14.699	*
14	F3*F4	−7.960	*	−10.532
	Intercept	10.796	5.040	16.506

The symbol "*" marks the variables that are not incorporated in the equations.
Note: Annual SST: Annual sea surface temperature. Winter SST: Average sea surface temperature of caloric winter season. Summer SST: Average sea surface temperature of caloric summer season.

(Imbrie and Kipp, 1971), but we constrained the regression by using a "Best Subsets Regression" (MINITAB program) criterion for selecting terms to enter into the equation (Table 2).

We calculated MAT SST for the 694 coretop and MD972151 downcore samples using the same 694 coretops as the calibration data set. The MAT procedure used in this study is the same as that reported in Prell (1985), which introduces a dissimilarity coefficient (the squared chord distance) to measure the dissimilarity between coretop–coretop or coretop–downcore paired species percentage data. Higher values of the squared chord distance indicate higher dissimilarity. The squared chord distance also has the effect of amplifying the signals of less dominant species and reducing the influence of dominant species abundances. MAT SST estimates of coretop or downcore samples are thus calculated from the weighted average of observed SST values of 10 best analog samples. We applied a cut-off dissimilarity value of 0.4, as suggested by Prell (1985), for coretop samples that did not qualify to be included in an SST estimation. Therefore, in some cases,

the number of best analogs used to calculate the SST estimate was less than 10.

The SIMMAX approach (SIMMAX28-1900) adopted here is a revised version which has been applied in a small set of SCS coretop samples (Pflaumann and Jian, 1999); this routine calculates the scalar product of the normalized faunal assemblage vectors as a similarity index. This similarity index ranges from 0 to 1, with 0 indicating complete dissimilarity and 1 indicating full similarity. Both coretop and downcore data sets were analyzed by the SIMMAX routine without any threshold in the search for similarity, and with scalar-product-weighted averaging of the SSTs of the 10 best analogs from the western Pacific 694 coretop sample data set. In this study, we also used the geographic distance-weighting procedure for SST estimation (Pflaumann et al., 1996). This is important as coretops of the western Pacific are unevenly distributed. The resultant output for the single core and time slices contains similarity, estimated annual average, and winter and summer SSTs, as well as the minima, maxima and standard deviations of the 10 best analogs for the three SST estimates, respectively.

RAM SST calibrations were performed for the first time on the western Pacific coretop data. The RAM technique (Waelbroeck et al., 1998) was proposed to improve the estimation accuracy and precision of the MAT, adopting the same dissimilarity coefficient (i.e. squared chord distance) but with two important modifications: (1) selecting only good analog coretop samples by examining the rate of increase in dissimilarity; and (2) remapping and interpolating the coretop fauna database into a more homogenous, evenly distributed space as a function of winter and summer SSTs. These modifications should make RAM useful for SST reconstructions in the western Pacific, where less evenly distributed and/or less well-preserved samples may limit the selection of good analogs for downcore estimates. In this study, a threshold value of the dissimilarity coefficient of 0.6 was applied to reject poor analog coretop samples. The 694 western Pacific coretop samples were expanded by 466 "virtual coretops" by the RAM remapping procedure (with a grid step of 0.4 °C and an interpolation radius of 0.5 °C) in a winter and summer SST, and annual environmental space, respectively. In this paper, we use the latest RAM02 software with the following parameters for all regional calibrations: initial $\alpha = 0.1$, $\beta = 10$, $\gamma = 0.25$ °C, $R = 0.3$ °C, as explained in Kucera et al. (this volume).

A fundamentally different approach for SST calibration, ANN uses sophisticated algorithms to search for a relationship between coretop fauna abundance distributions and SSTs. The general principles and architecture of a back propagation (BP) neural network and its application in reconstructions of past environmental conditions from assemblage counts have been described by Malmgren and Nordlund (1997) and Malmgren et al. (2001). A trained ANN can be best compared to a long, complicated, recurrent mathematical formula transforming species abundances into a desired variable(s). An ANN can successfully learn very complex, nonlinear relationships; this technique is not as dependent on the size, coverage and balance of the calibration dataset as are modern analog techniques (MAT, SIMMAX, RAM). Unfortunately, it is virtually impossible to interpret the meaning of the coefficients associated with individual neurons and thus to understand how the network assigns output variables values to unknown samples.

The networks used in this paper were trained on a database with counts of 28 species in 1111 Pacific coretops. The details of the training results are presented in Kucera et al. (2004). It is important to point out that the ANN SST estimation presented in this study was based on whole Pacific database training, which may reduce the comparability to the other techniques.

3. Comparisons of SST estimations

3.1. Modern SST calibrations

We calibrated (or trained) abundances of 28 species of planktic foraminifers in 694 coretops to annual average, winter, and summer SSTs following the procedures of the five individual SST estimation techniques (IKM, MAT, SIMMAX, RAM, ANN) described above. The calibration results are examined here by comparing scatter plots of observed vs. estimated SSTs (Fig. 2). The

success of the calibration can also be evaluated by calculating the correlation (R^2) between the observed and estimated SSTs as well as the root mean squared error (RMS) of the residuals of the estimated SSTs (Table 3). To simplify the presentation of our results, only the annual average SSTs estimated by these five different techniques are presented for comparison.

Based on their success in predicting observed SSTs from coretop samples, the different techniques can be divided into two distinct groups. For all SST estimations, the IKM technique produced by far the lowest correlation and the highest RMS value (Fig. 2; Table 3). The IKM showed relatively large scattering in observed vs. estimated SST plots and gave a magnitude of ~1.3–1.4 °C of the uncertainties of the estimates.

MAT, SIMMAX, RAM and ANN appeared to yield similarly accurate results. The relatively high correlations and small RMS values (Fig. 2; Table 3) produced by the four techniques indicated good predictability in estimating SSTs in the 694 coretop data. Applying these four techniques to the coretops gives somewhat lower estimation uncertainties of ~0.8–0.9 °C, which represents a significant improvement over the IKM. Among the four techniques, our results suggest also that RAM produces a much lower RMS value (~0.85 °C) than MAT (~0.97 °C), SIMMAX (~0.95 °C), and ANN (~0.96 °C). Kucera et al. (this volume) have pointed out that the low error values produced by RAM reflect the fact that the leaving-one-out validation procedure in the RAM02 software is implemented only after the two-dimensional interpolation and thus underestimates the full error rate. The difference is, however, negligible compared with the ~0.5 °C estimation uncertainty difference between the IKM and MAT/SIMMAX/RAM/ANN groups.

We also noticed that in estimating winter and summer SSTs, these five different techniques yielded results similar to those for estimating annual average SSTs (Table 3). The IKM showed large errors in estimating the SSTs, while the other four (MAT, SIMMAX, RAM, and ANN) exhibited smaller errors with better predictability than the IKM. All five techniques produced relatively larger uncertainties when estimating winter SSTs (1.2–1.7 °C) versus summer SSTs (0.7–1.3 °C).

3.2. Coretop evaluation of estimation biases

To evaluate potential biases in the calibration of the five SST estimation techniques, which might lead to inaccurate downcore estimates, we examined the relationship of SST residuals (the difference between estimated minus observed SSTs) with observed SST (Fig. 3), latitudinal distribution (Fig. 4), and water depth of coretops (Fig. 5). Techniques (MAT, SIMMAX, RAM) adopting dissimilarity (or similarity) indices for comparing coretop faunal samples were also

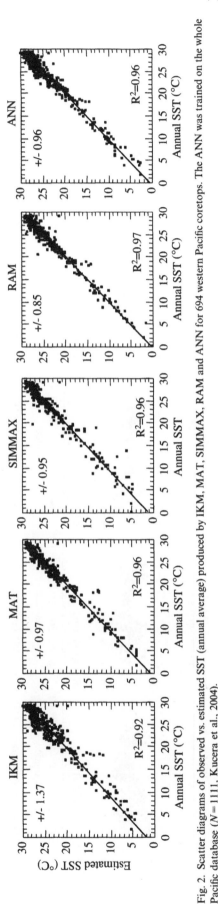

Fig. 2. Scatter diagrams of observed vs. estimated SST (annual average) produced by IKM, MAT, SIMMAX, RAM and ANN for 694 western Pacific coretops. The ANN was trained on the whole Pacific database ($N = 1111$, Kucera et al., 2004).

Table 3
Statistical results for IKM, MAT, SIMMAX, RAM, and ANN

Methods	Annual SST		Winter SST		Summer SST	
	R^2	RMS error	R^2	RMS error	R^2	RMS error
IKM	0.92	1.37	0.89	1.73	0.91	1.35
MAT	0.96	0.97	0.96	1.14	0.96	1.00
SIMMAX	0.95	0.96	0.96	1.12	0.96	0.90
RAM	0.97	0.85	0.95	1.17	0.98	0.72
ANN	0.96	0.96	0.94	1.27	0.98	0.74

Note: Annual SST: Annual sea surface temperature. Winter SST: Average sea surface temperature of caloric winter season. Summer SST: Average sea surface temperature of caloric summer season.

evaluated by examining the relationship of the SST residuals with different levels of dissimilarity (or similarity), since the lack of good analog samples in the coretop data set might also bias the SST estimates (Fig. 6).

No significant correlation ($R^2 \leqslant 0.01$) was found between SST residuals and observed SST, water depth, or latitudes given by the five different techniques. This suggests that the variables did not bias the calibration or training estimation results for SST. Although the correlation was not significant, IKM, MAT, and SIMMAX SST estimates appeared to be too high in SST $< 15\,^{\circ}$C and too low in SST $> 25\,^{\circ}$C (Fig. 3). These patterns suggest that these three techniques may tend to underestimate the full range of SST variability in downcore faunal records. In evaluating the bias patterns associated with observed SSTs, the RAM and ANN estimates exhibited a noticeable improvement at the cold and warm ends of the SST range. These are indications that RAM and ANN might provide better reconstructions in the full range of possible SST variation.

When we examined the latitudinal distribution of the SST residuals by the five different techniques (Fig. 4), it also appeared that IKM, MAT, and SIMMAX SST estimates have large uncertainties, mostly at latitudes poleward of 30° north and south. The large errors (primarily overestimates) expressed in high latitudes indicate that IKM, MAT, and SIMMAX techniques are more susceptible to lacking good analogs in high latitudes. The 694 coretop database used here might not represent faunal distribution patterns well at higher latitudes, due to relatively few published data and also poorer preservation of carbonate sediments at high latitudes of the northwest Pacific (Fig. 1). These three techniques appear to have been forced to use fauna factors or analogs of lower latitudes for estimating the high-latitude coretop SSTs. Several large SST errors in the MAT and SIMMAX estimates associated with large dissimilarity (Fig. 6) support the idea that the MAT and SIMMAX estimates were biased by the no-analog

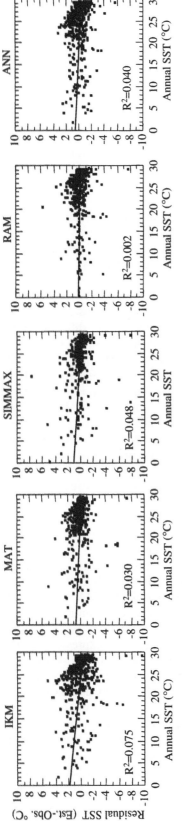

Fig. 3. Scatter diagrams of observed SST vs. ΔSST (estimated minus observed values) produced by IKM, MAT, SIMMAX, RAM and ANN for 694 western Pacific coretops.

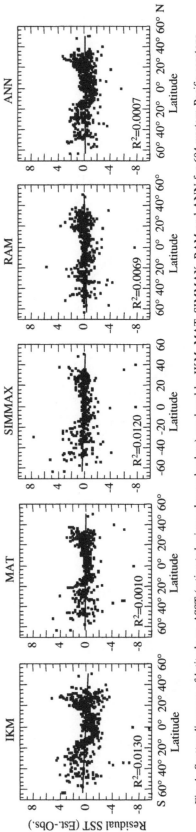

Fig. 4. Scatter diagrams of latitude vs. ΔSST (estimated minus observed values) produced by IKM, MAT, SIMMAX, RAM and ANN for 694 western Pacific coretops.

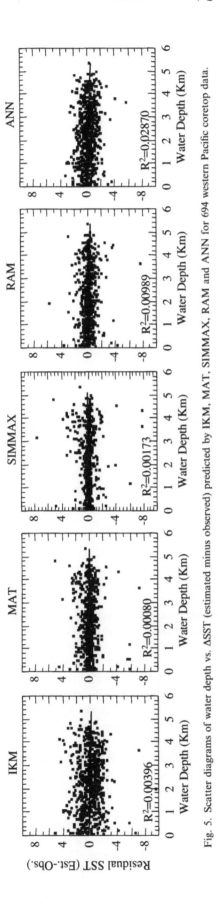

Fig. 5. Scatter diagrams of water depth vs. ΔSST (estimated minus observed) predicted by IKM, MAT, SIMMAX, RAM and ANN for 694 western Pacific coretop data.

conditions. In contrast, RAM and ANN appear to be more successful in estimating the high-latitude coretop SSTs, since there were only relatively small errors at the two ends (Fig. 3). For ANN, this could be in part due to the larger calibration database on which the neural networks were trained (Kucera et al., 2004).

Changes in different levels of carbonate preservation were previously thought to produce biases in SST estimates based on planktic foraminifer faunal assemblages (Thompson, 1976; Le, 1992; Miao and Thunell, 1994). Poor carbonate preservation might preferentially eliminate some species living primarily in warm surface water (such as *G. ruber* and *G. sacculifer*) with relatively thin and delicate tests. With the dissolution of these warm water species, SST estimate techniques might be biased toward colder estimates. The five different SST estimation techniques appear to be less affected by the dissolution problem; we observed no significant correlation between SST residuals and water depth, and no large error was observed from coretops located at deeper water depths (Fig. 5).

3.3. Downcore evaluation of estimation biases

We evaluated SST estimations by these five different techniques on a high-resolution planktic foraminifer fauna record of core MD972151 (8°43.73′N 109°52.17′E, water depth 1589 m) taken from the southern SCS during the 1997 IMAGES cruise (Chen et al., 1998). This record combines planktic foraminifer isotope stratigraphy, AMS C^{14} age model (Lee et al., 1999) and paleomagnetic stratigraphy (Lee, in preparation) with high-resolution planktic foraminifer fauna abundance data (Huang et al., 2002) and alkenone SST data (Huang et al., 1999). SST estimates based on the five above-described transfer function techniques are shown for a Holocene (0–4 kya) as well as an LGM (19–23 kya) window of the record, with comparisons to alkenone SST estimates and $\delta^{18}O$ variations (Fig. 7).

The observed annual average SST at this core site is 28.0 °C. All five techniques appear to have succeeded in reconstructing the absolute value of the observed SST based on the Holocene part of the record, although the stratigraphy of this core suggests that the age of the coretop sediments reaches ~1 kya. While applying SIMMAX and RAM to the downcore record, we excluded the coretop at the site MD972151. All five SST estimates fluctuated within a range of 27–29 °C and indicated a few small cooling events in the late Holocene window. While the MAT, SIMMAX, RAM and ANN SST estimates agreed well with each other and variations fell into a narrow range of ~27–28 °C, the IKM yielded warmer estimates by ~1–1.5 °C than all the other techniques and the alkenone SST. The IKM SST estimates may be biased by a low communality (~0.7) of

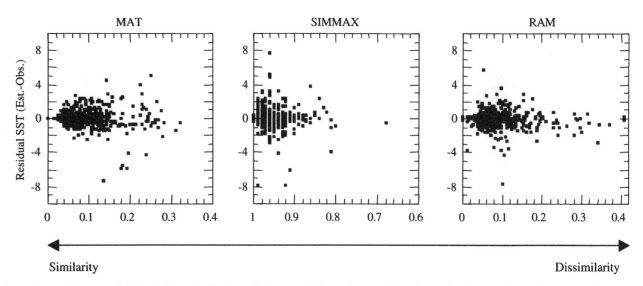

Fig. 6. Scatter diagrams of dissimilarity or similarity coefficients vs. ΔSST (estimated minus observed values) produced by MAT, SIMMAX and RAM for 694 western Pacific coretops.

the downcore samples (Fig. 7). The low communality indicates that the faunal assemblages in core MD972151 are not well explained by the factor model (Table 1) used in generating the transfer function. The appearance of local or high diversity of faunal assemblages in the southern SCS might not be represented well in the factor model of all western Pacific data. In contrast, all the other techniques appeared to overcome the problem, since they exhibited either low dissimilarity (or high similarity) or had a small standard deviation of estimates based on different partitions of the calibration database.

In the LGM window, all estimation techniques yielded much greater variability than in the Holocene. The IKM SST estimates remained within a range of 28–29 °C, but differed from the other techniques by +1.5–2 °C (Fig. 7). This appeared as "tropical stability" in glacial oceans if we accept the IKM SST estimate for the LGM in the southern SCS. In disagreement with the IKM, all other techniques produced much cooler SST estimates in the LGM. MAT, SIMMAX, RAM and ANN, as they did in the Holocene window, showed resemblance and consistently yielded LGM SSTs in the range of ~26.5–27 °C, although there are some indications of the presence of minor no-analog faunas in the LGM (relatively high dissimilarity and standard deviation) (Fig. 7). If these estimates are considered more reliable, then the faunal data would suggest ~1 °C cooling in the glacial southern SCS. In the LGM window, the alkenone SST shows a large offset of ~3 °C to the IKM and ~1–1.5 °C to the other techniques. As compared to the alkenone estimates of the Holocene, the organic geochemical method yields a ~2 °C glacial cooling in the southern SCS.

4. Discussion and conclusions

Our tests of five different SST estimation techniques against a western Pacific coretop data set indicated that better predictions could be obtained with MAT, SIMMAX, RAM and ANN than with IKM. The superiority of the other techniques over IKM when predicting coretop SSTs has been reported in a number of previous studies (Pflaumann et al., 1996; Ortiz and Mix, 1997; Waelbroeck et al., 1998; Malmgren et al., 2001). One source of the large errors in the IKM estimates might be the application of a factor model to faunal percentage data. A factor model is efficient in summarizing multivariate fauna data and translating that data into a few independent, oceanographically interpretable factor "assemblages". However, the composition of the factor assemblages might not reflect real ecological associations of the faunas in oceans. This discrepancy may have led to systematic biases in the IKM estimates. A newly revised version of IKM (Mix et al., 1999) where downcore faunas are included in the construction of the factor model appears to improve the predictability of IKM estimates, but this technique was not tested in this study.

MAT, SIMMAX and RAM obtained SST estimates from the observed values of SSTs from a set of coretop analogs, which were selected based on calculations of similarity/dissimilarity coefficients. The different coefficients adopted by these techniques may have resulted in slightly different estimation results. Since all similarity/dissimilarity coefficients adopted by the techniques have the common effect of amplifying the signals of less dominant species and of reducing the influence of dominant species abundances, we suspect that the use

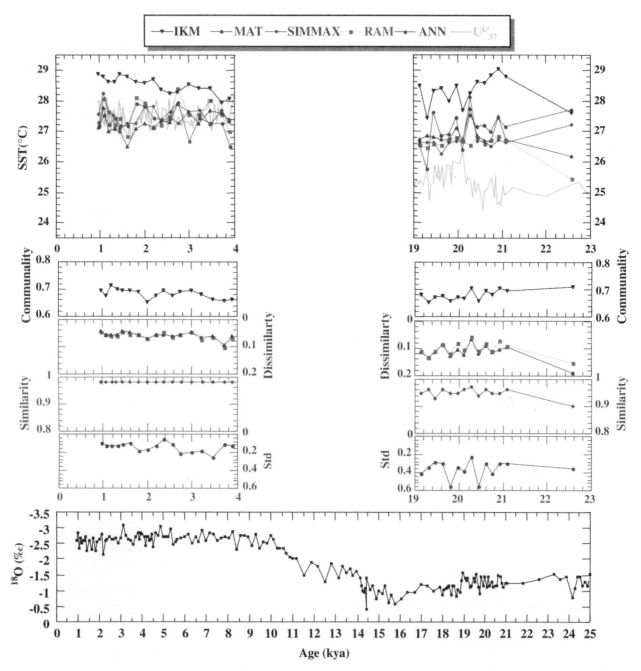

Fig. 7. Holocene and LGM SST reconstructions in core MD972151 (southern South China Sea) by IKM, MAT, SIMMAX, RAM and ANN and by planktic foraminifer isotope stratigraphy (Lee et al., 1999). The SST reconstructions presented here are compared with alkenone SST estimates from the same core (Huang et al., 1999).

of different coefficients is not a major factor affecting the accuracy of these estimations. While MAT only takes a simple average of the best 10 analogs, SIMMAX and RAM are considered to yield better estimations because they adopt more advanced weighting or truncating functions. In the SIMMAX estimations, a coretop sample from the downcore record was excluded from the analog searching because it would introduce a strong bias in the geographical distance weighting procedure (Pflaumann et al., 1996). The very good

estimation ability indicated by the RAM results is only partly due to the analog searching that was based on an interpolated, expanded database (coretop itself exclusive), which may have significantly improved the sparseness of the coretop data in the western Pacific. On the other hand, it may mainly result from the leaving-one-out technique which RAM applied after virtual samples were added, and which explains the generally lower prediction errors of this technique (Kucera et al., this volume).

The ANN technique offered the same high accuracy and precision in coretop SST estimations as RAM. In our study, ANN demonstrated an ability to learn the very complicated relationships between faunas and SSTs and to produce very accurate estimates. ANN, although its training procedures are extremely complex and time-consuming, provided the same potential as RAM for estimating highly accurate SSTs in the western Pacific. Future work will be needed to compare the estimation results of ANN with training on western Pacific coretop data only.

SST reconstructions based on alkenones (Pelejero et al., 1999) and Mg/Ca data (Lea et al., 2000; Stott et al., 2002; Visser et al., 2003; Rosenthal et al., 2003) suggest a general cooling of ~2–3 °C in the western tropical Pacific marginal seas or open ocean during the LGM. In contrast to the general agreement of LGM alkenone and Mg/Ca SST data, atmosphere–ocean coupled GCMs produce a relatively wide range of estimations of 1–6 °C for LGM cooling on the surface ocean of the western tropical Pacific (Weaver et al., 1999; Bush and Philander, 1999; Hewitt et al., 2001; Shin et al., 2003). Although different atmospheric and/or oceanic processes were incorporated in these GCMs, all modeled estimates of surface ocean temperatures in the western tropical Pacific were interpretable by specific climatic mechanisms. If we consider the faunal SST estimates to be more reliable when different techniques yield similar results, the downcore data of MD972151 from the southern SCS suggest a ~1 °C cooling of the western tropical Pacific during the LGM. This ~1 °C cooling is supported by MAT, SIMMAX, RAM and ANN (Fig. 7), and thus it is robust and also independent of the methodology of SST estimation techniques that were used.

The contribution of the SCS core to determining the magnitude of glacial cooling in the western Pacific assumes that the SCS is a good representative of tropical Pacific climate. However, recent studies (Kienast et al., 2001; Kiefer and Kienast, 2004) suggest that SST in the SCS appears strongly influenced by terrestrial and/or local climate conditions and therefore deviates from open ocean conditions in the western tropical Pacific. Nevertheless, if this ~1 °C glacial cooling is accepted, our study implies that much larger variations in sea surface salinity can be extracted from the planktic foraminifer oxygen isotope data which show ~1.6‰ difference between the Holocene and LGM (Fig. 7). This is consistent with the current though that large changes in hydrological cycles analogous to ENSO are a dominant feature of the tropical Pacific (Cane and Clement, 1999; Clement and Cane, 1999; Clement et al., 1999; Stott et al., 2002; Koutavas et al., 2002; Visser et al., 2003). Although a ~1 °C discrepancy exists between our faunal and alkenone SST estimates of the LGM in core MD972151, this discrepancy may result from differential sensitivity of the faunal and alkenone methods, or from different optimum growing seasons of planktic foraminifers and coccoliths, and/or complex sedimentation patterns in the southern SCS. In any case, our results provide a more conservative estimate of LGM cooling in the western tropical Pacific and will have to be considered in further assessments of surface ocean climate in the glacial tropics.

Acknowledgements

This research was supported by the National Science Council (NSC92-2611-M-019-016), Academia Sinica (Asian Paleoenvironmental Changes (APEC) Projects), and the National Taiwan Ocean University, Republic of China. We thank Thorsten Kiefer and Tim Barrows for their constructive reviews.

References

Barrows, T.T., Juggins, S., De Deckker, P., Thiede, J., Martinez, J.I., 2000. Sea-surface temperatures of the southwest Pacific Ocean during the last glacial maximum. Paleoceanography 15, 95–109.

Be, A.W.H., 1967. Foraminifera families: Globigerinide and Globorotaliidae. Fiches d'Idendification du Zooplancton. Fraser, J.H. Cons., Int. Explor. Mer, Charlottenlund: Sheet 118.

Beck, J.W., Edwards, R.L., Ito, E., Taylor, F.W., Recy, J., Rougerie, F., Joannot, P., Heinin, C., 1992. Sea-surface temperature from coral skeletal strontium/calcium ratios. Science 257, 644–647.

Bush, A.B.G., Philander, S.G.H., 1999. The climate of the last glacial maximum: results from a coupled atmosphere-ocean general circulation model. Journal of Geophysical Research 104, 24509–24525.

Cane, M., Clement, A.C., 1999. A role for the tropical Pacific coupled ocean-atmosphere system on Milankovitch and millennial timescales. Part II: Global Impacts, American Geophysical Union, pp. 373–384.

Chen, M.T., Ho, H.W., Lai, T.D., Zheng, L., Miao, Q., Shea, K.S., Chen, M.P., Wang, P., Wei, K.Y., Huang, C.Y., 1998. Recent planktonic foraminifers and their relationships to surface ocean hydrography of the south China sea. Marine Geology 146, 173–190.

CLIMAP Project Members, 1976. The surface of the ice-age Earth. Science 191, 1131–1137.

CLIMAP Project Members, 1981. Seasonal reconstructions of the Earth's surface at the last glacial maximum. Geological Society of America Map and Chart Series, MC-36, Geological Society of America, Boulder, CO.

Clement, A.C., Cane, M., 1999. A role for the tropical Pacific coupled ocean-atmosphere system on Milankovitch and millennial timescales. Part I: a modeling study of tropical Pacific variability, American Geophysical Union, pp. 363–371.

Clement, A.C., Seager, R., Cane, M.A., 1999. Orbital controls on the El Niño/Southern Oscillation and the tropical climate. Paleoceanography 14 (4), 441–456.

Coulbourn, W.T., Parker, F.L., Berger, W.H., 1980. Faunal and solution patterns of planktonic foraminifera in surface sediments of the North Pacific. Marine Micropaleontology 5, 329–399.

Guilderson, T., Fairbanks, R.G., Rubenstone, J.L., 1994. Tropical temperature variations since 20,000 years ago: modulating interhemispheric climate change. Science 263, 663–665.

Hewitt, C.D., Broccoli, A.J., Mitchell, J.F.B., Stouffer, R.J., 2001. A coupled model study of the last glacial maximum: was part of the North Atlantic relatively warm? Geophysical Research Letters 28, 1571–1574.

Huang, C.-Y., Wang, C.-C., Zhao, M., 1999. High-resolution Carbonate Stratigraphy of IMAGES Core MD972151 from South China sea. The Journal of Terrestrial, Atmospheric, and Oceanic Sciences 10, 225–238.

Huang, C.C., Chen, M.T., Lee, M.Y., Wei, W.Y., Huang, C.Y., 2002. Planktic foraminifer faunal sea surface temperature records of the past two glacial terminations in the South China Sea near Wan-An shallow (IMAGES core MD972151). Western Pacific Earth Sciences 2, 1–4.

Hutson, W.H., 1980. The Agulhas Current during the Late Pleistocene: analysis of modern faunal analogs. Science 207, 64–66.

Hutson, W.H., Prell, W.L., 1980. A paleoecological transfer function, FI-2, for Indian Ocean planktonic foraminifera. Paleontology 54, 381–399.

Imbrie, J., Kipp, N.G., 1971. A new micropaleontological method for quantitative paleoclimatology: application to a Late Pleistocene Caribbean core. In: Turekian, K.K. (Ed.), The Late Cenozoic Glacial Ages. Yale University Press, New Haven, pp. 71–181.

Kiefer, T., Kienast, M., 2004. The deglacial warming in the Pacific Ocean: a review with emphasis on Heinrich Event 1. Quaternary Science Reviews, this issue (doi: 10.1016/j.quascirev.2004.02.021).

Kiefer, T., Sarnthein, M., Erlenkeuser, H., Grootes, P.M., Roberts, A.P., 2001. North Pacific response to millennial-scale changes in ocean circulation over the last 60 kyr. Paleoceanography 16, 179–189.

Kienast, M., Steinke, S., Stattegger, K., Calvert, S.E., 2001. Synchronous tropical South China Sea SST change and Greenland warming during deglaciation. Science 291, 2132–2134.

Kipp, N.G., 1976. New transfer function for estimating past sea-surface conditions from sea-bed distribution of planktonic foraminiferal assemblages in the North Atlantic. Memoir of Geological Society of America 3–41.

Koutavas, A., Lynch-Stieglitz, J., Marchitto Jr., T.M., Sachs, J.P., 2002. El Niño-like pattern in ice age tropical Pacific sea surface temperature. Science 297, 226–230.

Kucera, M., Weinelt, M., Kiefer, T., Pflaumann, U., Hayes, A., Weinelt, M., Chen, M.-T., Mix, A.C., Barrows, T., Cortijo, E., Duprat, J., Waelbroeck, C., 2004. Reconstruction of sea-surface temperatures from assemblages of planktonic foraminifera: multi-technique approach based on geographically constrained calibration datasets and its application to glacial Atlantic and Pacific Oceans. Quaternary Science Reviews, this issue (doi: 10.1016/j.quascirev.2004.07.014).

Le, J., 1992. Palaeotemperature estimation methods: sensitivity test on two western equatorial Pacific cores. Quaternary Science Reviews 11, 801–820.

Lea, D.W., Pak, D.K., Spero, H.J., 2000. Climate impact of Late Quaternary equatorial Pacific sea surface temperature variations. Science 289, 1719–1724.

Lee, T.-Q., Environmental magnetic record of deep-sea core MD972151 from the southwestern South China Sea, in preparation.

Lee, M.-Y., Wei, K.-Y., Chen, Y.-G., 1999. High resolution oxygen isotope stratigraphy for the last 150,000 years in the southern South China Sea: core MD972151. The Journal of Terrestrial, Atmospheric, and Oceanic Sciences 10, 239–254.

Malmgren, B.A., Nordlund, U., 1997. Application of artificial neural networks to paleoceanographic data. Palaeogeography, Palaeoclimatology, Palaeoecology 136, 359–373.

Malmgren, B.A., Kucera, M., Nyberg, J., Waelbroeck, C., 2001. Comparison of statistical and artificial neural network techniques

for climating past sea surface temperatures from planktonic foraminifer. Paleoceanography 16, 1–11.

Miao, Q., Thunell, R.C., 1994. Glacial-Holocene carbonate dissolution and sea surface temperatures in the South China and Sulu Seas. Paleoceanography 9, 269–290.

Mix, A.C., Bard, E., Schneider, R., 2001. Environmental processes of the ice age: land, oceans, glaciers (EPILOG). Quaternary Science Reviews 20, 627–657.

Mix, A.C., Morey, A.E., Pisias, N.G., Hosterler, S.W., 1999. Foraminiferal faunal estimates of paleotemperature: circumventing the no-analog problem yields cool ice age trophics. Paleoceanography 14 (3), 350–359.

Ortiz, J.D., Mix, A.C., 1997. Comparison of Imbrie–Kipp transfer function and modern analog temperature estimates using sediment trap and core top foraminiferal faunas. Paleoceanography 12 (2), 175–190.

Parker, F.L., 1962. Planktonic foraminiferal species in Pacific sediments. Micropaleontology 8, 219–254.

Pelejero, C., Grimalt, J.O., Sarnthein, M., Wang, L., Flores, J.-A., 1999. Molecular biomarker record of sea surface temperature and climatic change in the South China Sea during the last 140,000 years. Marine Geology 156, 109–121.

Pflaumann, U., Jian, Z., 1999. Modern distribution patterns of planktonic foraminifera in the South China Sea and western Pacific: a new transfer technique to estimate regional sea-surface temperatures. Marine Geology 156, 41–83.

Pflaumann, U., Duprat, J., Pujol, C., Labeyrie, L.D., 1996. SIMMAX: a modern analog technique to deduce Atlantic sea surface temperatures from planktonic foraminifera in deep-sea sediments. Paleoceanography 11 (1), 15–35.

Prell, W.L., 1985. The stability of low-latitude sea-surface temperatures: and evaluation of the CLIMAP reconstruction with emphasis on the positive SST anomalies. United States Department of Energy, Office of Energy Research, TR025, US, Government Printing Office, vol. 1–2, pp. 1–60.

Prell, W., Martin, A., Cullen, J., Trend, M., 1999. The Brown University Foraminiferal Data Base. IGBP PAGES/World Data Center-A for Paleoclimatology Data Contribution Series # 1999-027.

Rind, D., Peteet, D., 1985. Terrestrial conditions at the last glacial maximum and CLIMAP sea-surface temperature estimates: are they consistent? Quaternary Research 24, 1–22.

Rosenthal, Y., Oppo, D.W., Linsley, B.K., 2003. The amplitude and phasing of climate change during the last deglaciation in the Sulu Sea, western equatorial Pacific, Geophysical Research Letters, 30(8), 2002GL016612.

Shin, S.-I., Liu, Z., Otto-Bliesner, B., Brady, E.C., Kutzbach, J.E., Harrison, S.P., 2003. A simulation of the last glacial maximum climate using the NCAR-CCSM. Climate Dynamics 20, 127–151.

Stott, L., Poulsen, C., Lund, S., Thunell, R., 2002. Super ENSO and global climate oscillations at millennial time scales. Science 297, 222–226.

Thiede, J., Nees, S., Schulz, H., De Deckker, P., 1997. Organic surface conditions recorded on the sea floor of the southwest Pacific Ocean through the distribution of foraminifers and biogenic silica. Palaeogeography, Palaeoclimatology, Palaeoecology 131, 207–239.

Thompson, P.R., 1976. Planktonic foraminiferal dissolution and the progress towards a Pleistocene equatorial Pacific transfer function. Journal of Foraminiferal Research 6, 208–227.

Thompson, P.R., 1981. Planktonic foraminifera in the western north Pacific during the past 150 000 years: comparison of modern and fossil assemblages. Palaeogeography, Palaeoclimatology, Palaeoecology 35, 241–279.

Ujiie, Y., Ujiie, H., 2000. Distribution and oceanographic relations of modern planktonic foraminifera in the Ryukyu Arc region, northwest Pacific Ocean. Journal of Foraminiferal Research 30 (4), 336–360.

Visser, K., Thunell, R., Stott, L., 2003. Magnitude and timing of temperature changes in the Indo-Pacific warm pool during deglaciation. Nature 421, 152–155.

Waelbroeck, C., Labeyrie, L., Duplessy, J.-C., Guiot, J., Labracherie, M., Leclair, H., Duprat, J., 1998. Improving past sea surface temperature estimates based on planktonic fossil faunas. Paleoceanography 13 (3), 272–283.

Webster, P.N., Streeten, N., 1978. Late Quaternary ice age climates of tropical Australia, interpretation and reconstruction. Quaternary Research 10, 279–309.

WOA, 1998. World Ocean Atlas 1998, version 2, http://www.nodc. noaa.gov/oc5/woa98.html. Technical report, National Oceanographic Data Center, Silver Spring, Maryland.

ELSEVIER

Quaternary Science Reviews 24 (2005) 1063–1081

QSR

Patterns of deglacial warming in the Pacific Ocean: a review with emphasis on the time interval of Heinrich event 1

T. Kiefer[a,*], M. Kienast[b,1]

[a]*Department of Earth Sciences, University of Cambridge, Downing Street, Cambridge CB2 3EQ, UK*
[b]*Woods Hole Oceanographic Institution, Woods Hole, MA, 02543, USA*

Received 10 September 2003; accepted 25 February 2004

Abstract

Based on a compilation of currently available records of past sea surface temperature (SST) variability, estimated from a variety of different proxies, the regional-scale deglacial SST development in the Pacific Ocean is tentatively classified into four endmember types. The subtropical and tropical Pacific is characterized by a continuous deglacial warming without marked interruption during the time interval of Heinrich event 1 (H1), whereas the subarctic North Pacific exhibits centennial-scale warm-cold oscillations during this time interval. SST records from marginal seas of the Pacific show a deglacial warming interrupted by a cooling event coeval with H1, followed by a marked Bølling SST increase. A single SST record from the southwestern subantarctic Pacific displays a continuous deglacial warming across H1 followed by an Antarctic Cold Reversal-type cooling during the Allerød. Thus, in contrast to the deglacial SST development in the Atlantic, which has been inferred to be overwhelmingly driven by the redistribution of heat through changes in the meridional overturning circulation (MOC), none of the open oceanic Pacific SST records reviewed here displays any obvious and/or dominant response to the reduction of the MOC and/or the reorganization of atmospheric circulation during H1. Within the limits of absolute chronologies, all tropical and subtropical Pacific SST records show an onset of deglacial warming at 19 ± 1 ka, coeval with the onset of the deglacial rise in sea level.
© 2004 Elsevier Ltd. All rights reserved.

1. Introduction

The transition from the last glacial maximum to the Holocene is punctuated by various high-frequency oscillations, for example the Bølling warm phase, the Heinrich event 1 (H1) and Younger Dryas (YD) cold periods, or the Antarctic Cold Reversal (ACR). These oscillations may not have been global in extent, and their geographic distribution is still a matter of considerable debate (see for example Broecker and Hemming, 2001; Clark et al., 2002). For the Atlantic Ocean an asynchronous development of deglacial SST

between the northern and southern hemisphere is suggested in some records (e.g., Rühlemann et al., 1999; Vidal et al., 1999; see Clark et al., 2002 for a comprehensive review), supporting the concept of the bipolar seesaw (Crowley, 1992; Broecker, 1998; Alley and Clark, 1999; Stocker, 2000; for a different view see Morgan et al., 2002). According to the seesaw hypothesis, a decrease/cessation of the southward flux of North Atlantic Deep Water is accompanied by a decrease in northward heat transport in the Atlantic, which, in turn, leads to an accumulation of heat in the tropical Atlantic, the South Atlantic, and the Southern Ocean. Conversely, a vigorous North Atlantic Deep Water flux extracts heat from these latter regions.

Irrespective of whether one accepts a northern (e.g., Alley et al., 2002) or a southern (e.g., Petit et al., 1999) hemisphere lead of the deglacial warming, or a synchroneity of both (e.g., Steig et al., 1998; Grootes

*Corresponding author. Tel.: +44-1223-333442; fax: +44-1223-333450

E-mail address: tkie02@esc.cam.ac.uk (T. Kiefer).

[1]Now at: Department of Oceanography, Dalhousie University, Halifax, NS, Canada.

et al., 2001), it is evident that the deglacial warming trend in the circum-North Atlantic realm was temporarily suppressed during H1. Thus, H1 is both associated with a significant cooling in the North Atlantic (Bond et al., 1992; Bard et al., 2000), with the most pronounced reduction of the MOC during the last 30 kyr (e.g. Sarnthein et al., 1994; McManus et al., 2004), as well as with a reorganization of atmospheric circulation (e.g., Porter and An, 1995; Wang et al., 2001; Rohling et al., 2003). As such, H1 represents a major perturbation of the climate system, and mapping its regional manifestation will further our understanding of mechanistic linkages between different components of the climate system, as well as of the communication of climate signals between distant regions.

Current model simulations of SSTs in the Pacific Ocean in response to a collapse of the MOC show conflicting results. For example, the simulations examined by Manabe and Stouffer (1995), Schiller et al. (1997); see also Mikolajewicz et al. (1997), and Vellinga and Wood (2002) all show a significant cooling in the North Pacific as a result of a MOC shutdown. In contrast, the simulations of Huang et al. (2000) as well as of Weaver et al. (1999) and Schmittner et al. (2003) predict a subtle warming throughout the Pacific. This discrepancy could be partly due to the fact that usually only snapshots in time are displayed in these presentations, aimed at highlighting features in the Atlantic Ocean rather than integrated averages of the dynamic response in the Pacific to a collapse of the MOC. Furthermore, all these modelling studies tend to be meltwater experiments on modern boundary conditions, and they are based on the a priori assumption that the climatic signal during H1 originates in the North Atlantic. All these models do agree, however, in predicting only very minor changes in tropical Pacific SSTs in response to a collapse of the MOC.

Here we review deglacial Pacific SST records in an attempt to evaluate the impact of H1 on the deglacial SST development in the Pacific Ocean (for a definition of H1 see below). We further aim to establish regional differences and commonalities of the deglacial SST rise in the Pacific Ocean, and to examine potential mechanistic linkages.

2. Approach

In this review, only SST records with high resolution and with independent radiocarbon chronologies are considered for defining deglacial SST patterns (Table 1, Fig. 1). Radiocarbon chronologies are essential because lead-lag relations with respect to planktonic foraminiferal $\delta^{18}O$ could be obscured by changes in sea surface salinities, which have been shown to be significant, particularly during the deglaciation (e.g. Rosenthal

et al., 2003). For the same reason, we do not include SST reconstructions based solely on foraminiferal $\delta^{18}O$. However, in order to increase the data coverage throughout the Pacific, and in line with the MARGO (Multiproxy Approach for the Reconstruction of the Glacial Ocean surface) strategy, we incorporate SST records from different proxies, namely alkenone unsaturation index, Mg/Ca ratio, and foraminiferal and dinocyst assemblages (Table 1, Fig. 1), even though this direct comparison of SST records derived from different proxies might impose significant distortions to our interpretations. This approach is justified, however, by the fact that a number of Pacific sites where deglacial SSTs have been reconstructed using different proxies show an overall agreement between all methods used (see further discussion below). Furthermore, none of our conclusions is based on a single record or records using a single SST proxy.

The SST records were adopted as published by the authors, making no attempt to harmonize the data towards any particular calibration or method. This approach imposes a focus on the main features of the deglacial rise in SST, thereby largely ignoring absolute temperatures or temperature gradients.

Given the fact that both the atmospheric transport of the temperature signal as well as the impact of changes in the strength of the MOC should be felt almost instantaneously around the globe we do not attempt to evaluate decadal to centennial lead-lag features or whether H1 is the result of the transmission of paleoclimate signals from the North Atlantic elsewhere or whether they are due to a common forcing ('joint dependency', Clark and Bartlein, 1995). Instead, we classify patterns of similar SST evolution across the deglaciation, thereby generally tolerating deviations in the timing of events on the order of several centuries to one millennium. This is considered to approximate the actual precision of most radiocarbon-based age constraints for Pacific sediment cores, which are limited by the (usually small) analytical error of the ^{14}C dates, errors and uncertainties in the paleo-reservoir ages (Sikes et al., 2000; Kovanen and Easterbrook, 2002; Kienast et al., 2003), and errors and uncertainties in the relationship between ^{14}C ages and calendar ages beyond the tree-ring calibration (Stuiver et al., 1998).

By definition, Heinrich events are restricted to the North Atlantic. Accordingly, any assessment of climate change outside the North Atlantic during or in response to Heinrich events has to resort to defining a time period that is equivalent to these North Atlantic events. Thus, for the purpose of this review H1 is defined as the time interval 17.5–14.7 ka. Even though this broad period possibly extends beyond the time of the actual deposition of detrital layers in the North Atlantic (François and Bacon, 1994; but see also Elliot et al., 1998; Clarke et al., 1999; Veiga-Pires and Hillaire-Marcel, 1999;

Table 1
Core sites in the Pacific with deglacial records of sea surface temperature (SST). Deglacial SST patterns are labelled 'A'–'D' as in Fig. 6. Sedimentation rates (SR) are given for the deglacial interval (ca 19–11 cal ka).

Core	Region	Latitude	Longitude	Water depth (m)	Age constraint	SR (cm/kyr)	SST Proxies	SST pattern	Reference
Core sites with deglacial SST records reviewed in this study									
PAR87A-10	Subarctic NE Pacific	54.36	−148.47	3664	^{14}C ages	11	Dinocysts, Radiolaria	C	de Vernal and Pedersen (1997); Sabin and Pisias (1996)
MD01-2416	Subarctic NW Pacific	51.27	167.73	2317	^{14}C ages/^{14}C plateau	10–34	Mg/Ca, PF	C	Kiefer et al. submitted
ODP Site 883	Subarctic NW Pacific	51.20	167.77	2385	^{14}C ages/^{14}C plateau	10–28	Mg/Ca, PF	C	Kiefer et al. submitted
JT96-09	Subarctic NE Pacific	48.91	−126.89	920	^{14}C ages	63	$U^{K'}_{37}$	C?	Kienast and McKay (2001)
W8709A-8TC	Subarctic NE Pacific	42.50	−127.68	3111	^{14}C ages	9	$U^{K'}_{37}$ radiolaria	C	Prahl et al. 1995; Sabin and Pisias (1996)
ODP 1019	California margin	41.68	−124.93	980	^{14}C ages	63	$U^{K'}_{37}$, Np coiling ratios	C?	Barron et al. (2003); Mangelsdorf et al. (2000); Mix et al. (1999)
ODP 1017	California margin	34.53	−120.89	955	^{14}C ages	19	$U^{K'}_{37}$	C	Seki et al. (2002); Mangelsdorf et al. (2000)
ODP 893	California margin	34.29	−120.04	575	^{14}C ages	170	PF	D	Hendy et al. (2002)
PC17	Subtropical Pacific	21.36	−158.19	503	^{14}C ages	2.5	$U^{K'}_{37}$	A	Lee et al. (2001); Lee and Slowey 1999
17940	South China Sea	20.12	117.38	1727	^{14}C ages/benthic δ^{18}O	21–76	$U^{K'}_{37}$, PF	D	Pelejero et al. (1999); Pflaumann and Jian (1999)
SO50-31KL	South China Sea	18.76	115.87	3360	^{14}C ages	10	$U^{K'}_{37}$, PF	D?	Huang et al. (1997b); Chen and Huang (1998)
SCS90-36	South China Sea	18.00	111.49	2050	^{14}C ages	6	$U^{K'}_{37}$, PF	D?	Huang et al. (1997a)
18252	South China Sea	9.23	109.38	1273	^{14}C ages	70	$U^{K'}_{37}$, PF	D	Kienast et al. (2001)
MD97-2151	South China Sea	8.73	109.87	1589	^{14}C ages	23	$U^{K'}_{37}$, PF	D	Huang et al. (2002)
MD97-2141	Sulu Sea	8.80	121.30	3633	^{14}C ages	22	Mg/Ca	A	Rosenthal et al. (2003)
MD98-2181	W tropical Pacific	6.30	125.83	2114	^{14}C ages	48	Mg/Ca	A	Stott et al. (2002)
17964	South China Sea	6.16	112.21	1556	^{14}C ages	46	$U^{K'}_{37}$	D	Pelejero et al. (1999)
18287	South China Sea	5.65	110.65	598	^{14}C ages	55	$U^{K'}_{37}$, PF, Mg/Ca	D	Steinke et al. (2001); Whitko et al. (2002)
TR163-19	E equatorial Pacific	2.26	−90.95	2348	^{14}C ages	4	Mg/Ca	A	Lea et al. (2000); Spero and Lea (2002)
V21-30	E equatorial Pacific	−1.22	−89.68	617	^{14}C ages	15	Mg/Ca	A	Koutavas et al. (2002)
ERDC-92	Central equat. Pacific	−2.23	157.00	1598	^{14}C ages	1–2.5	Mg/Ca	A	Palmer and Pearson (2003)
MD98-2162	W equatorial Pacific	−4.69	117.90	1855	^{14}C ages	33	Mg/Ca	A	Visser et al. (2003)
17748-2/ GeoB3302-1	SE Pacific	−33.22	−72.10	1498	^{14}C ages	13–66	$U^{K'}_{37}$	A	Kim et al. (2002)
MD97-2120	SW Pacific	−45.53	174.93	1210	^{14}C ages	13–22	Mg/Ca	B	Pahnke et al. (2003)
Core sites with lack of independent age control, very low resolution, or inferred dominance of localized SST variability									
TT39-PC12	NE Pacific	49.41	−128.19	2369	^{14}C ages on bulk organic C	ca. 15	Radiolaria		Sabin and Pisias (1996)
TT39-PC17	NE Pacific	48.23	−130.01	2795	^{14}C ages on bulk organic C	ca. 20	Radiolaria		Sabin and Pisias (1996)

Table 1 (continued)

Core	Region	Latitude	Longitude	Water depth (m)	Age constraint	SR (cm/kyr)	SST Proxies	SST pattern	Reference
V20-120	Subarctic NW Pacific	47.40	167.75	6216	radiolaria abundance	ca. 3	Radiolaria		Heusser and Morley (1997)
EW9504-17PC	NE Pacific	42.24	−125.89	2671	few ^{14}C ages	28	Radiolaria		Pisias et al. (2001)
W8709A-13PC	NE Pacific	42.12	−125.75	2712	^{14}C ages	21	Radiolaria, Np Coiling ratios		Pisias et al. (2001); Mix et al. (1999)
ODP 1019	NE Pacific	41.68	−124.93	989	^{14}C ages	63	Radiolaria		Pisias et al. (2001)
ODP 1020	California margin	41.00	−126.43	3038	benthic δ^{18}O	19	$U_{37}^{K'}$		Kreitz et al. (2000)
RC14-105	NW Pacific	39.68	157.55	5630	radiolaria abundance	ca. 3	Radiolaria		Heusser and Morley (1997)
RC14-99	NW Pacific	36.97	147.93	5652	radiolaria abundance	ca. 5	Radiolaria		Heusser and Morley (1997)
KT94-15 PC-9	Japan Sea	39.57	139.41	807	Correlation to KH-79-3 L-3	ca. 13	$U_{37}^{K'}$		Ishiwatari et al. (2001)
GH93 KI-5	Japan Sea	39.57	139.44	754	Correlation to KH-79-3 L-3	ca. 17	$U_{37}^{K'}$		Ishiwatari et al. (2001)
KH-79-3 L-3	Japan Sea	37.06	134.71	935	^{14}C ages, tephra	6.5	$U_{37}^{K'}$		Ishiwatari et al. (1999), (2001)
V1-80-P3	NE Pacific	38.43	−123.80	1600	^{14}C ages on bulk organic C	ca. 10	Radiolaria		Sabin and Pisias (1996)
L13-81-G138	NE Pacific	38.41	−123.97	2531	^{14}C ages on bulk organic C	ca. 8	Radiolaria		Sabin and Pisias (1996)
F8-90-G21	California margin	37.22	−123.24	1605	^{14}C ages	ca. 14	Radiolaria		Sabin and Pisias (1996)
EW9504-13PC	California margin	36.99	−123.27	2510	^{14}C ages	29	Radiolaria		Pisias et al. (2001)
ODP 1018	California margin	36.99	−122.72	2477	benthic δ^{18}O	36	$U_{37}^{K'}$		Mangelsdorf et al. (2000)
F2-92-P3	California margin	35.62	−121.60	799	^{14}C ages	22	Radiolaria		Sabin and Pisias (1996)
V1-81-G15	California margin	33.60	−120.42	1000	^{14}C ages	7	Radiolaria		Sabin and Pisias (1996)
ODP 893	California margin	34.29	−120.04	575	benthic δ^{18}O	170	$U_{37}^{K'}$		Herbert et al. (2001)
EW9504-09PC	California margin	32.86	−119.96	1194	benthic δ^{18}O	10	$U_{37}^{K'}$		Herbert et al. (2001)
ODP 1012	California margin	32.28	−118.38	1772	benthic δ^{18}O	16	$U_{37}^{K'}$		Herbert et al. (2001)
EW9504-03PC	California margin	32.07	−117.37	1245	benthic δ^{18}O	ca. 6	$U_{37}^{K'}$		Herbert et al. (2001)
C-4	Subtropical NW Pacific	33.15	137.70	3343	^{14}C ages	6	Mixed microfossils		Chinzei et al. 1987
KT92-17-St14	Subtropical NW Pacific	32.67	138.46	3252	^{14}C ages	12	$U_{37}^{K'}$		Sawada and Handa (1998)
KT92-17-St19	Subtropical NW Pacific	31.10	138.67	3280	^{14}C ages	14	$U_{37}^{K'}$		Sawada and Handa (1998)
KT92-17-St20	Subtropical NW Pacific	30.38	138.65	3280	^{14}C ages	8	$U_{37}^{K'}$		Sawada and Handa (1998)
V28-304	Subtropical NW Pacific	28.53	134.13	824	benthic δ^{18}O	ca. 5	Radiolaria		Heusser and Morley (1997)
RN95-PC1	East China Sea	32.08	129.00	676	one ^{14}C age, planktonic δ^{18}O	ca. 4	PF		Ujiié and Ujiié (2003)
RN94-PC3	East China Sea	30.93	131.85	1536	^{14}C ages	ca. 13	PF		Ujiié and Ujiié (2003)
RN95-PC3	East China Sea	30.83	128.17	500	^{14}C ages	14	PF		Ujiié and Ujiié (2003)
RN94-PC6	East China Sea	29.75	131.43	3031	one ^{14}C age	ca. 14	PF		Ujiié and Ujiié (2003)
DGKS9603	East China Sea	28.15	127.27	1100	^{14}C ages	13	PF		Li et al. (2001)
RN93-PC3	East China Sea	27.69	126.43	1292	^{14}C ages	ca. 17	PF		Ujiié and Ujiié (2003)
MD982193	East China Sea	27.40	126.27	1614	^{14}C ages	40	PF		Ujiié and Ujiié (2003)
RN93-PC8	East China Sea	24.56	123.75	1561	^{14}C ages	ca. 21	PF		Ujiié and Ujiié (2003)
RN93-PC12	East China Sea	24.02	124.43	2160	^{14}C ages	ca. 9	PF		Ujiié and Ujiié (2003)

Core	Region	Lat.	Long.	Depth	Age control	Res.	Proxy	Reference
RGS0487 BC9/ GC11	California margin	23.68	−111.08	381	benthic δ^{18}O	ca. 2	$U^{K'}_{37}$	Herbert et al. (2001)
LPAZ 21P	California margin	22.99	−109.47	624	benthic δ^{18}O	4	$U^{K'}_{37}$	Herbert et al. (2001)
PC20	Subtropical Pacific	21.34	158.17	640	correlated planktic δ^{18}O	2.5	$U^{K'}_{37}$, PF	Lee et al. (2001)
17938-2	South China Sea	19.79	117.54	2840	planktic δ^{18}O	ca. 15	PF	Chen et al. (1999)
17927-2	South China Sea	17.25	119.45	2804	planktic δ^{18}O	ca. 15	PF	Wang et al. (1999)
17954	South China Sea	14.80	111.53	1520	benthic δ^{18}O	6	$U^{K'}_{37}$, PF	Pelejero et al. (1999); Wang et al. (1999)
17961	South China Sea	8.51	112.33	1968	benthic δ^{18}O	11	$U^{K'}_{37}$	Pelejero et al. (1999)
KH92-1-5cBX	W tropical Pacific	3.53	141.87	2282	few ^{14}C ages	ca. 1	$U^{K'}_{37}$	Okhouchi et al. (1994)
ODP 806B	W equatorial Pacific	0.32	159.36	2520	planktic δ^{18}O	4	Mg/Ca	Lea et al. (2000)
W8402A-14GC	Central equat. Pacific	0.95	−138.95	4287	planktic δ^{18}O	ca. 2	$U^{K'}_{37}$	Lyle et al. (1992)
TR163-11	E equatorial Pacific	6.45	−85.82	1950	planktic δ^{18}O	2	PF	Martinez et al. (2003)
ODP677B	E equatorial Pacific	1.20	−83.74	3461	^{14}C age at onset of deglacial	6.5	PF	Martinez et al. (2003)
ODP506B	E equatorial Pacific	0.61	−86.09	2711	^{14}C age in LGM	5	PF	Martinez et al. (2003)
RC13-110	E equatorial Pacific	−0.10	−95.65	3231	benthic δ^{18}O	3	PF	Feldberg and Mix (2003)
TR163-38	E equatorial Pacific	−1.34	−81.58	2200	^{14}C age in LGM	7	PF	Martinez et al. (2003)
TR163-33	E equatorial Pacific	−1.91	−82.57	2230	^{14}C age in LGM	2	PF	Martinez et al. (2003)
ODP846B	E equatorial Pacific	−3.05	−90.49	3307	^{14}C age in LGM	3.5	PF	Martinez et al. (2003)
RR9702A-69TC	Southeastern Pacific	−16.01	−76.33	3228	^{14}C ages, benthic δ^{18}O	5	PF	Feldberg and Mix (2003)
Y71-6-12	Southeastern Pacific	−16.44	−77.56	2734	one ^{14}C age, benthic δ^{18}O	3	PF	Feldberg and Mix (2003)
RR9702A-63TC	Southeastern Pacific	−18.09	−79.04	2901	one ^{14}C age, benthic δ^{18}O	<1	–	Feldberg and Mix (2003)
TG7	SW Pacific	−17.25	78.11	3120	planktic δ^{18}O	1.5	$U^{K'}_{37}$	Calvo et al. (2001)
P69	SW Pacific	−40.40	178.00	2195	one ^{14}C age, tephra	30	PF	Nelson et al. (2000)
R657	SW Pacific	−42.53	179.49	1408	^{14}C ages	ca. 3	$U^{K'}_{37}$, PF	Sikes et al. (2002)
W268	SW Pacific	−42.85	178.98	980	^{14}C ages	ca. 1	$U^{K'}_{37}$	Sikes et al. (2002)
U939	SW Pacific	−44.53	179.50	1300	^{14}C ages	ca. 6	$U^{K'}_{37}$	Sikes et al. (2002)
U938	SW Pacific	−45.08	179.50	2700	^{14}C ages	ca. 8	$U^{K'}_{37}$, PF	Sikes et al. (2002)
E11-2	Subantarctic S Pacific	−56.00	−115.00	3109	planktic δ^{18}O	9	Mg/Ca	Mashiotta et al. (1999)

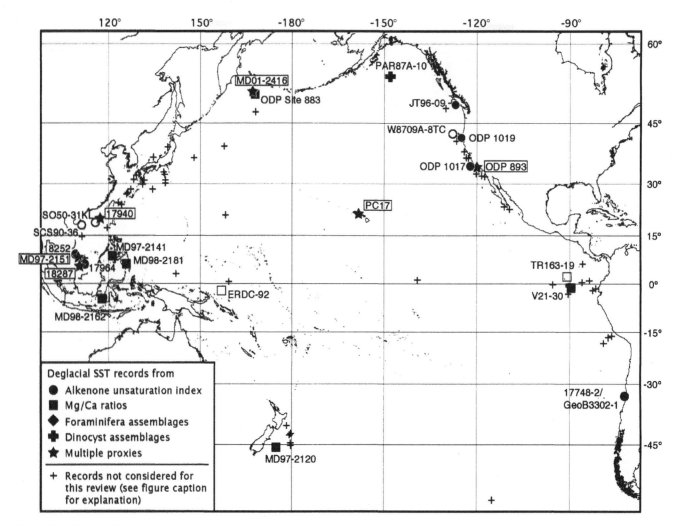

Fig. 1. Distribution of sites with deglacial SST records reviewed in this study (Table 1). Symbols indicate SST estimates derived from alkenones (circles), planktonic foraminiferal Mg/Ca ratios (squares), planktonic foraminiferal assemblages (diamonds), dinocyst assemblages (bold cross) or from multiple proxies (stars with framed core labels, see also Fig. 2). Regional SST patterns (see also Figs. 3–6) are inferred from records with radiocarbon-based age control and high resolution (full symbols), and supported by records with lower resolution (open symbols). Small crosses indicate sites not considered for this review because they lack an independent age control, and/or could not be assigned to a regional SST pattern due to very low resolution or inferred dominance of localized SST variability (see text for discussion).

Grousset et al., 2001), it corresponds to the time of reduced meridional overturning circulation (Sarnthein et al., 1994; McManus et al., 2004) and to the marked deglacial cooling in the North Atlantic region (Cacho et al., 1999; Bard et al., 2000), attributed to the 'Heinrich mode' (Sarnthein et al., 2001; Alley and Clark, 1999) of ocean circulation.

3. Assessment of the multi-proxy approach

Potentials and limitations of comparing different SST proxy records are assessed on the basis of six multi-proxy SST records from different climate regimes of the Pacific, the tropical South China Sea (cores 17940, 18287, and MD97-2151), the central subtropical North Pacific (core PC 17), the eastern mid-latitude Pacific

(Santa Barbara Basin, ODP Site 893), and the subarctic northwestern Pacific (core MD01-2416).

The only Pacific site with published SST estimates derived from foraminiferal census counts (Steinke et al., 2001), alkenone unsaturation ($U_{37}^{K'}$; Kienast et al., 2001; Steinke et al., 2001) and foraminiferal Mg/Ca ratios (Whitko et al., 2002) is in the southern South China Sea (site 18287; Fig. 2A). Here, $U_{37}^{K'}$, foraminiferal transfer function (FP-12E) and Mg/Ca estimates consistently record an average late glacial-Holocene SST difference of $2.2 \pm 0.5\,^{\circ}\mathrm{C}$. Despite an offset in the absolute SST values, the three methods also agree in recording the main SST features, that is a gradual warming during the later part of H1 (note that this core does not span the entire time interval of H1), an abrupt warming of at least $1\,^{\circ}\mathrm{C}$ at ca. 14.7 ka, and a relatively gentle warming towards a mid-Holocene SST maximum (Fig. 2A). An obvious difference between the three proxy records is the relative smoothness of the alkenone SST record as compared to the more variable foraminiferal Mg/Ca and

faunal records (Fig. 2A). This difference could also explain why a minor, albeit clearly detectable, cooling in the $U_{37}^{K'}$-record of 0.2–0.6 °C around 11–13 ka (Steinke et al., 2001) is poorly expressed in the foraminiferal transfer function and Mg/Ca records.

The high-frequency variability of the foraminiferal transfer function and Mg/Ca SSTs could be due to higher sensitivity of the methods, higher analytical noise or higher inherent errors of these approaches. Alternatively, the relatively smooth $U_{37}^{K'}$ record could be due to differential bioturbational mixing of alkenones and foraminifera (Bard, 2001), preferentially attenuating the alkenone record. The excellent agreement (in timing and amplitude) between all methods during the step-like Bølling warming at 14.7 cal ka, however, suggests that this effect is limited to a centimetre-scale. The sample-to-sample agreement, including a synchroneity of foraminiferal $\delta^{18}O$ and $U_{37}^{K'}$ SST, also attests to the autochthonous origin of the alkenones in the South China Sea cores. Whatever the cause(s) for the apparent discrepancy are, the high-frequency variability of foraminiferal transfer function and Mg/Ca SST estimates has the potential to obscure minor SST shifts of less than ~1 °C. For the purpose of this review, however, we note that all three methods of SST estimation clearly show a consistent SST development during the deglaciation.

Similarly, alkenones and a globally calibrated foraminiferal transfer function yield comparable patterns of the deglacial warming at site MD97-2151 in the southwestern South China Sea (Huang et al., 2002), despite differences in the estimated absolute glacial–interglacial SST amplitude (Fig. 2B). In particular, both estimates show a marked cooling during the time interval of H1 (Fig. 2B). Planktonic foraminifera and $U_{37}^{K'}$ also display a comparable deglacial SST development off Hawaii (Lee et al., 2001; see Fig. 2C), thus

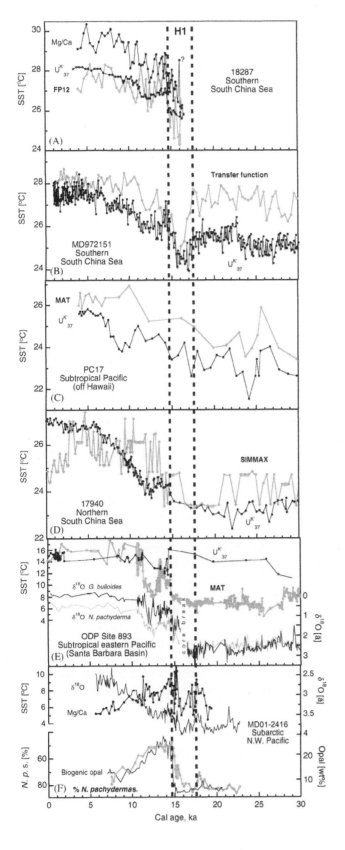

Fig. 2. Multi-proxy SST records from the Pacific (see Fig. 1). A: SST at site 18287 from alkenones, planktonic foraminiferal assemblages using a transfer function (Steinke et al., 2001), and Mg/Ca ratios of *Globigerinoides ruber* (Whitko et al., 2002; calibrated after Lea et al., 2000). (B) SST at site MD97-2151 from alkenones and planktonic foraminiferal assemblages using a foraminiferal transfer function (Huang et al., 2002); (C) SST at site PC17 from alkenones (Lee et al., 2001) and planktonic foraminiferal assemblages using a Modern Analog Technique (MAT; Lee and Slowey, 1999). (D) SST at site 17940 from alkenones (Pelejero et al., 1999) and from planktonic foraminiferal assemblages using the SIMMAX MAT (Pflaumann and Jian, 1999). (E) SST at ODP Site 893 from alkenones (Herbert et al., 2001) and planktonic foraminiferal assemblages using a MAT (Hendy and Kennett, 2000; I. Hendy, unpublished data), compared to $\delta^{18}O$ of *Globigerina bulloides* and *Neogloboquadrina pachyderma* (sinistral + dextral) (Hendy et al., 2002). (F) SST from core MD01-2416 estimated from Mg/Ca ratios of *Neogloboquadrina pachyderma* (sinistral), calibrated after Elderfield and Ganssen (2000), compared to $\delta^{18}O$ of the same species, *N. pachyderma* coiling ratios, and concentrations of biogenic opal (Kiefer et al., submitted).

further justifying the interchangable use of different SST proxies for the purpose of this study.

It is noted, however, that in the South China Sea, various other methods to derive SST estimates from foraminiferal abundances (e.g., SIMMAX, RAM and ANN) yield deviating SST estimates for core MD97-2151 (Chen et al., 2004) and core 18287 (Steinke et al., 2001). Correspondingly, at the northern South China Sea site 17940, a foraminiferal transfer function (SIMMAX) and alkenones yield quite different deglacial SST estimates. For example, an abrupt warming in the $U_{37}^{K'}$ record of ~1 °C at ca 14.7 ka (the Bølling warming) has no equivalent feature in the faunal record, but is instead bracketed by warmings at ~16 and ~13.7 ka (Fig. 2D; Pflaumann and Jian, 1999). Pflaumann and Jian (1999) argue that according to their different habitat the alkenone-producing coccolithophores and planktonic foraminifera record water temperatures from the euphotic zone and the entire upper water column, respectively. However, this deviation between the foraminiferal transfer function and the alkenone unsaturation SST estimates may also be introduced by a large uncertainty of the faunal estimates in the Pacific, arising from a limited number and diversity of coretop samples in the reference datasets (161 samples in Pflaumann and Jian, 1999), which form the basis of this particular SST calibration. This results in frequent poor-analog situations between coretop assemblages and the deglacial assemblages of cores 17940 (and 18287), and in a large sensitivity of the SST estimates to the transfer technique used (see Pflaumann and Jian (1999) and Steinke et al. (2001) for discussion). Thus, in the South China Sea, the discrepancy between different SST estimates appears to be more significant between different foraminiferal transfer techniques (MAT, SIMMAX, RAM, ANN; Steinke et al., 2001 and Chen et al., 2004) than between geochemical proxies (alkenones, Mg/Ca) and Imbrie–Kipp-type foraminferal transfer functions. Since the reason for this inconsistency between the different foraminiferal transfer techniques is not understood yet, we rely on the geochemical SST estimates in assigning a pattern of deglacial SST change in the South China Sea.

It is noted, however, that the timing of changes in the SIMMAX SST record from site 17940, particularly the inferred cold spell between ca. 20 and 16 ka, compares very well to the faunal SST estimates from the Okinawa Trough (Li et al., 2001), possibly indicating a regional signal, at least in foraminiferal abundances. In the Okinawa Trough, however, deglacial SST variability is interpreted to be significantly affected by changes in the geography related to variations in sea level (see the comprehensive review by Ujiié et al., 2003). Furthermore, studies by Brunner and Biscaye (1997) and Yamasaki and Oda (2003) indicate that at shallow water depths foraminiferal tests can be transported laterally, thus potentially obscuring foraminifera-based

SST reconstructions at sites of extreme sediment focusing, such as the Okinawa Trough.

In the Santa Barbara Basin alkenones and foraminiferal abundances yield different deglacial SST estimates (Hendy and Kennett, 2000; Herbert et al., 2001). Here, the foraminiferal SST estimates closely follow the planktonic $\delta^{18}O$ records throughout most of the deglaciation, whereas the $U_{37}^{K'}$ SST estimates show a marked cooling of ca 3 °C at the time of the Bølling warm period (Fig. 2E; note, however, that there are no alkenone SST data during the time interval of H1 or the YD). According to Hendy and Kennett (2000), the absolute SST estimates based on the planktonic foraminiferal assemblages in the Santa Barbara Basin are ambiguous due to significant differences (both due to foraminiferal preservation and water column structure) in faunal communities between the Santa Barbara Basin, which is dominated by eurythermal species, and sites in the California Current region and the North Atlantic used to calibrate the transfer methods. Comparing fossil assemblages from the size fraction > 125 μm with calibration data from the standard > 150 μm fraction added further imprecision on the order of ~0.5 °C (Hendy and Kennett, 2000). On the other hand, the alkenone SST estimates may be compromised by the inferred differential preservation of the di- and tri-unsaturated alkenones under oxic and anoxic bottom water conditions (Gong and Hollander, 1999), which would be particularly severe during the abrupt deglacial changes in bottom water oxygenation in the Santa Barbara Basin (Behl and Kennett, 1996). Furthermore, in their regional assessment of alkenone production along the California Margin, Herbert et al. (1998) note that several closely spaced samples from within the Santa Barbara Basin give a considerable range (1.2 °C) of estimated temperatures despite identical modern sea surface temperatures. These authors note, however, that there are significant SST gradients within the basin, which could explain the observed range of surface sediment SST estimates from within the Santa Barbara Basin. It is beyond the scope of this paper to resolve this discrepancy between the faunal and the $U_{37}^{K'}$ SST estimates for the Santa Barbara Basin. Accordingly, we adopt here the interpretation of Hendy et al. (2002) of a close link between the deglacial SST development in the Santa Barbara Basin and Greenland temperatures, which is also substantiated by records of the coiling ratio of *Neogloboquadrina pachyderma* from the same core (Hendy et al., 2002) and Mg/Ca SST estimates from a near-by site (MD02-2504, not shown) in the Santa Barbara Basin (Hill et al., 2003).

Another limitation of SST estimation based on faunal assemblages is revealed by comparing Mg/Ca temperatures and foraminiferal abundances from the subarctic Pacific (site MD01-2416; Fig. 2F; Kiefer et al., submitted). In this case, the foraminiferal census counts were not converted into SST estimates because no appropriate coretop calibration data set for the sub-arctic Pacific is currently available. Instead, the relative abundance of the (sub-)polar species *N. pachyderma* (left coiling; *Npl*%) is used as a semi-quantitative SST indicator (see also: Hendy et al., 2002; Elliott et al., 2001). In fact, Mg/Ca and *Npl*%, as well as $\delta^{18}O$ of *N. pachyderma* (left) in core MD01-2416 record the same SST features across the deglacial, which are three warm periods from 18.5 to 16.5 ka, at around 15.5 ka, and from 15–11 ka (Fig. 2F). In detail however, the records fail to agree across the 15–11 ka warm period, when high productivity is suggested by high concentrations of biogenic opal (Fig. 2F). The good (negative) correlation between biogenic opal and *Npl*% suggests that the faunal abundances are modulated by the high productivity conditions (Kiefer et al., submitted), thus cautioning the general use of *Npl*% as an SST indicator. In contrast, the Mg/Ca SST estimates appear to be unaffected by productivity changes and are hence used in this study.

The examples discussed here are certainly insufficient to conclusively rule out minor discrepancies between different SST proxies. In fact, these differences might well contain valuable information regarding variations in seasonality or in the thermal structure of the upper water column. Nevertheless, the records presented here demonstrate that SST estimates from alkenones, foraminiferal Mg/Ca ratios and foraminiferal assemblages generally agree in recording SST changes of more than ~1 °C (see also Bard (2000) for a similar conclusion). Furthermore, it appears that the distinction of different deglacial warming patterns throughout the Pacific (see below) is not affected by the choice of SST proxy.

4. Mapping common SST patterns

Bearing in mind the uncertainties related to the multi-proxy approach, we review in the following section the SST development during the deglaciation in currently available downcore records from the Pacific Ocean, focussing on the time interval of H1.

4.1. Continuous deglacial warming

The SST records compiled in Fig. 3 show a continuous deglacial warming, starting at around 19 ± 1 ka, which corresponds to the end of the Last Glacial Maximum (as defined by Mix et al., 2001). All these records have in common that there is no clear indication of a cessation of the deglacial SST increase during H1 (or the ACR or the YD), and that roughly half the deglacial warming occurs prior to the end of H1, i.e. prior to the onset of the Bølling warming. This pattern of continuous deglacial warming is characteristic

of the tropical and subtropical open Pacific (Figs. 3B–F), including the Sulu Sea (Fig. 3E), as well as of the Peru-Chile current off mid-latitude Chile (Fig. 3G) and possibly of the southeastern subantarctic Pacific (site E11-2; Mashiotta et al., 1999; not shown).

It is conceivable that the continuous warming trend observed in some of the lower resolution records or in records from lower sedimentation rate sites is merely due to unresolved SST variations or bioturbational attenuation of the signal, respectively. For example, two $U_{37}^{K'}$ SST records from sites in the South China Sea with low sedimentation rates (average 6–10 cm/kyr, Table 1) show only reduced deglacial SST oscillations characteristic of this marginal basin (see below). Instead, these records display a more or less continuous deglacial warming (Huang et al., 1997a,b; not shown). However, a number of sites from this group, particularly in the western equatorial Pacific, clearly have high enough resolution and sedimentation rates (Table 1) to resolve millennial-scale events such as H1, the ACR or the YD. The lack of any SST response during H1 suggests that the deglacial warming of this region was not affected by changes in the intensity of the MOC, and/or in the atmospheric circulation associated with H1. Furthermore, accepting that the first step in sea level rise occurred at ca 19 ka (see Lambeck and Chappell (2001) for review), the onset of the warming appears to be in phase with the initial melting of continental ice sheets.

4.2. The Antarctic type deglacial pattern

The Antarctic type SST pattern is similar to the continuous deglacial warming described above (i.e., temperatures increased gradually across H1), but punctuated by a moderate temperature decrease during the Allerød period, coeval with the ACR found in Antarctic ice cores (around 13–14 ka; Blunier et al. 1997). Because of the brevity of the ACR (~1000 years) and its relative low-amplitude signal it is difficult to unambiguously distinguish the Antarctic-type deglacial warming from the continuous warming pattern described above. Nevertheless, we adopt here the interpretation of Pahnke et al. (2003) of the deglacial Mg/Ca SST record of core MD97-2120 off New Zealand as showing an ACR (Fig. 3(H)). In contrast to this record, neither of the tropical or subtropical and subantarctic SE Pacific records show any clear indication of an ACR. However, as discussed above, this lack of an ACR signal could also be due to signal attenuation due to low sedimentation rates, most notably in the only other record from the subantarctic Pacific (site E11-2; Mashiotta et al., 1999) or due to low sampling frequency (e.g. site P69 off New Zealand; Nelson et al., 2000). Accordingly, the extent of the Antarctic-type signal cannot be mapped properly. However, the few SST records available suggest that, contrary to the Atlantic

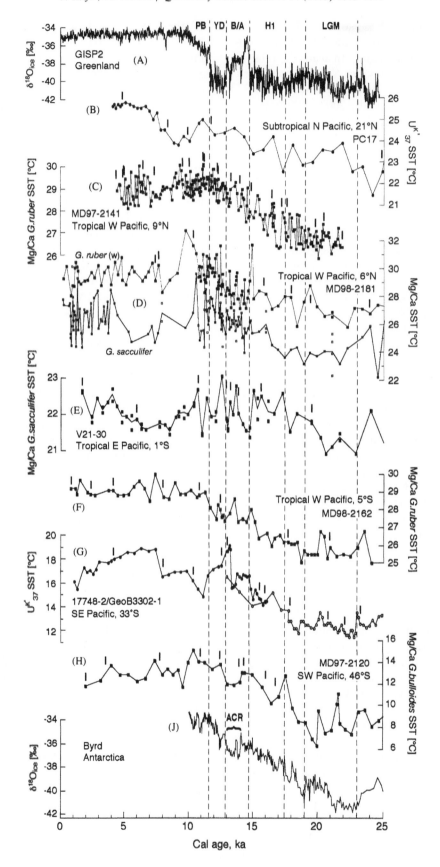

(e.g. Rühlemann et al., 1999), a link between SSTs and air temperatures over Antarctica is restricted to the region of circum-Antarctic atmospheric circulation. It is noted, however, that based on planktonic foraminiferal $\delta^{18}O$ records from the Pacific sector of the Southern Ocean and from the Great Australian Bight, respectively, Morigi et al. (2003) and Andres et al. (2003) postulate the occurrence of a cooling event coeval with the YD cold period in the Southern Ocean. If corroborated by independent SST estimates, these findings challenge the interpretation of Pahnke et al. (2003) of a close coupling of the South Pacific to Antarctic climate variability during the deglaciation.

4.3. Warm-cold oscillation during H1

Two SST records from the subarctic northwestern Pacific (sites ODP 883 and MD01-2416) and one record from the California margin (ODP Site 1017E) are characterized by a rapid warming leading into an SST maximum during the older part of H1 (ca. 18.5–16.5 ka), followed by a cooling during the younger part of H1 (ca. 16.5–15.5 ka) and another rapid warming towards the end of H1 (ca 15.2 ka; Fig. 4C, D, G). Whether or not the abrupt warming following this cold spell could be synchronous with the Bølling warming in the circum-North Atlantic realm will be discussed below. A similar pattern, albeit with lower resolution, is also evident in the SST record based on dinocyst assemblages from the subarctic northeastern Pacific (Fig. 4B). The only two other high-resolution SST records published from the North Pacific to date (sites JT96-09 and ODP 1019; Fig. 4E, F) do not cover the period of interest here. Nevertheless, it is noteworthy that both these $U_{37}^{K'}$ SST records show a warm spike preceding the inferred Bølling warming, possibly indicating that the warm-cold oscillation described above is characteristic of the entire North Pacific. On the other hand, both these records show a YD-type cooling, supporting a close coupling to North Atlantic climate variability during this climate event.

The abrupt warming at the termination of the late-H1 cold spell occurs at ca 15.5 ka at sites ODP 883 and MD01-2416 as well as at site ODP 1017. In core MD01-2416, the age of this warming is constrained by numerous, closely spaced AMS ^{14}C dates, which allow identification in the sediment core of the ^{14}C plateau

occurring from ca 12.4–12.7 ^{14}C ka (ca 15.3–14.35 cal. ka; Stuiver et al., 1998; Hughen et al., 2004). Thus, the dates of 12,460 ^{14}C years (reservoir-corrected) near the top and 12,470 ^{14}C years near the base of the radiocarbon plateau identified in the NW Pacific core MD01-2416 provide crucial age control points with centennial precision around 14.35 and 15.3 ka (Fig. 4C; Kiefer et al., submitted; Sarnthein et al., 2003). The fact that the ^{14}C dates defining the plateau occurs significantly after the SST maximum constitutes compelling evidence that the warming in the NW Pacific occurred several hundred years before the Bølling warming in Greenland at 14.7 cal. ka (Kiefer et al., submitted). This approach of using known features of the ^{14}C curve to establish age fixpoints is analogous to a recent study by Hajdas et al. (2003), who used the Younger Dryas radiocarbon plateau to constrain the timing of cold events in South America.

At ODP Site 1017A off California, the warming at 15.5 ka is delimited by three radiocarbon dates (see Fig. 4). The ^{14}C age closest to the abrupt SST increase (13,050 years at 300.5 cm corrected depth) yields two calibrated ages according to CALIB 4.3 (using a local reservoir age anomaly ΔR of 233 years; Kennett et al., 2000) of 14.37–14.14 and 15.17–14.64 ka (both 1σ), respectively. Choosing the former intercept places the midpoint of the abrupt warming at ca. 14.8 ka, which is indistinguishable from the age of the Bølling warming in the GISP2 ice core record. However, choosing the older intercept produces an age of ca. 15.3 ka, suggesting a lead of this warming with respect to the Bølling warming in the North Atlantic. Omitting this ^{14}C date from the age model (as per Kennett et al., 2000, and adopted by Seki et al., 2002), and interpolating between the two immediately adjacent ^{14}C dates (which have low 1σ calendar age ranges) corroborates the older age of the abrupt warming (Fig. 4G). Even though there does not seem to be any a priori reason to cull the age at 300.5 cm, assuming more or less constant sedimentation rates at this site, would further corroborate the older age of ~15.5 ka of this event.

The temperature fluctuation during H1 described here represents a departure from the simple antiphasing of the northeastern Pacific with respect to North Atlantic climate variability postulated to occur during Dansgaard–Oeschger events (Kiefer et al., 2001), and precludes a direct response of North Pacific SSTs to a

---

Fig. 3. SST records characterized by a continuous warming across the deglacial without major intermittent coolings, compared to the $\delta^{18}O$ records from ice cores in Greenland (A; GISP2; Stuiver and Grootes, 2000) and Antarctica (J; Byrd; Blunier et al., 1998). SST records are arranged North-South: (B) PC17 (Lee et al., 2001), (C) MD97-2141 (Rosenthal et al., 2003), (D) MD98-2181 on a recalculated age scale by linearly interpolating between calibrated ^{14}C ages; lines connect averages of multiple SST data (Stott et al., 2003), (E) V21-30 (Koutavas et al., 2003), (F) MD98-2162 (Visser et al., 2003), (G) 17748-2 and GeoB3302-1 (Kim et al., 2002), (H) MD97-2120 (Pahnke et al., 2003). The latter record shows a 1–2 °C ACR-type cooling during the Allerød. Dashes above SST records mark ^{14}C dates used for the age models, adopted from the original publications. PB = Preboreal, YD = Younger Dryas, B/A = Bølling/Allerød, H1 = Heinrich event 1, LGM = Last Glacial Maximum, ACR = Antarctic Cold Reversal. Symbols as in Fig. 1.

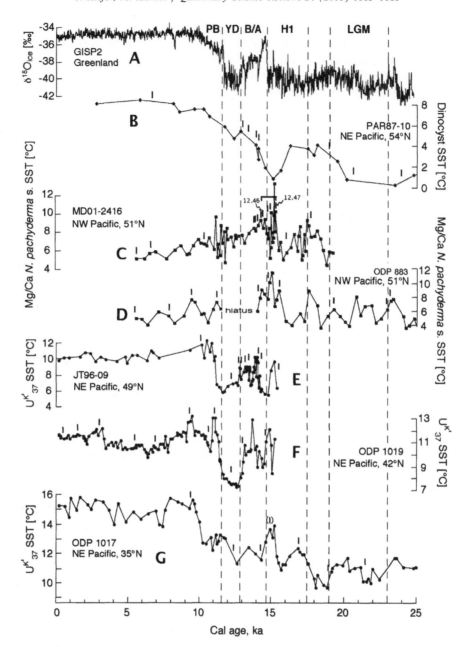

Fig. 4. SST records across the deglacial showing a warm-cold oscillation during H1 compared to the δ^{18}O record from Greenland ice core GISP2 (A; Stuiver and Grootes, 2000). SST records are arranged North-South: (B) PAR87-10 (de Vernal and Pedersen, 1997) on a recalibrated time scale, (C) MD01-2416 (Kiefer et al., submitted) with horizontal bar delineating the ^{14}C-Plateau at 12.4-12.7 ^{14}C ka used as a time marker for late H1/early Bølling (see text), (D) ODP Site 883 (Kiefer et al., submitted), (E) JT96-09 (Kienast and McKay, 2001), (F) ODP Site 1019 (Barron et al., 2003) on their untuned age scale, and (G) ODP Site 1017 (Seki et al., 2002). Dashes and abbreviations as in Fig. 3. The dash in brackets at ODP Site 1017 marks the position of the culled ^{14}C age discussed in the text. Numbers at MD01-2416 give the reservoir corrected ^{14}C ages defining the extent of the 14C plateau (see text). Symbols as in Fig. 1.

shut down of the MOC during H1. Even though the old age of the termination of this inter-H1 cold spell at ca. 15.5 ka is less than 800 years older than the GISP2 age of the Bølling warming (i.e., within the potential error of ^{14}C chronologies), we still suggest that it is a true feature because it is evident in three independent records, constrained by numerous AMS ^{14}C dates. Furthermore, we note that an analogous 'pre-Bølling warming' event

is also seen in records of *N. pachyderma* coiling ratios off Oregon and northern California (sites ODP 1019 and W8709A-13PC; Mix et al., 1999), and in the Santa Barbara Basin (Hendy et al., 2002). This observation also demonstrates the critical role of high resolution age control for revealing offsets in timing and duration of climate events, which otherwise might have been assumed to occur synchronously.

Due to the limited number of northern Pacific SST records, the mapping of the regional extent of this pattern is vague, in particular in the western and central Pacific. Differing SST patterns in the East China and South China Seas and near Hawaii define a maximum southern limit. At the east Pacific margin the southernmost site clearly reflecting the H1 warm-cold oscillation is ODP Site 1017 at 34.5°N. Thus, records of the warm-cold oscillation are currently restricted to the northern subarctic gyre and its southward extension, the California Current. This could be taken as an indication that the SST signal 'originates' somewhere in the subarctic Pacific and is then transmitted further across the North Pacific, possibly by surface currents.

4.4. The marginal sea type

The SST records of the marginal basins of the Pacific (South China Sea, Santa Barbara Basin) show a rapid warming synchronous (within the errors of radiocarbon chronology) with the Bølling warming in GISP2 (Fig. 5). This abrupt rise in SST is preceded by a SST minimum or a period of reduced warming during the time of H1. Both the low SSTs during H1 as well as the following rapid warming at ca. 14.7 ka are analogous to the GISP2 pattern (Fig. 5A) and suggest a strong coupling to the deglacial climate development of the North Atlantic region. This interpretation of the SST pattern in the South China Sea is not only corroborated by five independent $U_{37}^{K'}$ SST records from throughout the South China Sea that all show an abrupt warming at ca. 14.7 ka (Pelejero et al. 1999, Wang et al., 1999; Kienast et al., 2001; see Kienast et al., 2003, and Sarnthein et al., 2003, for a discussion of the ^{14}C dates of Wang et al. 1999; Huang et al., 2002) but also by the general agreement of multiple proxies (Steinke et al. 2001; Huang et al., 2002; Figs. 2A, B).

The interpretation of the Santa Barbara Basin record is more complicated. First, the interpretation of this record with respect to SST during H1 is compromised by a core break during part of the time interval of H1 (ca. 16.6–15.7 ka; Fig. 5B). Secondly, there is a significant discrepancy between the different deglacial SST estimates depending on the proxy used (see above). Nevertheless, the good agreement between foraminiferal abundance (Hendy and Kennett, 2000) and Mg/Ca (Hill et al., 2003) SST estimates suggests a close link of temperatures in this basin to the deglacial warming in Greenland. Some of the marine records described here do not show an actual cooling during H1 as evident in Greenland ice cores and North Atlantic sediment records but rather a slow rate of warming (Figs. 5B, C). Thus, we interpret these records as a combination of the North Atlantic type deglaciation and the continuous warming trend described above for the open low-latitude Pacific. Given the fact that both the Santa

Barbara Basin and particularly the South China Sea are surrounded by open ocean records that show very different patterns (see above), the most likely communication of these marginal basins and the North Atlantic is through the atmosphere (sensu Mikolajewicz et al., 1997).

The assertion of a deglacial SST pattern typical of marginal seas is further substantiated by an analogous discrepancy between the deglacial SST increase in the Cariaco Basin (Lea et al., 2003) compared to the open western tropical Atlantic (Rühlemann et al., 1999) and the Gulf of Mexico (Flower et al., 2004). Thus, the Cariaco Basin SST record shows a Greenland-type SST increase at the onset of the Bølling period, whereas the western tropical Atlantic and the Gulf of Mexico show a close coupling to Antarctic temperatures during the deglaciation, in line with the bipolar see-saw mechanism. Taken together, these records suggest that SSTs in marginal seas are influenced more significantly by continental/atmospheric variability than open ocean sites.

5. Discussion and conclusions

As more records become available, the generalized classification of the deglacial warming in the Pacific proposed here (Fig. 6) may be replaced by the realization of a regionally even more diverse and spatially more dynamic deglacial warming throughout the Pacific. Examples of this highly localized variability are available already from the Japan Sea, the East China Sea and the Kuroshio region, where sea level-related changes in the geographic and oceanographic setting and minor changes in the flow path of the Kuroshio warm current, respectively, have been shown to have profound effects on the deglacial warming on a very localized spatial scale (Sawada and Handa, 1998; Ishiwatari, 1999, 2001; Kim et al., 2000; Ujiié et al., 2003). Similarly, recent studies by Martinez et al. (2003), Koutavas and Lynch-Stieglitz (2003), and Feldberg and Mix (2003) demonstrate a high degree of complexity and regionality in the eastern equatorial Pacific. Here, SSTs are affected by the interaction of equatorial upwelling dynamics, including potential changes in the temperature of the source waters, with changes in the meridional SST gradients related to the intensity and position of the Intertropical Convergence Zone (ITCZ), and variable advection of water masses from the Peru Current. Finally, the northeast Pacific off California and Oregon is another example of high spatial heterogeneity during the deglaciation. Here, SSTs are postulated to be driven by an interaction of the intensity of the California Current and Countercurrent, extent of the North Pacific gyre, and the intensity and position of coastal and open ocean upwelling (Mix et al., 1999; Herbert et al., 2001;

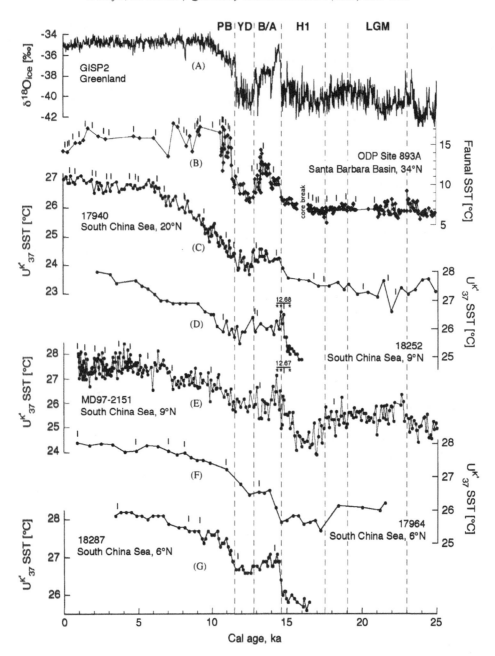

Fig. 5. SST records with 'Greenland-type' deglacial pattern compared to the $\delta^{18}O$ record from Greenland ice core GISP2 (A; Stuiver and Grootes, 2000). SST records are arranged North–South: (B) ODP Site 893 (Hendy and Kennett, 2000; I. Hendy, unpublished SST estimates from planktonic foraminiferal assemblages; Hendy et al., 2002)), (C) 17940 (Pelejero et al., 1999), (D) 18252 (Kienast et al., 2001), (E) MD97-2151 (Huang et al., 2002) on a recalibrated age scale after Lee et al. (1999), (F) 17964 (Pelejero et al., 1999), (F) 18287 (Steinke et al., 2001). Dashes and abbreviations as in Fig. 3. Symbols as in Fig. 1. Asterisks in D and E mark the three alternative intercept ages of the respective ^{14}C dates with the calibration curve of Stuiver et al. (1998), the average age of which (dash in between) was used for the age models.

Pisias et al., 2001; Hendy et al., 2002). In both of these cases, however, some of the inferred regionality of the deglacial warming could also be related to subtle discrepancies between the diverse proxies used to reconstruct SSTs at different sites within the same region (i.e., foraminiferal and radiolarian transfer functions, N. *pachyderma* coiling ratios, alkenones, foraminiferal Mg/Ca and planktonic foraminiferal $\delta^{18}O$).

It appears, however, that even the limited number of SST records available to date allows some key conclusions to be drawn about the deglacial warming in the Pacific:

(1) A close coupling of the deglacial SST increase in the Pacific to temperatures in the North Atlantic realm appears to be restricted to the marginal seas (Fig. 6), possibly also including some marginal sites along the northeastern Pacific. This calls for an at least northern

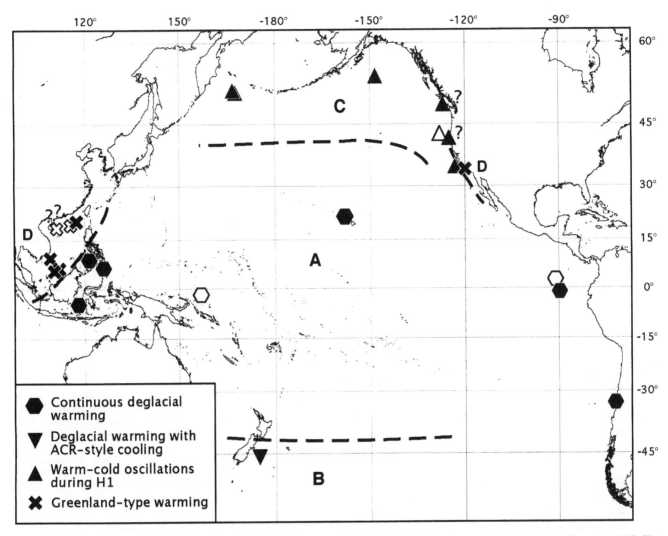

Fig. 6. Distribution of the four endmember-type deglacial SST patterns (A–D) in the Pacific: Continuous deglacial warming (hexagons, "A", Fig. 3B–G), deglacial warming with Antarctic Cold Reversal-style cooling (inverse triangle, "B", Fig. 3(H), warm-cold oscillations during Heinrich event 1 (triangles, "C", Fig. 4), and stepwise Greenland-type warming (crosses, "D", Fig. 5). Full and open symbols as in Fig. 1. The dashed grey lines outline the presumed regional extent of patterns A–D.

hemisphere-wide atmospheric teleconnection (sensu Mikolajewicz et al., 1997), which appears to be overwhelmed by different, and yet unexplained, processes governing the deglacial rise in SSTs in the more open oceanic settings.

(2) To the extent that the records presented here are representative of regional-scale deglacial warming patterns, there does not seem to be any obvious and/or dominant response of open ocean SSTs in the Pacific to the H1 reduction of the MOC or the associated cooling in the North Atlantic. This challenges some of the mechanisms that have been proposed to influence glacial-interglacial SST variability, particularly in the equatorial Pacific. For example, the persistent and pronounced cooling in the North Atlantic such as during H1 would increase the meridional gradients of SST and surface air temperature, resulting in intensified

trade winds and a south-ward migration of the ITCZ (Schiller et al., 1997). Both these atmospheric effects have been invoked to explain glacial-interglacial changes in SST along the equatorial upwelling zone, and in the eastern equatorial Pacific, respectively, and should thus lead to SST variations associated with H1 in this region as well. We concede, however, that the resolution and sensitivity of the presently available SST records from the eastern equatorial Pacific might not suffice to record any subtle SST variations linked to H1. Furthermore, a recent model study by Chiang et al. (2003) suggests that any displacement of the ITCZ due to changes in the intensity of the MOC could be offset by an opposing effect of North Atlantic sea ice cover on the position of the ITCZ. Finally, the inferred absence of an SST response during H1, particularly in the low latitude Pacific (Figs. 3 and 6), is generally consistent with many climate models that show only very minor changes in the tropical Pacific in response to a collapse of the MOC (e.g., Vellinga and Wood, 2002; Schmittner et al., 2003).

(3) The warm-cold oscillations in the northern North Pacific (Figs. 4 and 6) reveal a high degree of centennial-scale climate variability within the time interval of H1. For the North Atlantic region, a comparable variability within the time interval of H1 has been proposed to account for the variable origin and abundance of icebergs (Bond and Lotti, 1995; Elliot et al., 1998; Bard et al., 2000; Grousset et al., 2001), and for the fluctuating intensity of atmospheric polar circulation (Rohling et al., 2003). Nevertheless, regional temperatures remained invariably cold throughout H1 in the North Atlantic region (Bond et al., 1993; Bard et al., 2000; Stuiver and Grootes, 2000), indicating a high sensitivity of northern Pacific SSTs to yet unspecified climate perturbations during H1. Most notably, the internal SST variability during H1 in the subarctic northwest Pacific is decoupled from records of marine productivity from the same cores (Kiefer et al., submitted), suggesting that it is not primarily driven by changes in water column stratification and/or upwelling of sub-surface waters.

(4) The contrasting patterns of deglacial warming observed in various parts of the Pacific (Fig. 6) imply that both zonal and meridional gradients changed on centennial (or even shorter) timescales during the deglaciation. This will complicate any 'simple' reconstruction of meridional and/or zonal temperature gradients, as well as of inferred water vapour fluxes between the Pacific and Atlantic Oceans.

(5) The radiocarbon-dated SST records from the tropical and subtropical Pacific constrain the deglacial warming to have started at $19+/-1$ ka BP. Thus, the warming appears to be in phase with the initial melting of continental ice sheets.

Acknowledgements

This work was made possible through the generous sharing of data and manuscripts by a large number of colleagues. In particular we thank Ingrid Hendy for making available to us her unpublished faunal SST data of ODP Site 893. S. Kienast and T. F. Pedersen kindly provided a revised age model for core PAR87-10. The study benefited from discussions with and comments by S. Kienast, D. Lea, O. Marchal, D. Oppo, C. Pelejero, and M. Sarnthein. We owe our thanks to all MARGO participants for inspiration and exchange of data and ideas. The manuscript benefited from constructive reviews by P. Clark and an anonymous reviewer. Funding through the UK Natural Environmental Research Council (NER/A/S/2000/00493; TK) and a WHOI postdoctoral scholarship (Devonshire Foundation; MK) is gratefully acknowledged. This is WHOI contribution 10994.

References

Alley, R.B., Clark, U., 1999. The deglaciation of the northern hemisphere: a global perspective. Annual Review of Earth and Planetary Science 27, 149–182.

Alley, R.B., Brook, E.J., Anandakrishnan, S., 2002. A northern lead in the orbital band: north–south phasing of Ice-Age events. Quaternary Science Reviews 21, 431–441.

Andres, M.S., Bernasconi, S.M., McKenzie, J.A., Röhl, U., 2003. Southern Ocean deglacial record supports global Younger Dryas. Earth and Planetary Science Letters 216, 515–524.

Bard, E., 2000. Comparison of alkenone estimates with other temperature proxies. Geochemistry Geophysics Geosystems 2, 1–12.

Bard, E., Rostek, F., Turon, J.-L., Gendreau, S., 2000. Hydrological impact of Heinrich Events in the subtropical northeast Atlantic. Science 289, 1321–1324.

Bard, E., 2001. Paleoceanographic implications of the difference in deep-sea sediment mixing between large and fine particles. Paleoceanography 16, 235–239.

Barron, J.A., Heusser, L., Herbert, T., Lyle, M., 2003. High-resolution climatic evolution of coastal northern California during the past 16,000 years. Paleoceanography 18 10.1029/2002PA000768.

Behl, R.J., Kennett, J.P., 1996. Brief interstadial events in the Santa Barbara basin, NE Pacific, during the past 60 kyr. Nature 379, 243–246.

Blunier, T., Schwander, J., Stauffer, B., Stocker, T., Dällenbach, A., Indemühle, A., Tschumi, J., Chappellaz, J., Raynaud, D., Barnola, J.-M., 1997. Timing of the Antarctic Cold Reversal and the atmospheric CO_2 increase with respect to the Younger Dryas event. Geophysical Research Letters 24, 2683–2686.

Blunier, T., Chappellaz, J., Schwander, J., Dällenbach, A., Stauffer, B., Stocker, T.F., Raynaud, D., Jouzel, J., Clausen, H.B., Hammer, C.U., Johnsen, S.J., 1998. Asynchrony of Antarctic and Greenland climate change during the last glacial period. Nature 394, 739–743.

Brunner, C.A., Biscaye, P.E., 1997. Storm-driven transport of foraminifers from the shelf to the upper slope, southern Middle Atlantic Bight. Continental Shelf Research 17, 491–508.

Bond, G., Heinrich, H., Broecker, W., Labeyrie, L., McManus, J., Andrews, J., Huon, S., Jantschik, R., Clasen, S., Simet, C., Tedesco, K., Klas, M., Bonani, G., Ivy, S., 1992. Evidence for massive discharges of icebergs into the North Atlantic ocean during the last glacial period. Nature 360, 245–249.

Bond, G., Broecker, W., Johnsen, S., McManus, J., Labeyrie, L., Jouzel, J., Bonani, G., 1993. Correlations between climate records from North Atlantic sediments and Greenland ice. Nature 365, 143–147.

Bond, G.C., Lotti, R., 1995. Iceberg discharges into the North Atlantic on millennial time scales during the last glaciation. Science 267, 1005–1010.

Broecker, W.S., 1998. Paleocean circulation during the last deglaciation: a bipolar seesaw? Paleoceanography 13, 119–121.

Broecker, W.S., Hemming, S., 2001. Climate swings come into focus. Science 294, 2308–2309.

Cacho, I., Grimalt, J.O., Pelejero, C., Canals, M., Sierro, F.J., Flores, J.A., Shackleton, N., 1999. Dansgaard-Oeschger and Heinrich event imprints in Alboran Sea paleotemperatures. Paleoceanography 14, 698–705.

Calvo, E., Pelejero, C., Herguera, J.C., Palanques, A., Grimalt, J.O., 2001. Insolation independence of the southeastern Subtropical

Pacific sea surface temperature over the last 400 kyrs. Geophysical Research Letters 28, 2481–2484.

Chen, M.-T., Wang, C.-H., Huang, C.-Y., Wang, P., Wang, L., Sarnthein, M., 1999. A late Quaternary planktonic foraminifer faunal record of rapid climatic changes from the South China Sea. Marine Geology 156, 85–108.

Chen, M.-T., Huang, C.-C., Pflaumann, U., Waelbroeck, C., Kucera, M., 2005. Estimating glacial western Pacific sea-surface temperature: methodological overview and data compilation of surface sediment planktic foraminifer faunas. Quaternary Science Reviews this issue.

Chiang, J.C.H., Biasutti, M., Battisti, D.S., 2003. Sensitivity of the Atlantic Intertropical Convergence Zone to Last Glacial Maximum boundary conditions. Paleoceanography 18 10.1029/2003PA000916.

Chinzei, K., Fujioka, K., Kitazato, I., Koizumi, I., Oba, T., Oda, M., Okada, H., Sakai, T., Tanimura, Y., 1987. Postglacial environmental changes of the Pacific Ocean off the coasts of central Japan. Marine Micropaleontology 11, 273–291.

Clark, P.U., Bartlein, P.J., 1995. Correlation of late Pleistocene glaciation in the western United States with North Atlantic Heinrich events. Geology 23, 483–486.

Clark, P.U., Pisias, N.G., Stocker, T.F., Weaver, A.J., 2002. The role of the thermohaline circulation in abrupt climate change. Nature 415, 863–869.

Crowley, T.J., 1992. North Atlantic deep water cools the southern hemisphere. Paleoceanography 7, 489–497.

de Vernal, A., Pedersen, T.F., 1997. Micropaleontology and palynology of core PAR87A-10: A 23,000 year record of paleoenvironmental changes in the Gulf of Alaska, northeast North Pacific. Paleoceanography 12, 821–830.

Elderfield, H., Ganssen, G., 2000. Past temperature and $\delta^{18}O$ of surface ocean waters inferred from foraminiferal Mg/Ca ratios. Nature 405, 442–445.

Elliot, M., Labeyrie, L., Bond, G., Cortijo, E., Turon, J.-L., Tisnerat, N., Duplessy, J.-C., 1998. Millennial-scale iceberg discharges in the Irminger Basin during the last glacial period: relationship with the Heinrich events and environmental settings. Paleoceanography 13, 433–446.

Elliot, M., Labeyrie, L., Dokken, T., Manthe, S., 2001. Coherent patterns of ice rafted debris deposits in the nordic regions during the last glacial (10–60 ka). Earth and Planetary Science Letters 194, 151–163.

Feldberg, M.J., Mix, A.C., 2003. Planktonic foraminifera, sea surface temperatures, and mechanisms of oceanic change in the Peru and south equatorial currents, 0–150 ka BP. Paleoceanography 18 10.1029/2001PA000740.

Flower, B.P., Hastings, D.W., Hill, H.W., Quinn, T.M., 2004. Phasing of deglacial warming and Laurentide ice sheet meltwater in the Gulf of Mexico. Geology 32, 597–600.

Francois, R., Bacon, M.P., 1994. Heinrich events in the North Atlantic: radiochemical evidence. Deep-Sea Research I 41, 315–334.

Gong, C., Hollander, D.J., 1999. Evidence for differential degradation of alkenones under contrasting bottom water oxygen conditions: implication for paleotemperature reconstruction. Geochimica et Cosmochimica Acta 63, 405–411.

Grootes, P.M., Steig, E.J., Stuiver, M., Waddington, E.D., Morse, D.L., Nadeau, M.J., 2001. The Taylor Dome Antarctic ^{18}O record and globally synchronous changes in climate. Quaternary Research 56, 289–298.

Grousset, F.E., Cortijo, E., Huon, S., Herve, L., Richter, T., Burdloff, D., Duprat, J., Weber, O., 2001. Zooming in on Heinrich Layers. Paleoceanography 16, 240–259.

Hajdas, I., Bonani, G., Moreno, P.I., Ariztegui, D., 2003. Precise radiocarbon dating of late-glacial cooling in mid-latitude South America. Quaternary Research 59, 70–78.

Hendy, I.L., Kennett, J.P., 2000. Dansgaard-Oeschger cycles and the California Current System: planktonic foraminiferal response to rapid climate change in Santa Barbara Basin, Ocean Drilling Program hole 893A. Paleoceanography 15, 30–42.

Hendy, I.L., Kennett, J.P., Roark, E.B., Ingram, B.L., 2002. Apparent synchroneity of submillennial scale climate events between Greenland and Santa Barbara Basin, California from 30–10 ka. Quaternary Science Reviews 21, 1167–1184.

Herbert, T.D., Schuffert, J.D., Thomas, D., Lange, C., Weinheimer, A., Peleo-Alampay, A., Herguera, J.C., 1998. Depth and seasonality of alkenone production along the California margin inferred from a core top transect. Paleoceanography 13, 263–271.

Herbert, T.D., Schuffert, J.D., Andreasen, D., Heusser, L., Lyle, M., Mix, A., Ravelo, A.C., Stott, L.D., Herguera, J.C., 2001. Collapse of the California Current during glacial maxima linked to climate change on land. Science 293, 71–76.

Hill, T.M., Pak, D., Kennett, J.P., Lea, D.W., White, J., 2003. Abrupt climatic transitions: surface to intermediate-water response from high-resolution sediment records, Santa Barbara, California. Eos Transactions AGU, 84 (46), Fall Meeting Supplementary, Abstract OS31B-0210.

Huang, C.-C., Chen, M.-T., Lee, M.-Y., Wei, K.-Y., Huang, C.-Y., 2002. Planktic foraminifer faunal sea surface temperature records of the past two glacial terminations in the South China Sea near Wan-An Shallow (IMAGES core MD972151). Western Pacific Earth Sciences 2, 1–14.

Huang, C.-Y., Liew, P.-M., Meixun, Z., Chang, T.-C., Kuo, C.-M., Chen, M.-T., Wang, C.-H., Zheng, L.-F., 1997a. Deep sea and lake records of the Southeast Asian paleomonsoons for the last 25 thousand years. Earth and Planetary Science Letters 146, 59–72.

Huang, C.-Y., Wu, S.-F., Meixun, Z., Chen, M.-T., Wang, C.-H., Xia, T., Yuan, P.B., 1997b. Surface ocean and monsoon climate variability in the South China Sea since the last glaciation. Marine Micropaleontology 32, 71–94.

Huang, R.X., Cane, M.A., Naik, N., Goodman, P., 2000. Global adjustment of the thermocline in response to deepwater formation. Geophysical Research Letters 27, 759–762.

Hughen, K., Lehman, S., Southon, J., Overpeck, J., Marchal, O., Herring, C., Turnbull, J., 2004. ^{14}C activity and global carbon cycle changes over the past 50,000 years. Science 303, 202–207.

Ishiwatari, R., Yamada, K., Matsumoto, K., Houtatsu, M., Naraoka, H., 1999. Organic molecular and carbon isotopic records of the Japan Sea over the past 30 kyr. Paleoceanography 14, 260–270.

Ishiwatari, R., Houtatsu, M., Okada, H., 2001. Alkenone-sea surface temperature in the Japan Sea over the past 36 kyr: warm temperatures at the last glacial maximum. Organic Geochemistry 32, 57–67.

Kennett, J.P., Roark, E.B., Cannariato, K.G., Ingram, L., Tada, R., 2000. Latest Quaternary paleoclimatic and radiocarbon chronology, Hole 1017E, southern California margin. Proceedings of the Ocean Drilling Program, Scientific Results 167, 249–254.

Kiefer, T., Sarnthein, M., Erlenkeuser, H., Grootes, P., Roberts, A., 2001. North Pacific response to millennial-scale changes in ocean circulation over the last 60 ky. Paleoceanography 16, 179–189.

Kiefer, T., Sarnthein, M., Elderfield, H., Erlenkeuser, H., Grootes, P.M. Warmings in the far northwestern Pacific support pre-Clovis immigration to America during Heinrich event 1. Geology, submitted.

Kienast, M., Hanebuth, T.J.J., Pelejero, C., Steinke, S., 2003. Synchroneity of meltwater pulse 1a and the Bølling warming: New evidence from the South China Sea. Geology 31, 67–70.

Kienast, M., Steinke, S., Stattegger, K., Calvert, S.E., 2001. Synchronous tropical South China Sea SST change and Greenland warming during deglaciation. Science 191, 2132–2134.

Kienast, S.S., McKay, J.L., 2001. Sea surface temperatures in the subarctic northeast Pacific reflect millennial-scale climate oscillations during the last 16 kyrs. Geophysical Research Letters 28, 1563–1566.

Kim, J.-H., Schneider, R.R., Hebbeln, D., Müller, P.J., Wefer, G., 2002. Last deglacial sea-surface temperature evolution in the Southeast Pacific compared to climate changes on the South American continent. Quaternary Science Reviews 21, 2085–2097.

Kim, J.-M., Kennett, J.P., Park, B.-K., Kim, D.C., Kim, G.Y., Roark, E.B., 2000. Paleoceanographic change during the last deglaciation, East Sea of Korea. Paleoceanography 15, 254–266.

Koutavas, A., Lynch-Stieglitz, J., Marchitto, T.M., Sachs, J.P., 2002. El Niño-like pattern in ice age tropical Pacific sea surface temperature. Science 297, 226–230.

Koutavas, A., Lynch-Stieglitz, J., 2003. Glacial-interglacial dynamics of the Eastern Equatorial Pacific cold tongue-ITCZ system reconstructed from oxygen isotope records. Paleoceanography 18 10.1029/2003PA000894.

Kovanen, D.J., Easterbrook, D.J., 2002. Paleodeviations of radiocarbon marine reservoir values for the northeast Pacific. Geology 30, 243–246.

Kreitz, S.F., Herbert, T.D., Schuffert, J.D., 2000. Alkenone paleothermometry and orbital-scale changes in sea-surface temperature at site 1020, northern California margin. Proceedings of the Ocean Drilling Program, Scientific Results 167, 153–161.

Lea, D.W., Pak, D.K., Spero, H.J., 2000. Climate impact of late Quaternary equatorial Pacific sea surface temperature variations. Science 289, 1719–1724.

Lea, D.W., Pak, D.K., Peterson, L.C., Hughen, K.A., 2003. Synchroneity of tropical and high-latitude Atlantic temperatures over the last glacial termination. Science 301, 1361–1364.

Lee, K.E., Slowey, N.C., 1999. Cool surface waters of the subtropical North Pacific Ocean during the last glacial. Nature 397, 512–514.

Lee, K.E., Slowey, N.C., Herbert, T.D., 2001. Glacial SSTs in the subtropical North Pacific: A comparison of U^k_{37}, $\delta^{18}O$ and foraminiferal assemblage temperature estimates. Paleoceanography 16, 268–279.

Lee, M.-Y., Wei, K.-Y., Chen, Y.-G., 1999. High-resolution oxygen isotope stratigraphy for the last 150,000 years in the southern South China Sea. Journal of Terrestrial, Atmospheric and Oceanic Sciences 10, 239–254.

Li, T., Liu, Z., Hall, M.A., Berne, S., Saito, Y., Cang, S., Cheng, Z., 2001. Heinrich event imprints in the Okinawa Trough: evidence from oxygen isotope and planktonic foraminifera. Palaeogeography, Palaeoclimatology, Palaeoecology 176, 133–146.

Lyle, M.W., Prahl, F.G., Sparrow, M.A., 1992. Upwelling and productivity changes inferred from a temperature record in the central equatorial Pacific. Nature 355, 812–815.

Manabe, S., Stouffer, R.J., 1995. Simulation of abrupt climate change induced by freshwater input to the North Atlantic Ocean. Nature 378, 165–167.

Mangelsdorf, K., Güntner, U., Rullkötter, J., 2000. Climatic and oceanographic variations on the California continental margin during the last 160 kyr. Organic Geochemistry 31, 829–846.

Martinez, I., Keigwin, L., Barrows, T.T., Yokoyama, Y., Southon, J., 2003. La Niña-like conditions in the eastern equatorial Pacific and a stronger Choco jet in the northern Andes during the last glaciation. Paleoceanography 18 10.1029/2002PA000877.

Mashiotta, T.A., Lea, D.W., Spero, H.J., 1999. Glacial-interglacial changes in subantarctic sea surface temperature and $\delta^{18}O$-water using foraminiferal Mg. Earth and Planetary Science Letters 170, 417–432.

McManus, J.F., Francois, R., Gherardi, J.-M., Keigwin, L.D., Brown-Leger, S., 2004. Collapse and rapid resumption of Atlantic meridional circulation linked to deglacial climate changes. Nature 428, 834–837.

Mikolajewicz, U., Crowley, T.J., Schiller, A., Voss, R., 1997. Modelling teleconnections between the North Atlantic and North Pacific during the Younger Dryas. Nature 387, 384–387.

Mix, A.C., Lund, D.C., Pisias, N.G., Boden, P., Bornmalm, L., Lyle, M., Pike, J., 1999. Rapid climate oscillations in the Northeast Pacific during the last deglaciation reflect northern and southern hemisphere sources. In: Clark, P.U., Webb, R.S., Keigwin, L.D. (Eds.), Mechanisms of Global Climate Change at Millennial Time Scales. AGU Monograph. American Geophysical Union, Washington, DC, pp. 127–148.

Mix, A.C., Bard, E., Schneider, R., 2001. Environmental processes of the ice age; land, oceans, glaciers (EPILOG). Quaternary Science Reviews 20, 627–657.

Morgan, V., Delmotte, M., Ommen, T.V., Jouzel, J., Chappellaz, J., Woon, S., Masson-Delmotte, V., Raynaud, D., 2002. Relative timing of deglacial climate events in Antarctica and Greenland. Science 297, 1862–1864.

Morigi, C., Capotondi, L., Giglio, F., Langone, L., Brilli, M., Ravaioli, B.T., 2003. A possible record of the Younger Dryas event in deep-sea sediments of the Southern Ocean (Pacific sector). Palaeogeography, Palaeoclimatology, Palaeoecology 198, 265–278.

Nelson, C.S., Hendy, I.L., Neil, H.L., Hendy, C.H., Weaver, P.P.E., 2000. Last glacial jetting of cold waters through the Subtropical Convergence zone in the Southwest Pacific off eastern New Zealand, and some geological implications. Palaeogeography, Palaeoclimatology, Palaeoecology 156, 103–121.

Ohkouchi, N., Kawamura, K., Nakamura, T., Taira, A., 1994. Small changes in the sea surface temperature during the last 20,000 years: molecular evidence from the western tropical Pacific. Geophysical Research Letters 21, 2207–2210.

Pahnke, K., Zahn, R., Elderfield, H., Schulz, M., 2003. 340,000-year centennial-scale marine record of southern hemisphere climatic oscillation. Science 301, 948–952.

Palmer, M.R., Pearson, P.N., 2003. A 23,000-year record of surface water pH and pCO_2 in the western equatorial Pacific Ocean. Science 300, 480–482.

Pelejero, C., Grimalt, J.O., Heilig, S., Kienast, M., Wang, L., 1999. High-resolution U^K_{37} temperature reconstructions in the South China Sea over the past 220 kyr. Paleoceanography 14, 224–231.

Petit, J.R., Jouzel, J., Raynaud, D., Barkov, N.I., Barnola, J.-M., Basile, I., Bender, M., Chappellaz, J., Davis, M., Delaguye, G., Delmotte, M., Kotlyakov, V.M., Legrand, M., Lipenkov, V.Y., Lorius, C., Pepin, L., Ritz, C., Saltzmann, E., Stievenard, M., 1999. Climate and atmospheric history of the past 420,000 years from the Vostok ice core, Antarctica. Nature 399, 429–437.

Pflaumann, U., Jian, Z., 1999. Modern distribution patterns of planktonic foraminifera in the South China Sea and western Pacific: a new transfer technique to estimate regional sea surface temperatures. Marine Geology 156, 41–83.

Pisias, N.G., Mix, A.C., Heusser, L., 2001. Millennial scale climate variability of the northeast Pacific Ocean and northwest North America based on radiolaria and pollen. Quaternary Science Reviews 20, 1561–1576.

Porter, S.C., An, Z., 1995. Correlation between climate events in the North Atlantic and China during the last glaciation. Nature 375, 305–308.

Prahl, F.G., Pisias, N., Sparrow, M.A., Sabin, A., 1995. Assessment of sea surface temperature at 42°N in the California Current over the last 30,000 years. Paleoceanography 10, 763–773.

Rohling, E., Mayewski, P., Challenor, P., 2003. On the timing and mechanism of millennial-scale climate variability during the last glacial cycle. Climate Dynamics 20, 257–267.

Rosenthal, Y., Oppo, D.W., Linsley, B.K., 2003. The amplitude and phasing of climate change during the last deglaciation in the Sulu Sea, western equatorial Pacific. Geophysical Research Letters 30 10.1029/2002GL016612.

Rühlemann, C., Mulitza, S., Müller, P.J., Wefer, G., Zahn, R., 1999. Warming of the tropical Atlantic Ocean and slowdown of thermohaline circulation during the last deglaciation. Nature 402, 511–514.

Sabin, A.L., Pisias, N.G., 1996. Sea surface temperature changes in the northeastern Pacific Ocean during the past 20,000 years and their relationship to climate change in northwestern North America. Quaternary Research 46, 48–61.

Sarnthein, M., Winn, K., Jung, S.J.A., Duplessy, J.C., Erlenkeuser, H., Ganssen, G., 1994. Changes in east Atlantic deepwater circulation over the last 30,000 years: Eight time slice reconstructions. Paleoceanography 9, 209–267.

Sarnthein, M., Stattegger, K., Dreger, D., Erlenkeuser, H., Grootes, P.M., Haupt, B., Jung, S., Kiefer, T., Kuhnt, W., Pflaumann, U., Schäfer-Neth, C., Schulz, H., Schulz, M., Seidov, D., Simstich, J., van Kreveld, S., Vogelsang, E., Völker, A., Weinelt, M., 2001. Fundamental modes and abrupt changes in North Atlantic circulation and climate over the last 60 ky: concepts, reconstruction and numerical modeling. In: Schäfer, P., Ritzrau, W., Schlüter, M., Thiede, J. (Eds.), The northern North Atlantic: A Changing Environment. Springer, Berlin, pp. 365–410.

Sarnthein, M., Grootes, P.M., Kiefer, T., Kienast, M., Schulz, M., 2003. Mega-^{14}C plateau provides global age tie point for pre-Bølling DO event 1. Eos Transactions AGU 84 (46) Fall Meeting Supplement, Abstract GC12A-0156.

Sawada, K., Handa, N., 1998. Variability of the Kuroshio ocean current over the past 25,000 years. Nature 392, 592–595.

Schiller, A., Mikolajewicz, U., Voss, R., 1997. The stability of the North Atlantic thermohaline circulation in a coupled ocean-atmosphere general circulation model. Climate Dynamics 13, 325–347.

Schmittner, A., Saenko, O.A., Weaver, A.J., 2003. Coupling of the hemispheres in observations and simulations of glacial climate change. Quaternary Science Reviews 22, 659–671.

Seki, O., Ishiwatari, R., Matsumoto, K., 2002. Millennial climate oscillations in NE Pacific surface waters over the last 82 kyr: new evidence from alkenones. Geophysical Research Letters 29 10.1006/qres.2001.2235.

Sikes, E.L., Samson, C.R., Guilderson, T.P., Howard, W.R., 2000. Old radiocarbon ages in the southwest Pacific Ocean during the last glacial period and deglaciation. Nature 405, 555–559.

Sikes, E.L., Howard, W.R., Neil, H.L., Volkman, J.K., 2002. Glacial-interglacial sea surface temperature changes across the subtropical front east of New Zealand based on alkenone unsaturation ratios and foraminiferal assemblages. Paleoceanography 17 10.1029/2001PA000640.

Spero, H.J., Lea, D.W., 2002. The cause of carbon isotope minimum events on glacial terminations. Science 296, 522–525.

Steig, E.J., Brook, E.J., White, J.W.C., Sucher, C.M., Bender, M.L., Lehman, S.J., Morse, D.L., Waddington, E.D., Clow, G.D., 1998. Synchronous climate changes in Antarctica and the North Atlantic. Science 282, 92–95.

Steinke, S., Kienast, M., Pflaumann, U., Weinelt, M., Stattegger, K., 2001. A high-resolution sea-surface temperature record from the tropical South China Sea (16,500–3000 yr B.P.). Quaternary Research 55, 352–362.

Stocker, T.F., 2000. Past and future reorganizations in the climate system. Quaternary Science Reviews 19, 301–319.

Stott, L., Poulsen, C., Lund, S., Thunell, R., 2002. Super ENSO and global climate oscillations at millennial time scales. Science 297, 222–226.

Stuiver, M., Reimer, P.J., Bard, E., Beck, J.W., Burr, G.S., Hughen, K.A., Kromer, B., McCormac, F.G., van der Plicht, J., Spurk, M., 1998. INTCAL98 radiocarbon age calibration, 24,000-0 cal BP. Radiocarbon 40, 1041–1083.

Stuiver, M., Grootes, P.M., 2000. GISP2 oxygen isotope ratios. Quaternary Research 53, 277–284.

Ujiié, Y., Ujiié, H., Taira, A., Nakamura, T., Oguri, K., 2003. Spatial and temporal variability of surface water in the Kuroshio source region, Pacific Ocean, over the past 21,000 years: evidence from planktonic foraminifera. Marine Micropaleontology 940, 1–30.

Veiga-Pires, C.C., Hillaire-Marcel, C., 1999. U and Th isotope constraints on the duration of Heinrich events H0-H4 in the southeastern Labrador Sea. Paleoceanography 14, 187–199.

Vellinga, M., Wood, R.A., 2002. Global climatic impacts of a collapse of the Atlantic thermohaline circulation. Climatic Change 54, 251–267.

Vidal, L., Schneider, R.R., Marchal, O., Bickert, T., Stocker, T.F., Wefer, G., 1999. Link between the North and South Atlantic during the Heinrich events of the last glacial period. Climate Dynamics 15, 909–919.

Visser, K., Thunell, R., Stott, L., 2003. Magnitude and timing of temperature change in the Indo-Pacific warm pool during deglaciation. Nature 421, 152–155.

Wang, L., Sarnthein, M., Erlenkeuser, H., Grimalt, J., Grootes, P., Heilig, S., Ivanova, E., Kienast, M., Pelejero, C., Pflaumann, U., 1999. East Asian monsoon climate during the Late Pleistocene: high-resolution sediment records from the South China Sea. Marine Geology 156, 245–284.

Wang, Y.J., Cheng, H., Edwards, R.L., An, Z.S., Wu, J.Y., Shen, C.-C., Dorale, J.A., 2001. A high-resolution absolute-dated late Pleistocene monsoon record from Hulu Cave, China. Science 294, 2345–2348.

Weaver, A.J., 1999. Millennial timescale variability in ocean/climate models. In: Webb, R.S., Clark, P.U., Keigwin, L.D. (Eds.), Mechanisms of Global Climate Change at Millennial Time Scales. Geophysical Monograph. American Geophysical Union, Washington, DC, pp. 285–300.

Whitko, A.N., Hastings, D.W., Flower, B.P., 2002. Past sea surface temperatures in the tropical South China Sea based on a new foraminiferal Mg calibration. MARSci MARSci.2002.01.020101.

Yamasaki, M., Oda, M., 2003. Sedimentation of planktonic foraminifera in the East China Sea: evidence from a sediment trap experiment. Marine Micropaleontology 49, 3–20.

Quaternary Science Reviews 24 (2005) 1083–1093

Perspectives on mapping the MARGO reconstructions by variogram analysis/kriging and objective analysis

Christian Schäfer-Neth*, André Paul, Stefan Mulitza

DFG Research Center Ocean Margins and Department of Geosciences, University of Bremen, D-28334 Bremen, Germany

Received 8 January 2004; accepted 21 June 2004

Abstract

Paleo-data are not always useful in their original scattered distribution: For many numerical modeling issues and for display and comparison, gridded versions that provide meaningful estimates for under-sampled regions are a must. We constructed a test data set with a spatial resolution identical to the Multi-proxy Approach for the Reconstruction of the Glacial Ocean (MARGO) samples from the World Ocean Atlas temperatures and assess the performance of (i) variogram estimation and kriging and (ii) the Levitus objective analysis in reconstructing the original data. The two methods complement each other with respect to the facility of application and the quality of the results. Kriging requires a careful parameter adjustment but delivers the smallest deviation from the original data (1.22 °C in the global average), whereas the Levitus analysis provides a fast and efficient tool for checking the samples from different proxy data against each other during the compilation of the final MARGO database, at the expense of a slightly higher error (1.56 °C).

1. Introduction

Every major advance in the reconstruction of paleooceanographic data, such as by the CLIMAP (1981) and GLAMAP (Sarnthein et al., 2003) projects, has triggered numerous studies that employed the new data in the form of regularly gridded fields. Common applications are the forcing of ocean and atmosphere models and the validation of coupled climate models, but gridding is already necessary for displaying and comparing data. However, interpolating the scattered paleo data to a regular grid is no straightforward task, and unsuitable methods may easily generate artifacts that will mislead any further studies. By compiling not only species counts but in addition other available temperature proxies, and by improving the spatial sampling density, Multi-proxy Approach for the Reconstruction of the Glacial Ocean (MARGO) marks a

new step towards a better reconstruction of the glacial sea surface. Again, it is desirable to produce regular grids from the new seasonal data. Using a test data set based on modern temperatures but with a spatial resolution identical to the MARGO data set, we compared two different interpolation methods: variogram analysis and kriging (Deutsch and Journel, 1992), and the objective analysis by Levitus (1982). According to the results of this comparison, variogram analysis and kriging is best suited for the final gridding of the new MARGO temperature reconstructions, which underlines a remark by Wunsch (1996, Chapter 5.4), that "Kriging [...] deserves more oceanographic attention".

2. Test data set

For comparing the different interpolation techniques, we adopted the following strategy:

We constructed three test data sets (annual mean, January–March and July–September) from the unanalyzed World Ocean Atlas 1998 temperature data (WOA,

*Corresponding author. Tel.: +49-421-218-7188; fax: +49-421-218-7040.

E-mail address: csn@uni-bremen.de (C. Schäfer-Neth).

0277-3791/$ - see front matter © 2004 Elsevier Ltd. All rights reserved.
doi:10.1016/j.quascirev.2004.06.017

1998) for 10 m depth. Except for an $1° \times 1°$ averaging, these temperatures have not been treated by any interpolation, specifically, they are not biased due to some inherent a priori "information" from any of the methods we discuss here. Apart from the high southern latitudes during austral winter, these data provide an almost complete global coverage which facilitates a comparison with the gridding results. Furthermore, the analyzed 10 m WOA data have been used to calibrate the transfer techniques throughout the groups participating in MARGO.

The WOA temperatures were averaged to every even longitude and latitude, that is, the $2° \times 2°$ grid used by the CLIMAP Project Members (1981), that we chose as the target for the present interpolation exercise. The choice of a 2° spacing is a compromise; in the areas of densest sampling, a gridding to 1° or even 0.5° would be feasible, but where data are sparse, 5° would be a more reasonable distance. At this stage of a preliminary test for a gridding of the actual MARGO data, the 2° grid served as a test bed, which can be easily refined or coarsened by changing a few parameters of the employed programs, depending on future requirements and data availability. To construct an input data set for the present study, we binned the original MARGO core locations into 2° longitude/latitude squares and sampled the 2° WOA data set at these points (Fig. 1). In this way, the test data and the original MARGO data are of identical resolution.

After interpolating from the test points back to the $2° \times 2°$ grid, we computed the absolute differences between the original and the gridded fields as a measure of interpolation skill, used to assess the interpolation techniques.

3. Interpolation/gridding methods

An interpolation of paleooceanographic data from the core locations to a regular grid faces three main problems: (i) There are areas with dense spatial sampling in contrast to others that are practically void of data. In the case of the MARGO data set, examples for both situations can be found in the northern North Atlantic Ocean and the subtropical South Pacific Ocean, respectively (Fig. 1). These extremes in sampling density require an interpolation method that is able to fill in the large gaps, but does not discard the small-scale information present in the densely sampled regions. Ideally, when estimating the values in the under-sampled areas, the interpolation scheme should use the information on small-scale variability and the spatial correlations of the samples that is available from the well-sampled areas. The use of such information would facilitate extrapolations beyond the areas sampled by MARGO, e.g., into the polar regions. (ii) The spatial

temperature variability changes itself quite considerably between the different ocean basins. For example, it is very high at oceanic fronts but very low in the centers of the subtropical gyres. These regional changes should be accounted for in the mapping process. If there is other related knowledge available, such as information on the ocean currents, it might be sensible to include this information as well. (iii) The reconstructions at individual core sites may differ greatly with respect to their error margins. If these can be quantified, the gridding method should use this information for weighting the samples and computing the possible errors of the gridded data.

If the gridding is intended as an interface between geological samples and ocean circulation models, the more straightforward interpolation methods such as nearest neighbor interpolation, splines, and triangulation are not well suited for a number of reasons. Triangulation interpolates with linear distance weights within triangles that are constructed between the sample locations, and the gridded data set emerges as a surface of connected triangular planar tiles. Thus, the value at a given point is determined only by the values of the three surrounding vertices, and potentially useful information from farther samples is discarded. In case of highly uneven sample spacings this is clearly not desirable as small-scale characteristics of the interpolated field are abruptly removed or distorted at the transition from dense to sparse sampling. Another problem arises from the piecewise planarity of the gridded field because it places the horizontal gradients at the edges of the interpolating triangles (Taylor et al., 2004). These locations can be very different from the real-world gradients, and an ocean circulation model driven with a field containing such patterns would develop unrealistic fronts and currents. Spline interpolation avoids these sharp wrinkles, but suffers from a tendency to generate local bulges or depressions around detached sample locations, thereby generating unrealistic current loops in ocean models. Nearest neighbor schemes with an inverse distance weighting often induce pronounced gradients halfway between the sample points. As with triangulation, the location of these gradients is dictated entirely by the arrangement of the samples and not by the physical processes that underly the sampled values. Besides being insufficient for forcing ocean-only models, interpolated fields that contain these artifacts cannot be drawn on when results from coupled ocean–atmosphere models are to be assessed. Furthermore, these methods are poor extrapolators because the extrapolated field is dominated by the spatial derivatives of the data set at the very boundary of the sampled area, which results in a high probability of overshoots.

This situation can be greatly improved if the gridding scheme propagates information from the sample locations to the grid points in a more sophisticated manner,

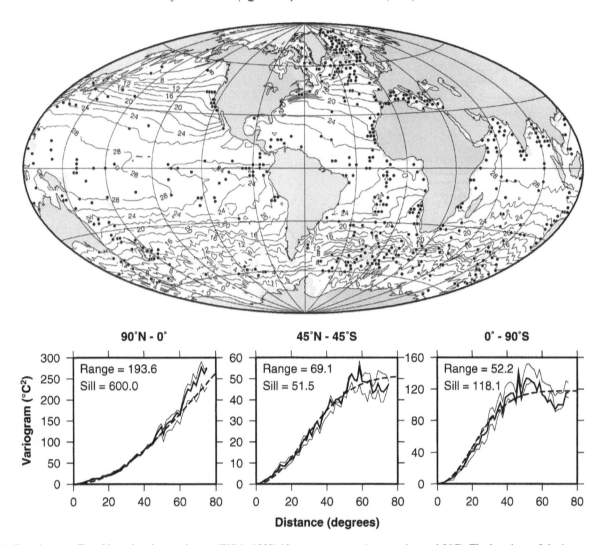

Fig. 1. Test data set. Top: Unanalyzed annual mean WOA (1998) 10 m temperatures (contour interval 2 °C). The locations of the input test data points (dots) were obtained by binning the MARGO core sites into 2° × 2° boxes. The geographical grid spacing is 30° and the central meridian is located at 60°W. The total number of data points is 530 (507 for the January–March data and 482 for July–September, not shown here). Light grey shading indicates gaps in the WOA data set. Bottom: Empirical (heavy solid) and modeled (dashed) modified Gaussian variograms for the three latitude belts. For comparison, the empirical variograms for the winter and summer data are indicated by the thin lines. Note the different scales of the vertical axes. The sill of the variogram for the northern belt cannot be identified from the empirical variograms. It was set to the indicated value based on a comparison of the empirical variograms's inflection points.

implicitly accounting for the underlying physical processes that determined the sample data. This is even more desirable if the interpolation is applied to paleo-data that are representative for a background climatic state different from today. One possibility to incorporate the physical constraints would be to assimilate the sample data into a coupled atmosphere–ocean model. In our associated paper (Paul and Schäfer-Neth, 2004) we discuss an example of this approach, based on the same test data set. Here, we compare variogram analysis and kriging (VAK hereafter) and the objective analysis devised by Levitus (1982, LOA). We demonstrate that these approaches are able to cope with uneven sampling density, and to reasonably extrapolate if required.

3.1. Variogram analysis and kriging

Variogram analysis and kriging (VAK) is a weighted-average interpolation method that adjusts the averaging weights according to the spatial variability of the sample data. (The text books by Deutsch and Journel (1992) and Wackernagel (2003) give an excellent introduction into the subject.) VAK is performed in two steps. First, the spatial variability of the sample data set is quantified by mapping the variance of paired sample data against the distance between the data points. For this purpose, the total distance mapped is subdivided into *lags* of equal width. The resulting curve, the so-called *empirical variogram*, typically shows low variances at short distances and eventually reaches some saturation value

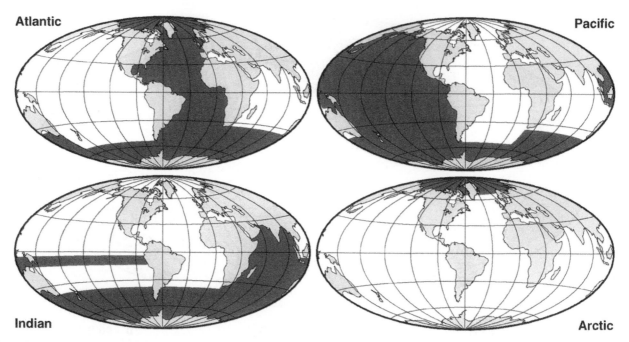

Fig. 2. Communication masks (grey shade) for the four major oceans (masks for the marginal seas were employed as well). Pairs of data points were allowed to contribute to the empirical variograms only if both points belong to a given mask. Likewise, the kriging and the Levitus estimate for a gridded point was computed only from those samples that share at least one ocean mask with the grid point.

for large distances (Fig. 1, bottom panels, solid lines). To feed this information into the kriging interpolation, an analytical *variogram model* is fitted to the empirical curve (dashed lines). Most of the commonly used models, such as the modified Gaussian that is employed here, are determined by two parameters: a *sill* that represents the highest variance for a large data point separation, and a *range* that denotes the distance at which the variogram model attains its sill. Therefore, the range is a measure of the distance over which the data are correlated. Because an empirical variogram does not necessarily start with zero variance at small distances, the variogram model may have to be adjusted by a constant offset, termed the *nugget effect*. In the second step, the actual kriging interpolation, the data values for a given set of locations are estimated, in our case, on the regular $2° \times 2°$ grid. For this purpose, kriging calculates the averaging weights according to the variogram model from the distances between the sample points and the location of a desired estimate. This weighting method minimizes the *kriging variance*, that is, the error variance of the estimates.

In this way, VAK accounts for the spatial variability present in the sampled temperature data. Given the fact that these variations are the result of ocean currents, mixing, and sea surface heat exchange, information on these processes is implicitly introduced into the inter-polated field. Conceivably, if the global ocean is considered, there are regions that are closely related to each other because of the physical processes, whereas the linkages between others are much weaker. Therefore

it is not appropriate to use all possible data pairs for estimating the empirical variograms. We implemented a simple representation of these restrictions of inter-basin exchanges into the GSLIB variogram estimation routine "gamv2" (Deutsch and Journel, 1992, available at www.gslib.com) by employing communication masks (Fig. 2) that inhibit a pairing of data points that are unlikely to influence each other in the real ocean. These masks are used identically in the LOA method (see below). Another feature of the global temperature data set are the great differences of the spatial variability between different regions of the ocean. To account for this, "gamv2" was further revised to compute separate empirical variograms for three overlapping zonal belts that extend from the south pole to the equator, from $45°S$ to $45°N$, and from the equator to the north pole, respectively. A finer subdivision of the data set into overlapping $90° \times 90°$ boxes of the globe yielded only minor differences between the variograms for a given latitude (not shown here). Thus, we judge the belt-wise split as being sufficient, albeit necessary, as borne out by the great differences between the empirical variograms (Fig. 1, bottom, solid lines). A third modification was applied to "gamv2" to enable the use of spherical coordinates. This is required by the zonal cyclicity of the data set and the convergence of the meridians in high latitudes.

Parallel changes of the code were included in the GSLIB kriging routine "okb2d" that we employed for this study. The kriging was carried out separately for the three belts, and we combined the resulting belt-wise fields by weighted averaging in the overlapping regions. The weights increased linearly from zero at a belt's southern and northern limits to one at its central latitude. The modified codes of "gamv2" and "okb2d", the communication masks, and the test data set may be obtained from www.palmod.uni-bremen.de/~csn.

There are other geostatistical packages that could be modified in a similar manner: *GSTAT* (Pebesma and Wesseling, 1998, available at www.gstat.org) provides a graphical interface and additional variogram models that could be more appropriate under some circumstances. For example, if a data set with cyclic boundaries like ours should be analyzed over zonal distances much larger than 90°, the periodic model might be a sensible choice. GSTAT also allows a masking of given areas by to user-defined polygons, comparable to the communication masks we use here. However, GSTAT is programmed for cartesian coordinates only and would thus require additional work to make it applicable to a sphere. Its code structure is much more complex than that of GSLIB. *Spherekit* (available at www.ncgia.ucs-b.edu/pubs/spherekit) is already designed for spherical coordinates and uses a graphical front-end, but provides no masking or clipping of specified regions. Spherekit relies on specific versions of GMT,[1] Tcl/Tk,[2] and NetCDF[3] that have to be installed as well. The great advantage of GSLIB is that it comes in plain FORTRAN without any requirements except a compiler, and that code modifications can be easily implemented.

One might consider whether a realistic geodetic coordinate system could be more appropriate than the spherical one. Given the World Geodetic System 84, the earth's oblateness amounts to 1:298.257, and using spherical coordinates for kriging introduces a 0.3% anisotropy between the meridional and zonal directions. This is only a minor error and can be safely neglected in the context of the present application.

3.2. Levitus objective analysis

The interpolation by Levitus (1982, LOA) comprises three stages and starts with an initial guess by placing all 2° input data into their respective boxes on the 2° × 2° grid and filling the other boxes with the zonal mean of the input data for a given latitude. In the second stage, the first guess gets iteratively improved by adding the distance-weighted differences between input data and first guess to the grid points, using a hierarchy (Levitus, 1982) of predefined search radii (in our case 14°, 11°, 8°, and 5°). As already noted in the previous section, we

modified the original Levitus method by applying communication masks (Fig. 2) to prevent a spreading of information across the continents during the second stage, and all distances were computed within spherical coordinates. In a final step, the interpolated field is smoothed by the subsequent application of a median and a 3 × 3 point Shapiro (1971) filter. If there is no input data available for a grid point within the search radius, LOA preserves the first guess like the method of Reynolds and Smith (1999).

4. Results

4.1. Variogram analysis and kriging

We computed the empirical variograms for the three zonal belts omni-directionally with 50 lags of 1.5° width, that is, over a total distance of 75°. In the southern and the equatorial belt, the variograms show a well-defined Gaussian shape that reaches sills exceeding values of $100\,°C^2$ and $50\,°C^2$, respectively, at a range of roughly 50°–60° (Fig. 1, bottom panels, heavy solid lines). In the northern belt, no sill is visible. For a given latitude belt, both sills and ranges do not differ very much between the annual and seasonal data sets (thin lines) and we restrict our discussion to the annual-mean data.

Because of the general shape of the empirical variograms, it is advisable to use a Gaussian variogram model for kriging. However, Gaussian models are almost horizontal for small distances, which leads to many almost-zero entries in the coefficient matrix of the kriging equations. Even if the matrix remains invertible, there is a tendency to generate small-scale noise in the kriged field that can be reduced by adding a tiny artificial nugget effect (Englund and Sparks, 1991; Ababou et al., 1994; Pannatier, 1996). According to Wackernagel (2003, Chapter 16), a modified version of the Gaussian model,

$$\gamma(h) = c \cdot \left[1 - \exp\left(-\frac{h^p}{a^p}\right)\right], \qquad (1)$$

provides a much cleaner solution to this problem, where c and a denote the model's sill and range, and h represents the spatial distance. Instead of $p = 2$, the exponent of the standard Gaussian model, the modified model employs a reduced value, $p < 2$, that must be adjusted together with a and c when fitting the model to the empirical variogram. The fits were performed as follows: Direct least-square fits of model and empirical variograms resulted in values of $p > 2$, which violates the assumed variogram model. Therefore we varied p stepwise over the range $1.50 \leqslant p \leqslant 1.95$ and calculated a and c by least-square fitting the model for a given p. To narrow down the possible range of p, we visually inspected the fit of modeled and empirical variogram, and the smoothness of the resulting kriged SST field. For all exponents $p \leqslant 1.7$, the resulting model slopes emerged distinctly smaller than the slope of the empirical data. If p was chosen $\geqslant 1.9$, the interpolated SST data were prone to the undesired small-scale disturbances. These findings leave $1.7 < p < 1.9$ as a suitable range, and we performed our analyses with $p = 1.8$.

The *practical range* of this modified Gaussian model, that is, the distance at which the model attains 95% of its sill, is indicated in Fig. 1 and Table 1 for all three belts, and given by

$$a_{95} = 3^{1/p} \cdot a \approx 1.84 \cdot a. \qquad (2)$$

As stated above, the empirical variogram for the northern belt shows no visible sill, caused by the large temperature difference between samples from the equatorial and Arctic Oceans. Given its curvature, a power variogram model given by

$$\gamma(h) = c \cdot h^a \qquad (3)$$

seems to be more appropriate in this case, where the exponent a is constrained to $0 < a < 2$. For values very close to 2, the power model is horizontal in its beginning, just as the standard Gaussian model, and generates the same small-scale noise in the gridded field. Unfortunately, in our case the fitted power model has an exponent of $a = 1.994$, and the power model is no suitable choice. Therefore we used the modified Gaussian model with an assumed sill of $600\,°C^2$. This is five times the sill found for the southern belt ($118\,°C^2$), corresponding to the ratio of the values that both empirical variograms assume at their inflection points, namely $\approx 60\,°C^2$ (southern belt) and $\approx 300\,°C^2$ (northern, Fig. 1), and consistent with the $\approx 25-30\,°C$ SST contrast between the equatorial and arctic regions. The range of the model was determined by least-square fitting as for the other latitude belts.

Kriging was then carried out with an omnidirectional search radius of $50°$ and a maximum of 40 data points to be included in the estimation of a given grid point. The small-scale features are well represented in the gridded result (Fig. 3, top), especially, if the areas of strong gradients (see above) are considered. There is a general agreement between dense spatial sampling

Table 1
Parameters of the variograms fitted to the annual mean SST data (see Fig. 1, bottom)

Latitude belt	Practical range (deg)	Sill ($°C^2$)	Data points
North	193.6	600.0	267
Equatorial	69.1	51.5	354
South	52.2	118.1	277

The sill of the northern belt's variogram model represents no fit result but was directly set (see text).

(Fig. 1) and low kriging variance (Fig. 4). As a measure of the skill of the interpolation method, we chose the absolute difference between gridded and original data as displayed in Fig. 5; its per-ocean averages are listed in Table 2. The global mean difference is $1.22\,°C$ for the annual data and 1.24 and $1.28\,°C$ for northern Winter and Summer, respectively. By and large, low kriging variances coincide with small differences between the gridded and the original data (Fig. 5, top). However, there are discrepancies. The frontal systems associated with the Gulf Stream, the Kuroshio, the Brazil Current and the Agulhas retroflection, that are clearly visible in the original data, are hardly present in the gridded fields. These regions with strong oceanic fronts are the most difficult ones to interpolate, as can be seen from the differences between gridded and original field that exceed $5\,°C$ (Fig. 5, top). In case of the three currents, this comes as no surprise because none of these regions is adequately sampled. In the Gulf Stream and the Brazil Current, there are only two or three points along the respective current's axis, and there is literally nothing in the Kuroshio. This poor sampling leads to an underestimation of spatial variability, especially across the currents, and the interpolation scheme does not know anything about the existence of the front. The only remedy to this problem would be a better sampling. For the region of the Agulhas retroflection, the sampling is quite dense, but the original data show high variability on a very small spatial scale that cannot be preserved by the interpolation, which is evident from the undulating pattern in the difference field south of Africa. If this were to be changed, the original data set had to be subdivided further into smaller regions (Schäfer-Neth and Paul, 2003) and the VAK be carried out on a smaller scale. In the Arctic Ocean, there are two regions where the kriged field differs considerably from the WOA data set: Along the Siberian coast the gridded SSTs fall well below the freezing point, whereas they are too high in the Barents Sea (Fig. 5). The cooling is due to a tendency of Gaussian (and almost-Gaussian) variogram models to extrapolate beyond the sample data range near the boundaries of the area (Wackernagel, 2003). Although there are no boundaries on a sphere, the coast of Siberia is farthest away from the sample data points, simply because of the presence of the large Eurasian land mass. This problem can easily be overcome by including some additional points in the data-void Arctic Ocean with a prescribed freezing-point temperature. Indeed, if the gridded field was cut off at the freezing point, the mean absolute temperature error would drop from $3.5\,°C$ to about $1.8\,°C$ (Table 2) in the Arctic Ocean. The unrealistic warming occurs because the steep and high variogram model fitted for the northern latitude belt, that is, the entire northern hemisphere, is not fully appropriate for the highest northern latitudes where there are only small

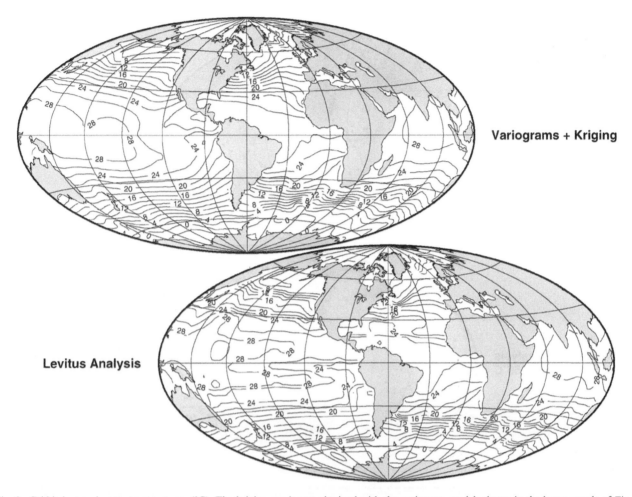

Fig. 3. Gridded annual mean temperatures (°C). The kriging result was obtained with the variogram models shown in the lower panels of Fig. 1 (dashed curves) and used up to 40 sample data within a maximum search radius of 50° around each grid point. The Levitus interpolation employed all available samples in a staggered hierarchy of four search radii of 14°, 11°, 8°, and 5°.

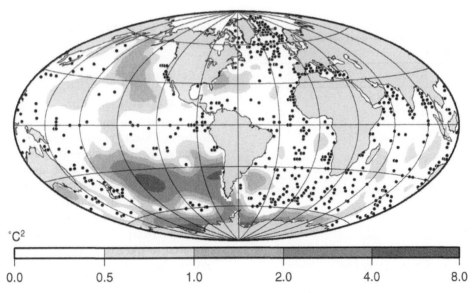

Fig. 4. Kriging variance for the interpolation of the annual WOA (1998) temperatures (°C^2) and locations of the sample points. The low values of the variance in areas of denser spatial sampling indicate that the variogram models were appropriately chosen.

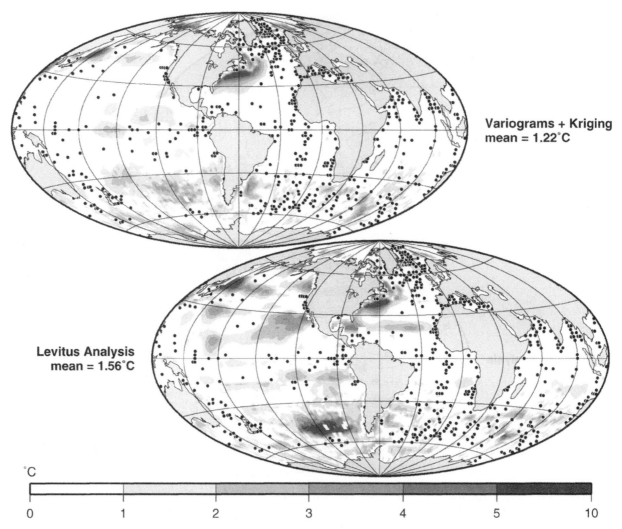

Fig. 5. Absolute difference between the interpolated and the unanalyzed annual temperatures (°C) and sample locations. In contrast to the kriging variance, the differences are not necessarily small in regions of dense spatial sampling and become large especially near oceanic fronts that emerge less sharp with both methods.

Table 2
Annual mean absolute differences (°C) between the original and the interpolated fields (see Fig. 5)

Ocean	Kriging	Levitus
Atlantic	1.29	1.40
Pacific	1.19 (1.15)	1.75
Indian	0.93	1.04
Arctic	3.52 (1.69)	1.84
Global	1.22 (1.15)	1.56

Values in parentheses result if all SSTs below −1.8 °C are set to this value in the kriged field.

temperature variations. In the present configuration, the Barents Sea is too strongly influenced by the higher mid- and low-latitude temperatures. The solution would be to further subdivide the northern belt and consider variograms of a shorter total range, and consequently

a lower sill, in the Arctic Ocean (Schäfer-Neth and Paul, 2003).

4.2. Levitus objective analysis

As can be seen from Fig. 3 (bottom), the LOA is characterized by a tendency to generate zonally oriented structures. Since many regions of the ocean are dominated by zonal current systems, the LOA yields fairly realistic results in these areas. For example, the 26 °C isoline in the southern Indian Ocean, the 8–10 °C belt in the southern Atlantic Ocean, and the 24–26 °C tongue in the equatorial Pacific Ocean show only small deviations from the original data (Fig. 5, bottom). Compared to the 28 °C line of the original (Fig. 1, top) field, the latter structure is somewhat over-emphasized. However, in areas with more meridional flows and low spatial sampling density, the zonal bias of the LOA may introduce unrealistic features such as the meanders of the 22 and 24 °C isolines in the northern subtropical Atlantic Ocean and the squeezing of the 20–24 °C area off Baja California. Naturally, these pattern shifts result in large differences of 3–4 °C between the WOA data and the LOA interpolation (Fig. 5, bottom). The meridional temperature gradient, which is present in the original data from the North Pacific Ocean at about 40°N, becomes split into two weaker gradients around 30°N and 45°N. There, the LOA yields the highest absolute differences, beyond 5 °C, between the original and the interpolated data, except from the poorly sampled South Pacific Ocean and the Gulf Stream region for which the LOA fails to reproduce the location of the frontal system. This is reflected by the absolute difference between the original and the Levitus-interpolated data, which is highest for the Pacific Ocean (Table 2). It should be noted that the strength of the temperature gradient across the Gulf Stream is quite close to the original one, despite its displacement. In some cases of dense sampling at the coasts, the LOA preserves small-scale features very well, as exemplified by the 14° line west of North America and the almost straight 26° isoline west of Australia. The global mean differences relative to WOA amount to 1.56 °C in the annual mean and to 1.49 and 1.71 °C for Winter and Summer.

5. Discussion

According to the mean errors for the Atlantic Ocean (Table 2), the two methods are of comparable quality if the sampling density is high. However, if there are void regions wider than 20–30°, such as the subtropical North Atlantic Ocean, considerable differences may occur, as borne out by the meanders and loops of the 24° and 26° isolines in that area (Fig. 3), and by the considerably different absolute errors for the Pacific Ocean (Table 2).

5.1. Variogram analysis and kriging

To some extent, VAK resembles the optimal interpolation technique devised by Reynolds and Smith (1999). Their approach employs horizontal correlation lengths derived from the input data set to estimate gridding weights, quite similar to the variograms that can be viewed as an inverse measure of spatial correlation. Applied to satellite data, this method can rely on a much higher sampling density which enables an automatic fitting of the correlation length scales for any point of the globe. This is not generally possible for the presently available paleo-temperatures. The method of Reynolds and Smith accounts for the errors of the individual samples by adjusting the interpolation weights. This would be beneficial for the gridding of

the paleo-data and could in principle be accomplished by VAK. However, the statistical error of the individual temperature estimates is about ±1 °C for all MARGO reconstructions and thus provides no additional information. Hence, the kriging variances obtained by VAK for every gridded point do not reflect the quality of the data, but solely the spatial sampling density and the appropriateness of the variogram models fitted to the empirical variograms (Fig. 4). Nevertheless, the kriging variances provide an excellent indicator as to the quality of the interpolation, which will increase if the variance is minimized by a careful adjustment of the model variogram parameters. It should be noted that the temperatures reconstructed for a given core site by different methods or from different proxies may vary by much more than ±1 °C. If it can be ruled out in the future that this reflects different seasons of plankton growth or varying depth habitats, these deviations can be accounted for in the gridding procedure. With optimally tuned variogram parameters, VAK yields a very good interpolation over the data-void areas in-between the well-sampled ones, and reasonable extrapolations for the polar regions.

5.2. Levitus objective analysis

The LOA provides a fast method that can easily be repeated upon the arrival of new or revised samples, which is a clear advantage during the compilation of the data set and for the comparison of the individual temperature reconstruction methods. The characteristic zonal features produced by this method are due to the initialization with the zonal mean values of the sample data set. Once initialized, the field is updated at all grid points that are nearer to any of the sample points than the longest of the search radii, which is set to 14° in the present case. In the subtropical North Atlantic Ocean, the loops of the 22–24° isolines extend over a meridional distance corresponding to this radius. For the northern Pacific Ocean, which is void of samples over distances of 40–50°, this search radius implies that the LOA does not update the first guess at all. A clear indication for this is the split of the front associated with the North Pacific Current (Fig. 3, bottom), caused by the uneven meridional distribution of the samples that enter the initial guess for the North Pacific Ocean. Constructing the initial guess from the per-basin zonal means would improve this situation. These findings might depend on the search radius lengths and we repeated the Levitus interpolation with doubled and halved influence radii. Longer radii broadened the gradients and diminished the preference for zonally oriented structures, but did not remedy the split of the front in the north Pacific Ocean. The general widening of the gradients yielded even greater differences between the interpolated and the WOA data at the frontal areas. In the under-sampled parts of the southern Pacific Ocean, the longer influence radii caused higher interpolated SSTs and as well a larger difference to WOA. The global mean difference was in this case 1.91 °C. With the shorter radii, the global mean error turned out to be only slightly lower, namely 1.87 °C. This similar difference was caused by too sharp gradients and an enhancement of the zonal structures. Thus, the original choice of 14°, 11°, 8°, and 5° constitutes the optimum, at least for a uniform application to the entire global grid. A local adaptation of the search radii to the changing sampling density should considerably improve the performance of the LOA. Regardless of the search radius, the requirement of a suitable first guess poses a problem outside the sampled area, that is, in the polar regions. Since there are no samples, the interpolation starts with the values of the northern- and southernmost samples, hence with temperatures that are much too high for the polar ocean. The subsequent iterative process partly adjusts this, but the final temperature distribution around Antarctica resembles the original one only very coarsely. To resolve this problem, additional sample points are necessary. If there is independent evidence for an ice cover, artificial "samples" with prescribed freezing point temperature are sufficient.

The subsequent Shapiro (1971) and median filters that employ a stencil of 3 × 3 grid points imply a reduction of the method's effective resolution by a factor of two, but compared to the VAK result, this does not flatten out the gradients in the interpolated fields. In fact, the VAK gradients are weaker and less realistic than those of LOA.

6. Summary

(1) The results of VAK are fairly reasonable even in under-sampled areas and when extrapolating poleward of the sampled regions. VAK fails only near strong oceanic fronts. Given the similar accuracy of the input data, the kriging variances reflect only the sampling density and provide no quality assessment for individual samples. However, minimizing the variance by tuning the variogram parameters generally reduces the difference between original and interpolated data set. A more sophisticated regional grouping of the sample data could avoid empirical variograms that show no sill, thereby opening a chance of automating the parameter adjustment. Then, VAK could be used as a convenient data-in/grid-out 'black-box' without the elaborate fitting of the variogram models.

(2) LOA is easily applied and yields good results in areas of dense sampling, especially in the well-sampled intermediate and high latitudes of the Atlantic Ocean, and in regions where zonal ocean currents match the method's preferential zonal spreading of information. This preference may cause artifacts where ocean currents are not predominantly zonal. For extrapolation beyond the sampled region, additional tie points are necessary, this would already be the case during the determination of the first guess. An adaptation of the influence radius to the local sampling density could lead to a better match of the interpolated and the original field.

(3) For gridding a global data set, all computations of distances and distance-related parameters must be carried out using spherical coordinates. The input data set should be subdivided into regions of similar spatial data dependencies. For VAK, this avoids mismatches between actual and modeled variability. For LOA, this allows search radii and an initial guess that are both adjusted to the spatial sampling density. Ocean communication masks are highly advisable, because they suppress unrealistic influences of the samples across the continental barriers.

(4) Given the different performances of both methods, we regard LOA as appropriate for work that is 'under way', for example, for the discussion of the (in part) large variations between the SST samples from different proxies and methods. If this discussion has eventually led to a consistent compilation of samples, VAK should be used for the gridding of the final MARGO SST data. Since both methods are problematic poleward of the sampled area, additional samples will greatly improve the results. These samples could be derived from proxy-based estimates of the seasonal ice covers.

Acknowledgements

This research was funded by the Deutsche Forschungsgemeinschaft (DFG) as part of the DFG Research Center "Ocean Margins" of the University of Bremen, No. RCOM0186, and by the BMBF (DEKLIM E grant 01LD0019 to S.M.). The thorough and helpful comments by Martin Weinelt and an anonymous referee prompted us to clarify the manuscript in general, and in particular to put the geostatistical analyses on a consistent basis.

References

Ababou, R., Bagtzoglou, A.C., Wood, E.F., 1994. On the condition number of covariance matrices arising in kriging, estimation, and simulation of random fields. Mathematical Geology 26 (1), 99–133.

CLIMAP Project Members, 1981. Seasonal reconstructions of the Earth's surface at the Last Glacial Maximum. Geological Society of America, Map and Chart Series MC-36, pp. 1–18.

Deutsch, C.V., Journel, A.G., 1992. GSLIB, Geostatistical Software Library and User's Guide. Oxford University Press, New York, Oxford.

Englund, E., Sparks, A., 1991. GEO-EAS 1.2.1 User's Guide. Technical report, United States Environmental Protection Agency.

Levitus, S., 1982. Climatological atlas of the World Ocean. NOAA Prof. Paper No. 13, 173pp.

Pannatier, Y., 1996. VARIOWIN: Software for Spatial Data Analysis in 2D. Springer, Berlin, Heidelberg.

Paul, A., Schäfer-Neth, C., 2004. How to combine sparse proxy data and coupled climate models. Quaternary Science Reviews, this issue, doi:10.1016/j.quascirev.2004.05.010.

Pebesma, E.J., Wesseling, C.G., 1998. GSTAT: a program for geostatistical modelling, prediction and simulation. Computers and Geosciences 24 (1), 17–31.

Reynolds, R.W., Smith, T.M., 1999. Improved global sea surface temperature analyses using optimum interpolation. Journal of Climate 7, 929–948.

Sarnthein, M., Gersonde, R., Niebler, S., Pflaumann, U., Spielhagen, R., Thiede, J., Wefer, G., Weinelt, M., 2003. Preface: Glacial atlantic ocean mapping (GLAMAP-2000). Paleoceanography 18, doi:10.1029/2002PA00769.

Schäfer-Neth, C., Paul, A., 2003. The Atlantic Ocean at the Last Glacial Maximum: 1. Objective mapping of the GLAMAP sea-surface conditions. In: Wefer, G., Mulitza, S., Ratmeyer, V. (Eds.), The South Atlantic in the Late Quaternary: Reconstruction of Mass Budget and Current System. Springer, Berlin, Heidelberg, pp. 531–548.

Shapiro, R., 1971. The use of linear filtering as a parameterization of atmospheric diffusion. Journal of Atmospheric Sciences 28, 523–531.

Taylor, S.P., Haywood, A.M., Valdes, P.J., Sellwood, B.W., 2004. An evaluation of two spatial interpolation techniques in global sea-surface temperature reconstructions: last Glacial Maximum and Pliocene case studies. Quaternary Science Reviews 23, 1041–1151.

Wackernagel, H., 2003. Multivariate Geostatistics, 3rd ed. Springer, Berlin, Heidelberg.

WOA, 1998. World ocean atlas 1998, http://www.nodc.noaa.gov/oc5/woa98.html. Technical report, National Oceanographic Data Center, Silver Spring, MD.

Wunsch, C., 1996. The Ocean Circulation Inverse Problem. Cambridge University Press, New York.

Quaternary Science Reviews 24 (2005) 1095–1107

How to combine sparse proxy data and coupled climate models

André Paul*, Christian Schäfer-Neth

DFG Research Center Ocean Margins and Department of Geosciences, University of Bremen, D-28334 Bremen, Germany

Received 2 February 2004; accepted 15 May 2004

Abstract

We address the problem of reconstructing a global field from proxy data with sparse spatial sampling such as the MARGO (multi-proxy approach for the reconstruction of the glacial ocean surface) SST (sea-surface temperature) and $\delta^{18}O_c$ (oxygen-18/oxygen-16 isotope ratio preserved in fossil carbonate shells of planktic foraminifera) data. To this end, we propose to 'assimilate' these data into coupled climate models by adjusting some of their parameters and optimizing the fit. In particular, we suggest to combine a forward model and an objective function that quantifies the misfit to the data. Because of their computational efficiency, earth system models of intermediate complexity are particularly well-suited for this purpose. We used one such model (the University of Victoria Earth System Climate Model) and carried out a series of sensitivity experiments by varying a single model parameter through changing the atmospheric CO_2 concentration. The unanalyzed World Ocean Atlas SST and the observed sea-ice concentration served as present-day targets. The sparse data coverage as implied by the locations of 756 ocean sediment cores from the MARGO SST database was indeed sufficient to determine the best fit. As anticipated, it turned out to be the 365 ppm experiment. We also found that the 200 ppm experiment came surprisingly close to what is commonly expected for the Last Glacial Maximum ocean circulation. Our strategy has a number of advantages over more traditional mapping methods, e.g., there is no need to force the results of different proxies into a single map, because they can be compared to the model output one at a time, properly taking into account the different seasons of plankton growth or varying depth habitats. It can be extended to more model parameters and even be automated.

1. Introduction

The multi-proxy approach for the reconstruction of the glacial ocean surface (MARGO) project will produce various sea-surface temperature (SST) reconstructions from different proxies as well as a reconstruction of $\delta^{18}O_c$ (oxygen-18/oxygen-16 isotope ratio preserved in fossil carbonate shells of planktic foraminifera). It is planned to provide these data at the core locations (Fig. 1) as well as in the form of regularly gridded fields and paleo-maps (see the companion paper by Schäfer-Neth et al., 2005). Such paleo-maps are not only useful for displaying and discussing the data, but also for forcing ocean or atmosphere models.

Examples for simulations of the ocean at the Last Glacial Maximum (LGM) based on the CLIMAP Project Members (1981) SST reconstruction or modifications thereof are given by Seidov et al. (1996), Winguth et al. (1999), Schäfer-Neth and Paul (2001) and Paul and Schäfer-Neth (2003, 2004). In our previous work (Paul and Schäfer-Neth, 2003, 2004) we use the GLAMAP SST reconstruction (Sarnthein et al., 2003a) for the Atlantic Ocean, which as compared to CLIMAP is characterized by 1–2 °C colder tropics and seasonally ice-free Nordic Seas.

On the one hand, classical surface mapping methods of paleo-climate proxy variables make insufficient or no use of (1) the specific properties of each proxy (e.g., different seasons of plankton growth or varying depth habitats) and (2) the dynamical constraints of the ocean circulation, or, more broadly, the coupled ocean–sea-ice–atmosphere system.

*Corresponding author. Tel.: +49-421-218-7189; fax: +49-421-218-7040.

E-mail address: apau@palmod.uni-bremen.de (A. Paul).

0277-3791/$ - see front matter © 2004 Elsevier Ltd. All rights reserved.
doi:10.1016/j.quascirev.2004.05.010

MARGO core locations

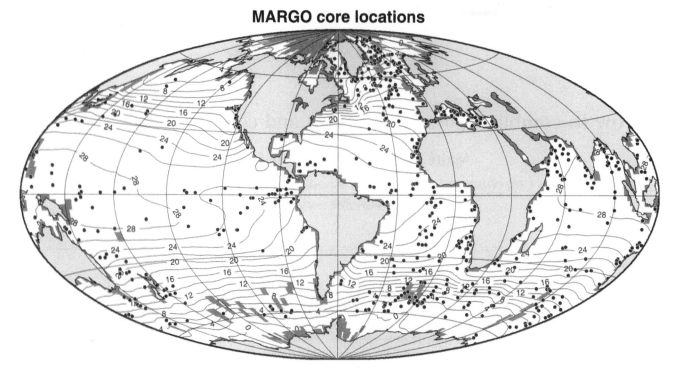

Fig. 1. Unanalyzed annual mean WOA (1998) 10 m temperature (contours at an interval of 2 °C) and distribution of the 756 ocean sediment cores from the MARGO (multi-proxy approach for the reconstruction of the glacial ocean surface) SST (sea-surface temperature) database (circles). The WOA temperature data was smoothed with a cosine arch filter of 2000 km width. Dark gray shading indicates gaps in the original data set. The central meridian of the Hammer equal-area projection is located at 60°W, and the line interval of the geographical grid is 30° (the same map projection is also used in Fig. 2).

On the other hand, coupled climate models do not require the proxy data for forcing and allow for a data–model comparison directly at the core locations (Broccoli and Marciniak, 1996), but their LGM results, although all physically plausible in some sense, are radically different among each other (Mix, 2003). For example, there is no agreement on whether the LGM meridional overturning circulation (MOC) was weaker or stronger than today. Furthermore, these models simulate a wide range of sea-surface temperatures. More generally, they agree with some paleo-data, but are in conflict with some other paleo-data.

We therefore suggest to combine coupled climate models and sparse proxy data and construct an SST map that accommodates best the information obtained from paleo-proxies with the physical constraints of the climate system.

More recently any such data–model combination has been termed 'data assimilation' (Wunsch, 1996), with the general goal to use data in order to improve the performance of numerical models (Hargreaves and Annan, 2002). Here it means finding a set of parameter values such that the model equilibrium solution is best compatible with observational (or proxy-) data.

This method is not entirely new. Even the most comprehensive coupled climate model contains para-

meterizations of processes that either have not been understood yet from first principles, or that cannot be resolved because of their spatial or temporal scales, and there is some freedom in tuning the associated model parameters. This freedom is used to optimize the fit to present-day climate data. Basically, what we have in mind is to tune a climate model to fit paleo-proxy data.

An earth system model of intermediate complexity is particularly well-suited for this purpose, because it is computationally efficient and can be used to carry out more than one LGM experiment. To illustrate our point of 'assimilating' reconstructed proxy-data into a coupled climate model, we performed a series of experiments with one such model and used the same test data set for present-day SST as in Schäfer-Neth et al. (2005), complemented by a second data set for observed sea-ice concentration. We varied a single model parameter (through changing the atmospheric CO_2 concentration) and left all boundary conditions unchanged. Our strategy could be easily extended to, e.g., the MARGO reconstructions of SST and sea-ice extent, by designing an LGM experiment according to the Paleo-Modelling Intercomparison Project (PMIP) 2 recommendations (http://www-lsce.cea.fr/pmip2/).

2. Methods

We used a coupled climate model, combined it with an explicit objective function that quantified the misfit to the data, selected a parameter which presumably affected the results significantly, performed a series of sensitivity experiments and evaluated the objective function for discrete values of the selected parameter.

2.1. Coupled climate model

We chose the 'UVic Earth System Climate Model' version 2.6, which consists of the Modular Ocean Model (MOM) 2 of the Geophysical Fluid Dynamics Laboratory (GFDL) (Pacanowski, 1996), coupled to an atmospheric energy–moisture balance model (Fanning and Weaver, 1996) and a sea-ice model. We used the sea-ice model in its standard form, in which the thermodynamics is based on the zero-heat capacity formulation by Semtner (1976), together with the lateral growth and melt parameterization by Hibler (1979), and the dynamics is elastic-viscous-plastic (Hunke and Dukowicz, 1997). This particular model version allows for the advection of moisture by monthly winds prescribed from the NCEP reanalysis climatology (Kalnay et al., 1996) and is described in detail by Weaver et al. (2001). The 280 and 365 ppm restart files were kindly provided by Michael Eby (pers. comm.), which enabled us to initialize the model from near-steady states. All experiments were integrated for at least 1000 years, by which time the global air-sea heat flux averaged over 10 years was between 0.009 and 0.135 $W\,m^{-2}$. This corresponded to a residual trend in the global annual-mean ocean temperature between 0.02 and 0.30 °C per 1000 years.

2.2. Objective function

An objective (or 'cost') function typically characterizes an 'inverse problem' (Wunsch, 1996). Here we combine it with a 'forward model' like the UVic coupled climate model.

The purpose of an objective function F is to provide a measure for the misfit between data and model, and hence it involves the observations (or proxy data) $\mathbf{x}$, the corresponding model output $\mathbf{x}'$ and the model parameters $\mathbf{p}$:

$$F(\mathbf{x}, \mathbf{x}', \mathbf{p}) = \sum_{n=1}^{N} \sum_{l=1}^{L} f_{ln}(\mathbf{x}, \mathbf{x}', \mathbf{p}). \qquad (1)$$

The components f_{ln} of the objective function are taken as sum of squares (e.g., Jentsch, 1991; Hargreaves and Annan, 2002):

$$f_{ln}(\mathbf{x}, \mathbf{x}', \mathbf{p}) = \sum_{i} \left(\frac{x_{iln} - x'_{iln}}{g_{iln}} \right)^2. \qquad (2)$$

In our case, the number of variables $N = 2$, where $n = 1$ stands for temperature and $n = 2$ for sea-ice concentration. Furthermore, the number of seasons $L = 2$, where $l = 1$ stands for January–February–March (JFM) and $l = 2$ for July–August–September (JAS). The sum on i is over all grid cells that contain data. In particular, $x_{il,n=1}$ refers to the World Ocean Atlas unanalyzed temperature data for 10 m depth (WOA, 1998) and $x_{il,n=2}$ to the Atmospheric Model Intercomparison Project (AMIP) 2 sea-ice concentration data (Taylor et al., 2000).

An important issue is the choice of the weighting factors g_{iln} that are to give each component f_{ln} an approximately equal weight (Jentsch, 1991). We set them such that the objective function became the sum of the individual root-mean square (RMS) seasonal errors, each normalized by the corresponding RMS seasonal contrast of the 365 ppm experiment and squared. The components with $n = 2$ that referred to sea-ice concentration were further multiplied by a factor of 0.12, which is the ratio of the area affected by the AMIP 2 sea-ice concentration data (43.1×10^6 km^2) and the area covered by the annual-mean WOA unanalyzed SST data (361.7×10^6 km^2).

2.3. Selected parameter

To illustrate our technique, we applied a globally uniform climate forcing

$$Q = \Delta F_{2x} \ln \frac{C}{C_0}$$

to the energy balance at the top of the atmosphere by directly reducing the outgoing longwave radiation (Weaver et al., 2001). Here C is a prescribed atmospheric CO_2 concentration, C_0 some reference level ($C_0 = 350$ ppm) and $\Delta F_{2x} = 5.77$ $W\,m^{-2}$ corresponds to a specified radiative forcing of 4 $W\,m^{-2}$ for a doubling of CO_2, as estimated from calculations with radiative transfer models (Ramanathan et al., 1987; Hartmann, 1994).

2.4. Series of sensitivity experiments

Usually, the model parameters $\mathbf{p} = (p_1, \ldots)$ would be adjusted within their uncertainty ranges. In our simple example, we tuned the climate forcing Q over a wide range through changing the CO_2 concentration between 200 and 560 ppm (Table 1). This range is much larger than the uncertainty of the radiative transfer scheme, which has been calibrated against satellite and shipboard measurements by Fanning and Weaver (1996).

2.5. Evaluation of the objective function

'Assimilating' (paleo-) data into the coupled climate model now means to evaluate the objective function for

Table 1
List of experiments with atmospheric CO_2 concentration, radiative forcing and equilibrium temperature response (in terms of the global mean surface-air and ocean temperatures, T_a and T_o)

Experiment	pCO_2 (ppm)	Q (W m^{-2})	T_a (°C)	T_o (°C)
1	200	−3.23	11.1	3.3
2	280	−1.28	12.9	3.6
3	365	0.24	14.2	4.2
4	450	1.45	15.2	4.6
5	560	2.67	16.2	5.1

Experiment 5 was a transient experiment, initialized with an atmospheric CO_2 concentration of 280 ppm. The atmospheric CO_2 concentration was increased by 1% per year until the final value of 560 ppm was reached. At this time (i.e., after 70 years of model integration), the global mean surface-air and ocean temperatures were $T_a = 14.9$ °C and $T_o = 3.8$ °C.

discrete values of the selected parameter and search for a minimum. With our choice of the objective function, this is equivalent to the 'least squares method'. To facilitate the comparison with the more traditional mapping methods discussed by Schäfer-Neth et al. (2005), we interpolated the model output as well as the two test data sets (the unanalyzed WOA SST and AMIP2 sea-ice concentration) to a $2° \times 2°$ grid. We computed two types of RMS errors for SST: The first one was based on all unanalyzed WOA data and referred either to the annual mean or the two seasons JFM and JAS. The second one was determined from the annual mean of the unanalyzed WOA data restricted to those $2° \times 2°$ latitude–longitude grid cells that contained at least one of the 756 ocean sediment cores from the MARGO SST database. Furthermore, in presenting or results, we used the same Hammer equal-area map projection as in Schäfer-Neth et al. (2005) (cf. Fig. 1).

3. Results

The simulated annual-mean SST for intermediate atmospheric CO_2 concentrations was generally too cold in the Atlantic Ocean and generally too warm in the Indian and Pacific Oceans (Fig. 2). Regional positive anomalies indicate that the Gulf Stream separated from the coast too far south, and that the water that reached the surface in the major coastal upwelling areas was too warm.

The subtropical and subpolar fronts were reproduced by the coupled climate model, albeit with smaller SST gradients than observed (not shown). In the 365 ppm experiment, the North Atlantic Ocean turned out to be too cold, because the sea ice extended too far equatorwards (cf. Figs. 3 and 4). This in turn was caused by the fact that the NADW was formed slightly too far south (cf. Fig. 5).

For the 365 ppm experiment, the global mean surface-air temperature (Table 1) was close to present-day observations (e.g. 13.84 °C according to the NCEP reanalysis climatology, Kalnay et al., 1996). Comparing the global mean surface-air temperatures for the 280 and 560 ppm experiments, the equilibrium temperature response for a doubling of the atmospheric CO_2 concentration with respect to preindustrial times was 3.3 °C, which is in the range simulated by more comprehensive coupled climate models.

With respect to the present-day climatology, too much sea ice was simulated especially in the Nordic Seas, but also in the Southern Ocean (Figs. 3 and 4). The sea-ice concentration in the 200 ppm experiment as compared to the 365 ppm experiment showed a large expansion during Northern Hemisphere winter down to 50°N (Fig. 3). In contrast, there was still an ice-free region off southern Norway during Northern Hemisphere summer (Fig. 4). In both cases, the simulated sea-ice margins roughly corresponded to the GLAMAP reconstruction (Gersonde et al., 2003; Paul and Schäfer-Neth, 2003; Sarnthein et al., 2003b). During Southern Hemisphere winter, sea ice vastly expanded in the Drake Passage and Atlantic and the Indian sectors of the Southern Ocean.

Fig. 5 shows the MOC in the Atlantic Ocean for LGM and modern atmospheric CO_2 concentrations. The deep circulation is represented by two cells with centers at 1500 and 3500 m depth (in the 200 ppm experiment) and 1500 and 4000 m depth (in the 365 ppm experiment). In the 200 ppm experiment, the formation of North Atlantic Deep Water (NADW) takes place in two latitude bands between 40–50°N and 60–70°N, and the outflow of NADW to the Southern Ocean is 2 Sv ($1 Sv = 1 \times 10^6$ m^3 s^{-1}). In contrast, in the 365 ppm experiment, the formation of NADW is concentrated in the 60–70°N latitude band and the outflow of NADW to the Southern Ocean is 12 Sv. The cooling of the tropical SST in the 200 ppm experiment with respect to the 365 ppm experiment amounts to about 3 °C (not shown).

The RMS annual-mean error of SST as a function of latitude for the global ocean was generally smallest for the 365 ppm experiment (left column in Fig. 6), in accordance with the global area-weighted RMS annual-mean errors (cf. Table 2, which can be directly compared to the results presented by Schäfer-Neth et al., 2005). However, in the Atlantic Ocean, the RMS error was smaller for the 450 ppm experiment than for the 365 ppm experiment, while in the Pacific and Indian Oceans, the 280 ppm experiment performed best. The same basic pattern was born out by the unanalyzed WOA data restricted to the MARGO core locations (Table 3 and right column in Fig. 6).

The RMS seasonal SST errors were also generally smallest for the 365 ppm experiment (Table 4). With

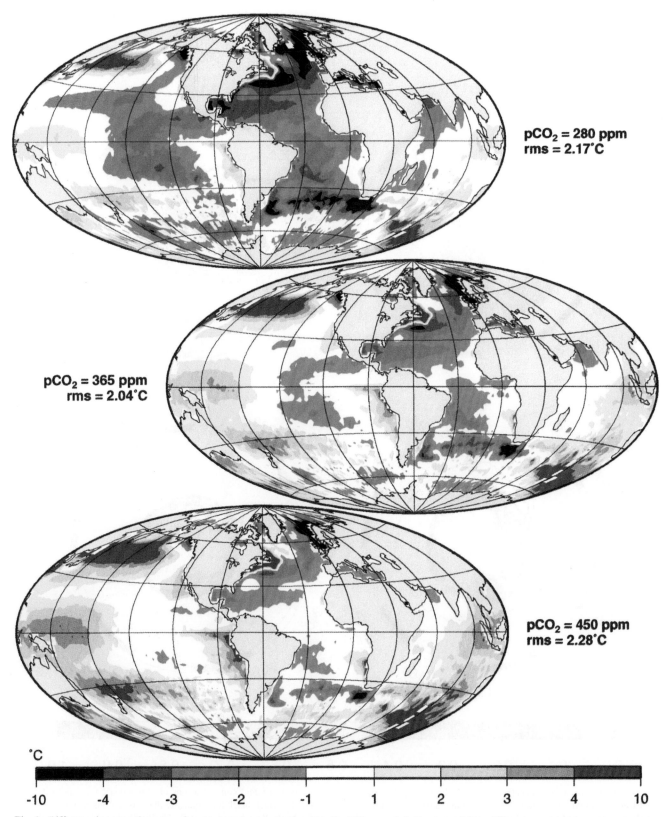

Fig. 2. Difference between the sea-surface temperature as simulated by the UVic coupled climate model for different concentrations of atmospheric CO_2 (all other boundary conditions were unchanged) and the unanalyzed annual-mean WOA (1998) 10 m temperature shown in Fig. 1, in units of °C. Top: 280 ppm. Center: 365 ppm. Bottom: 450 ppm.

JFM sea-ice concentration

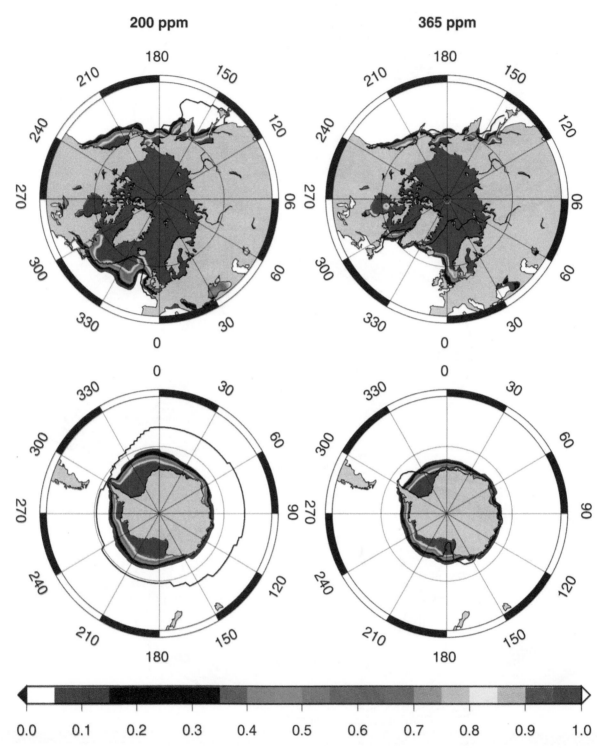

Fig. 3. Simulated sea-ice concentration for JFM (January–February–March) for two different concentrations of atmospheric CO_2, but otherwise unchanged boundary conditions. Left: 200 ppm (value appropriate for the LGM). The thick black lines indicate the sea-ice boundary based on the CLIMAP and GLAMAP reconstructions (Gersonde et al., 2003; Paul and Schäfer-Neth, 2003; Sarnthein et al., 2003b). Right: 365 ppm (present-day value). In the Northern Hemisphere, the thick black line indicates the 50% contour of the AMIP 2 observed climatology for the period of 1979–2001 (Taylor et al., 2000). A 50% contour was used by Sarnthein et al. (2003b) for calibrating their method of reconstructing past sea-ice extent. In the Southern Hemisphere, the thick black line is the 15% contour of the AMIP 2 climatology. A 15% contour is commonly used for indicating present-day sea-ice extent. The polar stereographic map projection (also used in Fig. 4) extends to 40°N in the Northern Hemisphere and 40°S in the Southern Hemisphere, and the line interval of the geographical grid is 30°.

JAS sea-ice concentration

200 ppm

365 ppm

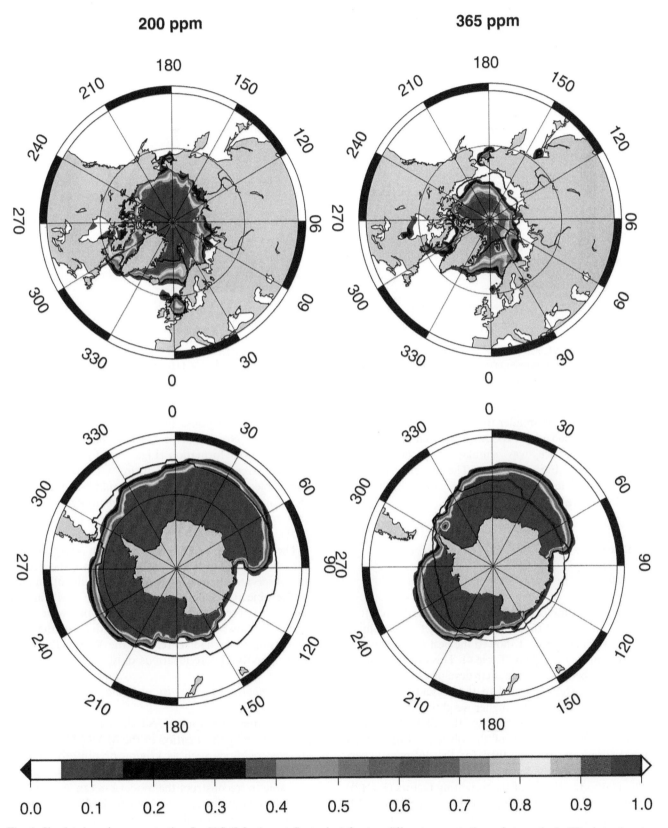

Fig. 4. Simulated sea-ice concentration for JAS (July–August–September) for two different concentrations of atmospheric CO_2, but otherwise unchanged boundary conditions. Left: 200 ppm (value appropriate for the LGM). Right: 365 ppm (present-day value). The thick black lines have a similar meaning as in Fig. 3.

Atlantic Ocean Meridional Overturning Streamfunction

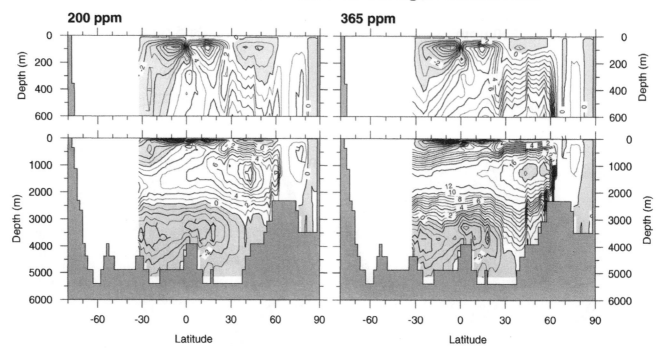

Fig. 5. Vertical meridional overturning streamfunction as simulated by the UVic coupled climate model for two different concentrations of atmospheric CO_2, but otherwise unchanged boundary conditions, in units of Sv ($1\,Sv = 1 \times 10^6\,m^3\,s^{-1}$). Left: 200 ppm. Right: 365 ppm.

respect to sea-ice concentration, the RMS seasonal errors tended to be smaller for the 450 ppm experiment than for the 365 ppm experiment, which reflects that too much sea ice was simulated in the 365 ppm experiment (cf. Figs. 3 and 4). The objective function still attained its minimum value for the 365 ppm experiment (Table 4).

4. Discussion

4.1. Optimum fit

The explicit use of an objective function allowed for a concise measure of the misfit to the target. Accordingly, the 365 ppm experiment showed the best agreement with the present-day test data (Table 4). This did not come as a surprise, because during their development all climate models are carefully tuned to the present-day climate. However, we note that a higher weighting of the sea-ice concentration data relative to the SST data could produce a smaller value of the objective function for the 450 ppm experiment than for the 365 ppm experiment. This shows the sensitivity of the objective function (Eq. (1)) to the choice of the weighting factors (Eq. (2)).

Furthermore, in the 365 ppm experiment the RMS annual-mean SST error reached its minimum value because of a balanced representation of the Atlantic and Indo-Pacific Oceans ('harmony of errors', cf. Table 2

and Fig. 2). The RMS error also concealed large anomalies in regions such as the northern North Atlantic or North Pacific Oceans that are important for deep water formation (Figs. 2 and 6c–f). We could possibly circumvent this problem by choosing a 'minimax' objective function, which would force any minimization procedure to focus on regions with largest data–model discrepancies (LeGrand and Alverson, 2001).

The pre-industrial CO_2 concentration (280 ppm) gave already a noticeably larger deviation from the present-day observations than the 365 ppm experiment, in terms of the objective function as well as the RMS annual-mean SST error; similarly, the 450 ppm experiment. Table 3 shows that we would arrive at the same conclusion if we had to rely on the WOA (1998) data at the MARGO core locations only.

4.2. Comparison to classical mapping methods

Compared to mapping SST data from a density of points given by the location of the MARGO cores using classical methods such as kriging or objective analysis (Schäfer-Neth et al., 2005), the coupled climate model was able to 'reconstruct the modern ocean surface' with an only slightly lower accuracy (the RMS annual-mean error with respect to all available unanalyzed WOA data is 2.04 °C as compared to 1.22 °C for variogram analysis/kriging and 1.56 °C for objective analysis).

Basin Root-Mean Square Error

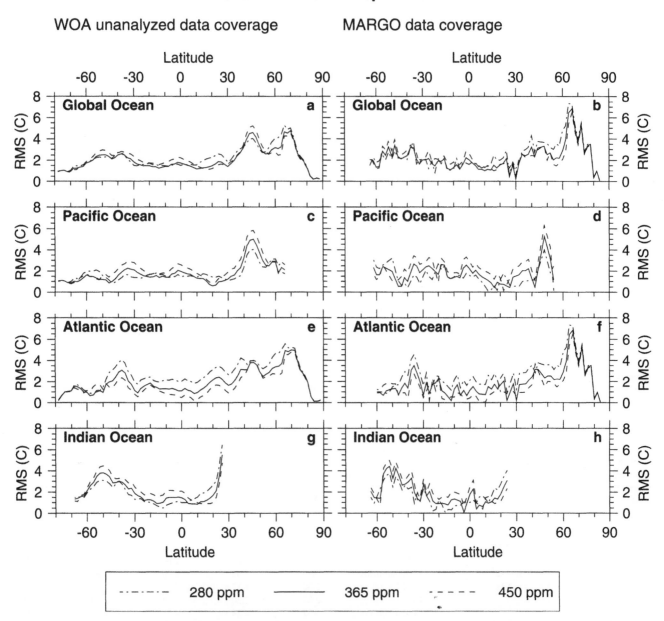

Fig. 6. Zonal-mean distribution of the RMS annual-mean difference between the sea-surface temperature as simulated by the UVic coupled climate model for different concentrations of atmospheric CO_2 (280, 365 and 450 ppm) and the unanalyzed 10 m temperature of the World Ocean Atlas (WOA, 1998), in units of °C, for the global ocean as well as the Pacific, Atlantic and Indian Oceans. Left column: With respect to all available unanalyzed WOA data. Right column: With respect to the unanalyzed WOA data at the MARGO core locations only.

Furthermore, the 'reconstructed' sea-surface conditions are consistent with the fundamental equations of the coupled climate model. Spatial and temporal correlations of the different regions are implicit in these equations. At the chosen resolution of the ocean model, information on ocean currents is exploited and frontal systems are preserved. As an advantage of a coupled climate model (also as compared to an ocean-only model subject to restoring boundary conditions) the sea-ice distribution can be simulated and compared to an independent set of observations. This enabled us to add sea-ice concentration to the objective function.

In addition, there is no need for a sophisticated gridding and mapping of the proxy data at the core locations for other uses than displaying or comparing it. In particular, it is not necessary to extrapolate the proxy data into areas where no sediment cores can be taken, e.g., areas covered by sea ice today, or to make assumptions about seasons during which there is only very little plankton growth and no significant imprint on

Table 2
RMS annual-mean SST differences between the simulated and observed fields at all WOA (1998) unanalyzed data locations (binned into 2° longitude/latitude squares)

Ocean	Atmospheric CO_2 concentration (ppm)				
	200	280	365	450	560
Atlantic	4.29	2.89	2.18	1.83	1.78
Pacific	2.26	1.72	1.85	2.32	2.95
Indian	2.24	1.96	2.23	2.66	3.20
Mediterranean	6.20	4.16	3.10	2.32	1.56
Global	2.97	2.17	2.04	2.28	2.73

Table 3
RMS annual-mean SST differences between the simulated and observed fields restricted to those $2° \times 2°$ latitude–longitude grid cells that contained at least one ocean sediment core from the MARGO SST database

Ocean	Atmospheric CO_2 concentration (ppm)				
	200	280	365	450	560
Atlantic	4.63	2.98	2.34	2.10	2.10
Pacific	1.91	1.55	1.93	2.51	3.20
Indian	2.48	2.07	2.25	2.64	3.16
Mediterranean	6.19	4.10	3.03	2.25	1.48
Global	3.55	2.46	2.24	2.39	2.75

Table 4
Hemispheric RMS seasonal errors (model–data) and seasonal contrasts (model only)

Variable	Hemisphere	Season	Atmospheric CO_2 concentration (ppm)				
			200	280	365	450	560
Sea-surface temperature	NH	JFM	4.13	2.97	2.67	2.72	3.01
		JAS	3.77	2.69	2.64	2.77	3.25
		Model contrast	4.85	4.92	4.90	4.99	4.97
	SH	JFM	2.02	1.87	2.22	2.71	3.27
		JAS	2.81	2.20	2.01	2.20	2.70
		Model contrast	4.32	4.34	4.35	4.27	4.17
Sea-ice concentration	NH	JFM	0.21	0.15	0.14	0.13	0.13
		JAS	0.15	0.13	0.12	0.12	0.12
		Model contrast	0.19	0.18	0.20	0.21	0.18
	SH	JFM	0.11	0.12	0.07	0.07	0.07
		JAS	0.26	0.24	0.19	0.14	0.14
		Model contrast	0.32	0.31	0.28	0.24	0.21
Objective function			2.27	1.35	1.24	1.41	1.90

Here 'NH' stands for 'Northern Hemisphere' and 'SH' for 'Southern Hemisphere'. The value of the objective function is the sum of the individual RMS seasonal differences, each normalized by the corresponding RMS seasonal contrast of the 365 ppm experiment and squared. The components related to sea-ice concentration are further weighted by the ratio of the area affected by the AMIP 2 sea-ice concentration data and the area covered by the annual-mean WOA unanalyzed SST data (which is 0.12, see Methods).

the sedimentary record is left, e.g., winter in the Southern Ocean. The proxy data is only used where and when it is available.

Finally, there is no need to force the results of different proxies into a single map, no matter how desirable such a map would be for other purposes. The results from the different proxies used in MARGO can be as radically different as those from different coupled climate models. As an advantage of our strategy, they can be compared to the model output one at a time, properly taking into account the different seasons of plankton growth or varying depth habitats. In this way the discrepancies among the different proxies can be addressed and clarified.

A disadvantage of using a coupled climate model for assimilating proxy data is that it is computationally much more demanding than classical mapping methods. Another problem is the ambiguity in minimizing the objective function, which is hidden in the usual forward problem. Model errors might compensate: We might adjust one or a few parameters considered uncertain or critical to match the paleo-data, while actually other parameters may be in cause. To distinguish between multiple solutions, corresponding to multiple sets of model parameters, ultimately requires the computation of the joint 'probability density function' (PDF) of parameter values (Hargreaves and Annan, 2002).

4.3. The 200 ppm experiment

In many ways the 200 ppm experiment came surprisingly close to what is commonly expected from an experiment subject to full LGM boundary conditions. First, it satisfies 'a widespread, if not universal, belief that the LGM circulation was weaker than today', which however, has not been firmly established from plaeotracer data yet (Wunsch, 2003). Second, far from the direct influence of the ice sheets in the high northern latitudes, the main cause of glacial cooling must have been lower levels of atmospheric greenhouse gas concentrations. With a given climate sensitivity of $4 \, W \, m^{-2}$ for a doubling of the atmospheric CO_2 concentration, the tropical cooling in the UVic coupled climate model turned out to be $3 \, °C$, which is in accordance with evidence from planktic fauna and flora as well as oxygen isotope measurements (e.g., Crowley, 2000; Schäfer-Neth and Paul, 2004). Third, sea-ice cover vastly expanded, but the Nordic Seas were partly ice-free during summer (Sarnthein et al., 2003b).

The severe reduction of the MOC in the 200 ppm experiment as compared to the 365 ppm experiment is the result of a change in the subtle balance of thermal and haline buoyancy forcing in the Atlantic Ocean. According to the analysis by Schmittner et al. (2002), changes in the atmospheric hydrologic cycle dominate changes in the surface heat flux. We found that these changes caused a reduced convection intensity, decreased SST and reduced evaporation in the North Atlantic Ocean, which in a positive feedback loop led to a further reduction of the MOC. As a result, south of $65 \, °N$, evaporation decreased much more than precipitation, which is reflected in the meridional gradients of

sea-surface density (Paul and Schäfer-Neth, 2003) and depth-integrated steric height (Schmittner et al., 2002).

To determine absolute LGM circulation rates, additional proxy data is required. Passive, steady-state tracer data alone (such as $\delta^{13}C$) do not suffice, but must be coupled with a 'clock' (LeGrand and Wunsch, 1995). This could be provided by a reconstruction of the LGM density field with accuracy and spatial sampling adequate to infer the paleo-geostrophic shear (Lynch-Stieglitz et al., 1999), or with well-distributed measurements of a radioactive tracer such as ^{14}C (Meissner et al., 2003) or the $^{231}Pa/^{230}Th$ ratio (Yu et al., 1996; Marchal et al., 2000). The absolute strength and associated stability of the glacial circulation are important because they set the stage for understanding rapid climate changes such as the Dansgaard–Oeschger or Heinrich events.

5. Outlook

Regarding the 'assimilation' of real proxy data, our objective function could be adapted to compare the different MARGO SST proxies to the model output one at a time, properly taking into account the different seasons of plankton growth or varying depth habitats. Instead of the observed sea-ice concentration, we could use the reconstructed sea-ice extent for average LGM winter and summer conditions, as depicted in Figs. 3 and 4. By including an isotopic cycle into the UVic coupled climate model, simulating the oxygen isotope ratios $\delta^{18}O_w$ and $\delta^{18}O_c$ and comparing the outcome to the MARGO $\delta^{18}O_c$ reconstruction, we could even exploit the information on the LGM density field that is implicit in the oxygen isotope data (see Discussion). Ideally, we would aim for a good fit to the LGM as well as the present-day data.

The next step would be to extend our method to more than one model parameter, choose a grid of discrete values and search on this grid for a minimum of the objective function. We would select a restricted number of model parameters which are either poorly known or may affect the results significantly (Jentsch, 1991). Parameters related to radiation (e.g., the planetary and atmospheric emissivities, scattering coefficients and planetary albedo, cf. Fanning and Weaver (1996)) dominate the globally averaged climate, while parameters related to dynamics (the coefficients of horizontal diffusion and advection in the atmosphere, the horizontal and vertical diffusion coefficients in the ocean) are mainly important for the redistribution of heat and moisture, and hence the climatic gradients (Jentsch, 1991).

The manual search for the optimum fit could in principle be automated by using inverse methods such as, e.g., nudging, a Kalman filter or the adjoint method

(Wunsch, 1996). Recent examples for the application of inverse methods to paleoceanographic problems are given by Winguth et al. (1999, inverse physical-biogeochemical ocean model), LeGrand and Alverson (2001, inverse ocean box model), Wunsch (2003, inverse dynamical ocean model) and Grieger and Niebler (2003, semi-inverse ocean model).

Each of these inverse methods has its own strengths and weaknesses, e.g., nudging a climate model to data violates such general principles as the conservation of heat and salt, whereas the adjoint method requires finding the inverse ('adjoint') of a complex forward model, which is an extremely tedious and time-consuming task.

Furthermore, a coupled climate model is nonlinear by nature, and the objective function needs no longer have a unique minimum, but 'may come to resemble a chaotic function' with many nearby, or distant, minima, and 'hills, plateaus, and valleys' inbetween (Wunsch, 1996, p. 386). These multiple minima produce predictions that come all very close to the target observations.

A method that overcomes such problems is the Monte Carlo Markov Chain method (Hargreaves and Annan, 2002) based on a Bayesian approach to parameter estimation and the Metropolis-Hastings algorithm. It does not require finding the adjoint and yields the solution to the inverse problem in terms of an estimate for the joint posterior probability density function (PDF), instead of an unique optimum solution. However, at present, this method can only be applied to very efficient climate models. To sample the PDF of the model parameter space of the UVic coupled climate model in its standard form would go beyond the presently available computational resources and require to severely degrade its horizontal and vertical resolution.

6. Conclusion

We addressed the question whether or not a global SST field can be reconstructed from data available at the MARGO core locations. In response to this question, we found that

(1) combining a forward model like the UVic coupled climate model and the use of an objective function quantifies the misfit to the target data in a concise manner,
(2) the sparse MARGO data coverage is indeed sufficient to determine the optimum fit,
(3) the accuracy is comparable to that of classical mapping methods such as kriging or objective analysis.

The strategy of 'assimilating' sparse proxy data (e.g., the MARGO SST reconstruction) into a coupled climate model is free from many restrictions imposed on classical mapping methods by limited sampling density. Finally, we note that our 200 ppm experiment came surprisingly close to what is commonly expected from an experiment subject to full LGM boundary conditions.

Acknowledgements

We thank the participants of the second MARGO workshop (15–17 September 2003, Vilanova i la Geltrú, Spain) for many stimulating discussions. Furthermore, we thank Michael Eby and the Climate Modelling Group at the University of Victoria for providing us with a copy of the UVic coupled climate model. Finally we thank three anonymous referees, as well as Michael Schulz, for their very constructive comments on our original manuscript. This research was funded by the Deutsche Forschungsgemeinschaft (DFG) as part of the DFG Research Center 'Ocean Margins' of the University of Bremen, No. RCOM0185.

References

Broccoli, A.J., Marciniak, E.P., 1996. Comparing simulated glacial climate and paleodata: a reexamination. Paleoceanography 11, 3–14.

CLIMAP Project Members, 1981. Seasonal reconstructions of the Earth's surface at the last glacial maximum. Geological Society of America, Map and Chart Series MC-36, 1–18.

Crowley, T., 2000. CLIMAP SSTs re-revisited. Climate Dynamics 16, 241–255.

Fanning, A.F., Weaver, A.J., 1996. An atmospheric energy–moisture balance model: climatology, interpentadal climate change, and coupling to an ocean general circulation model. Journal of Geophysical Research 101 (D10), 15,111–15,128.

Gersonde, R., Abelmann, A., Brathauer, U., Cortese, G., Fütterer, D., Grobe, H., Niebler, H.-S., Segl, M., Sieger, R., Zielinski, U., 2003. Last glacial maximum sea surface temperature and sea ice extent in the Southern Ocean (Atlantic-Indian sector). a multiproxy approach. Paleoceanography 18, doi 10.1029/2002PA00773.

Grieger, B., Niebler, S., 2003. Glacial South Atlantic surface temperatures interpolated with a semi-inverse ocean model. Paleoceanography 18, doi 10.1029/2002PA000773.

Hargreaves, J.C., Annan, J.D., 2002. Assimilation of paleo-data in a simple Earth system model. Climate Dynamics 19, 371–381, doi 10.1007/s00382-002-0241-0.

Hartmann, D.L., 1994. Global Physical Climatology. Academic Press, San Diego, 411pp.

Hibler, W.D., 1979. A dynamic-thermodynamic sea-ice model. Journal of Physical Oceanography 9, 815–846.

Hunke, E.C., Dukowicz, J.K., 1997. An elastic-viscous-plastic model for sea ice dynamics. Journal of Physical Oceanography 27, 1849–1867.

Jentsch, V., 1991. An energy balance climate model with hydrological cycle. 1. Model description and sensitivity to internal parameters. Journal of Geophysical Research 96 (D9), 17169–17179.

Kalnay, E., Kanamitsu, M., Kistler, R., Collins, W., Deaven, D., Gandin, L., Iredell, M., Saha, S., White, G., Woolen, J., Zhu, Y., Chelliah, M., Ebisuzaki, W., Higgins, W., Janowiak, J., Mo, K.C., Ropelewski, C.A.L., Reynolds, R., Jenne, R., 1996. The NCEP/NCAR reanalysis project. Bulletin of the American Meteorological Society 77, 437–471.

LeGrand, P., Alverson, K., 2001. Variations in atmospheric CO_2 during glacial cycles from an inverse ocean modeling perspective. Paleoceanography 16, 604–616.

LeGrand, P., Wunsch, C., 1995. Constraints from paleotracer data on the North Atlantic circulation during the last glacial maximum. Paleoceanography 6, 1011–1045.

Lynch-Stieglitz, J., Curry, B., Slowey, H., 1999. Weaker Gulf Stream in the Florida Straits during the last glacial maximum. Nature 402, 644–648.

Marchal, O., François, R., Stocker, T.F., Joos, F., 2000. Ocean thermohaline circulation and sedimentary $^{231}Pa/^{230}Th$ ratio. Paleoceanography 15, 625–641.

Meissner, K.J., Schmittner, A., Weaver, A.J., Adkins, J.F., 2003. The ventilation of the North Atlantic Ocean during the last glacial maximum—a comparison between simulated and observed radiocarbon ages. Paleoceanography 18, doi 10.1029/2002PA00762.

Mix, A.C., 2003. Chilled out in the ice-age Atlantic. Nature 425, 32–33.

Pacanowski, R.C.E., 1996. MOM 2. Documentation, user's guide and reference manual. Technical Report 3.2, GFDL Ocean Group, GFDL, Princeton, New Jersey.

Paul, A., Schäfer-Neth, C., 2003. Modeling the water masses of the Atlantic Ocean at the last glacial maximum. Paleoceanography 18, doi 10.1029/2002PA000783.

Paul, A., Schäfer-Neth, C., 2004. The Atlantic Ocean at the last glacial maximum: 2. Reconstructing the current systems with a global ocean model. In: Wefer, G., Mulitza, S., Ratmeyer, V. (Eds.), The South Atlantic in the Late Quaternary: Reconstruction of Material Budgets and Current Systems. Springer, Berlin, Heidelberg, pp. 549–583.

Ramanathan, V., Callis, L., Cess, R., Hansen, J., Isaksen, I., Kuhn, W., Lacis, A., Luther, F., Mahlman, J., Reck, P., Schlesinger, M., 1987. Climate-chemical interactions and effects of changing atmospheric trace gases. Reviews of Geophysics 25, 1441–1482.

Sarnthein, M., Gersonde, R., Niebler, S., Pflaumann, U., Spielhagen, R., Thiede, J., Wefer, G., Weinelt, M., 2003a. Preface: glacial Atlantic Ocean mapping (GLAMAP-2000). Paleoceanography 18, doi 10.1029/2002PA00769.

Sarnthein, M., Pflaumann, U., Weinelt, M., 2003b. Past extent of sea ice in the northern North Atlantic inferred from foraminiferal paleotemperature estimates. Paleoceanography 18, doi 10.1029/2002PA00771.

Schäfer-Neth, C., Paul, A., 2001. Circulation of the glacial Atlantic: a synthesis of global and regional modeling. In: Schäfer, P., Ritzrau, W., Schlüter, M., Thiede, J. (Eds.), The northern North Atlantic: a changing environment. Springer, Berlin, Heidelberg, pp. 441–462.

Schäfer-Neth, C., Paul, A., 2004. The Atlantic Ocean at the last glacial maximum: 1. Objective mapping of the GLAMAP sea-surface conditions. In: Wefer, G., Mulitza, S., Ratmeyer, V. (Eds.), The South Atlantic in the Late Quaternary: Reconstruction of Material Budgets and Current Systems. Springer, Berlin, Heidelberg, pp. 531–548.

Schäfer-Neth, C., Paul, A., Mulitza, S., 2005. Perspectives on mapping the MARGO reconstructions by variogram analysis/kriging and objective analysis. Quaternary Science Reviews, this issue.

Schmittner, A., Meissner, K.J., Eby, M., Weaver, A.J., 2002. Forcing of the deep ocean circulation in simulations of the last glacial maximum. Paleoceanography 17, 26–35.

Seidov, D., Sarnthein, M., Stattegger, K., Prien, R., Weinelt, M., 1996. North Atlantic ocean circulation during the last glacial maximum and a subsequent meltwater event: a numerical model. Journal of Geophysical Research C 101, 16,305–16,332.

Semtner, A.J., 1976. A model for the thermodynamic growth of sea ice in numerical investigations of climate. Journal of Physical Oceanography 6, 379–389.

Taylor, K., Williamson, D., Zwiers, F., 2000. The sea surface temperature and sea ice concentration boundary conditions for AMIP II simulations, http://www-pcmdi.llnl.gov/pcmdi/pubs/ab60.html. Technical Report, Program for Climate Model Diagnosis and Intercomparison, Lawrence Livermore National Laboratory.

Weaver, A.J., Eby, M., Wiebe, E.C., Bitz, C.M., Duffy, P.B., Ewen, T.L., Fanning, A.F., Holland, M.M., MacFadyen, A., Matthews, H.D., Meissner, K.J., Saenko, O., Schmittner, A., Wang, H., Yoshimori, M., 2001. The UVic earth system climate model: model description, climatology, and applications to past, present and future climates. Atmosphere-Ocean 39, 361–428.

Winguth, A.M.E., Archer, D., Duplessy, J.C., Maier-Reimer, E., Mikolajewicz, U., 1999. Sensitivity of paleonutrient tracer distributions and deep-sea circulation to glacial boundary conditions. Paleoceanography 14, 304–323.

WOA, 1998. World ocean atlas 1998, http://www.nodc.noaa.gov/oc5/woa98.html. Technical Report, National Oceanographic Data Center, Silver Spring, MD.

Wunsch, C., 1996. The Ocean Circulation Inverse Problem. Cambridge University Press, New York, 442pp.

Wunsch, C., 2003. Determining paleoceanographic circulations, with emphasis on the last glacial maximum. Quaternary Science Reviews 22, 371–385.

Yu, E.-F., Francois, R., Bacon, M.P., 1996. Similar rates of modern and last-glacial ocean thermohaline circulation inferred from radiochemical data. Nature 379, 689–694.

List of Forthcoming Papers

Context

G.A. Schmidt, A.N. LeGrande, The Goldilocks abrupt climate change event

Research and Review papers

W.F. Ruddiman, Cold climate during the closest Stage 11 analog to recent Millennia

R.B. Alley, A.M. Ágústsdóttir, The 8k Event: cause and consequences of a major Holocene abrupt climate change

B.P. Onac, I. Viehmann, J. Lundberg, S.-E. Lauritzen, C. Stringer and V. Popi, U–Th ages constraining the Neanderthal footprint at Vârtop Cave, Romania

G.H. Denton, R.B. Alley, G.C. Comer, W.S. Broecker, The role of seasonality in abrupt climate change

G.A. Milne, A.J. Long and S.E. Bassett, Modelling Holocene relative sea-level observations from the Caribbean and South America

M.J. Bentley, D.A. Hodgson, J.A. Smith and N.J. Cox, Relative sea level curves for the South Shetland Islands and Marguerite Bay, Antarctic Peninsula

M. Nakada and H. Inoue, Rates and causes of recent global sea-level rise inferred from long tide gauge data records

M. Presti, L. De Santis, G. Brancolini and P.T. Harris, Continental shelf record of the East Antarctic Ice Sheet evolution: seismo-stratigraphic evidence from the George V Basin

C.J. Proctor, P.J. Berridge, M.J. Bishop, D.A. Richards and P.L. Smart, Age of Middle Pleistocene fauna and Lower Palaeolithic industries from Kent's Cavern, Devon

P. McNeil, L.V. Hills, B. Kooyman and S.M. Tolman, Mammoth tracks indicate a declining Late Pleistocene population in southwestern Alberta, Canada

A. Blundell and K. Barber, A 2800-year palaeoclimatic record from Tore Hill Moss, Strathspey, Scotland: the need for a multi-proxy approach to peat-based climate reconstructions

J. Sun, Long-term fluvial archives in the Fen Wei Graben, central China, and their bearing on the tectonic history of the India–Asia collision system during the Quaternary

M. Antón, A. Galobart and A. Turner, Co-existence of scimitar-toothed cats, lions and hominins in the European Pleistocene. Implications of the post-cranial anatomy of *Homotherium latidens* (Owen) for comparative palaeoecology

M.L. Filippi and M.R. Talbot, The palaeolimnology of northern Lake Malawi over the last 25 ka based upon the elemental and stable isotopic composition of sedimentary organic matter

J.J. Lowe, Quaternary Science Reviews/Elsevier Prize for Top Student, 2004, University of London M.Sc. Degree in Quaternary Science, Royal Holloway and University College

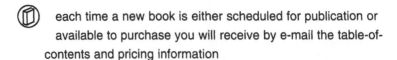

Printed and bound by CPI Group (UK) Ltd, Croydon, CR0 4YY

08/05/2025

01864797-0001